T0299976

PHYSIOLOGY OF MOLLUSCS

Volume 1

A Collection of Selected Reviews

PHYSIOLOGY OF MOLLUSCS

Volume 1

A Collection of Selected Reviews

Edited by

Saber Saleuddin, PhD
Spencer Mukai, PhD

Apple Academic Press Inc.	Apple Academic Press Inc.
3333 Mistwell Crescent	9 Spinnaker Way
Oakville, ON L6L 0A2	Waretown, NJ 08758
Canada	USA

©2017 by Apple Academic Press, Inc.
First issued in paperback 2021
No claim to original U.S. Government works

ISBN 13: 978-1-77-463526-1 (pbk)
ISBN 13: 978-1-77-188572-0 (hbk)

Physiology of Molluscs: A Collection of Selected Reviews (2-volume set)
International Standard Book Number-13: 978-1-77188-408-2 (hardback)
International Standard Book Number-13: 978-1-315-20748-3 (CRC/Taylor & Francis ebook)
International Standard Book Number-13: 978-1-77188-409-9 (AAP ebook)
Physiology of Molluscs, Volume 1: A Collection of Selected Reviews
International Standard Book Number-13: 978-1-77188-572-0 (hardback)
International Standard Book Number-13: 978-1-315-20712-4 (CRC/Taylor & Francis ebook)
International Standard Book Number-13: 978-1-77188-574-4 (AAP ebook)
Physiology of Molluscs, Volume 2: A Collection of Selected Reviews
International Standard Book Number-13: 978-1-77188-573-7 (hardback)
International Standard Book Number-13: 978-1-315-20711-7 (CRC/Taylor & Francis ebook)
International Standard Book Number-13: 978-1-77188-575-1 (AAP ebook)

Library and Archives Canada Cataloguing in Publication

Physiology of molluscs : a collection of selected reviews / edited by Saber Saleuddin, PhD, Spencer Mukai, PhD.

Includes bibliographical references and indexes.
Issued in print and electronic formats.
ISBN 978-1-77188-572-0 (volume 1 : hardcover).--ISBN 978-1-77188-574-4 (volume 1 : pdf)
1. Bivalves--Physiology. I. Saleuddin, Saber, editor II. Mukai, Spencer, 1962-, editor

QL430.6.P59 2016	594'.4	C2016-905658-9	C2016-905659-7

Library of Congress Cataloging-in-Publication Data

Names: Saleuddin, Saber, editor. | Mukai, Spencer, editor.
Title: Physiology of molluscs : a collection of selected reviews / editors, Saber Saleuddin, PhD, Spencer Mukai, PhD.
Description: New Jersey : Apple Academic Press, Inc., 2016- | Includes bibliographical references and index.
Identifiers: LCCN 2016037722 (print) | LCCN 2016038776 (ebook) | ISBN 9781771884082 (set : hardcover : alk. paper) | ISBN 9781771884099 (ebook)
Subjects: LCSH: Mollusks--Physiology.
Classification: LCC QL431.2 .P49 2016 (print) | LCC QL431.2 (ebook) | DDC 594--dc23
LC record available at https://lccn.loc.gov/2016037722

Apple Academic Press also publishes its books in a variety of electronic formats. Some content that appears in print may not be available in electronic format. For information about Apple Academic Press products, visit our website at **www.appleacademicpress.com** and the CRC Press website at **www.crcpress.com**

ABOUT THE EDITORS

Saber Saleuddin, PhD

Saber Saleuddin, PhD, is a University Professor Emeritus of the Department of Biology at York University in Toronto, Ontario, Canada. Dr. Saleuddin received his early education in Bangladesh. He received his doctorate in molluscan zoology from the University of Reading in the UK. After an NRC Research Fellowship at the University of Alberta, studying biomineralization in molluscs, he continued his research on biomineralisation in the laboratory of Karl Wilbur at Duke University. Though offered a position at Duke, he accepted a faculty appointment at York University in Canada, where he taught for 37 years. The university recognized his outstanding contributions to research, teaching, and administration by honoring him as a University Professor. He has published more than a hundred papers in international journals and has co-edited three books on molluscan physiology. He served as co-editor of the *Canadian Journal of Zoology* for 18 years and was president of the Canadian Society of Zoologists, from whom he was awarded the Distinguished Service Medal.

Spencer Mukai, PhD

Spencer Mukai, PhD, is currently an instructor and technician at York University's Glendon College campus (Toronto, Ontario, Canada), where he is facilitating the implementation of a new biology undergraduate teaching laboratory. Dr. Mukai's research interests are in the neuroendocrine regulation of reproduction, growth, and osmoregulation in molluscs. He has published in and served as reviewer for national and international journals. After receiving his BSc and PhD from the Department of Biology, York University, Dr. Mukai has spent time as a postdoctoral fellow and research associate as well as an instructor at York University's Keele campus. He has demonstrated labs in invertebrate physiology and zoology for many years and has taught a variety of courses, including invertebrate physiology and endocrinology, animal physiology, environmental physiology, histology, human physiology, parasitology, introductory biology, ecology, and conservation biology.

DEDICATION TO OUR MENTORS

Professor Alastair Graham, FRS
1906–2000

Professor Alastair Graham, born in Edinburgh, was one of the most distinguished molluscan biologists. Professor Graham started his teaching career at Sheffield University. After 20 years of teaching and administrative duties, including the Chair of Zoology and Dean of the Faculty at Birkbeck College in London, he took up the Chair of Zoology at the University of Reading in 1952 and later served as Deputy Vice Chancellor. During his career he published many papers in collaboration with Dr. Vera Fretter. One of their most important publications is the classic *British Prosobranch Molluscs.* He received many accolades for his scholarship, including a DSc and a Fellowship of the Royal Society. Professor Graham was also the editor of *The Journal of Molluscan Studies* for many years.

Duke Professor Karl M. Wilbur
1912-1994

Professor Wilbur was born in New York, and following his doctoral degree at the University of Pennsylvania, he joined the Zoology Department of Duke University in 1946. He became a James B. Duke Professor in 1961. His major interest in research was the physiology of mineralization, primarily in molluscs. Professor Wilbur was an eminent cell physiologist. In addition to many scientific papers, Professor Wilbur is best remembered for coediting the classical volume *The Physiology of Mollusca* and being Editor-in-Chief of the series *The Mollusca,* published by then Academic Press. One of Professor Wilbur's long-time collaborators was Professor Norimitsu Watabe of the University of South Carolina. Together they published many articles that made a significant advancement to the field of biomineralization.

CONTENTS

LIST OF CONTRIBUTORS

Jean Alupay
Department of Linguistics, University of Southern California, Los Angeles, CA, USA. E-mail: alupay@ usc.edu.

Ryan A. Bell
Ottawa Hospital Research Institute, Ottawa, ON, Canada.

Thierry Caquet
INRA, UAR 1275 Research Division Ecology of Forests, Grasslands and Freshwater Systems, Champenoux, France, E-mail: thierry.caquet@rennes.inra.fr.

Eric S. Clelland
Bamfield Marine Sciences Center 100 Pachena Rd., Bamfield, BC, Canada V0R 1B0; Tel.: +1 25 728 3301x255; E-mail: research@bamfieldmsc.com.

Marie-Agnès Coutellec
INRA, UMR 0985 Ecology and Ecosystem Health, Rennes, France, E-mail: marie-agnes.coutellec@ rennes.inra.fr.

Bernard M. Degnan
School of Biological Sciences, The University of Queensland, St. Lucia, Queensland 4072, Australia.

Alexander Fedosov
A. N. Severtzov Institute of Ecology and Evolution, Russian Academy of Science, Leninsky prospect, 33, Moscow 119071, Russia.

Helga E. Guderley
Département de biologie, Université Laval, Québec, QC, Canada G1T 2M7, E-mail: Helga.Guderley@ bio.ulaval.ca.

Matthew J. Harrington
Department of Biomaterials, Max Planck Institute of Colloids and Interfaces, Potsdam 14424, Germany.

John-Douglas Matthew Hughes
Faculty of Medicine, University of Ottawa, Ottawa, ON, Canada K1H 8M5.

Julita S. Imperial
Department of Biology, University of Utah, Salt Lake City, UT, United States.

Yuri Kantor
A. N. Severtzov Institute of Ecology and Evolution, Russian Academy of Science, Leninsky prospect, 33, Moscow 119071, Russia.

Kevin M. Kocot
School of Biological Sciences, The University of Queensland, St. Lucia, Queensland 4072, Australia.

Jennifer Mather
Department of Psychology, University of Lethbridge, Lethbridge, AB, Canada T1K 6T5.

Carmel McDougall
School of Biological Sciences, The University of Queensland, St. Lucia, Queensland 4072, Australia.

Baldomero M. Olivera
Department of Biology, University of Utah, Salt Lake City, UT, United States.

Christopher J. Ramnanan
Faculty of Medicine, University of Ottawa, Ottawa, ON, Canada K1H 8M5, E-mail: cramnana@uottawa.ca.

Antje Reinecke
Department of Biomaterials, Max Planck Institute of Colloids and Interfaces, Potsdam 14424, Germany.

Kenneth Simkiss
The University of Reading, Whiteknights Park, Reading RG6 6UR, UK.

Isabelle Tremblay
Département de biologie, Université Laval, Québec, QC, Canada G1T 2M7.

Nicole B. Webster
Department of Biological Sciences, University of Alberta, Edmonton, AB, Canada T6G 2E9; Tel.: +1 780 492 5751; E-mail: nwebster@ualberta.ca.

LIST OF ABBREVIATIONS

ABC	ATP-binding cassette
ABO	accessory boring organ
AMPs	antimicrobial peptides
BaPH	BaP-hydroxylase activity
DEB	dynamic energy budget
DOPA	3,4-dihydroxyphenylalanine
ECOD	ethoxycoumarin-O-deethylase activity
EDTA	ethylenediamine tetracetic acid
EPH	epoxide hydrolases
ER	endoplasmic reticulum
FREPs	fibrinogen-related proteins
G6PDH	glucose-6-phosphate dehydrogenase
GPX	glutathione peroxidase
GPX	glutathione peroxidase
LAAOs	l-amino acid oxidases
LPO	lipid peroxidation
MAPKs	mitogen-activated protein kinases
MRP	multidrug resistance-associated protein
MTs	metallothioneins
NMO	menadione reductase
OCLTT	capacity-limited thermal tolerance
PDH	pyruvate dehydrogenase
PKC	protein kinase C
ROS	reactive oxygen species
SOD	superoxide dismutase
TMP-1	thread matrix protein-1
UBF	upstream binding factor
XRD	X-ray diffraction

PREFACE

The first comprehensive treatment on the physiology of molluscs was published in two volumes, edited by K. M. Wilbur and C. M. Yonge in 1964 and 1966. Almost 20 years later, a landmark compendium in multiple edited volumes on the biology of molluscs was published between 1983 and 1988. This series dedicated two volumes (volumes 4 and 5) to review papers on molluscan physiology. K. M. Wilbur was the editor-in-chief of this important series. The volumes in 1964 and 1966 and those in the 1980s were all published by then Academic Press.

The only review series on selected aspects of molluscan physiology since the 1980s was a special volume of the *Canadian Journal of Zoology*, published in 2013, which was edited by Saber Saleuddin. As luck would have it, we were approached by Apple Academic Press in 2014 to edit another volume dedicated to molluscan physiology, which we enthusiastically agreed to undertake.

With the rapid development of cutting-edge proteomic, molecular biological, and cellular imaging techniques, our understanding of molluscan physiology, specifically in the areas of neurobiology, reproductive biology, and shell formation, has increased exponentially over the last several years. Therefore, we felt that compiling an edited volume of review papers was warranted, and we hope that this book will serve as an important resource for researchers, professors, and students.

Editing a review series is a daunting task. The major challenge of such an endeavor is not what areas we could cover but how to deal with topics where we were unable to find excellent contributors. Thus, the titles and areas of research included in this book are our personal choices based on availability of contributors and their willingness to write within the allotted time frame. Furthermore, in certain fields of physiology, such as osmoregulation and defense mechanisms, we felt that the fields have not advanced significantly enough to warrant reviews. To partially compensate for not covering certain fields, we have included two papers previously published in the *Canadian Journal of Zoology*. The only instructions we gave to contributing authors is that the coverage be comprehensive, with a brief introduction, present knowledge highlighting the significant recent findings, and finally, provide suggestions about future directions in the context of recent developments.

We are indebted to friends and colleagues around the globe who have kindly contributed to this volume. During the months of writing, rewriting, and editing, the authors have been unfailingly cooperative in all we have requested them to do. We gratefully thank the appraisers who provided an immense service by providing critical appraisal and evaluation of each paper. Each revised paper was so much better following the evaluation reports. The fact that this service is given freely attests to the generosity of our colleagues.

We had expected that a single volume should suffice, but as the project developed it became apparent we needed two volumes. In grouping papers for the two volumes, we tried to ensure that the majority of papers in each volume complemented each other and were aimed at specific readers. Thus, Volume 1 is on shell structure, mineralization, the dynamics of calcium transport, shell drilling, byssus proteins, locomotion, and reproduction. Volume 2 includes reviews on the neural mechanisms of learning, reproductive behavior, responses to environmental stress and hormones, and neurotransmitters. We believe that the reviews included in these two volumes make a significant contribution to our understanding not only of molluscan physiology but also the physiology of animals in general.

We are grateful to Sandra Jones Sickels, Ashish Kumar, and Rakesh Kumar of Apple Academic Press for their invaluable guidance and support not only at the planning stages, but also during the editing and printing processes. Finally, we are grateful to the Canadian Science Publishing of Ottawa for allowing us to reprint two papers from the *Canadian Journal of Zoology.*

CHAPTER 1

DEVELOPING PERSPECTIVES ON MOLLUSCAN SHELLS, PART 1: INTRODUCTION AND MOLECULAR BIOLOGY

KEVIN M. KOCOT[1], CARMEL MCDOUGALL, and BERNARD M. DEGNAN

[1]*Present Address: Department of Biological Sciences and Alabama Museum of Natural History, The University of Alabama, Tuscaloosa, AL 35487, USA; E-mail: kmkocot@ua.edu*

School of Biological Sciences, The University of Queensland, St. Lucia, Queensland 4072, Australia

CONTENTS

ABSTRACT

Molluscs (snails, slugs, clams, squid, chitons, etc.) are renowned for their highly complex and robust shells. Shell formation involves the controlled deposition of calcium carbonate within a framework of macromolecules that are secreted by the outer epithelium of a specialized organ called the mantle. Molluscan shells display remarkable morphological diversity, structure, and ornamentation; however, the physiological mechanisms underlying the evolution and formation of the shell are just beginning to be understood. Examination of genes expressed in the mantle and proteins incorporated into the shell suggests that the genetic program underlying shell fabrication is rapidly evolving. This includes lineage-specific integration of conserved, ancient gene families into the mantle gene regulatory network and the evolution of genes encoding proteins with novel repetitive motifs and domain combinations, which results in the expression of markedly different shell matrix protein repertoires in even closely-related molluscs. Here, we review the molecular physiology of shell formation with emphasis on the protein components that are particularly rapidly evolving. Nonprotein components such as chitin, other polysaccharides, and lipids are also reviewed. The high degree of novelty in molluscan biomineralized structures is discussed with emphasis on topics of recent interest including the image-forming aragonitic eye lenses of chiton shells and shell pigments. Finally, unanswered questions including some dealing with basic concepts such as the homology of the nacreous shell layers of gastropods and bivalves are discussed.

1.1 INTRODUCTION

Biomineralization is the process by which living organisms convert ions in solution into solid minerals (Simkiss & Wilbur, 1989). The great success of molluscs can be attributed in part to their ability to secrete calcareous skeletal structures with evidence for molluscan biomineralization extending back to the late Precambrian (Runnegar, 1996). All eight major lineages of Mollusca produce calcified exoskeletons, in the form of shells (such as those produced by bivalves, gastropods, and *Nautilus*) or sclerites (spines, scales, etc. produced by chitons and aplacophorans). However, secondary reduction or loss of the shell has occurred in several lineages (e.g., Kröger et al., 2011; Wägele & Klussmann-Kolb, 2005). In this chapter, we begin the discussion of molluscan biomineralization physiology with an emphasis on recent insights on the molecular biology of shell formation from studies using evolutionary developmental, comparative genomic/transcriptomic, and proteomic approaches. We highlight the importance of comparative studies in understanding the principles of biomineralization and a need for more such studies that include representatives from all lineages of Mollusca.

1.1.1 DIVERSITY AND STRUCTURE OF MOLLUSCAN EXOSKELETONS

With forms as disparate as the familiar garden snail, "headless" filter feeding bivalves, tiny meiofaunal worms, and giant squid, there is extreme variation in morphology among the eight major lineages of Mollusca (Haszprunar et al., 2008). Figure 1.1 shows the current consensus of molluscan phylogeny based on recent studies (Kocot et al., 2011; Smith et al., 2011; Vinther et al., 2012) with an exemplar of each major lineage. These are Polyplacophora (chitons), Caudofoveata (=Chaetodermomorpha), Solenogastres (=Neomeniomorpha), Monoplacophora, Gastropoda (snails and slugs), Bivalvia (clams, scallops, oysters, etc.), Cephalopoda (octopuses, squids, and *Nautilus*), and Scaphopoda (tusk shells). Despite the disparity in morphology among the major lineages of Mollusca, the majority of species rely on mineralized exoskeletons in the form of a shell and/or sclerites. Molluscan exoskeletons provide physical defense, support, and, in some species, desiccation resistance (Carefoot & Donovan, 1995; Fishlyn & Phillips, 1980; reviewed by Furuhashi et al., 2009). Examination of the diversity of form and structure of molluscan exoskeletons quickly reveals the great diversity that has evolved (Fig. 1.2).

Exoskeletons of extant molluscs are layered structures that contain calcium carbonate, proteins, glycoproteins, polysaccharides, and lipids. In many shelled molluscs, the mineralized layers are often covered by an entirely organic outer layer (the cuticle or periostracum). Mineralized layers are composed predominantly of calcium carbonate (as aragonite, calcite, or rarely vaterite) with a small fraction of protein and polysaccharides (reviewed by Furuhashi et al., 2009; Marin et al., 2013). A number of different shell microstructures may occur in mineralized layers of molluscan shells (Chateigner et al., 2000). These are generally classified as (1) prismatic microstructures with mutually parallel, adjacent prism-shaped crystals that do not strongly interdigitate along their mutual boundaries, (2) nacreous microstructures with laminar polygonal to rounded tablets arranged in broad sheets, (3) crossed or crossed lamellar microstructures with sheets of thin, parallel rods, and (4) homogeneous microstructures with aggregations of irregularly shaped crystallites with a granular appearance (Chateigner et al., 2000; see Bandel, 1990; Carter & Clark, 1985 for detailed discussions of shell microstructure). Of these, the prismatic and nacreous microstructures are the best studied. The prismatic layer is resistant to crack propagation and puncture (Eichhorn et al., 2005; Li & Nardi, 2004; Su et al., 2004), whereas the nacreous layer is best known for being more ductile and fracture resistant (Chateigner et al., 2000; Li et al., 2006). We refer the reader to Chateigner et al. (2000) for high-quality scanning electron micrographs of each of these different microstructure types.

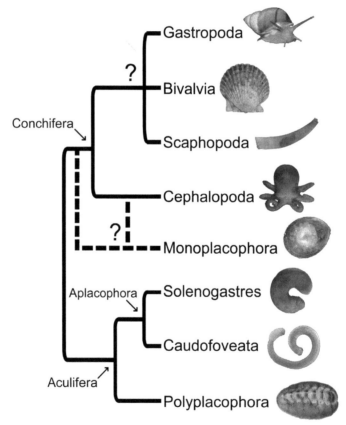

FIGURE 1.1   Current consensus of evolutionary relationships among the major lineages of Mollusca as inferred by Kocot et al. (2011), Smith et al. (2011), and Vinther et al. (2012). Photos are not to scale. Photo of *Argopecten* (Bivalvia) by Dan Speiser. Photo of *Chaetoderma* (Caudofoveata) by Christiane Todt. Photo of *Laevipilina* (Monoplacophora) by Greg Rouse and Nerida Wilson. (Used with permission.)

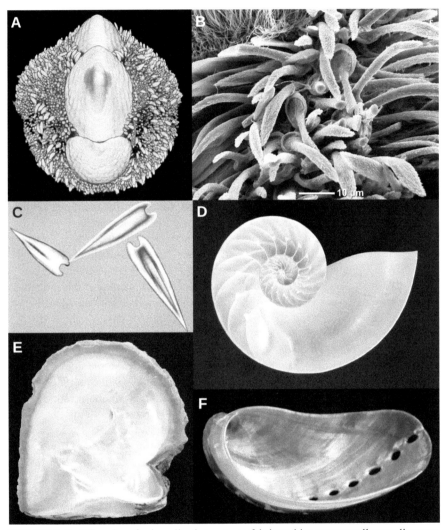

FIGURE 1.2 Diversity of mineralized structures fabricated by extant molluscan lineages. A. Micro CT scan of a juvenile specimen of *Cryptoplax larvaeformis* (Polyplacophora) showing anterior shell valves and sclerites. Specimen is approximately 1-cm wide. Photo by Jeremy Shaw. B. Scanning electron micrograph (SEM) of sclerites of *Macellomenia schanderi* (Solenogastres). C. Micrograph of sclerites of an undescribed species of *Falcidens* (Caudofoveata) from New Zealand illuminated with polarized light. Smallest sclerite is approximately 100 μm in length. D. Laterally bisected shell of *Nautilus* (Cephalopoda). E. Shell of the pearl oyster *Pinctada maxima* (Bivalvia). F. Shell of the abalone *Haliotis asinina* (Gastropoda).

Polyplacophora is a clade of slug-like molluscs that are dorsally protected by eight serially arranged shells (=valves) and a thick, fleshy girdle bearing calcareous sclerites. The shells of polyplacophorans, or chitons as they are commonly called, typically consist of four layers (Haas, 1972, 1976, 1981; summarized by Kaas & Van Belle, 1985; Fig. 1.3). The outermost layer is the cuticle, which is sometimes called the periostracum or "properio-stracum," as it differs from conchiferan periostracum in composition (reviewed by Haas, 1981; Saleuddin & Petit, 1983). This thin, transparent layer covers the tegmentum, which is the dorsally visible part of the shell. The tegmentum of chiton shells is quite different from that of shell layers observed in conchiferan shells as it contains calcium carbonate as well as substantial amounts of organic material (mostly polysaccharides). Calcified layers of conchiferan shells typically have some, but relatively very little organic material (see below; reviewed by Eernisse & Reynolds, 1994). The tegmentum is typically sculptured, and may be pigmented (e.g., Sigwart & Sirenko, 2012). Below the tegmentum is the articulamentum. This shell layer contains less organic material and is "somewhat nacreous" (Haas, 1981). In most chitons (but not the basal Lepidopleurida), the articula-mentum forms insertion plates, which project into the surrounding leathery girdle to anchor the shells in place. The hypostracum, which is also a predominantly calcareous layer, underlies the articulamentum. This layer differs from the articulamentum by having significantly less organic mate-rial and a different microstructure (see below). Finally, the myostracum, which lies below the hypostracum, is a modified hypostracum that serves for attachment of muscles. The girdle or mantle, which surrounds the shells, is covered with the same glycoproteinaceous cuticle material that covers the shells (Beedham & Trueman, 1968; Kniprath, 1981) and bears many calcareous sclerites. Chiton sclerites vary in morphology from fine, scale-like structures to large pronounced spines (Haas, 1981). The calcar-eous layers of chiton shells and sclerites are composed of aragonite (Carter & Hall, 1990; Haas, 1981; Treves et al., 2003). The crystalline structure of chiton shells has been explored in relatively few taxa. In those that have been studied, the tegmentum is formed by rods of spherulitic sectors. The hypostracum is composed of crossed lamellae with bundles of crystals. Unlike conchiferans (see below), the hypostracum crystallographic c-axis coincides with the bisectrix of these crossing fibers (Haas, 1981). Because of the unique microstructure of the hypostracum and unusual composi-tion and structure of the chiton tegmentum, it has been hypothesized that chiton shells are not strictly homologous to the shells of other conchiferans (Eernisse & Reynolds, 1994; Furuhashi et al., 2009; Haas, 1981; Schel-tema, 1993; reviewed by Kocot, 2013).

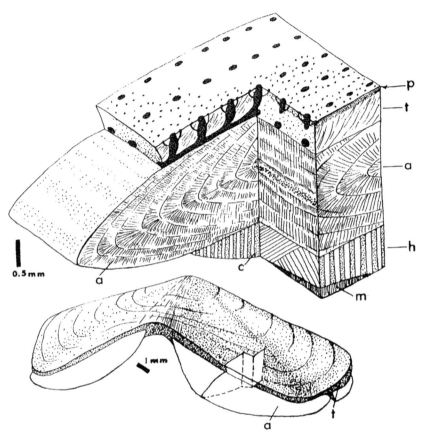

FIGURE 1.3 Structure of a chiton-shell valve. Above: Whole shell valve with cut-out region corresponding to enlargement below. Below: Enlargement showing shell layers. Abbreviations: a, articulamentum; c, crossed lamellar structure of hypostracum; h, hypostracum; m, myostracum; pp, properiostracum (cuticle); t, tegmentum. Modified from Haas (1976).

Caudofoveata (=Chaetodermomorpha) and Solenogastres (=Neomeniomorpha), collectively called Aplacophora, are worm-shaped, shell-less molluscs (reviewed by Todt et al., 2008; Todt, 2013). Although traditionally viewed as basal, plesiomorphic molluscs (see Salvini-Plawen & Steiner, 2014 and references therein), recent molecular phylogenetic studies (Kocot et al., 2011; Smith et al., 2011; Vinther et al., 2012) have grouped Aplacophora + Polyplacophora in a clade called Aculifera (Scheltema, 1993; Fig. 1.1). Examination of fossil paleoloricate "chitons" (Sutton & Sigwart, 2012; Sutton et al., 2012) has led to the interpretation that aplacophorans are derived from chiton-like ancestors that secondarily lost their shells (Sutton

& Sigwart, 2012; Sutton et al., 2012; Vinther et al., 2012; Vinther, 2014, 2015). Developmental studies have also been cited as evidence for a chiton-like ancestor of Aplacophora (Scheltema & Ivanov, 2002). Although extant aplacophorans lack shells, most of the body surface is covered with a glyco-proteinaceous cuticle and a dense coat of calcareous sclerites. Although the sclerites of the burrowing caudofoveates are relatively uniform, there is great variation in the morphology of solenogaster sclerites. Solenogaster sclerites may be solid or hollow and can exhibit a variety of shapes, such as needles, scales, hooks, and paddles, just to name a few (García-Álvarez & Salvini-Plawen, 2007). Presence of scale-like sclerites in the putatively early branching solenogaster order Pholidoskepia (Salvini-Plawen, 2003) and observation of scale-like sclerites in larvae and early juvenile solenogaster species that later develop hollow needles (Okusu, 2002; Todt & Kocot, 2014) suggests that scale-like sclerites (as also found in Caudofoveata) are plesio-morphic for Aplacophora (Salvini-Plawen, 2003). Aplacophoran spicules are composed of aragonite (Rieger and Sterrer 1975; Scheltema & Ivanov, 2002, 2004), with the long axis of the crystals aligned with the long axis of the spicules (reviewed by Ehrlich, 2010).

Monoplacophora is a small group of around 30 described species of single-shelled molluscs that mostly live in the deep sea (reviewed by Hasz-prunar & Ruthensteiner, 2013; Haszprunar, 2008; Lindberg, 2009). Some authors prefer the more specific name Tryblidia for the extant Monopla-cophora because several extinct "monoplacophorans" are of uncertain phylo-genetic affinity. Most monoplacophorans have a thin outer periostracum, a prismatic shell layer with large quadrangular or hexagonal prisms, and an inner nacreous layer (Erben et al., 1968; Hedegaard & Wenk, 1998; Meen-akshi et al., 1970; Wingstrand, 1985). However, in *Veleropilina*, *Rokopella*, and *Micropilina*, the prismatic layer is apparently absent (see Haszprunar & Ruthensteiner, 2013 for discussion) and the outer shell layer is composed of smooth or granular material with unknown microstructure (presumably homogeneous; Checa et al., 2009; Cruz et al., 2003; Marshall, 2006; Warén & Hain, 1992).

Scaphopods are marine burrowing microcarnivores with a conical shell that is open at both ends. The shell grows from the anterior end and is removed at the posterior end to allow for increased water flow into the mantle cavity as the animal grows (de Paula & Silveira, 2009). Some species produce "tubes" or "pipes" from the posterior mantle margin (Hebert, 1986; Shimek, 1989). The shell may bear longitudinal or, rarely, annular ribs. Generally, scaphopods have a trilayered shell organization similar to that of gastropods and bivalves. The organic periostracum may be thick but typically it is very thin or completely eroded in adult animals, probably due to their sand burrowing activity. An outer, very thin crystalline prismatic layer with tightly packed crystals is present in the majority of species of the order Gadilida giving these species a polished appearance. The inner-most shell layer is a complex, crossed-lamellar layer, which may have a regular or irregular structure (Steiner, 1995; Reynolds & Okusu, 1999). The shell is composed of aragonite (Bøggild, 1930).

Cephalopoda includes the extant nautiloids, octopods, vampyropods, and decabrachians (cuttlefish, squid, and *Spirula*) as well as a rich diversity of fossil forms (reviewed by Kröger et al., 2011; Young et al., 1998). Among the living cephalopods, only members of Nautiloidea have retained an external shell as adults, whereas others have reduced or (more-or-less) completely lost their shell. Cephalopod shell structure and the general mechanisms of shell formation in this group were reviewed by Bandel (1990) and Budelmann et al. (1991). In Nautiloidea, the most plesiomorphic extant cephalopod lineage, the thick, external shell is aragonitic with prismatic, spherulitic, and nacreous configurations. In *Nautilus*, internal chambers of the shell are used for buoyancy control; an osmotic gradient is established by active transport of salts to the space between the mantle tissue and the shell. This allows for the extraction of liquid from the hollow chamber and inward diffusion of gas (reviewed in detail by Budelmann et al., 1991). Most of the extant diversity of Cephalopoda is dominated by taxa with internalized and usually highly reduced shells (Birchall & Thomas, 1983; Hunt and El Sherief, 1990; Sousa Reis & Fernandes, 2002). The pelagic cephalopod *Spirula* has a calcified internal shell similar to that of *Nautilus*, which is also used for buoyancy control. Cuttlefish (e.g., *Sepia*) also use their internal shell for this function. Here, the shell is not coiled with relatively few large chambers, but contains small chambers with many flat, subdivided chambers subdivided by serially arranged organic membranes. Most other cephalopods (e.g., octopus and squid) have completely uncalcified, chitinous vestiges of the shell.

The filter- or deposit-feeding bivalves are easily recognized by their characteristic hinged shell. Shell structure and mineralogy within the group are highly variable (Kobayashi & Samata, 2006). The Lower-Middle Cambrian protobranch bivalve *Pojetaia runnegari* is the oldest known bivalve fossil. It seems to have had a single-layer shell with a prismatic microstructure that was deposited onto an organic periostracum (Runnegar & Pojeta, 1985). The pearl oysters (Pterioida) are perhaps the best-studied bivalve molluscs with respect to biomineralization, due to their economic importance. Pearl oysters exhibit the condition observed in most bivalves; they have a shell with an inner nacreous layer, a middle prismatic layer, and an outer protein-aceous layer.

Gastropoda is the most species-rich class of Mollusca. There is a great diversity of shell organization and microstructures within this clade. In the well-studied vetigastropod *Haliotis*, the shell consists of three layers: an outer organic periostracum (that is often eroded in adults), a prismatic layer made up of needle-shaped crystals enveloped by an organic sheath, and a nacreous layer consisting of aragonitic tablets surrounded and perfused by thin organic matrix (summarized by Marie et al., 2010). Adult patello-gastropods such as *Lottia* have a shell consisting of five layers (Mann et al., 2012; Marie et al., 2013; Suzuki et al., 2010). The outer-most layer is primarily calcite with a mosaic organization whereas the remaining layers are composed of prismatically arranged crystals of aragonite (Marie et al., 2013; Suzuki et al., 2010). Crossed lamellar shell microstructure is wide-spread in other gastropods (Dauphin & Denis, 2000).

1.1.2 MANTLE TISSUE

Mantle tissue (=pallial tissue; Fig. 1.4) is responsible for the secretion of molluscan shells and sclerites. The mantle forms and isolates a chamber from the external environment (see Simkiss Chapter 2 of this volume) and secretes an organic matrix of polysaccharides (e.g., chitin) and protein, which is presumed to be the site of calcium carbonate crystal nucleation (reviewed by Addadi et al., 2006; Furuhashi et al., 2009; Wilbur & Saleuddin, 1983; Wilbur, 1972). Mantle tissue morphology and the process of shell formation in general are most well-known in bivalves and gastropods. In these taxa, there are conserved cellular and morphogenetic movements that initiate larval shell secretion. Larval shell formation begins at the end of gastrula-tion, with the differentiation and local thickening of a group of ectodermal cells in the post-trochal dorsal region (the shell gland or shell field). These cells elongate and then invaginate transitorily to form the shell gland, which is analogous to the adult mantle and responsible for the secretion of larval shell. The periphery of the shell gland produces an extracellular lamella—the organic periostracum—that will serve as the site of calcium carbonate deposition (Bielefeld & Becker, 1991; Cather, 1967; Hohagen & Jackson, 2013; Kniprath, 1981). Later, the shell gland flattens and grows into the more recognizable adult mantle epithelium (Jackson et al., 2007; Kniprath, 1977, 1980, 1981).

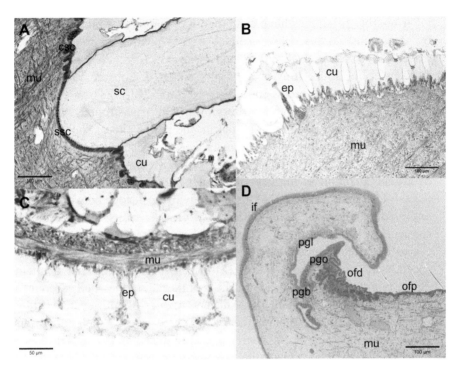

FIGURE 1.4 Histological sections of molluscan mantle tissues. A. Sclerite secretion in *Acanthopleura gemmata* (Polyplacophora). B. Various stages of sclerite secretion and lifting through cuticle in *Cryptoplax larvaeformis*. C. Epidermal papillae, cuticle, and voids from decalcified sclerites in the thick cuticle of *Proneomenia custodiens* (Solenogastres). D. Mantle tissue of *Haliotis asinina* (Gastropoda). Specimen prepared by Kathryn Green. Abbreviations: csc, cuticle secreting cells; cu, cuticle; ep, epidermal papillae; inner fold of mantle; mu, muscle; ofd, distal part of outer fold; ofp, proximal part of outer fold; pgb, base of the periostracal groove ; pgl, periostracal groove; pgo, outer fold of the periostracal groove; sc, sclerite; ssc, sclerite secreting cells.

Larval conchiferans (e.g., gastroods, bivalves, scaphopods) typically have a discreet shell gland that secretes the periostracum at its distal edge. The mantle tissue and the periostracum form the crystallization chamber where calcium is deposited adjacent to the periostracum. In contrast to conchiferans, chiton shells are secreted underneath a thin layer of cuticle (the same material that covers the entire dorsum; "properiostracum" *sensu* Haas, 1981) by a broad "plate field" (Kniprath, 1980; reviewed by Eernisse & Reynolds, 1994). This dramatic difference in shell formation mode has led some workers to question the homology of chiton shells to those of conchiferans (reviewed by Kocot, 2013; see below).

A number of studies have examined the anatomy of bivalve (reviewed by Morse and Zardus, 1997; see also Acosta-Salmón & Southgate, 2006; Checa, 2000; Fang et al., 2008) and gastropod (e.g., Fleury et al., 2008; Jackson et al., 2006; Jolly et al., 2004; Kapur & Gibson, 1967; McDougall et al., 2011; Sud et al., 2002; Werner et al., 2013; Zylstra et al., 1978) mantle tissue. Bivalve mantle differs from that of gastropods in some key ways. Most notably, the mantle margin, the active site of shell formation, in bivalves has three folds or grooves whereas gastropods generally only have two (Kniprath, 1978; Zylstra et al., 1978). However, this may be an over-generalization as the keyhole limpet *Diodora* sp. mantle margin has three folds (Budd et al., 2014). In adult bivalves, a ridge between the outer and median fold defines the periostracal groove, which secretes the periostracum. This outer organic shell layer is secreted from basal cells with a greatly infolded apical cell membrane or, in the case of *Crassostrea*, a specialized "periostracum gland" (Morrison, 1993). The outer epithelium of the mantle (i.e., the surface of the mantle facing the shell) secretes the calcified layers of the shell. Here, different zones of cells secrete different types of layers. In bivalves with a typical three-layered shell consisting of periostracum, prismatic, and nacreous layers, the epithelial cells that secrete the prismatic shell layer are columnar (Carriker, 1992) and distal to the those that secrete nacre, which are cuboidal (Fang et al., 2008; Sudo et al., 1997).

The sclerite-bearing epidermis of chitons (Haas, 1976; Kniprath, 1981) and aplacophorans (Kingsley et al., 2012; Woodland, 1907) contains calcium carbonate-secreting cells, cuticle-secreting cells, and papillae (reviewed by Ehrlich, 2010). In most chitons, an epithelium of columnar cells secretes calcium carbonate portion of the sclerite while marginal cells containing many vesicles secrete the cuticular covering of the sclerite (Haas, 1981, Fig. 1.4A). Sclerite secretion in the chiton *Cryptoplax* (Fig. 1.4B) is similar but, because this species has a relatively thick cuticle, sclerites must be pushed up through the cuticle. This appears to be achieved by growth of mantle cells (possibly papillae) that subsequently "retreat." This process is similar to what has been observed in proneomeniid (and other) solenogaster aplacophorans (e.g., Woodland, 1907), which also have a thick cuticle (Fig. 1.4C). Sclerite secretion in the solenogaster aplacophoran *Helicoradomenia* is similar to that of chitons except just one cell secretes the calcareous portion of the sclerite (as is the case in *Proneomenia*) and no special cell elongation is needed to push the sclerite through the relatively thin cuticle of this species (Kingsley et al., 2012).

1.2 INSIGHTS FROM GENOMICS, TRANSCRIPTOMICS, AND PROTEOMICS

At the time of writing this chapter, well-annotated genomes were publicly available from only three molluscs: *Lottia gigantea* (Simakov et al., 2013), *Pinctada fucata* (Takeuchi et al., 2012), and *Crassostrea gigas* (Zhang et al., 2012). However, advances in high-throughput sequencing (reviewed by Metzker, 2010) have made it possible for researchers to deeply sequence the transcriptomes of biological samples as small as a single cell (e.g., Hashimshony et al., 2012). Studies applying such an approach to the study of molluscan mantle tissue have provided new insight into the genes expressed in mantle and their interactions. Recent phylogenomic studies addressing molluscan evolutionary relationships have also contributed a significant amount of transcriptome data (González et al., 2015; Kocot et al., 2011; Smith et al., 2011; Zapata et al., 2014). Similarly, proteomic tools make it possible to identify the proteins and peptides incorporated into mineralized structures (e.g., Mann & Edsinger-Gonzales, 2014; Mann & Jackson, 2014; Mann et al., 2012). Here, we summarize recent studies that have employed such approaches to improve understanding of the molecular physiology of molluscan biomineralization.

1.2.1 DIFFERENT GENE REPERTOIRES

Several studies have used transcriptomic approaches to identify the biomineralization gene repertoires of bivalves including *Pinctada* (pearl oysters; Fang et al., 2011; Gardner et al., 2011; Huang et al., 2013; Jackson et al., 2010; Jones et al., 2014; Joubert et al., 2010; Kinoshita et al., 2011; McGinty et al., 2012; Shi et al., 2013; Zhao et al., 2012), *Mytilus* (mussels; Freer et al., 2014; Hüning et al., 2013), *Pecten* (Artigaud et al., 2014), *Hyriopsis* (Bai et al., 2010, 2013), and *Laternula* (Clark et al., 2010; Sleight et al., 2015) and gastropods including *Haliotis* (abalone; Jackson et al., 2006, 2007, 2010), *Patella* (Werner et al., 2013) *Cepaea* (Mann & Jackson, 2014). However, relatively few comparative studies have been performed (Jackson et al., 2010). By directly comparing the transcriptome of nacre-forming cells in a bivalve (*Pinctada* maxima) and gastropod (*Haliotis asinina*), Jackson et al. (2010) found tremendous differences in these two mantle transcriptomes, with less than 10% of the genes expressed in the nacre-secreting cells having significant similarity. Of these, most could be identified as being involved in processes other than biomineralization. Notably, *P. maxima* had high representation of genes annotated with lyase activity due to the abundant expression of two alpha carbonic anhydrase (CA) genes. Alpha CAs have previously been shown to be involved in biomineralization in various metazoan taxa (Horne et al., 2002; Jackson et al., 2007; Miyamoto et al., 1996; Moya et al., 2008; Wilbur & Saleuddin, 1983).

In order to focus on genes likely involved in the patterning of the nacreous layer of these animals' shells, Jackson et al. (2010) identified gene products that possessed a signal peptide (indicating an extracellular [secreted] protein) from each gene set. From *H. asinina* they identified 129 sequences and from *P. maxima* they identified 125 sequences that bear a signal peptide. When these "secretomes" were searched against each other and a variety of databases, the authors found that the majority were unique; 95 (74%) and 71 (57%) of the putative secreted proteins in *H. asinina* and *P. maxima*, respectively, shared no similarity with sequences in GenBank's nonredundant protein database or EST databases, or the genome of the patellogastropod *Lottia gigantea*. Of the 54 *P. maxima*-secreted products that shared similarity with a previously described sequence, 12 of these were previously identified as bivalve-specific biomineralization proteins (McDougall et al., 2013; Yano et al., 2006; Zhang et al., 2006; Aguilera et al., 2014 manuscript in preparation). Interestingly, only six novel *H. asinina* proteins and one novel *P. maxima* secreted protein shared similarity with proteins encoded by the *Lottia* genome, suggesting rapid evolution of lineage-specific biomineralization gene repertoires.

Proteomic studies have also shed light on differences among molluscan lineages in the molecular physiology of biomineralization (e.g., Joubert et al., 2010; Liao et al., 2015; Mann & Jackson, 2014; Mann et al., 2012; Marie et al., 2011; Marie et al., 2013; Pavat et al., 2012). Marie et al. (2011) observed that the shell protein repertoire of the mussel *Mytilus edulis* is partly similar to that of other bivalves (i.e., *Pinctada*), but also shares few similarities with that of the gastropod *Haliotis*. Also, Marie et al. (2013) examined the proteins incorporated into the shell of the patellogastropod *Lottia gigantea*. Similar to the results of Jackson et al. (2010), who used a transcriptomic approach, the shell matrix protein (SMP) repertoire of *Lottia* was found to be more similar to that of the bivalve *Pinctada* than to that of the vetigastropod *Haliotis*. Given the fundamental crystallographic differences between the limpet and abalone shells (e.g., presence/absence of nacre and crossed lamellae), these results might suggest that the secretome of the abalone mantle is relatively derived. These works highlight the importance of comparative studies for elucidating the evolution of the molluscan biomineralization toolkit.

To this end, Mann and Jackson (2014) characterized the transcriptome and shell matrix proteome of another gastropod, the common grove snail *Cepaea nemoralis*. Interestingly, the shell proteome was dominated by novel proteins with no known protein domains. Specifically, 31 out of the 59 identified shell proteins (52.5%) were completely unknown. Comparison of the *C. nemoralis* shell proteome to shell proteomes of five molluscan species (*Crassostrea gigas*, *L. gigantea*, *H. asinina*, *P. maxima*, and *P. margaritifera*) revealed 28 of 59 *C. nemoralis* proteins (47.5%) that shared similarity with one or more proteins in shell proteomes of the other species. Interestingly, only one *C. nemoralis* protein had high similarity to one of the 94 proteins in the shell of *H. asinina* and only 34 were similar to proteins (631 in total) in the *L. gigantea* shell proteome. Taken together, these studies indicate that the SMPs directing shell formation in bivalves and gastropods (and even among lineages of gastropods) are markedly different.

1.2.2 COMMON PRINCIPLES

Recent comparative studies have revealed a surprising diversity in the genetic toolkits used in shell secretion by different molluscs. However, there are underlying common principles. All shell- and/or sclerite-forming molluscs use specialized cellular machinery located in the mantle tissue to actively concentrate and secrete calcium carbonate into a closed-off space formed by the mantle and an organic matrix. The shell matrix, which consists of proteins, glycoproteins, chitin, and other polysaccharides, has been shown to be very important in determining the structure of the resulting shell (reviewed by Furuhashi et al., 2009; Marin et al., 2008, 2013).

1.2.2.1 STRUCTURAL PROTEINS

Earlier hypotheses of mollusc shell formation focused on the presence of an extrapallial fluid (e.g., Wilbur & Saleuddin, 1983). However, most contemporary views of biomineralization refer to a protein–polysaccharide gel rather than a fluid (Addadi et al., 2006; Marin et al., 2013) and view certain SMPs in this gel as the site of nucleation (Evans, 2008). Marin et al. (2008, 2013) and Evans (2008) reviewed the structure, function, and evolution of molluscan shell proteins. Structural proteins are by far the best-known component of the molluscan shell matrix. These proteins appear to function in promoting (Kim et al., 2004, 2006) or inhibiting (Kim et al., 2006; Mann et al., 2007;

Michenfelder et al., 2003) crystallization of aragonite or calcite and modulating the morphology of the structures that are produced (Evans, 2008).

1.2.2.1.1 Acidic Shell Proteins

Highly acidic proteins have been implicated in the biomineralization of many organisms, and molluscs are no exception. The organic matrix of bivalve, gastropod, and polyplacophoran shells contains a high proportion of acidic amino acids – particularly aspartate, one of two amino acids that possess a negative charge (the other acidic amino acid, glutamate, is much less common; Hare, 1963; Piez, 1961; Simkiss, 1965). This amino acid bias is reflected in a number of notably acidic characterized SMPs, including MSP1 (pI 3.2; Sarashina & Endo, 2001), Aspein (pI 1.45; Tsukamoto et al., 2004), Caspartin (Marin et al., 2005), Calprismin (Marin et al., 2005), MPP1 (pI 1.21; Samata et al., 2008), Pif (which is cleaved to produce two acidic peptides with pI's of 4.99 and 4.65; Suzuki et al., 2013), and the Asprich family (pI 3.1; Gotliv et al., 2005). Additionally, many other SMPs contain short acidic domains, such as N16/Pearlin (Samata et al., 1999), AP7 and AP24 (Michenfelder et al., 2003), some Shematrin proteins (Yano et al., 2006), and Silkmapin (Liu et al., 2015). Recent transcriptomic and proteomic studies have confirmed that the presence of acidic proteins is a common theme in molluscan shells, and have indicated that many more proteins of this nature await characterization (e.g., Jackson et al., 2010; Mann & Jackson, 2014; Marie et al., 2013).

That acidic proteins directly interact with positively charged calcium ions is well-accepted, but their true function within the shell matrix is not completely understood. In the context of in-vitro assays, acidic peptides have been demonstrated to trigger crystal nucleation via the concentration of calcium ions (Hare, 1963), or to control polymorph selection by interacting with and restricting growing crystal step-edges (Michenfelder et al., 2003). The first characterized acidic matrix proteins were isolated from calcitic layers and caused the precipitation of calcite in vitro (Falini et al., 1996; Marin et al., 2005; Takeuchi et al., 2008), prompting speculation that they were involved in the selection of this particular crystal polymorph. Subsequently, acidic proteins were also identified from aragonitic shell layers (Fu et al., 2005; Suzuki et al., 2009) indicating that the role of these proteins is not restricted to a particular $CaCO_3$ polymorph. Recent research has found that acidic proteins (or mimics thereof) can trigger the formation and stabilization of amorphous calcium carbonate (Politi et al., 2007; Smeets et al., 2015), which is thought to be the initial phase of biomineralization in molluscan and other systems (reviewed by Marin et al., 2008; Weiner & Addadi, 2011).

Marie et al. (2007) examined the physical properties of the SMP repertoire of the (freshwater) unionid bivalve *Unio pictorum* using trifluoromethanesulfonic acid-induced deglycosylation. Two-dimensional (2D) gel electrophoresis analysis of the SMPs before and after deglycosylation showed that the SMPs are heavily glycosylated. Glycosylation imparts an acidic pH to SMPs. The sulfated sugar moiety bound to these proteins (Crenshaw & Ristedt, 1976; Marxen & Becker, 1997; Simkiss, 1965) appears to impart a calcium-binding activity, which is weakened by deglycosylation (Marie et al., 2007). A similar calcium-binding activity has been observed in a vertebrate calcified tissue-associated glycoprotein (Ganss & Hoffman, 1993). Calcium-binding activity imparted by saccharides is also known in echinoderms (Farach-Carson et al., 1989) and has been suspected among mollusc shell components (Samata, 1990) previously.

1.2.2.1.2 Basic Shell Proteins

While the acidic protein fraction has been included in models of biomineralization as a major element (e.g., Addadi et al., 2006), the role of basic proteins has generally been overlooked. Basic proteins (or proteins with basic domains) have the potential to interact either directly with carbonate ions, or with other acidic macromolecules within the organic matrix. The existence of basic proteins has been revealed via 2D gel electrophoresis of SMPs from a number of taxa (Furuhashi et al., 2010; Marie et al., 2007; Marie et al., 2009; Pavat et al., 2012), and a growing number of proteins with a predicted basic pI have been characterized, including Lustrin A (Shen et al., 1997), Prisilkin (Kong et al., 2009), PFMG3 (Wang et al., 2011), Periostracin (Waite et al., 1979), Perlucin (Weiss et al., 2000), Perlustrin (Weiss et al., 2000), Perlwapin (Treccani et al., 2006), and Perlinhibin (Mann et al., 2007). In pearl oysters, two gene families encoding basic proteins, the lysine (K)-rich mantle proteins (KRMPs; McDougall et al., 2013; Zhang et al., 2006) and Shematrins (McDougall et al., 2013; Yano et al., 2006), are among the most highly expressed genes in the mantle (Jackson et al., 2010; Kinoshita et al., 2011) and are major components of the shell matrix, particularly within the prismatic layer (Marie et al., 2012). The level of expression of these proteins indicates that they may function in providing the framework of the organic matrix via interactions mediated by basic domains.

1.2.2.1.3   Silk Proteins and Other Repetitive Low-complexity Domain-containing Proteins

A particularly striking feature of SMPs is the preponderance of repetitive, low-complexity domains found within them. For example, of 39 proteins identified in the *Lottia* shell matrix identified by Marie et al. (2013), 13 were repetitive low complexity domain-containing (RLCD) proteins; likewise, 4 out of 14 and 23 out of 83 proteins from abalone shells (Marie et al., 2010) and pearl oyster shells (Marie et al., 2012), respectively, were found to possess RLCDs.

In many cases these RLCD domains contain a high proportion of glycine and alanine residues (e.g., McDougall et al., 2013), explaining why these amino acids were found to be highly abundant in amino acid analyses of shell matrices (Hare, 1963; Piez, 1961; Simkiss, 1965). This particular amino acid composition and the detection of an X-ray diffraction pattern suggestive of a beta-sheet structure drew researchers to liken this component of SMPs to spider silk fibroins, which have similar characteristics (Weiner & Hood, 1975; Weiner & Traub, 1980), and silk-like proteins became a central tenet of the model proposed for molluscan biomineralization (Weiner and Traub, 1984). Subsequent research demonstrated that the beta-sheet diffraction pattern probably originated from chitin within the matrix rather than the silk-like proteins themselves, which are likely to exist in a disordered state and form a hydrogel-like structure (Addadi et al., 2006; Falini et al., 2003; Levi-Kalisman et al., 2001). Interestingly, spider silk fibroins exist in a disordered state within silk glands prior to being extruded in a fibrous form (Hijirida et al., 1996).

Structural disorder of matrix proteins is rapidly becoming a widely recognized feature of biomineralized structures in many taxa and, interestingly, is associated with biased amino acid compositions and protein repetitiveness (Kalmar et al., 2012). Therefore, the presence of RLCDs in biomineralization-associated proteins may reflect their tendency to adopt an intrinsically disordered conformation. A survey of 39 molluscan aragonite-associated proteins revealed that all possessed a disordered region and that many were associated with aggregation motifs (Evans, 2012). Proteins of this type are likely responsible for assembling the framework of shell organic matrices.

Interestingly, RLCD-containing protein-encoding genes seem to be fast-evolving. For example, the pearl oyster shematrin gene family includes at least eight orthology groups that differ by the gain, loss, and shuffling of motifs (McDougall et al., 2013; Fig. 1.5). This high rate of evolution is likely due to the instability of repetitive sequences (Sezutsu & Yukuhiro, 2000).

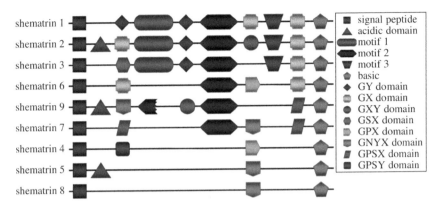

FIGURE 1.5   Schematic representation of sequence motifs in shematrin genes from pearl oysters. Modified from McDougall et al. (2013).

1.2.2.1.4   Modularity

Many SMPs exhibit a modular architecture with each module (i.e., protein domain) having distinct functionality. The most well documented examples of modular SMPs correspond to nacrein and Lustrin A. Nacrein contains a CA domain that is interrupted by the insertion of a RLCD rich in Gly and Asn. CAs have previously been shown to be involved in biomineralization in various metazoans (Horne et al., 2002; Jackson et al., 2007; Miyamoto et al., 1996; Moya et al., 2008). This RLCD region has been proposed to regulate the activity of the CA domain, acting as an inhibitor of the precipitation of calcium carbonate (Miyamoto et al., 2005). Lustrin A is the most complex multimodular SMP discovered so far and is characterized by numerous proline-, cysteine-, and GS-domains. The C-terminus domain of lustrin A exhibits high similarity with several protease inhibitors (Shen et al., 1997; Gaume et al., 2014). Although most SMPs do not exhibit sequence similarity with known proteins, many proteins contain, in addition to RLCDs, enzymatic domains such as peroxidase, CA, tyrosinase, or glycosidase domains. For example, *Lottia gigantea* CA-2 contains Asp- and Glu-rich domains in its C-terminus (Marie et al., 2013).

1.2.2.2   CHITIN AND OTHER POLYSACCHARIDES

Currently, understanding of chitin and other polysaccharides and their function in molluscan shells lag behind that of proteins. Proteins have been found in every type of molluscan shell analyzed so far, but whether chitin and/or other polysaccharides are present in all molluscan shells/sclerites is unclear. Furuhashi et al. (2009) provided a detailed review on the understanding of chitin and its role in molluscan shells. The few analyses of the polysaccharides in molluscan shells performed so far suggest that molluscs exhibit different sugar signatures (Marie et al., 2007 2009; Pavat et al., 2012). Chitin has been reliably identified in the shells of at least some bivalves, gastropods, and cephalopods but details on the structure and polymorphism (α- vs. β-chitin) are wanting. A number of different approaches have been used to detect chitin in molluscan shells, but these tests may also produce false positives or provide inaccurate pictures of chitin structure in the presence of other molecules. For example, Calcofluor White binds to chitin as well as certain acidic proteins (Albani et al., 1999, 2000). Inferences with respect to chitin network structure may be inaccurate due to nonspecific binding of such stains to molecules other than chitin. Infrared spectroscopy has also been used to test for chitin presence but insoluble proteins may confound results from this approach. Furuhashi et al. (2009) advocate the use of fluorescence probes with chitin-binding proteins (e.g., GFP-tagged chitin binding protein) and infrared spectroscopy before and after treatment with chitinase more specific tools for detection of chitin than stains such as Calcofluor White. Using the latter approach, they demonstrated the presence of both neutral polysaccharides and chitin in the cuticle of an unidentified solenogaster, the shell plates and sclerites of the chiton *Acanthopleura japonica*, the shells of the bivalves *Pinctada fucata* and *Atrina japonica*, the gastropod *Haliotis discus*, and the cephalopod *Nautilus* sp.

Much of our knowledge on chitin in mollusc shells is thanks to transcriptomic and proteomic approaches. In an attempt to understand the molecular basis underlying shell formation, Aguilera (2014) analyzed the mantle transcriptome of eight bivalve and three gastropod species. This study found over-representation of proteins with polysaccharide-binding domains within the mantle transcriptomes. These include chitin-binding Periotrophin-A, chitin-binding domain, chitinases II, chitinase-insertion domain, polysaccharide deacetylase, and galactose-binding domain-like, among others. In addition, Mann and Jackson (2014) described several *C. nemoralis* shell proteins that have high similarity with other molluscan shell-forming proteins. These include two chitin-binding domain-containing proteins. Further, they also found a protein with a chitin-binding Periotrophin-A domain and a chitinase in most of the sampled gastropods and bivalves. This emphasizes the importance of chitin in shell formation in at least these taxa (Falini and Fermani, 2004; Weiss et al., 2006).

1.2.2.3 LIPIDS

Lipids have long been known to be a minor constituent of the organic molecules found in mollusc shells (Wilbur & Simkiss, 1968). Cobabe and Pratt (1995) investigated the lipid content of the shells of *Arca zebra*, a heterotrophic bivalve, *Codakia orbicularis*, a bivalve that hosts chemoautotrophic bacteria, as well as several fossil bivalves (1.4 myo). They found that lipids comprise between 300 and 700 ppm of the total shell weight and did not vary with trophic strategy. This shell–lipid suite is dominated by cholesterol, fatty acids (recovered as fatty acid methyl esters), ketones, phytadienes, and, in some cases, alkanes. Samata and Ogura (1997) showed that lipids are present in the nacreous layer of *Pinctada fucata* and Rousseau et al. (2006) showed that lipids are present in the nacreous layer of *Pinctada margaritifera*. More recently, Farre and Dauphin (2009) examined in detail the lipid composition of the pteriomorph bivalves *Pinctada margaritifera* and *Pinna nobilis*. The shells of these bivalves contain polar lipids (phospholipids), sterols (cholesterol), triglycerides (triolein), fatty acids (oleic acid), steroids (stearyl oleate), and waxes. In the nacreous layer, the most abundant lipid components are apolar waxes, free fatty acids, and very polar lipids. Steroids and sterols are represented in lesser amounts and there are only traces of triglycerides. The situation is similar in the prismatic layer except fatty acids are lacking whereas triolein is more abundant in the prismatic layer than the nacreous layer. The physiological function of lipids in molluscan biomineralization is unclear. Extracted phospholipids have the ability to bind calcium ions (Isa & Okazaki, 1987), so they may be involved in calcification.

1.3 NOVELTY IN MOLLUSCAN BIOMINERALIZATION

Perhaps the most fascinating aspect of molluscan biomineralization is the degree of novelty it encompasses at all levels of organization: from the genes and proteins controlling the process, to the diversity of microarchitectures represented, through to the myriad of structures generated. This highly evolvable system is reflected in some astonishing innovations within the molluscan phylum, such as image-forming aragonite lenses found in chiton-shell plates (Speiser et al., 2011, 2014) and the exquisite paper-thin brood chambers of argonauts, which are often mistaken to be true shells but are, in fact, secreted from specialized webs at the tips of the arms of the female and held on to via suckers (Finn, 2013). Novelty can also be generated by the loss or reduction of structures, as seen in many cephalopods and opisthobranchs.

Structure aside, incredible diversity can also be seen in the coloration incorporated into molluscan shells and in the minerals from which the structures are composed. Some of these phenomena are explored further below.

1.3.1   REDUCTION OR LOSS OF THE SHELL

Many gastropods, particularly terrestrial and marine slugs, have reduced, internalized, or completely lost the shell. Why would these animals give up the safety afforded to them by the shell? In the terrestrial realm, loss of the shell is likely an evolutionary response to calcium limitation (Solem, 1974, 1978; South, 1992). In the marine realm, this secondary reduction or loss of the shell usually coincides with the sequestration or production of toxic chemical compounds that make these animals noxious or toxic (Derby et al., 2007; Wägele & Klussmann-Kolb, 2005). For example, the shell-less nudibranch *Glossodoris quadricolor* feeds on the sponge *Latrunculia magnifica and sequesters from it the* ichthyotoxic substance latrunculin B. It is thought that this compound then protects it from predation by fish (Mebs, 1985).

Interestingly, in some sea slugs that have secondarily lost shells, subdermal calcareous sclerites are produced (e.g., Brenzinger et al., 2013; Jörger et al., 2010; Schrödl & Neusser, 2010). Subepidermal, calcareous spicules are present in the meiofaunal gastropod taxa Acochlidia, Rhodopemorpha, and potentially *Platyhedyle* (Saccoglossa). Here, they are considered as an adaptation to the interstitial habitat, probably serving to stabilize certain body parts during movements through the interstices (Jörger et al., 2008). Many larger sea slugs such as nudibranchs also have internalized calcareous spicules. Here, these structures are often spiny and are thought to serve a defensive purpose (Penney, 2006; Thompson, 1960). Whether the production of these spicules is governed by a similar process to that in the Aculifera is unknown.

Further, the shelled deep-sea scaly foot gastropod (Neomphalida) has a foot covered in sclerites, which are noncalcified but contain iron as pyrite and greigite (Chen et al., 2015; Warén et al., 2003). Little is known about the physiology underlying the formation of these structures.

1.3.2   COLORATION OF MOLLUSCAN SHELLS

The natural beauty of seashells never fails to attract the attentions of beachgoers, young and old alike, and has done so since early human history (d'Errico et al., 2005). Part of this attraction stems from the stunning array of shapes that molluscan shells exhibit, and part from the often bright or ornately patterned coloration that they possess. The role of coloration in molluscan shells is not well understood; in some cases, the patterning quite effectively camouflages the organism against their habitat; however, in many molluscs this is not the case. Given that many molluscs with colored shells do not have image-forming eyes, reproduce via broadcast spawning, or remain buried in sediment for the extent of the life of the organism, the extravagant patterns are unlikely to serve as a signal to conspecifics (Bauchau, 2001). The fundamental role of coloration has been hypothesized to be as a means to dispose of waste products of metabolism (Comfort, 1951), to increase shell strength (Cain, 1988), or as a means to provide positional information to the mantle (Bauchau, 2001); however, support for all three of these theories is lacking.

The mechanisms underlying the production of color in molluscan shells are diverse (Aguilera et al., 2014; Barnard & De Waal, 2006; Comfort, 1951; Hedegaard et al., 2006). Pigments can be found within the proteinaceous periostracum that covers the outer surface of the shell, and also within the calcified layers themselves (Budd et al., 2014; Needham, 1975). Numerous types of pigments have been identified from molluscan shells, including pyrroles (bilins and porphyrins), polyenes (including carotenoids), and melanins (Barnard & De Waal, 2006; Comfort, 1951; Hedegaard et al., 2006). In some cases, these pigments appear closely associated with protein shell components; however, other species do not appear to use protein-associated pigmentation mechanisms (Mann & Jackson, 2014). Some molluscs do not use pigments to create their coloration at all—they have evolved shell microstructures which produce structural color, that is, color generated through the interference of reflected wavelengths of light from thin films. The most notable example of structural color in molluscs is mother-of-pearl; the architecture of nacre tablets in species such as pearl oysters and abalone results in a stunning display of reflected colors (Rayleigh, 1923; Snow et al., 2004; Webster & Anderson, 1983) that is likely to be the byproduct of an architecture that has been optimized for shell strength. However, there are examples of structural color that have clearly evolved to serve a function in their own right, such as the striking iridescent blue lines found on the shell of the limpet *Patella pellucida*. In this species, the shell ultrastructure maximizes the intensity of blue reflection from the stripes, possibly to mimic the bright blue coloration of toxic nudibranchs found in the same habitat (Li et al., 2015).

Very little is known about how shell coloration and patterning is controlled at the molecular level. A number of studies have demonstrated that pigmentation in species displaying intraspecific variation follows Mendelian patterns of inheritance (Evans et al., 2009; Gantsevich et al., 2005; Liu et al., 2009; Luttikhuizen & Drent, 2008), indicating a genetic basis for pigmentation in these molluscs. Evidence for a genetic basis also comes from studies on the juvenile abalone, *Haliotis asinina*. In this species, the expression of the *sometsuke* gene maps precisely with areas of red pigmentation on the shell, and the corresponding protein has been isolated from the shell itself (Jackson et al., 2006; Marie et al., 2010). *Sometsuke* has not been identified in any other molluscan shell proteome, indicating that it may be restricted to abalone.

It appears that the incredible diversity of coloration seen within molluscan shells is reflected in the complexity underlying it. The coloration can be generated by a diversity of pigments (or by no pigment at all!), can fulfill a broad range of functions, and is likely controlled by a number of different molecular processes. The lack of common principles indicates that shell coloration, like many aspects of biomineralization, likely evolved many times independently across the phylum.

1.3.3   CHITON SHELL EYES

In most chitons, the tegmentum is permeated by sensory structures called esthetes, which have a variety of sensory and possibly secretory functions (e.g., Eernisse & Reynolds, 1994; Speiser et al., 2011). In Schizochitonidae and Chitonidae esthetes may be capped with an ocellus that includes a lens (reviewed by Eernisse & Reynolds, 1994). Speiser et al. (2011) recently used electron probe X-ray microanalysis and X-ray diffraction to show that the chiton *Acanthopleura granulata* has shell eyes with the first aragonite lenses ever discovered. These eyes appear to be used to sense shadows produced by a would-be predator passing over the animal. Further, it appears that the eye structure results in two different refractive indices that are hypothesized to be optimal for function when the animal is submersed in water at high tide and exposed to air at low tide, respectively.

1.4   CONCLUSIONS AND OPEN QUESTIONS

1.4.1   MORE COMPARATIVE STUDIES NEEDED

Numerous recent studies have employed high throughput sequencing and proteomic approaches to improve our understanding of the process of biomineralization in molluscs. However, the vast majority of these studies have focused on economically important gastropods and bivalves. Currently, high quality genomes are available only from gastropods and bivalves (Simakov et al., 2013; Takeuchi et al., 2012; Zhang et al., 2012), although comparable data from cephalopods are forthcoming (Albertin et al., 2012). For obvious reasons, high quality genomic resources from other lineages of Mollusca would be highly beneficial toward understanding the evolution of the physiological mechanisms responsible for biomineralization.

Scaphopods are of particular interest because of their apparent close relationship to gastropods and bivalves (Kocot et al., 2011; Smith et al., 2011; Vinther et al., 2012). Because gastropods and bivalves are economically and ecologically important and well-studied with respect to biomineralization, comparative work in Scaphopoda has important bearing on studies in these two groups. In particular, given the apparent differences in biomineralization between gastropods and bivalves (e.g., Jackson et al., 2010), data from Scaphopoda would help clarify if either gastropods or bivalves are derived with respect to biomineralization or if the process is as highly variable across Mollusca in general (as suspected). Some very detailed studies have addressed scaphopod development (Wanninger & Haszprunar, 2001, 2002, 2003), but little is known about their biomineralization and limited genomic resources are available (Kocot et al., 2011; Smith et al., 2011).

Although relatively hard to obtain (but see Wilson et al., 2009), Monoplacophora would be another very interesting group to study due to its antiquity (Haszprunar, 2008; Haszprunar & Ruthensteiner, 2013; Lindberg, 2009). Genome sequencing of the monoplacophoran *Laevipilina antarctica* is currently underway (M. Schrödl, personal communication).

Because Aculifera (Aplacophora + Polyplacophora) is sister to all other extant molluscs (Kocot et al., 2011; Smith et al., 2011), studies of this group would provide important evolutionary context for molluscan biomineralization. Although many aplacophorans live in deep and/or polar habitats, some species are relatively easily accessible and have been successfully spawned in the laboratory (Okusu, 2002; Todt & Wanninger, 2010). In particular, solenogaster aplacophorans produce a phenomenal array of diverse sclerite types (reviewed by García-Álvarez & Salvini-Plawen, 2007). How these structures are achieved is a mystery, but their morphology is likely regulated by the same type of organic matrix found in shelled molluscs. Deeper and wider taxon sampling in comparative studies of biomineralization will help to understand the essential requirements for the production of mineralized structures in the Mollusca.

1.4.2   ARE SHELLS AND SCLERITES PRODUCED BY DIFFERENT MOLLUSCAN LINEAGES HOMOLOGOUS?

Although aculiferan (chiton + aplacophoran) sclerites, chiton valves, and conchiferan shells are all extracellular calcareous secretions of the mantle, structural, and developmental differences suggest that these features are not strictly homologous (Eernisse & Reynolds, 1994; Furuhashi et al., 2009; Haas, 1981; Scheltema, 1993; reviewed by Kocot, 2013). Specifically, the lack of a true periostracum, periostracal groove, and a differentiated larval shell-secreting epithelium (shell gland) in chitons distinguishes their shell structure and formation from that of the conchiferans. Further, developmental studies have shown that chiton shells are secreted by postrochal (2d) cells (Heath, 1899; Henry et al., 2004) during development. These cells (Conklin, 1897; Lillie, 1895), but sometimes also other micromere lineages (2a, 2b, 2c, and sometimes 3c), form the conchiferan shell gland (Damen & Dictus, 1994; Render, 1997). Interestingly, chiton sclerite-secreting cells arise from postrochal (2a, 2c, 3c, and 3d) as well as pretrochal cells (1a and 1d), suggesting that chiton sclerites are not strictly homologous to chiton or conchiferan shells (no cell lineage studies have been conducted in aplacophorans). Hence, the gene regulatory networks and physiological mechanisms that produce these structures may differ significantly.

There is also some question regarding the homology of shell layers within the Conchifera. The debate centers on nacre, which is found in bivalve, gastropod, cephalopod, and monoplacophoran lineages (Chateigner et al., 2000). Although generally similar, there are fundamental differences in mineralogy between the taxa; bivalves and monoplacophorans possess "sheet nacre' (tablets arranged in a brick-like pattern) with alignment of all three axes of the aragonite tablets (bivalves) or a randomly oriented *a* axis (monoplacophorans), gastropods and cephalopods possess "columnar nacre" (tablets stacked upon each other), with the *c*-axis of the tablet perpendicular to the surface of the shell and the *a* and *b* axes aligned within a stack (gastropods) or alignment of all three axes (cephalopods) (Chateigner et al., 2000; Meldrum & Cölfen, 2008). These differences, and the strikingly different nacre building gene sets that underlie them (Jackson et al., 2010), call in to question the assumption of homology of nacre in conchiferan taxa and has bearing on our understanding of the evolution of biomineralization in molluscs. Whether the other shell layers are similarly divergent between molluscan classes remains to be investigated.

1.4.3   WHAT DOES IT ALL MEAN?

Even with high quality genomic resources spanning the diversity of Mollusca, more data does not mean more understanding. However, genomic resources will continue to provide profound insight into physiological processes such as biomineralization. There is a growing need for implementation of advanced analytical techniques looking at gene family evolution (Aguilera, 2014; De Bie et al., 2006; Domazet-Lošo et al., 2007) and gene networks (Shannon et al., 2003; Smoot et al., 2011). Further, more "traditional" techniques with a much longer history of use in the field of physiology (see Simkiss Chapter 2 of this volume) should not be forgotten in the "-omics" era. Such comparative studies will undoubtedly continue to improve understanding of the complex physiological process of molluscan biomineralization.

KEYWORDS

- **biomineralization**
- **shell**
- **periostracum**
- **mantle**
- **silk**
- **RLCD**

REFERENCES

Acosta-Salmón, H.; Southgate, P. C. Wound Healing after Excision of Mantle Tissue from the Akoya Pearl Oyster, *Pinctada fucata*. *Comp. Biochem. Physiol., A: Mol. Integr. Physiol.* **2006,** *143*(2), 264–268.

Addadi, L.; Joester, D.; Nudelman, F.; Weiner, S. Mollusk Shell Formation: A Source of New Concepts for Understanding Biomineralization Processes. *Chem.—Eur. J.* **2006,** *12*(4), 980–987.

Aguilera, F.; McDougall, C.; Degnan, B. M. Evolution of the Tyrosinase Gene Family in Bivalve Molluscs: Independent Expansion of the Mantle Gene Repertoire. *Acta Biomater.* **2014,** *10*(9), 3855–3865.

Aguilera F. Investigation of Gene Family Evolution and the Molecular Basis of Shell Formation in Molluscs. Ph.D. Thesis, The University of Queensland: Brisbane, Australia, 2014.

Albani, J. R.; Sillen, A.; Coddeville, B.; Plancke, Y. D.; Engelborghs, Y. Dynamics of Carbohydrate Residues of α 1-Acid Glycoprotein (orosomucoid) followed by Red-edge Excitation Spectra and Emission Anisotropy Studies of Calcofluor White. *Carbohydr. Res.* **1999,** *322*(1), 87–94.

Albani, J. R.; Sillen, A.; Plancke, Y. D.; Coddeville, B.; Engelborghs, Y. Interaction between Carbohydrate Residues of α 1-acid Glycoprotein (Orosomucoid) and Saturating Concentrations of Calcofluor White. A Fluorescence Study. *Carbohydr. Res.* **2000,** *327* (3), 333–340.

Albertin, C. B.; Bonnaud, L.; Brown, C. T.; Crookes-Goodson, W. J.; da Fonseca, R. R.; Di Cristo, C.; Dilkes, B. P.; Edsinger-Gonzales, E.; Freeman, Jr., R. M.; Hanlon, R. T. Cephalopod Genomics: A Plan of Strategies and Organization. *Stand. Genomic Sci.* **2012,** *7*(1), 175.

Artigaud, S.; Thorne, M. A.; Richard, J.; Lavaud, R.; Jean, F.; Flye-Sainte-Marie, J.; Peck, L. S.; Pichereau, V.; Clark, M. S. Deep Sequencing of the Mantle Transcriptome of the Great Scallop *Pecten maximus. Mar. Genomics* **2014,** *15*, 3–4.

Bai, Z.; Yin, Y.; Hu, S.; Wang, G.; Zhang, X.; Li, J. Identification of Genes Potentially Involved in Pearl Formation by Expressed Sequence Tag Analysis of Mantle from Freshwater Pearl Mussel (*Hyriopsis cumingii* Lea). *J. Shellfish Res.* **2010,** *29*(2), 527–534.

Bai, Z.; Zheng, H.; Lin, J.; Wang, G.; Li, J. Comparative Analysis of the Transcriptome in Tissues Secreting Purple and White Nacre in the Pearl Mussel *Hyriopsis cumingii. PLoS ONE* **2013,** *8*(1), e53617.

Bandel, K. Cephalopod Shell Structure and General Mechanisms of Shell Formation. In *Skelet. Biominer. Patterns Process. Evol. Trends*; 1990; pp 97–115. http://onlinelibrary. wiley.com/doi/10.1029/SC005p0097/pdf.

Barnard, W.; De Waal, D. Raman Investigation of Pigmentary Molecules in the Molluscan Biogenic Matrix. *J. Raman Spectrosc.* **2006,** *37*, 342–352.

Bauchau, V. Developmental Stability as the Primary Function of the Pigmentation Patterns in Bivalve Shells. *Belg. J. Zool.* **2001,** *131*(Suppl. 2), 23–28.

Beedham, G. E.; Trueman, E. R. The Cuticle of the *Aplacophora* and its Evolutionary Significance in the Mollusca. *J. Zool.* **1968,** *154*(4), 443–451.

Bielefeld, U.; Becker, W. Embryonic Development of the Shell in *Biomphalaria glabrata* (Say). *Int. J. Dev. Biol.* **1991,** *35*, 121–131.

Birchall, J. D.; Thomas, N. L. On the Architecture and Function of Cuttlefish Bone. *J. Mater. Sci.* **1983,** *18*(7), 2081–2086.

Bøggild, O. B. The Shell Structure of the Molluscs D. Kgl. Danske Vidensk. *Selsk. Skrifter. Naturvidensk. og Math* **1930,** *9*, 230–326.

Brenzinger, B.; Padula, V.; Schrödl, M. Insemination by a Kiss? Interactive 3D-microanatomy, Biology and Systematics of the Mesopsammic cephalaspidean Sea Slug *Pluscula cuica* Marcus, 1953 from Brazil (Gastropoda: Euopisthobranchia: Philinoglossidae). *Org. Divers. Evol.* **2013,** *13*(1), 33–54.

Budd, A.; McDougall, C.; Green, K.; Degnan, B. M. Control of Shell Pigmentation by Secretory Tubules in the Abalone Mantle. *Front. Zool.* **2014,** *11*, 62.

Budelmann, B. U.; Riese, U.; Bleckmann, H. *Structure, Function, Biological Significance of the Cuttlefish "lateral lines."* In 1st International Symposium on the Cuttlefish Sepia; Boucaud-Camou, E., Ed.; Centre dePublications de l'Universite de Caen: Caen 1991; pp 201–209.

Cain, A. J. The Scoring of Polymorphic Colour and Pattern Variation and its Genetic Basis in Molluscan Shells. *Malacologia* **1988,** *28*(1–2), 1–15.

Carefoot, T. H.; Donovan, D. A. Functional Significance of Varices in the Muricid Gastropod *Ceratostoma foliatum. Biol. Bull.* **1995,** *189*(1), 59–68.

Carriker, M. R. Prismatic Shell Formation in Continuously Isolated (*Mytilus edulis*) and Periodically Exposed (*Crassostrea virginica*) extrapallial Spaces: Explicable by the Same Concept. *Am. Malacol. Bull.* **1992,** *9*, 193–197.

Carter, J. G.; Hall, R. M. Polyplacophora, Scaphopoda, Archaeogastropoda, and Paragastropoda (Mollusca). In *Skeletal Biomineralization: Patterns, Processes and Evolutionary Trends*; Carter, J. G., Ed.; Van Nostrand Reinhold, New York, 1990; Vol. 2 *Atlas and Index*, pp 29–31.

Carter, J. G.; Clark, G. R. Classification and Phylogenetic Significance of Molluscan Shell Microstructure. In *Molluscs, Notes for a Short Course;* Bottjer, D. J.; Hickman, C. S.; Ward, P. D.; Broadhead, T. W., Eds.; University of Tennessee, Department of Geological Sciences Studies in Geology, 1985.

Cather, J. N. Cellular Interactions in the Development of the Shell Gland of the Gastropod, *Ilyanassa. J. Exp. Zool.* **1967,** *166*, 205–223.

Chateigner, D.; Hedegaard, C.; Wenk, H. Mollusc Shell Microstructures and Crystallographic Textures. *J. Struct. Geol.* **2000,** *22*(11–12), 1723–1735.

Checa, A. A New Model for periostracum and Shell formation in Unionidae (Bivalvia, Mollusca). *Tissue Cell* **2000,** *32*(5), 405–416.

Checa, A. G.; Sánchez-Navas, A.; Rodríguez-Navarro, A. Crystal Growth in the Foliated Aragonite of Monoplacophorans (Mollusca). *Cryst. Growth Des.* **2009,** *9*(10), 4574–4580.

Chen, C.; Copley, J. T.; Linse, K.; Rogers, A. D.; Sigwart, J. How the Mollusc got its Scales: Convergent Evolution of the Molluscan Scleritome. *Biol. J. Linn. Soc.* **2015,** *114*(4), 949–954.

Clark, M.; Thorne, M.; Vieira, F.; Cardoso, J.; Power, D.; Peck, L. Insights into Shell Deposition in the Antarctic Bivalve *Laternula elliptica*: Gene Discovery in the Mantle Transcriptome using 454 Pyrosequencing. *BMC Genomics* **2010,** *11*(1), 362.

Cobabe, E. A.; Pratt, L. M. Molecular and Isotopic Compositions of Lipids in Bivalve Shells: A New Prospect for Molecular Paleontology. *Geochim. Cosmochim. Acta* **1995,** *59*(1), 87–95.

Comfort, A. The Pigmentation of Molluscan Shells. *Biol. Rev.* **1951,** *26*(3), 285–301.

Conklin, E. G. The Embryology of *Crepidula. J. Morphol.* **1897,** *13*, 1–226.

Crenshaw, M. A.; Ristedt, H. The Histochemical Localization of Reactive Groups in Septal Nacre from *Nautilus pompilius* L. In *The Mechanisms of Mineralization in the Invertebrates and Plants*; 1976; pp 355–367.

Cruz, R.; Weissmüller, G.; Farina, M. Microstructure of Monoplacophora (Mollusca) Shell Examined by Low-voltage Field Emission Scanning Electron and Atomic Force Microscopy. *Scanning* **2003,** *25*(1), 12–18.

Damen, P.; Dictus, W. J. A. G. Cell Lineage of the Prototroch of *Patella vulgata* (Gastropoda, Mollusca). *Dev. Biol.* **1994,** *162*(2), 364–383.

Dauphin, Y.; Denis, A. Structure and Composition of the Aragonitic Crossed Lamellar Layers in Six Species of Bivalvia and Gastropoda. *Comp. Biochem. Physiol., A. Mol. Integr. Physiol.* **2000,** *126*(3), 367–377.

De Bie, T.; Cristianini, N.; Demuth, J. P.; Hahn, M. W. CAFE: A Computational Tool for the Study of Fene Family Evolution. *Bioinformatics* **2006,** *22*(10), 1269–1271.

d'Errico, F.; Henshilwood, C.; Vanhaeren, M.; van Niekerk, K. *Nassarius kraussianus* Shell Beads from Blombos Cave: Evidence for Symbolic Behaviour in the Middle Stone Age. *J. Hum. Evol.* **2005,** *48*(1), 3–24.

de Paula, S. M.; Silveira, M. Studies on Molluscan Shells: Contributions from Microscopic and Analytical Methods. *Micron* **2009,** *40*(7), 669–690.

Derby, C. D.; Kicklighter, C. E.; Johnson, P. M.; Zhang, X. Chemical Composition of Inks of Diverse Marine Molluscs Suggests Convergent Chemical Defenses. *J. Chem. Ecol.* **2007,** *33*(5), 1105–1113.

Domazet-Lošo, T.; Brajković, J.; Tautz, D. A Phylostratigraphy Approach to Uncover the Genomic History of Major Adaptations in Metazoan Lineages. *Trends Genet.* **2007,** *23*(11), 533–539.

Eernisse, D. J.; Reynolds, P. D. Polyplacophora. In *Microscopic Anatomy of Invertebrates*; Harrison, F. W.; Kohn, A. J., Eds.; Wiley-Liss: New York, 1994; Vol. 5, pp 55–110.

Ehrlich, H. Molluscs Spicules. *Biol. Mater. Mar. Orig.* **2010,** 211–242.

Ehrlich, H. Chitin and Collagen as Universal and Alternative Templates in Biomineralization. *Int. Geol. Rev.* **2010,** *52*(7–8), 661–699.

Eichhorn, S. J.; Scurr, D. J.; Mummery, P. M.; Golshan, M.; Thompson, S. P.; Cernik, R. J. The Role of Residual Stress in the Fracture Properties of a Natural Ceramic. *J. Mater. Chem.* **2005,** *15*(9), 947–952.

Erben, H. K.; Flajs, G.; Siehl, A. Über die Schalenstruktur von Monoplacophoren. *Verlag der Akademie der Wissenschaften und der Literatur*; in Kommission bei F. Steiner, Wiesbaden, 1968, 1.

Evans, J. S. "Tuning in" to Mollusk Shell Nacre- and Prismatic-associated Protein Terminal Sequences. Implications for Biomineralization and the Construction of High Performance Inorganic–Organic Composites. *Chem. Rev.* **2008,** *108*(11), 4455–4462.

Evans, S.; Camara, M.; Langdon, C. Heritability of Shell Pigmentation in the Pacific Oyster, *Crassostrea gigas*. *Aquaculture* **2009,** *286*(3), 211–216.

Evans, J. S. Aragonite-associated Biomineralization Proteins are Disordered and contain Interactive Motifs. *Bioinformatics* **2012,** *28*(24), 3182–3185.

Falini, G.; Albeck, S.; Weiner, S.; Addadi, L. Control of Aragonite or Calcite Polymorphism by Mollusk Shell Macromolecules. *Science* **1996,** *271*(5245), 67–69.

Falini, G.; Weiner, S.; Addadi, L. Chitin–Silk Fibroin Interactions: Relevance to Calcium Carbonate Formation in Invertebrates. *Calcif. Tissue Int.* **2003,** *72*(5), 548–554.

Falini, G.; Fermani, S. Chitin Mineralization. *Tissue Eng.* **2004,** *10*(1–2), 1–6.

Fang, Z.; Feng, Q.; Chi, Y.; Xie, L.; Zhang, R. Investigation of Cell Proliferation and Differentiation in the Mantle of *Pinctada fucata* (Bivalve, Mollusca). *Mar. Biol.* **2008,** *153*(4), 745–754.

Fang, D.; Xu, G.; Hu, Y.; Pan, C.; Xie, L.; Zhang, R. Identification of Genes Directly Involved in Shell formation and their Functions in Pearl Oyster, *Pinctada fucata*. *PLoS ONE* **2011,** *6*(7), e21860.

Farach-Carson, M. C.; Carson, D. D.; Collier, J. L.; Lennarz, W. J.; Park, H. R.; Wright, G. C. A Calcium-binding, Asparagine-linked Oligosaccharide is Involved in Skeleton formation in the Sea Urchin Embryo. *J. Cell Biol.* **1989,** *109*(3), 1289–1299.

Farre, B.; Dauphin, Y. Lipids from the Nacreous and Prismatic Layers of Two Pteriomorpha Mollusc Shells. *Comp. Biochem. Physiol., B: Biochem. Mol. Biol.* **2009,** *152*(2), 103–109.

Finn, J. K. Taxonomy and Biology of the Argonauts (Cephalopoda: Argonautidae) with Particular Reference to Australian Material. *Molluscan Res.* **2013,** *33*(3), 143–222.

Fishlyn, D. A.; Phillips, D. W. Chemical Camouflaging and Behavioral Defenses against a Predatory Seastar by Three Species of Gastropods from the Surfgrass *Phyllospadix* Community. *Biol. Bull.* **1980,** *158*(1), 34–48.

Fleury, C.; Marin, F.; Marie, B.; Luquet, G.; Thomas, J.; Josse, C., Serpentini, A.; Lebel, J. M. Shell Repair Process in the Green Ormer *Haliotis tuberculata*: A Histological and Micro-structural Study. *Tissue Cell* **2008**, *40*(3), 207–218.

Freer, A.; Bridgett, S.; Jiang, J.; Cusack, M. Biomineral Proteins from *Mytilus edulis* Mantle Tissue Transcriptome. *Mar. Biotechnol.* **2014**, *16*(1), 34–45.

Fu, G.; Valiyaveettil, S.; Wopenka, B.; Morse, D. E. $CaCO_3$ Biomineralization: Acidic 8-kDa Proteins isolated from Aragonitic Abalone Shell Nacre can Specifically Modify Calcite Crystal Morphology. *Biomacromolecules* **2005**, *6*(3), 1289–1298.

Furuhashi, T.; Schwarzinger, C.; Miksik, I.; Smrz, M.; Beran, A. Molluscan Shell Evolution with Review of Shell Calcification Hypothesis. *Comp. Biochem. Physiol.: B Biochem. Mol. Biol.* **2009**, *154*(3), 351–371.

Furuhashi, T.; Miksik, I.; Smrz, M.; Germann, B.; Nebija, D.; Lachmann, B.; Noe, C. Comparison of Aragonitic Molluscan Shell Proteins. *Comp. Biochem. Phys., B* **2010**, *155*(2), 195–200.

Gantsevich, M.; Tyunnikova, A.; Malakhov, V. The Genetics of Shell Pigmentation of the Mediterranean Mussel *Mytilus galloprovincialis* Lamarck, 1819 (Bivalvia, Mytilida). *Dokl. Biol. Sci.* **2005**, *404*(1), 370–371.

García-Álvarez, O.; Salvini-Plawen, L. Species and Diagnosis of the Families and Genera of Solenogastres (Mollusca). *Iberus* **2007**, *25*(2), 73–143.

Ganss, B.; Hoffmann, W. Calcium Binding to Sialic Acids and its Effect on the Conformation of Ependymins. *Eur. J. Biochem.* **1993**, *217*(1), 275–280.

Gardner, L.; Mills, D.; Wiegand, A.; Leavesley, D.; Elizur, A. Spatial Analysis of Biomineralization Associated Gene Expression from the Mantle Organ of the Pearl Oyster *Pinctada maxima*. *BMC Genomics* **2011**, *12*(1), 455.

Gaume B.; Denis, F.; Van Wormhoudt, A.; Huchette, S.; and Jackson, D. J. Characterization and Expression of the Biomineralising Gene Lustrin A During Shell Formation of the European Abalone *Haliotis tuberculata*. *Comp. Biochem. Physiol., B* **2014**, *169*, 1–8.

Gotliv, B.-A.; Kessler, N.; Sumerel, J. L.; Morse, D. E.; Tuross, N.; Addadi, L.; Weiner, S. Asprich: A Novel Aspartic Acid-rich Protein Family from the Prismatic Shell Matrix of the Bivalve *Atrina rigida*. *ChemBioChem* **2005**, *6*(2), 304–314.

González, V. L.; Andrade, S. C.; Bieler, R.; Collins, T. M.; Dunn, C. W.; Mikkelsen, P. M.; Taylor, J. D.; Giribet, G. A Phylogenetic Backbone for Bivalvia: An RNA-Seq Approach. *Proc. R. Soc. B: Biol. Sci.* **2015**, *282*(1801), 20142332.

Haas, W. Untersuchungen über die Mikro- und Ultrastruktur der Polyplacophorenschale. *Biomineralization* **1972**, *5*, 1–52.

Haas, W. Observations on the Shell and Mantle of the Placophora. In *The Mechanisms of Mineralization in the Invertebretaes and Plants*; University of South Carolina Press: Columbia, 1976; pp 389–402.

Haas, W. Evolution of Calcareous Hardparts in Primitive Molluscs. *Malacologia* **1981**, *21*(1–2), 403–418.

Hare, P. E. Amino Acids in the Proteins from Aragonite and Calcite in the Shells of *Mytilus californianus*. *Science* **1963**, *139*(3551), 216–217.

Hashimshony, T.; Wagner, F.; Sher, N.; Yanai, I. CEL-Seq: Single-cell RNA-Seq by Multiplexed Linear Amplification. *Cell. Rep.* **2012**, *2*(3), 666–673.

Haszprunar, G. *Monoplacophora (Tryblidia)*. In *Phylogeny and Evolution of the Mollusca*; Ponder, W. F.; Lindberg, D. L., Eds.; University of California Press: Berkeley and Los Angeles, 2008; pp 97–104.

Haszprunar, G.; Schander, C.; Halanych, K. Relationships of Higher Molluscan Taxa. In *Phylogeny and Evolution of the Mollusca*; Ponder, W. F.; Lindberg, D. L., Eds.; University of California Press: Berkeley and Los Angeles, 2008; pp 19–32.

Haszprunar, G.; Ruthensteiner, B. Monoplacophora (Tryblidia)-some Unanswered Questions. *Am. Malacol. Bull.* **2013**, *31*(1), 189–194.

Heath, H. Development of *Ischnochiton*. *Zool. Jahrbuecher Abt. Fuer Anat. Ontog. Tiere* **1899**, *12*, 567–656.

Hebert, A. Reproductive Behavior and Anatomy of Three Central Californian Scaphopods. Master's Thesis, California State University: Hayward, 1986.

Hedegaard, C.; Wenk, H. Microstructure and Texture Patterns of Mollusc Shells. *J. Molluscan Stud.* **1998**, *64*, 133–136.

Hedegaard, C.; Bardeau, J. -F.; Chateigner, D. Molluscan Shell Pigments: An *in situ* Resonance Raman Study. *J. Molluscan Stud.* **2006**, *72*(2), 157–162.

Henry, J. Q.; Okusu, A.; Martindale, M. Q. The Cell Lineage of the Polyplacophoran, *Chaetopleura apiculata*: Variation in the Spiralian Program and Implications for Molluscan Evolution. *Dev. Biol.* **2004**, *272*(1), 145–160.

Hijirida, D. H.; Do, K. G.; Michal, C.; Wong, S.; Zax, D.; Jelinski, L. W. ¹³C NMR of *Nephila clavipes* Major Ampullate Silk Gland. *Biophys. J.* **1996**, *71*(6), 3442–3447.

Hohagen J.; Jackson., D. J. An Ancient Process in a Modern Molluscs: Early Development of the Shell in *Lymnaea stagnalis*. *BMC Dev. Biol.* **2013**, *13*, 27.

Horne, F.; Tarsitano, S.; Lavalli, K. L. Carbonic Anhydrase in Mineralization of the Crayfish Cuticle. *Crustaceana* **2002**, *75*(9), 1067–1081.

Huang, X. D.; Zhao, M.; Liu, W. G.; Guan, Y. Y.; Shi, Y.; Wang, Q.; Wu, S. Z.; He, M. X. Gigabase-scale Transcriptome Analysis on Four Species of Pearl Oysters. *Mar. Biotechnol.* **2013**, *15*(3), 253–264.

Hüning, A. K.; Melzner, F.; Thomsen, J.; Gutowska, M. A.; Krämer, L.; Frickenhaus, S.; Rosenstiel, P.; Pörtner, H.-O.; Philipp, E. E.; Lucassen, M. Impacts of Seawater Acidification on Mantle Gene Expression Patterns of the Baltic Sea Blue Mussel: Implications for Shell Formation and Energy Metabolism. *Mar. Biol.* **2013**, *160*(8), 1845–1861.

Hunt, S.; El Sherief, A. A Periodic Structure in the "Pen" Chitin of the Squid *Loligo vulgaris*. *Tissue Cell* **1990**, *22*(2), 191–197.

Isa, Y.; Okazaki, M. Some Observations on the Ca^{2+}-binding Phospholipid from Scleractinian Coral Skeletons. *Comp. Biochem. Physiol., B: Comp. Biochem.* **1987**, *87*(3), 507–512.

Jackson, D. J.; McDougall, C.; Green, K.; Simpson, F.; Wörheide, G.; Degnan, B. M. A Rapidly Evolving Secretome Builds and Patterns a Sea Shell. *BMC Biol.* **2006**, *4*, 40.

Jackson, D. J.; Wörheide, G.; Degnan, B. M. Dynamic Expression of Ancient and Novel Molluscan Shell Genes during Ecological Transitions. *BMC Evol. Biol.* **2007**, *7*(1), 160.

Jackson, D. J.; McDougall, C.; Woodcroft, B.; Moase, P.; Rose, R. A.; Kube, M.; Reinhardt, R.; Rokhsar, D. S.; Montagnani, C.; Joubert, C.; Piquemal, D.; Degnan, B. M. Parallel Evolution of Nacre Building Gene Sets in Molluscs. *Mol. Biol. Evol.* **2010**, *27*(3), 591–608.

Jörger, K. M.; Neusser, T. P.; Haszprunar, G.; Schrödl, M. Undersized and Underestimated: 3D Visualization of the Mediterranean Interstitial Acochlidian Gastropod *Pontohedyle milaschewitchii* (Kowalevsky, 1901). *Org. Divers. Evol.* **2008**, *8*(3), 194–214.

Jörger, K. M.; Stöger, I.; Kano, Y.; Fukuda, H.; Knebelsberger, T.; Schrödl, M. On the Origin of Acochlidia and Other Enigmatic Euthyneuran Gastropods, with Implications for the Systematics of Heterobranchia. *BMC Evol. Biol.* **2010**, *10*(1), 323.

Jolly, C.; Berland, S.; Milet, C.; Borzeix, S.; Lopez, E.; Doumenc, D. Zona Localization of Shell Matrix Proteins in Mantle of *Haliotis tuberculata* (Mollusca, Gastropoda). *Mar. Biotechnol.* **2004,** *6*(6), 541–551.

Jones, D. B.; Jerry, D. R.; Khatkar, M. S.; Moser, G.; Raadsma, H. W.; Taylor, J. J.; Zenger, K. R. Determining Genetic Contributions to Host Oyster Shell Growth: Quantitative Trait Loci and Genetic Association Analysis for the Silver-lipped Pearl Oyster, *Pinctada maxima*. *Aquaculture* **2014,** *434*, 367–375.

Joubert, C.; Piquemal, D.; Marie, B.; Manchon, L.; Pierrat, F.; Zanella-Cléon, I.; Cochennec-Laureau, N.; Gueguen, Y.; Montagnani, C. Transcriptome and Proteome Analysis of *Pinctada margaritifera* Calcifying Mantle and Shell: Focus on Biomineralization. *BMC Genomics* **2010,** *11*(1), 613.

Kaas, P.; Van Belle, R. *A. Monograph of Living Chitons. Vol. 1. Order Neoloricata: Lepidopleurina*, Brill: Leiden, 1985.

Kalmar, L.; Homola, D.; Varga, G.; Tompa, P. Structural Disorder in Proteins Brings Order to Crystal Growth in Biomineralization. *Bone* **2012,** *51*(3), 528–534.

Kapur, S. P.; Gibson, M. A. A Histological Study of the Development of the Mantle-edge and Shell in the Freshwater Gastropod, *Helisoma duryi eudiscus* (Pilsbry). *Can. J. Zool.* **1967,** *45*(6), 1169–1181.

Kim, I. W.; Morse, D. E.; Evans, J. S. Molecular Characterization of the 30-AA N-terminal Mineral Interaction Domain of the Biomineralization Protein AP7. *Langmuir* **2004,** *20*(26), 11664–11673.

Kim, I. W.; Collino, S.; Morse, D. E.; Evans, J. S. A Crystal Modulating Protein from Molluscan Nacre that Limits the growth of Calcite *in vitro*. *Cryst. Growth Des.* **2006,** *6*, 1078.

Kingsley, R.; Froelich, J.; Marks, C.; Spicer, L.; Todt, C. Formation and Morphology of Epidermal Sclerites from a Deep-sea Hydrothermal Vent Solenogaster *Helicoradomenia* sp. (Solenogastres, Mollusca). *Zoomorphology* **2012,** *132*(1), 1–9.

Kinoshita, S.; Wang, N.; Inoue, H.; Maeyama, K.; Okamoto, K.; Nagai, K.; Kondo, H.; Hirono, I.; Asakawa, S.; Watabe, S. Deep Sequencing of ESTs from Nacreous and Prismatic Layer Producing Tissues and a Screen for Novel Shell Formation-related Genes in the Pearl Oyster. *PLoS ONE* **2011,** *6*(6), e21238.

Kniprath, E. Ontogeny of Shell Field in *Lymnaea stagnalis*. *Wilhelm Rouxs Arch. Dev. Biol.* **1977,** *181*(1), 11–30.

Kniprath, E. Growth of the Shell-field in *Mytilus* (Bivalvia). *Zool. Scr.* **1978,** 7, 119–120.

Kniprath, E. Larval Development of the Shell and the Shell Gland in *Mytilus* (Bivalvia). *Wilhelm Rouxs Arch. Dev. Biol.* **1980,** *188*(3), 201–204.

Kniprath, E. Ontogeny of the Molluscan Shell Field: A Review. *Zool. Scr.* **1981,** *10*(1), 61–79.

Kobayashi, I.; Samata, T. Bivalve Shell Structure and Organic Matrix. *Mater. Sci. Eng. C.* **2006,** *26*(4), 692–698.

Kocot, K. M.; Cannon, J. T.; Todt, C.; Citarella, M. R.; Kohn, A. B.; Meyer, A.; Santos, S. R.; Schander, C.; Moroz, L. L.; Lieb, B. Phylogenomics Reveals Deep Molluscan Relationships. *Nature* **2011,** *477*(7365), 452–456.

Kocot, K. M. Recent Advances and Unanswered Questions in Deep Molluscan Phylogenetics. *Am. Malacol. Bull.* **2013,** *31*(1), 1–14.

Kong, Y.; Jing, G.; Yan, Z.; Li, C.; Gong, N.; Zhu, F.; Li, D.; Zhang, Y.; Zheng, G.; Wang, H.; Xie, L.; Zhang, R. Cloning and Characterization of Prisilkin-39, a Novel Matrix Protein Serving a Dual Role in the Prismatic Layer Formation from the Oyster *Pinctada fucata*. *J. Biol. Chem.* **2009,** *284*(16), 10841–10854.

Kröger, B.; Vinther, J.; Fuchs, D. Cephalopod Origin and Evolution: A Congruent Picture Emerging from Fossils, Development and Molecules. *BioEssays* **2011**, *8*(33), 602–613.

Levi-Kalisman, Y.; Falini, G.; Addadi, L.; Weiner, S. Structure of the Nacreous Organic Matrix of a Bivalve Mollusk Shell Examined in the Hydrated State Using Cryo-TEM. *J. Struct. Biol.* **2001**, *135*(1), 8–17.

Li, X.; Nardi, P. Micro/nanomechanical Characterization of a Natural Nanocomposite Material—the Shell of Pectinidae. *Nanotechnology* **2004**, *15*(1), 211.

Li, X.; Xu, Z. H.; Wang, R. *In situ* Observation of Nanograin Rotation and Deformation in Nacre. *Nano Lett.* **2006**, *6*(10), 2301–2304.

Li, L.; Kolle, S.; Weaver, J.; Ortiz, C.; Aizenberg, J.; Kolle, M. A Highly Conspicuous Mineralized Composite Photonic Architecture in the Translucent Shell of the Blue-rayed Limpet. *Nat. Commun.* **2015**, *6*, 6322.

Liao, Z.; Bao, L. F.; Fan, M. H.; Gao, P.; Wang, X. X.; Qin, C. L.; Li, X. M. In-depth Proteomic Analysis of Nacre, Prism, and Myostracum of *Mytilus* Shell. *J. Proteomics* **2015**, *122*, 26–40.

Lillie, F. R. The Development of the Unionidae. *J. Morphol.* **1895**, 10, 1–100.

Lindberg, D. R. Monoplacophorans and the Origin and Relationships of Mollusks. *Evol. Educ. Outreach* **2009**, *2*(2), 191–203.

Liu, X.; Wu, F.; Zhao, H.; Zhang, G.; Guo, X. A Novel Shell Color Variant of the Pacific abalone *Haliotis discus hannai* Ino Subject to Genetic Control and Dietary Influence. *J. Shellfish Res.* **2009**, *28*(2), 419–424.

Liu, X.; Dong, S.; Jin, C.; Bai, Z.; Wang, G.; Li, J. Silkmapin of *Hyriopsis cumingii*, a Novel Silk-like Shell Matrix Protein Involved in Nacre Formation. *Gene* **2015**, *555*(2), 217–222.

Luttikhuizen, P.; Drent, J. Inheritance of Predominantly Hidden Shell Colours in *Macoma balthica* (L.) (Bivalvia: Tellinidae). *J. Molluscan Stud.* **2008**, *74*(4), 363–371.

Mann, K.; Siedler, F.; Treccani, L.; Heinemann, F.; Fritz, M. Perlinhibin, a Cysteine-, histidine-, and Arginine-rich miniprotein from abalone (*Haliotis laevigata*) Nacre, Inhibits In Vitro Calcium Carbonate Crystallization. *Biophys. J.* **2007**, *93*(4), 1246–1254.

Mann, K.; Edsinger-Gonzales, E.; Mann, M. In-depth Proteomic Analysis of a Mollusc Shell: Acid-soluble and Acid-insoluble Matrix of the Limpet *Lottia gigantea. Proteome Sci.* **2012**, *10*(1), 28.

Mann K.; Edsinger-Gonzales, E. The *Lottia gigantea* Shell Proteome: Re-analysis Including MaxQuant iBAQ Quantitation and Phosphoproteome Analysis. *Proteome Sci.* **2014**, *12*, 28.

Mann, K.; Jackson, D. J. Characterization of the Pigmented Shell-forming Proteome of the Common Grove Snail *Cepaea nemoralis. BMC Genomics* **2014**, *15*(1), 249.

Marie, B.; Luquet, G.; Pais De Barros, J.-P.; Guichard, N.; Morel, S.; Alcaraz, G.; Bollache, L.; Marin, F. The Shell Matrix of the Freshwater Mussel *Unio pictorum* (Paleoheterodonta, Unionoida). Involvement of Acidic Polysaccharides from Glycoproteins in Nacre Mineralization. *FEBS J.* **2007**, *274*(11), 2933–2945.

Marie, B.; Marin, F.; Marie, A.; Bédouet, L.; Dubost, L.; Alcaraz, G.; Milet, C.; Luquet, G. Evolution of Nacre: Biochemistry and Proteomics of the Shell Organic Matrix of the Cephalopod *Nautilus macromphalus. ChemBioChem* **2009**, *10*(9), 1495–1506.

Marie, B.; Marie, A.; Jackson, D. J.; Dubost, L.; Degnan, B. M.; Milet, C.; Marin, F. Proteomic Analysis of the Organic Matrix of the Abalone *Haliotis asinina* Calcified Shell. *Proteome Sci.* **2010**, *8*, 54

Marie B.; Le Roy, N.; Zanella-Cleon, I.; Becchi, M.; Marin, F. Molecular Evolution of Mollusc Shell Proteins: Insights from Proteomic Analysis of the Edible Mussel *Mytilus. J. Mol. Evol.* **2011**, *72*, 531–546.

Marie, B.; Joubert, C.; Tayalé, A.; Zanella-Cléon, I.; Belliard, C.; Piquemal, D.; Cochennec-Laureau, N.; Marin, F.; Gueguen, Y.; Montagnani, C. Different Secretory Repertoires Control the Biomineralization Processes of Prism and Nacre Deposition of the Pearl Oyster Shell. *Proc. Natl. Acad. Sci.* **2012,** *109*(51), 20986–20991.

Marie, B.; Jackson, D. J.; Ramos-Silva, P.; Zanella-Cléon, I.; Guichard, N.; Marin, F. The Shell-forming Proteome of *Lottia gigantea* Reveals both Deep Conservations and Lineage-specific Novelties. *FEBS J.* **2013,** *280*(1), 214–232.

Marin, F.; Amons, R.; Guichard, N.; Stigter, M.; Hecker, A.; Luquet, G.; Layrolle, P.; Alcaraz, G.; Riondet, C.; Westbroek, P. Caspartin and Calprismin, Two Proteins of the Shell Calcitic Prisms of the Mediterranean Fan Mussel *Pinna nobilis. J. Biol. Chem.* **2005,** *280*(40), 33895–33908.

Marin, F.; Luquet, G.; Marie, B.; Medakovic, D. Molluscan Shell Proteins: Primary Structure, Origin, and Evolution. *Curr. Top. Dev. Biol.* **2008,** *80*, 209–276.

Marin, F.; Marie, B.; Benhamada, S.; Silva, P.; Le Roy, N.; Guichard, N.; Wolf, S.; Montagnani, C.; Joubert, C.; Piquemal, D. 'Shellome': Proteins Involved in Mollusk Shell Biomineralization-diversity, Functions. In *Recent Advances in Pearl Research*; 2013; pp 149–168. https://www.researchgate.net/profile/Benjamin_Marie2/publication/235752273_'Shel lome'_Proteins_Involved_in_Mollusc_Shell_Biomineralization_Diversity_Functions/ links/02bfe51320782ecb0c000000.pdf.

Marshall, B. A. Four New Species of Monoplacophora (Mollusca) from the New Zealand Region. *Molluscan Res.* **2006,** *26*(2), 61–68.

Marxen, J. C.; Becker, W. The Organic Shell Matrix of the Freshwater Snail *Biomphalaria flabrata. Comp. Biochem. Physiol., B: Biochem. Mol. Biol.* **1997,** *118*(1), 23–33.

McDougall, C.; Green, K.; Jackson, D. J.; Degnan, B. M. Ultrastructure of the Mantle of the Gastropod *Haliotis asinina* and Mechanisms of Shell Regionalization. *Cells Tissues Organs* **2011,** *194*(2), 103.

McDougall, C.; Aguilera, F.; Degnan, B. M. Rapid Evolution of Pearl Oyster Shell Matrix Proteins with Repetitive, Low-complexity Domains. *J. R. Soc. Interface* **2013,** *10*(82), 20130041.

McGinty, E. L.; Zenger, K. R.; Jones, D. B.; Jerry, D. R. Transcriptome Analysis of Biomineralisation-related Genes within the Pearl Sac: Host and Donor Oyster Contribution. *Mar. Genomics* **2012,** *5*, 27–33.

Mebs, D. Chemical Defense of a Dorid Nudibranch, *Glossodoris quadricolor*, From the Red Sea. *J. Chem. Ecol.* **1985,** *11*(6), 713–716.

Meldrum, F. C.; Cölfen, H. Controlling Mineral Morphologies and Structures in Biological and Synthetic Systems. *Chem. Rev.* **2008,** *108*(11), 4332–4432.

Meenakshi, V. R.; Harpe, P. E.; Watabe, N.; Wilbur, K. M.; Menzies, R. J. Ulstrastructure, Histochemistry and Amino Acid Composition of the Shell of *Neopilina. Sci. Rep. Southeast Pac. Exp.* **1970,** *2*, 1–12.

Metzker, M. L. Sequencing Technologies—The Next Generation. *Nat. Rev. Genet.* **2010,** *11*(1), 31–46.

Michenfelder, M.; Fu, G.; Lawrence, C.; Weaver, J. C.; Wustman, B. A.; Taranto, L.; Evans, J. S.; Morse, D. E. Characterization of Two Molluscan Crystal-modulating Biomineralization Proteins and Identification of Putative Mineral Binding Domains. *Biopolymers* **2003,** *70*(4), 522–533.

Miyamoto, H.; Miyashita, T.; Okushima, M.; Nakano, S.; Morita, T.; Matsushiro, A. A Carbonic Anhydrase from the Nacreous Layer in Oyster Pearls. *Proc. Natl. Acad. Sci.* **1996,** *93*(18), 9657–9660.

Miyamoto, H.; Miyoshi, F.; Kohno, J. The Carbonic Anhydrase Domain Protein Nacrein is Expressed in the Epithelial Cells of the Mantle and Acts as a Negative Regulator in Calcification in the Mollusc *Pinctada fucata*. *Zool. Sci.* **2005**, *22*(3), 311–315.

Morrison, C. M. Histology and Cell Ultrastructure of the Mantle and Mantle Lobes of the Easter Oyster *Crassostrea virginica* (Gmelin)—A Summary Atlas. *Am. Malacol. Bull.* **1993**, *10*(1), 1–24.

Morse, M. P.; Zardus, J. D. Chapter 2: Bivalvia. In *The Microscopic Anatomy of Invertebrates*; Harrison, F. W.; Kohn, A. J., Eds.; Wiley-Liss: New York, 1997; Vol. 6A *Mollusca II*, pp 7–118.

Moya, A.; Tambutté, S.; Bertucci, A.; Tambutté, E.; Lotto, S.; Vullo, D.; Supuran, C. T.; Allemand, D.; Zoccola, D. Carbonic Anhydrase in the Scleractinian Coral *Stylophora pistillata* Characterization, Localization, and Role in Biomineralization. *J. Biol. Chem.* **2008**, *283*(37), 25475–25484.

Needham, A. The Zoochromes of Helicid Shells. *Naturwissenschaften* **1975**, *62*(4), 183–184.

Okusu, A. Embryogenesis and Development of *Epimenia babai* (Mollusca Neomeniomorpha). *Biol. Bull.* **2002**, *203*(1), 87–103.

Pavat, C.; Zanella-Cléon, I.; Becchi, M.; Medakovic, D.; Luquet, G.; Guichard, N.; Alcaraz, G.; Dommergues, J. -L.; Serpentini, A.; Lebel, J. -M.; Marin, F. The Shell Matrix of the Pulmonate Land Snail *Helix aspersa maxima*. *Comp. Biochem. Phys., B:* **2012**, *161*(4), 303–314.

Penney, B. K. Morphology and Biological Roles of Spicule Networks in *Cadlina luteomarginata* (Nudibranchia, Doridina). *Invertebr. Biol.* **2006**, *125*(3), 222–232.

Piez, K. A. Amino Acid Composition of Some Calcified Proteins. *Science* **1961**, *134*(3482), 841–842.

Politi, Y.; Mahamid, J.; Goldberg, H.; Weiner, S.; Addadi, L. Asprich Mollusk Shell Protein: In Vitro Experiments Aimed at Elucidating Function in $CaCO_3$ Crystallization. *CrysEngComm* **2007**, *9*(12), 1171–1177.

Rayleigh, L. Studies of Iridescent Colour, and the Structure Producing it—II. Mother-of-pearl. *Proc. R. Soc. Lond., A: Mater.* **1923**, *102*(719), 674–677.

Render, J. Cell Fate Maps in the *Ilyanassa obsoleta* Embryo beyond the Third Division. *Dev. Biol.* **1997**, *189*(2), 301–310.

Reynolds, P. D.; Okusu, A. Phylogenetic Relationships among Families of the Scaphopoda (Mollusca). *Zool. J. Linn. Soc.* **1999**, *126*(2), 131–154.

Rieger, R. M.; Sterrer, W. New Spicular Skeletons in Turbellaria, and the Occurrence of Spicules in Marine Meiofauna. *J. Zool. Syst. Evol. Res.* **1975**, *13*(4), 207–278.

Rousseau, M.; Bédouet, L.; Lati, E.; Gasser, P.; Le Ny, K.; Lopez, E. Restoration of Stratum Corneum with Nacre Lipids. *Comp. Biochem. Physiol., B: Biochem. Mol. Biol.* **2006**, *145*(1), 1–9.

Runnegar, B.; Pojeta, J. Origin and Diversification of the Mollusca. In *The Mollusca*; Clarke, M. R.; Wilbur, K. M.; Trueman, E. R., Eds.; 1985; Vol. 10, pp 1–57.

Runnegar, B. Early Evolution of the Mollusca: The Fossil Record. In *Origin and Evolutionary Radiation of the Mollusca*; Taylor, J. D., Ed.; Oxford University Press: Oxford, 1996; p 77.

Saleuddin, A. S. M.; Petit, H. P. The Mode of Formation and the Structure of the Periostracum. In *The Mollusca*; Academic Press: New York, 1983; Vol. 5, pp 199–234.

Salvini-Plawen, L. On the Phylogenetic Significance of the Aplacophoran Mollusca. *Iberus* **2003**, *21*(1), 67–97.

Salvini-Plawen, L. V.; Steiner, G. The Testaria Concept (Polyplacophora + Conchifera) Updated. *J. Nat. Hist.* **2014**, *48*(45–48), 2751–2772.

Samata, T. Ca-binding Glycoproteins in Molluscan Shells with Different Types of Ultrastructure. *Veliger* **1990**, *33*(2), 190–201.

Samata, T.; Ogura, M. First Finding of Lipid Component in the Nacreous Layer of *Pinctada fucata*. *J. Foss. Res.* **1997**, *30*, 66.

Samata, T.; Hayashi, N.; Kono, M.; Hasegawa, K.; Horita, C.; Akera, S. A New Matrix Protein Family Related to the Nacreous Layer Formation of *Pinctada fucata*. *FEBS Lett.* **1999**, *462*(1–2), 225–229.

Samata, T.; Ikeda, D.; Kajikawa, A.; Sato, H.; Nogawa, C.; Yamada, D.; Yamazaki, R.; Akiyama, T. A Novel Phosphorylated Glycoprotein in the Shell Matrix of the Oyster *Crassostrea nippona*. *FEBS J.* **2008**, *275*(11), 2977–2989.

Sarashina, I.; Endo, K. The Complete Primary Structure of Molluscan Shell Protein 1 (MSP-1), an Acidic Glycoprotein in the Shell Matrix of the Scallop *Patinopecten yessoensis*. *Mar. Biotechnol.* **2001**, *3*(4), 362–369.

Scheltema, A. H. Aplacophora as Progenetic Aculiferans and the Coelomate Origin of Mollusks as the Sister Taxon of Sipuncula. *Biol. Bull.* **1993**, *184*(1), 57–78.

Scheltema, A. H.; Ivanov, D. L. An Aplacophoran Postlarva with Iterated Dorsal Groups of Spicules and Skeletal Similarities to Paleozoic Fossils. *Invertebr. Biol.* **2002**, *121*(1), 1–10.

Scheltema, A. H.; Ivanov, D. L. Use of Birefringence to Characterize Aplacophora sclerites. *Veliger* **2004**, *47*(2), 153–156.

Schrödl, M.; Neusser, T. P. Towards a Phylogeny and Evolution of Acochlidia (Mollusca: Gastropoda: Opisthobranchia). *Zool. J. Linn. Soc.* **2010**, *158*(1), 124–154.

Sezutsu, H.; Yukuhiro, K. Dynamic Rearrangement within the *Antheraea pernyi* Silk Fibroin Gene is Associated with Four Types of Repetitive Units. *J. Mol. Evol.* **2000**, *51*(4), 329–338.

Shannon, P.; Markiel, A.; Ozier, O.; Baliga, N. S.; Wang, J. T.; Ramage, D.; Amin, N.; Schwikowski, B.; Ideker, T. Cytoscape: A Software Environment for Integrated Models of Biomolecular Interaction Networks. *Genome Res.* **2003**, *13*(11), 2498–2504.

Shen, X.; Belcher, A. M.; Hansma, P. K.; Stucky, G. D.; Morse, D. E. Molecular Cloning and Characterization of Lustrin A, a Matrix Protein from Shell and Pearl Nacre of *Haliotis rufescens*. *J. Biol. Chem.* **1997**, *272*(51), 32472–32481.

Shi, Y.; Yu, C.; Gu, Z.; Zhan, X.; Wang, Y.; Wang, A. Characterization of the Pearl Oyster (*Pinctada martensii*) Mantle Transcriptome Unravels Biomineralization Genes. *Mar. Biotechnol.* **2013**, *15*(2), 175–187.

Shimek, R. L. Shell Morphometrics and Systematics: A Revision of the Slander, Shallow-water Genus *Cadulus* of the North-eastern Pacific (Scaphopoda: Gadilida). *Veliger* **1989**, *30*, 213–221.

Sigwart, J. D.; Sirenko, B. I. Deep-sea Chitons from Sunken Wood in the West Pacific (Mollusca: Polyplacophora: Lepidopleurida): Taxonomy, Distribution, and Seven New Species. *Zootaxa* **2012**, *3195*, 1–38.

Simakov, O.; Marletaz, F.; Cho, S.-J.; Edsinger-Gonzales, E.; Havlak, P.; Hellsten, U.; Kuo, D.-H.; Larsson, T.; Lv, J.; Arendt, D.; et al. Insights into Bilaterian Evolution from Three Spiralian Genomes. *Nature* **2013**, *493*(7433), 526–531.

Simkiss, K. The Organic Matrix of the Oyster Shell. *Comp. Biochem. Physiol.* **1965**, *16*, 427–435.

Simkiss, K.; Wilbur, K. M. *Biomineralization: Cell Biology and Mineral Deposition*, Academic Press: San Diego, 1989.

Sleight, V. A.; Thorne, M. A.; Peck, L. S.; Clark, M. S. Transcriptomic Response to Shell Damage in the Antarctic Clam, *Laternula elliptica*: Time Scales and Spatial Localisation. *Mar. Genomics* **2015**, *20*, 45–55.

Smeets, P. J. M.; Cho, K. R.; Kempen, R. G. E.; Sommerdijk, N. A. J. M.; De Yoreo, J. J. Calcium Carbonate Nucleation Driven by Ion Binding in a Biomimetic Matrix Revealed by *in situ* Electron Microscopy. *Nat. Mater.* **2015**, *14*(4), 394–399.

Smith, S. A.; Wilson, N. G.; Goetz, F. E.; Feehery, C.; Andrade, S. C. S.; Rouse, G. W.; Giribet, G.; Dunn, C. W. Resolving the Evolutionary Relationships of Molluscs with Phylogenomic Tools. *Nature* **2011**, *480*(7377), 364–367.

Smoot, M. E.; Ono, K.; Ruscheinski, J.; Wang, P.-L.; Ideker, T. Cytoscape 2.8: New Features for Data Integration and Network Visualization. *Bioinformatics* **2011**, *27*(3), 431–432.

Snow, M.; Pring, A.; Self, P.; Losic, D.; Shapter, J. The Origin of the Color of Pearls in Iridescence from Nano-composite Structures of the Nacre. *Am. Miner.* **2004**, *89*(10), 1353–1358.

Solem, A. *The Shell Makers*. John Wiley & Sons: New York, 1974.

Solem, A. *Classification of the Land Mollusca*, In *Pulmonates* Fretter, V.; Peake, J., Eds.; Academic Press: New York, 1978; Vol. 2A, pp 49–97.

Sousa Reis, C.; Fernandes, R. Growth Observations on *Octopus vulgaris* Cuvier, 1797 from the Portuguese Waters: Growth Lines in the Vestigial Shell as Possible Tools for Age Determination. *Bull. Mar. Sci.* **2002**, *71*(2), 1099–1103.

South, A. *Terrestrial Slugs: Biology, Ecology, and Control*. Chapman and Hall: London, 1992.

Speiser, D. I.; Eernisse, D. J.; Johnsen, S. A Chiton Uses Aragonite Lenses to form Images. *Curr. Biol.* **2011**, *21*(8), 665–670.

Speiser, D. I.; DeMartini, D. G.; Oakley, T. H. The Shell-eyes of the Chiton *Acanthopleura granulata* (Mollusca, Polyplacophora) Use Pheomelanin as a Screening Pigment. *J. Nat. Hist.* **2014**, *48*(45–48), 2899–2911.

Steiner, G. Larval and Juvenile Shells of Four North Atlantic Scaphopod Species. *Am. Malacol. Bull.* **1995**, *11*, 87–98.

Su, X.-W.; Zhang D.-M.; Heuer, A. H. Tissue Regeneration in the Shell of the Giant Queen Conch, *Strombus gigas*. *Chem. Mater.* **2004**, *16*(4), 581–593.

Sud D.; Poncet, J. M.; Saihi, A.; Lebel, J. M.; Doumenc, D.; Boucaud-Camou, E. A Cytological Study of the Mantle Edge of *Haliotis tuberculata* L. (Mollusca, Gastropoda) in Relation to Shell Structure. *J. Shellfish Res.* **2002**, *21*, 201–210.

Sudo, S.; Fujikawa, T.; Nagakura, T.; Ohkubo, T.; Sakaguchi, K.; Tanaka, M.; Nakashima, K.; Takahashi, T. Structures of Mollusc Shell Framework Proteins. *Nature* **1997**, *387*(6633), 563–564.

Sutton, M. D.; Sigwart, J. D. A Chiton Without a Foot. *Palaeontology* **2012**, *55*(2), 401–411.

Sutton, M. D.; Briggs, D. E. G.; Siveter, D. J.; Siveter, D. J.; Sigwart, J. D. A Silurian Armoured Aplacophoran and Implications for Molluscan Phylogeny. *Nature* **2012**, *490*(7418), 94–97.

Suzuki, M.; Kameda, J.; Sasaki, T.; Saruwatari, K.; Nagasawa, H.; Kogure, T. Characterization of the Multilayered Shell of a Limpet, *Lottia kogamogai* (Mollusca: Patellogastropoda), Using SEM-EBSD and FIB-TEM Techniques. *J. Struct. Biol.* **2010**, *171*(2), 223–230.

Suzuki, M.; Iwashima, A.; Kimura, M.; Kogure, T.; Nagasawa, H. The Molecular Evolution of the pif Family Proteins in Various Species of Mollusks. *Mar. Biotechnol.* **2013**, *15*(2), 145–158.

Suzuki, M.; Saruwatari, K.; Kogure, T.; Yamamoto, Y.; Nishimura, T.; Kato, T.; Nagasawa, H. An Acidic Matrix Protein, pif, is a Key Macromolecule for Nacre Formation. *Science* **2009**, *325*(5946), 1388–1390.

Takeuchi, T.; Sarashina, I.; Iijima, M.; Endo, K. In Vitro Regulation of CaCO(3) Crystal Polymorphism by the Highly Acidic Molluscan Shell Protein Aspein. *FEBS Lett.* **2008,** *582*(5), 591–596.

Takeuchi, T.; Kawashima, T.; Koyanagi, R.; Gyoja, F.; Tanaka, M.; Ikuta, T.; Shoguchi, E.; Fujiwara, M.; Shinzato, C.; Hisata, K. Draft Genome of the Pearl Oyster *Pinctada fucata*: A Platform for Understanding Bivalve Biology. *DNA Res.* **2012,** *19*(2), 117–130.

Thompson, T. E. The Development of *Neomenia carinata* Tullberg (Mollusca Aplacophora). *Proc. R. Soc., B: Biol. Sci.* **1960,** *153,* 263–278.

Todt, C.; Okusu, A.; Schander, C.; Schwabe, E. Solenogastres, Caudofoveata, and Polyplacophora. In *Phylogeny and Evolution of the Mollusca*; Ponder, W. F.; Lindberg, D. L., Eds.; University of California Press: Berkeley and Los Angeles, 2008; pp 71–96.

Todt, C.; Wanninger, A. Of Tests, Trochs, Shells, and Spicules: Development of the Basal Mollusk *Wirenia argentea* (Solenogastres) and its Bearing on the Evolution of Trochozoan Larval Key Features. *Front. Zool.* **2010,** *7*(6).

Todt, C. Aplacophoran Mollusks—Still Obscure and Difficult? *Am. Malacol. Bull.* **2013,** *31*(1), 1–7.

Todt, C.; Kocot, K. M. New Records for the Solenogaster *Proneomenia sluiteri* (Mollusca) from Icelandic Waters and Description of *Proneomenia custodiens sp. n. Pol. Polar Res.* **2014,** *35*(2), 291–310.

Treccani, L.; Mann, K.; Heinemann, F.; Fritz, M. Perlwapin, an Abalone Nacre Protein with Three Four-disulfide Core (whey acidic protein) Domains, Inhibits the Growth of Calcium Carbonate Crystals. *Biophys. J.* **2006,** *91*(7), 2601–2608.

Treves, K.; Traub, W.; Weiner, S.; Addadi, L. Aragonite Formation in the Chiton (Mollusca) Girdle. *Helv. Chim. Acta* **2003,** *86,* 1101–1112.

Tsukamoto, D.; Sarashina, I.; Endo, K. Structure and Expression of an Unusually Acidic Matrix Protein of Pearl Oyster Shells. *Biochem. Biophys. Res. Co.* **2004,** *320*(4), 1175–1180.

Vinther, J.; Sperling, E. A.; Briggs, D. E.; Peterson, K. J. A Molecular Palaeobiological Hypothesis for the Origin of Aplacophoran Molluscs and their Derivation from Chiton-like Ancestors. *Proc. R. Soc., B: Biol. Sci.* **2012,** *279*(1732), 1259–1268.

Vinther, J. A. Molecular Palaeobiological Perspective on Aculiferan Evolution. *J. Nat. Hist.* **2014,** *48*(45–48), 2805–2823.

Vinther, J. The Origins of Molluscs. *Palaeontology* **2015,** *58*(1), 19–34.

Waite, J.; Herbert, A.; Saleuddin, S. M.; Andersen, S. O. Periostracin—A Soluble Precursor of Sclerotized Periostracum in *Mytilus edulis* L. *J. Comp. Physiol.* **1979,** *130*(4), 301–307.

Wang, X.; Liu, S.; Xie, L.; Zhang, R.; Wang, Z. *Pinctada fucata* Mantle Gene 3 (PFMG3) Promotes Differentiation in Mouse Osteoblasts (MC3T3-E1). *Comp. Biochem. Phys., B* **2011,** *158*(2), 173–180.

Wanninger, A.; Haszprunar, G. The Expression of an Engrailed Protein During Embryonic Shell Formation of the Tusk-shell, *Antalis entalis* (Mollusca, Scaphopoda). *Evol. Dev.* **2001,** *3*(5), 312–321.

Wanninger, A.; Haszprunar, G. Muscle Development in *Antalis entalis* (Mollusca, Scaphopoda) and its Significance for Scaphopod Relationships. *J. Morphol.* **2002,** *254*(1), 53–64.

Wanninger, A.; Haszprunar, G. The Development of the Serotonergic and FMRF-amidergic Nervous System in *Antalis entalis* (Mollusca, Scaphopoda). *Zoomorphology* **2003,** *122*(2), 77–85.

Warén, A.; Hain, S. *Laevipilina antarctica and Micropilina arntzi*, Two New Monoplacophorans from the Antarctic. *Veliger* **1992,** *35,* 165–176.

Warén, A.; Bengtson, S.; Goffredi, S. K.; Van Dover, C. L. A Hot-vent Gastropod with Iron Sulfide Dermal Sclerites. *Science* **2000**, *302*(5647), 1007.

Wägele, H.; Klussmann-Kolb, A. Opisthobranchia (Mollusca, Gastropoda)—More than just Slimy Slugs. Shell Reduction and its Implications on Defence and Foraging. *Front. Zool.* **2005**, *2*(1), 1–18.

Webster, R.; Anderson, B. W. *Gems, Their Sources, Descriptions and Identification.* Butterworths: London, 1983.

Weiner, S.; Addadi, L. Crystallization Pathways in Biomineralization. *Annu. Rev. Mater. Res.* **2011**, *41*, 21–40.

Weiner, S.; Hood, L. Soluble Protein of the Organic Matrix of Mollusk Shells: A Potential Template for Shell Formation. *Science* **1975**, *190*(4218), 987–989.

Weiner, S.; Traub, W. X-Ray Diffraction Study of the Insoluble Organic Matrix of Mollusk Shells. *FEBS Lett.* **1980**, *111*(2), 311–316.

Weiner, S.; Traub, W. Macromolecules in Mollusc Shells and their Functions in Biomineralization. *Philos. Trans. R. Soc., B* **1984**, *304*(1121), 425.

Weiss, I. M.; Kaufmann, S.; Mann, K.; Fritz, M. Purification and Characterization of Perlucin and Perlustrin, Two New Proteins from the Shell of the Mollusc *Haliotis laevigata*. *Biochem. Biophys. Res. Co.* **2000**, *267*(1), 17–21.

Werner, G. D.; Gemmell, P.; Grosser, S.; Hamer, R.; Shimeld, S. M. Analysis of a Deep Transcriptome from the Mantle Tissue of *Patella vulgata* Linnaeus (Mollusca: Gastropoda: Patellidae) Reveals Candidate Biomineralising Genes. *Mar. Biotechnol.* **2013**, *15*(2), 230–243.

Wilbur, K. M. Shell Formation in Mollusks. In *Chemical Zoology*; Florkin, M.; Scheer, B. T., Eds.; Academic Press: New York and London, 1972; Vol. 7, *Mollusca*, pp 103–145.

Wilbur, K. M.; Saleuddin, A. S. M. Shell Formation. In *The Mollusca*; Saleuddin, A. S. M.; Wilbur, K. M., Ed.; Academic Press: New York, 1983; Vol. 4; pp 235–287.

Wilbur, K. M.; Simkiss, K. Calcified Shells. *Compr. Biochem.* **1968**, *26*, 229–295.

Wilson, N. G.; Huang, D.; Goldstein, M. C.; Cha, H.; Giribet, G.; Rouse, G. W. Field Collection of *Laevipilina hyalina* McLean, 1979 from Southern California, the Most Accessible Living monoplacophoran. *J. Molluscan Stud.* **2009**, *75*(2), 195–197.

Wingstrand, K. G. On the Anatomy and Relationships of Recent Monoplacophora. *Galathea Rep.* **1985**, *16*, 7–94.

Woodland, W. Studies in Spicule Formation. *Q. J. Microsc. Sci.* **1907**, *51*, 31.

Yano, M.; Nagai, K.; Morimoto, K.; Miyamoto, H. Shematrin: A Family of Glycine-rich Structural Proteins in the Shell of the Pearl Oyster *Pinctada fucata*. *Comp. Biochem. Physiol., B: Biochem. Mol. Biol.* **2006**, *144*(2), 254–262.

Young, R. E.; Vecchione, M.; Donovan, D. T. The Evolution of Coleoid Cephalopods and their Present Biodiversity and Ecology. *S. Afr. J. Mar. Sci.* **1998**, *20*(1), 393–420.

Zapata, F.; Wilson, N. G.; Howison, M.; Andrade, S. C.; Jörger, K. M.; Schrödl, M.; Goetz, F. E.; Giribet, G.; Dunn, C. W. Phylogenomic Analyses of Deep Gastropod Relationships Reject Orthogastropoda. *Proc. R. Soc., B: Biol. Sci.* **2014**, *281*(1794), 20141739.

Zhang, C.; Xie, L.; Huang, J.; Liu, X.; Zhang, R. A Novel Matrix Protein Family Participating in the Prismatic Layer Framework Formation of Pearl Oyster, *Pinctada fucata*. *Biochem. Biophys. Res. Co.* **2006**, *344*(3), 735–740.

Zhang, G.; Fang, X.; Guo, X.; Li, L.; Luo, R.; Xu, F.; Yang, P.; Zhang, L.; Wang, X.; Qi, H.; et al. The Oyster Genome Reveals Stress Adaptation and Complexity of Shell Formation. *Nature* **2012**, *490*(7418), 49–54.

Zhao, X.; Wang, Q.; Jiao, Y.; Huang, R.; Deng, Y.; Wang, H.; Du, X. Identification of Genes Potentially Related to Biomineralization and Immunity by Transcriptome Analysis of Pearl Sac in Pearl Oyster *Pinctada martensii*. *Mar. Biotechnol.* **2012,** *14*(6), 730–739.

Zylstra, U.; Boer, H. H.; Sminia, T. Ultrastructure, Histology, and Innervation of the Mantle Edge of the Freshwater Pulmonate snails *Lymnaea stagnalis* and *Biomphalaria pfeifferi*. *Calcif. Tissue Res.* **1978,** *26*(1), 271–282.

CHAPTER 2

DEVELOPING PERSPECTIVES ON MOLLUSCAN SHELLS, PART 2: CELLULAR ASPECTS

KENNETH SIMKISS

The University of Reading, Whiteknights Park, Reading RG6 6UR, UK

CONTENTS

ABSTRACT

The biological components involved in shell formation are considered in relation to the anatomical compartments, the extrapallial fluid, crystal nucleation, and mineral/matrix interactions. Four main theories of shell formation are considered in relation to calcium transport, organic frameworks and amorphous minerals, facilitated diffusion, exosomes, and molluscan calcium cells. The approach is to review the standard models and to indicate where alternative developments are required.

2.1   INTRODUCTION AND APPROACH

Collecting, studying, and admiring molluscan shells is a major study for many biologists and some hobbyists. Most natural history museums have their own collections, and many libraries have some books on the subject. Put the word conchology into any search system or booksellers catalog and you will be given access to a fund of information in books costing anywhere from £5 to £20,000 (Sowerby's 1812 edition). If your interests are wider and extend to the biology of molluscs, the word to explore is malacology and there you will also find a wealth of information on the Internet.

So, why write reviews to describe the models and concepts that are now in the literature databases? Greenhalgh's (2010) booklet "How to read a paper" suggests that a scientific review should be centered on the formulation of a problem, require a search for "evidence-based" studies, involve a test of hypotheses, and only accept good methodology that reflects the design of the research. These are clearly the basics of science, but those who work on molluscan shells and those who review their findings only rarely put them in the clear context of raising questions and testing hypotheses.

In an attempt to bring some order into a variety of examples of biomineralization as diverse as corals, sponges, molluscs, and crabs, Karl Wilbur (1980) identified three basic levels of complexity. These were (1) mineral formation that occurred intracellularly within cells as in coccolithophorids, (2) extracellular mineralization induced by single cells such as skeleton deposition in sea urchin larvae, and (3) mineralization by a sheet of cells as in the epithelia of molluscs. This classification presented an interesting proposal because it completely ignored phylogeny and emphasized some sort of "systems analysis." Wilbur was, however, not alone in this approach of searching for some basic components in what might support a fundamental scheme. A number of scientists have always used what is known as the "August Krogh Principle" which suggests that for many biological problems there may be an organism on which it can be most conveniently studied (Krebs 1975). Many biologists would respond to this approach with the retort "but only if you believe in comparative physiology," while the chemists would probably issue a warning that "it shows you are at last concerned with basic processes."

It must be acknowledged that molluscs are not always the most convenient organisms on which to investigate a process. The shell is a remarkably good defense against many types of investigation, and gastropods and bivalves often do seem to work very slowly. There are, therefore, some aspects of biomineralization where it may be helpful to compare molluscan activities with some other examples such as eggshell formation in birds to fill this need. The average rate of shell formation in a mollusc involves a calcium flux of about 10^{-6} mol/cm^2/h, whereas the formation of the avian eggshell involves a calcium flux of about 10^{-5} mol/cm^2/h. These examples are toward the two extremes of the rates of calcification, but both illustrate what Wilbur would call epithelial biomineralization. Both systems are also capable of inducing the formation of crystals of calcium carbonate. Both do this in an enclosed space using an organic substrate and both types of organism seem to be able to control the mineralogy of the deposits (i.e., whether it is calcite and/or aragonite based). Both examples of mineral formation will stop if you add inhibitors of the enzyme carbonic anhydrase and both use cellular control systems to turn the mineralization systems on and off.

2.2   HOW TO BEGIN?

There are eight main classes of mollusc ranging from chitons to squids, but this review will only concentrate on the gastropods and bivalves. They live in three types of habitat (marine, freshwater, and terrestrial) and produce shells of one or two main minerals (calcite and/or aragonite). Even so, many of the experiments that have been performed would have been easier to discuss if the systems had involved less types of test organisms from fewer and more stable environments.

Most scientists will go through some sort of check list before they design their experiments. This is because it is almost impossible to interpret experiments if there are too many variables and that makes it very difficult to claim to be testing major hypotheses without making major mistakes. There are about 85,000 species of molluscs in the world today that have had roughly 545 million years of evolutionary divergence. Compare that with one surviving species of human and only 2 million years of change.

2.3 IDENTIFYING THE BIOLOGICAL COMPONENTS

The requirements for shell formation in molluscs involve an enclosed space associated with cells that can secrete and accumulate specific ions within those spaces. These ions are then moved from the liquid to the solid phase and converted into inorganic crystalline minerals that are initiated, or nucleated, and held together in organic matrices.

Figure 2.1 is a generalized drawing of a bivalve shell. It is not drawn to scale and is intended to be used to identify the biological components with which it may be involved. In a slightly provocative way, it answers Greenhalgh's first requirement. It has been drawn to help the formulation of at least five problems. So now we need to make it evidence-based.

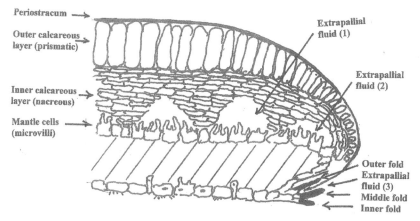

FIGURE 2.1 A vertical section through a bivalve shell and its mantle tissue. There are three folds in the mantle tissue edge. The space between the outer and middle folds form the periostracal gland that secretes the periostracum as a sheet that covers the shell. Note that the inner extrapallial fluid (1) is further away from the environmental water and contains most of the fluid while the outer extrapallial fluid (2) is smaller and in closer contact with the microvillus cells. The space in the periostracal gland also contains extrapallial fluid, and it should be noted that there is a specific basal cell at the base of the gland (3). All three fluids are involved in secreting and maintaining the shell. (The illustration is not to scale nor based on any species.)

On the outside of the shell is an impermeable sheet of denatured protein called the periostracum. This covers and protects the layers of crystals of calcium carbonate (calcite and/or aragonite) that are embedded in thin layers of soluble and insoluble matrix that form the shell. On the inside of the shell layer is a small gap called the extrapallial space. This is filled with an extrapallial fluid that is secreted by the epidermal sheet of mantle cells. From this it is possible to derive five questions. These are first, where are the anatomical compartments where the minerals are deposited? Second, what are the extrapallial fluids? Third, how do you nucleate and grow crystals? Fourth, what do shell proteins do? And finally, are there any theories about how this all fits together?

2.3.1 QUESTION 1: WHERE ARE THE ANATOMICAL COMPARTMENTS?

Molluscs have a remarkably versatile epidermis that covers their body. It is referred to as the mantle or pallium where it covers the viscera. This layer of "skin" can be used for functions as diverse as locomotion, respiration, and osmoregulation but, unlike the human skin, it is not covered in a layer of dead cells. There are basically three types of cell in the molluscan epidermis, namely, "mucus" cells, ciliated cells, and cells bearing microvilli on their apical surface (Simkiss, 1988). There is considerable diversification in the function and distribution of these cell types over the body but there is one particular region, the mantle, that is involved in generating the shell on the outside of the molluscs body.

The shell-forming epithelium of the mantle has been described in detail for a number of bivalve molluscs (McElwain & Bullard, 2014) and gastropods (Zylstra et al., 1978; McDougall et al., 2011). Where there are two different layers of the mineralized shell, the outermost region is usually made up of long crystals of prismatic calcite that are embedded in a sclerotized organic matrix and covered on the outside by a sheet of protein called the periostracum. Beneath the prismatic layer is the inner layer often composed of aragonite arranged like a stack of plates to form the nacreous layer. The whole shell is impregnated with a matrix of carbohydrates, lipids, and proteins that are trapped both within (i.e., intracrystalline) and between (i.e., intercrystalline) layers of inorganic minerals.

There has been considerable interest in the way that these layers of organic matrix might initiate the nucleation of the inorganic crystals. There are two theories involved, that is, the heteroepitaxial system where crystals form at selected sites on the matrix or the mineral bridge model where superimposed holes in the organic sheets permit a single crystal to pass through a number of stacked layers. This bridge system would result in one crystal carrying a continual mineral lattice through a number of matrix layers (Nudelman, 2015). There are some difficulties with this theory if different minerals occur where the nacreous and prismatic layers meet. There are, however, spacial differences in gene expression in some mantle tissues (Gardener et al., 2011) where different secretory processes are associated with prism and nacre deposition (Marie et al., 2012).

The organic matter of the shell that holds the crystals together is generally credited with at least three other functions. First, it may determine the type of mineral that is deposited in the shell; second, when bound together with the mineral layer, it produces a composite material of increased strength; and third, it may be involved with the organization of the mineral layers that are variously described as homogenous, prismatic, foliated, crossed lamellar, or other complex structural forms (Watabe, 1988). Clearly, the physical properties of the shell are influenced by both their composite form and their spatial arrangements, so that the strength of the shell differs significantly from its individual components. This is reflected physically in terms of the modulus of tensile strength, compression, bending strength, and hardness (Table 2.1). The mineralogy and the composite structure have contributed important physical properties that have undoubtedly influenced the types of protection that the shell has evolved in different kinds of molluscs (Currey, 1988, 1999; Vincent, 1982).

TABLE 2.1   Strength of Molluscan Shell Materials (Currey, 1988; Vincent, 1982).

Structure	Tension (MPa)	Compression (MPa)	Bending (MPa)	Hardness (kg mm^{-2})
Prismatic	60	250	140	162
Nacreous	80	380	220	168
Crossed lamella	40	250	100	250
Homogeneous	30	250	80	–

The shell of the adult mollusc is covered on its outer surface by a sheet of protein (the periostracum) and on the inner surface by the mantle epithelium. At the outer edge of this layer the mantle becomes folded, typically into two or three ridges that are referred to as the outer (nearest to the shell), middle, and inner mantle folds. It is the space between the outer and middle folds that secretes the periostracum, and this region is then referred to as the periostracal groove and by attaching to the outer region of the shell it converts the extrapallial space into a closed compartment (Fig. 2.1). This seal may occasionally be broken, particularly if a gastropod retracts into its shell (Nakahara, 1991). A number of attempts have been made with dyes and other fluids to confirm that there is a functional seal, keeping the extrapallial fluid away from the external environment, and for the purposes of this review, it will be considered normally to be intact. As the mollusc grows, the periostracum continues to be secreted and it is toughened by a denaturing of the proteins with a quinone "tanning" type of process (Gordon & Carriker, 1980; Waite, 1983). The structure and functions of the periostracum are so surprising that they deserve a small section on their own.

The structure of the periostracal groove in the primitive "living fossil" *Neotrigonia* has recently been described by Checa et al. (2014). As in all bivalves, there is a basal cell or group of cells at the bottom of this groove and it is these cells that secrete the protein sheet or pellicle that rapidly darkens as it becomes cross linked or tanned into a harder secretion. This is the periostracum, and as it continues to be formed, the outer surface becomes coated in a glycoprotein and starts to develop calcified nodules or "bosses" of calcite until it eventually emerges from the groove. The remarkable feature of *Neotrigonia* is that all the normal activities of shell formation, that is, secretion of protein, tanning, coating with glycoprotein and synthesis of crystals have already started while it is still in this simple groove of the epidermis. Within that space the epidermal cells also extend their microvilli by a distance of about 100 nm, and the whole process continues as the periostracum develops the layer of prismatic crystals that will eventually cover the mantle. Descriptions of some of the variants of these processes in other molluscs have been explored in detail by Saleuddin and Petit (1983) and are discussed in detail by Kocot et al. (Chapter 1, this volume).

So there are now three possible compartments that might be involved in shell formation, namely the larger inner extrapallial space between the shell and the mantle cells; the smaller outer extrapallial region that is so much closer to the microvilli of the mantle cells that they may contact the crystal nuclei; and the space between the outer and middle folds of the mantle edge where the periostracum is formed in what is then referred to as the periostracal groove.

2.3.2   QUESTION 2: WHAT IS EXTRAPALLIAL FLUID?

The extrapallial fluid is the film of liquid that receives all the secreted products of the mantle cells that are required to form and maintain the shell. It is present in all three of the compartments so the question arises as to whether there are one, two, or three varieties or states of this fluid.

There are relatively few analyses of the inorganic ions in these fluids as they lie beneath the shell and form only a thin layer of fluid (Table 2.2). The composition of this solution in marine bivalves shows a striking similarity to sea water whereas the fluid from freshwater molluscs is more dilute and clearly secreted from the cells of an osmoregulating animal. What one might have expected is that the fluids in contact with the calcareous shell would have been in equilibrium with the solubility of calcium carbonate or, if the shell was still being formed, it might be slightly supersaturated. The solubility of calcium carbonate minerals is normally expressed as the product of the concentrations of calcium and carbonate ions and is referred to as the solubility product or K_{sp}. This value differs for the different forms of calcium carbonate. Thus calcite, the least soluble of these crystals has a log K_{sp} value of −8.42, aragonite has a value of −8.22, vaterite −7.60 with amorphous calcium carbonate being very variable but the most soluble. Unfortunately, the value of K_{sp} is not a particularly useful concept in biological situations because it tends to vary with both the size of the crystals and the time taken for them to equilibrate with the solution (Williams, 1976). Even more important, it depends on the ionized rather than the total concentrations of the inorganic components in the fluids. Thus, when Misogianes and Chasteen (1979) looked at the concentration of calcium in the extrapallial fluid of the mussel *Mytilus edulis* they recorded a total level of 9.8 mM, but that included calcium bound to five proteins so that 84.7% of the calcium was complexed to small molecules and only 15.3% of the calcium was represented as the free ion.

TABLE 2.2　Concentrations of Ions in Extrapallial Blood or Sea Water Samples (Crenshaw, 1972; Moura et al., 2003).

Species	Fluid	Na (mM)	K (mM)	Mg (mM)	Ca (mM)	Cl (mM)
	Fresh water					
Anodonta cygnea	Extrapallial	15.5	0.46	4.3	3.14	17.1
Anodonta cygnea	Blood	14.8	0.36	4.4	2.95	17.1
	Sea water	427	18.0	53.0	9.2	49.6
Mytilus edulis	Extrapallial	442	9.5	58.0	10.7	477
Crassostrea virginica	Extrapallial	441	9.4	57.0	10.5	480

When these experiments were originally performed the main object of the analyses was to determine if the extrapallial fluid was supersaturated with calcium carbonate as this would suggest that the shell might be forming crystals out of this fluid. The results are, however, of limited value unless a specific mineral is formed. For example, the amorphous form of calcium carbonate has no fixed lattice and is only found when nucleation fails and growth of the solid phase becomes less organized. If these conditions persist the first solid material is likely to be amorphous, but that might then slowly transform into the more insoluble states of vaterite, aragonite, and finally calcite. This somewhat empirical situation is referred to as the Ostwald Lussac law of ripening (Williams, 1989) and it implies that the most soluble form of a polymorphic material may be the first solid phase to emerge. Given this situation, it is difficult to interpret the results of what an extrapallial fluid should contain, although there are always other details in a good experiment. When Crenshaw (1972) collected some extrapallial fluid samples, he used an indwelling catheter. He then noticed that if the bivalve closed its valves it was followed, after about 15 min, by a rise in the calcium level and a fall in the pH of the extrapallial fluid. This appeared to be due to a decline in the oxygen available to the bivalve and led to the accumulation of succinic acid. This situation persisted until the oyster reopened its valves and it suggested that the calcareous minerals of the molluscan shell were buffering the extrapallial fluid during the period of hypoxia (Crenshaw & Neff, 1969).

Most inorganic shells of molluscs are composed of the calcium salts of the main inorganic buffers that are found in the body, that is, the phosphates and carbonates. The fundamental equation for forming such calcareous crystals is $Ca^{2+} + HCO_3^- = CaCO_3 + H^+$. It is a proton releasing reaction and typically results in acidosis that should also be detectable at the site of shell formation unless these ions are being involved in other processes that displace the equilibrium.

It may be concluded that the composition of the extrapallial fluid may vary with ongoing activities so that chemical concepts, such as the solubility products of various crystal deposits, are difficult to apply in biological situations. As a result a different approach for assessing the activities of the ions in the extrapallial fluid(s) was attempted by using an *in vitro* preparation to investigate ion movements by measuring the electrical potential across this tissue (Istin & Kirschner, 1968). That work detected a large difference of about 50 mV across the tissue, but it was the mantle surface facing the shell that was positively charged. If this persisted *in vivo*, it would make it more difficult to accumulate calcium onto the shell, and it unfortunately appeared to be chloride ions rather than calcium ions that were crossing the epithelium. The possible effect of bicarbonate ions appeared to have been largely neglected. Subsequent work on mantle tissues from both freshwater and marine clams gave similar electrical potential differences with the shell side still positively charged. What had been expected was that the existence of an electrical potential across the mantle tissue would provide evidence for the transfer of ions onto or into the shell. It was subsequently discovered that the results were mainly due to the diffusion of calcium ions and chloride ions along concentration gradients and that any charge effects were actually mediated through an effect on other ions (Sorenson et al., 1980). This was an interesting but not very helpful contribution in trying to understand the movement of charged ions across a membrane that might have been secreting calcium and carbonate ions. Perhaps the most detailed study of the electrophysiology of the mantle epithelium was subsequently undertaken by Coimbra et al. (1988) using the freshwater bivalve *Anodonta cygnea*. The approach used both intact as well as stripped epidermal cells. Radioisotopes were used to track particular ions, while microelectrodes measured specific ion activities and millivolt potentials that could be modified with short circuit currents or enzyme inhibitors. This detailed work showed that as long as there were identical solutions on the two sides of the epithelium there was virtually no potential, but if the concentration of calcium ions was reduced on the shell-facing side a positive charge developed. There was no evidence for involving a calcium ion pump but the shell side of the epithelium had a large permeability for calcium ions, that is, roughly 10 times that of sodium, potassium, or chloride. They suggested a sodium/potassium ATPase system linked to bicarbonate produced by carbonic anhydrase might be one part of a transport system providing calcium and carbonate for shell formation. The epithelium was found to be clearly sensitive to carbon dioxide levels leading to speculations about the acid/base balance of the tissue (Moura et al., 2003). This aspect of biomineralization will be considered again in Section 2.4.1.

This now seems to be one of the situations when it might be helpful, to know the type of membrane potentials that exist in other organisms that form mineralized shells. Such measurements by Hurwitz et al. (1970) were taken from electrodes across the shell gland of a laying bird. The mucosal surface was about 10 mV negative, relative to the serosal layer, that is, the opposite orientation to that of the mollusc mantle.

So far, this review has considered four frustrating experiments. The search for a compartment that contained the ions necessary to form a shell revealed three possible compartments, and it was not certain that the extra-pallial fluid that they contained was supersaturated enough to produce a calcium carbonate mineral. The mantle epithelium that produced the fluid also did not seem to be actively pumping calcium ions into the extrapallial space. There are a number of possible explanations for these results. The most obvious is that shell formation may be an intermittent event in the mollusc or it may be easily shut down by the experimental procedures. A second set of suggestions would be that the initial hypotheses were wrong. The electron micrographs of the mantle of the bivalve *Pinctada radiata* showed that the epithelial cells were within less than 1 μm of the aragonite shell (Nakahara, 1991). Any fluid in this area would probably not be the same as the bulk extrapallial fluid that has been extracted for analysis and it therefore represents a separate functioning compartment. In addition, some samples of extrapallial fluid in *Mercenaria mercenaria* appeared to contain blood proteins suggesting that there may be leakage between the extrapallial fluid and the blood under experimental conditions (Yin et al., 2005). The design of the research in this work and the methodologies that were used forestalled the initial test of hypotheses about these extrapallial fluid(s).

Clearly, there are more details to be studied in interpreting the many functions in which the extrapallial fluid may be involved. Thus, for example, Hattan et al. (2001) purified a calcium-binding protein that represented over 50% of the organic matter in the extrapallial fluid of *M. edulis* and that appeared to regulate the growth of calcium carbonate crystals.

2.3.3   IDENTIFYING THE CHEMICAL COMPONENTS

Many of the approaches to understanding chemical reactions have been formulated by simplifying or controlling the conditions under which the studies have been undertaken. As such, they are often difficult to apply in a quantitative way to biological situations that are full of interfering components and very variable time scales.

2.3.4   QUESTION 3: HOW DO MOLLUSCS NUCLEATE AND GROW CRYSTALS?

A saturated solution is one that has dissolved a solid, such as a crystalized mineral, until it is unable to dissolve any more. A supersaturated solution is one that contains that little bit more. It is apparent straight away that there is a problem in those definitions, and if you wait long enough, the solution will revert to being a saturated one. The situation has a time and energy requirement and biological systems manipulate them both.

Calcium pumps are present in most cells and they control the cyto-plasmic content of this ion at a very low level. The enzyme carbonic anhy-drase is also present either in solution or attached to a variety of proteins. It is, therefore, relatively easy to envisage a fluid that becomes supersaturated with calcium and carbonate ions that may persist for some time in that state as there are a number of energy barriers involved in the formation of crystal lattice structures from hydrated ions.

In the absence of a preexisting surface, the formation of a new phase (such as a crystal of calcite) from another phase (such as an extrapallial fluid) would be called homogeneous nucleation. It is what many biologists hope to observe but most chemists doubt if it ever occurs. The problem is that very small clusters of ions have extremely large surfaces in relation to their volume so that they tend to dissolve again before they can grow. Perhaps even more important is the fact that when an ion passes from solu-tion into a lattice structure, it may have to shed some bound water so that the free energy of the new phase is less than that of the solvated phase. One of the ways to facilitate this is to provide a preexisting surface. This is known as heterogeneous nucleation and it occurs extensively in biological systems that incorporate either preexisting crystal lattices, or organic molecules with charged sites. Unfortunately, it is extremely difficult to relate model systems to experiments with biological matrices. As an example, it is possible to extract water soluble proteins from many biological shells but it is not clear whether this is a protein that is derived from within crystals or from between crystals. There is similarly an insoluble matrix. Is this the same as a soluble matrix except that it has been attached to a structural protein or is it entirely different? Do the extraction methods using acids or water define the proper-ties of subsequent tests? Is the system hydrophobic or hydrophilic before or after extraction and are the subsequent tests based on isolated groups or flexible molecules. Do calcium or carbonate ions influence the reactive properties of the matrix proteins before, during or after mineralization, that is, do the ions fold the proteins? There is a tendency in molluscan research to regard the mineralized matrix as a single entity whereas sections of a decalci-fied matrix may show a separate nucleation matrix followed by a spherulite-containing matrix followed by a columnar matrix as a temporal sequence from the initiation to the completion of the mineralized structures. Searching for nucleation sites from within shells is clearly a very complex system in biological materials but there are some excellent reviews of some of these very demanding experiments (Mann, 1989, 2001; Wheeler and Sikes 1989).

There is one other aspect of mineral formation that should be noted. If the level of supersaturation of the saline continues to rise, and there are only poorly organized or sparse nucleation sites then amorphous, rather than crystalline, minerals are likely to form. In that situation, as has already been explained, the calcium carbonate minerals will be produced in the sequence amorphous > vaterite > aragonite > calcite (Ostwald Lussac's law), whereas the series of insolubility is calcite > aragonite > vaterite > amorphous (Mann, 1986; da Silva & Williams, 1991).

It is apparent that these two series contain the possibility of a paradox but it all depends on the local conditions that dictate how and when calcium carbonate forms. And that directs you back to the aqueous conditions that exist around and within molluscan cells.

2.3.5   QUESTION 4: WHAT DO SHELL PROTEINS DO?

An important review of the main matrix proteins from the outer layer of the molluscan shell was published by Zhang and Zhang (2006). These proteins had been isolated and identified, together with some comments on their structure and suggestions as to their role in shell formation. They reported results for 13 of the main proteins in the nacreous layers of 2 species of *Haliotis*, 2 species of *Pinctada*, 2 species of *Biomphalarisa*, and 1 of *Pinna*, together with results for 10 proteins from the calcite layer of *Mytilus*, *Pinctada*, *Crassostrea*, *Painopecten*, *Adamussium*, *Atrina*, and *Pinna*. There were also eight proteins that occurred in both layers of shell. For the moment, it is the potential properties that were attributed to these proteins. These included some structural features, some sites that might block the various crystal faces of particular minerals, some that favored particular planes, some that induced step functions that might encourage crystal growth, some with strongly acidic groups, and some with enzyme sequences similar to carbonic anhydrase (i.e., nacreins) that can induce or act as negative regulators. It is obvious that with the right properties in the right places the shell matrix proteins could interact with the shell minerals in a wide range of ways.

This was a great start but a further major change in the study of matrix proteins occurred in 2011 when the first whole genome of a mollusc, the limpet *Lottia gigantea* was released into the public domain. This made it possible to produce a combination of proteomic and transcriptomic studies of the acid soluble and acid insoluble matrix proteins. Thus, for example, the *Lottia* shell matrix contained three proteins with strong similarities to nacreins, one of which also contained a secretion signal. The presence of a chitin framework with silk-like and strongly acidic proteins also suggested a key structural component for shell formation (Mann et al., 2012). The subsequent revelations of molecular biology that have been exposed as the genomes and protein repertoires of various molluscs have been discussed by Kocot et al. (Chapter 1, this volume).

A recent analysis of gene expression in different mantle areas and especially in the periostracal groove have confirmed and extended the structural continuity between the three folds at the mantle edge and two regions of shell formation, that is, the columnar and nacreous layers. The outer fold of the mantle is confirmed as being responsible for much of the secretory activities of the periostracum groove while the middle fold is a sensory organelle and the inner fold is involved with muscular activities (Gardener et al., 2011). One of the recurrent problems in understanding the biomineralization processes of the prismatic and nacreous layers of the shell was also studied by Marie et al. (2012) who were able to track "prismatic proteins" to the mantle edge and "nacreous proteins" to the central mantle pallium. This confirmed the finding that different secretory repertoires were producing prismatic calcite or nacreous aragonite presumably by transcriptional regulation of a single genome. This would similarly explain how mantle cells were able to transiently change the type of mineral they induced when repairing a damaged shell. What came as a major surprise, however, was the disparity between the genes in the nacre forming cells in the mantle of a bivalve (*Pinctada maxima*) and a gastropod (*Haliotis asininae*) that had so little in common that Jackson et al. (2010) concluded that they must be an example of convergent evolution rather than a closely related genome.

There are several large groups pursuing these molecular approaches in order to get an understanding of the evolution of the proteins in the molluscan shell (Jackson et al., 2006; Marin et al., 2008), for influencing studies on the of deposition of prism and nacre in the pearl oyster shell (Marie et al., 2012) for identifying the role of the calcifying mantle and shell of the pearl oyster (Joubert et al., 2010) and transcriptomes for the mantle tissue of *Mytilus edulis* (Freer et al. 2014). These samples have been used to identify which genes are turned on and off and which proteins are released at various times and at different sites. There are clearly some very effective interactions between switching on certain genes to form matrix proteins and inducing specific crystal forms (Belcher et al., 1996). Most molluscan shells are, however, composed of mainly calcium-based minerals, and, like the genes and proteins, these inorganic components will also be controlled by cells.

2.3.6   QUESTION 5: HOW IS CALCIUM TRANSPORTED TO FORM THE SHELL?

In 1971, E. W. Sutherland won the Nobel Prize for his discovery of the second messenger. It was soon to become a major component of cell biology as it explained how primary signals such as nerve impulses or water soluble hormones could pass across the cell membrane into the cytoplasm and trigger a controlled response (Clapham, 2007). One of the most common of these second signals involves the release of calcium ions inside the cell so that Berridge et al. (2000) would write about "the versatility and universality of calcium signalling." It soon became apparent, however, that intracellular levels of ionic calcium were maintained at about 100 nM whereas extracellular calcium was around 1 mM. There is a gradient of roughly 10,000 times between the high calcium activity levels outside and the extremely low levels inside a cell whereas even a small tenfold increase of ionic calcium inside a cell could be lethal. This implies that it would be very difficult to transport calcium ions through the cells of an epithelium such as the mantle tissue of a mollusc. There seemed to be only three ways around this difficulty, namely, to move the calcium around the cells, to encapsulate the calcium in a membrane bound vesicle, or to transport it in a non-ionic form.

2.3.6.1   THE PARACELLULAR ROUTE

If calcium ions are transported around the cells of an epithelium they require spaces between the cells together with some form of control over the types of materials that could use them. These intercellular channels provide what is termed the paracellular route. A variety of "tight" and "gap" junctions between cells are known but in terms of restricting access and controlling the passage of ions they do not seem to be very selective. These spaces are, however, strongly advocated as providing the route taken by calcium ions in a number of vertebrate studies (Hoenderop et al., 2005) including the intestinal absorption of calcium in humans (Khanal & Nemere, 2008). This route was also suggested by Neff (1972) whose ultrastructural study of the mantle of the clam exposed intercellular deposits of calcium between the mantle cells. These results were extended by Bleher and Machado (2004) who added some data on the rate of passage of ions along these intercellular routes.

2.3.6.2   THE VESICULAR ROUTE

There are a variety of vesicles in the cytoplasm of many cells, some of which are formed by an infolding of the plasma membrane (endocytosis) and some that are derived from intracellular organelles such as multivesicular bodies, lysosomes, Golgi complex, and endoplasmic reticulum systems (Simkiss, 2015). Many of these vesicles may be associated with microtubules and deliver their contents by exocytosis out of the cell, thereby providing the possibility of a vesicular route for calcium transport. By these means, a large number of different types of vesicle may be involved with the movement of inclusions into, out of, or around the cell. In an interesting calculation, Addadi et al. (2006) found that it would require 10^5 times the volume of a saturated calcium carbonate solution to move the same mass of solid aragonite through a cell. Clearly, transporting some materials in the solid state could have a lot of benefits over handling very large volumes of solution.

2.3.6.3   CALCIUM BINDING

The third suggestion for moving calcium ions through a cell without raising the calcium level of the cytoplasm would be to bind the ions to a protein. The recent proteome reports on molluscan proteins have revealed a large number of calcium-binding proteins in both shells and mantle samples. At the present time, there is only limited information on the strength of the binding of calcium by these proteins or on their relationships to other groups of proteins.

2.4   FOUR THEORIES

This section poses the final Greenhalgh (2010) question. Is it possible from the information in this review to propose a coherent theory that provides some testable hypotheses on how shells are formed? Remember that the motivations for much of the current work on molluscan shells were initiated by objectives that ranged from measuring the environmental impact of global warming to understanding evolution, and from fundamental science to commercial exploitation. They have exploited new technologies such as proteomes, transcriptomes, biocomposites, amorphous minerals, the moulding of crystal products, and even the promise of *in vitro* pearl culture. Despite that, it is possible to derive four possible attempts to explain molluscan shell formation from the viewpoint of cell physiology.

The first approach is based on the electrophysiological data obtained from the freshwater bivalve *Anodonta cygnea*. The second provides a basis for converting amorphous calcium carbonate into an organized crystalline structure. This theory does not provide explanations for the cellular accumulation of the ions or their transport to the mineralization sites. The third theory deals largely with calcium transport across epithelia and the acid–base consequences of mineral formation. It does not involve itself with nucleation, crystal growth, or matrix interactions. The fourth theory involves the genetic and transcriptome approaches to a variety of stress adaptations in the oyster and compares shell formation with shell repair. All four theories are shown in an abbreviated form in Figure 2.2.

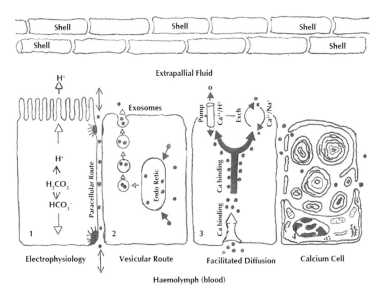

FIGURE 2.2   Diagram illustrating four possible systems involving the movement of ions at sites of shell formation (calcium ions are shown as black circles). (1) The paracellular route between two epithelial cells permits the diffusion of calcium ions into the extracellular fluid (Bleher & Machado, 2004). Intracellular ion movements produce an acid/base response resulting in the acidification of the extracellular fluid and possible release of calcium deposits (Machado et al., 1990). (2) The vesicular route delivering amorphous calcium carbonate to the extrapallial fluid (Addadi et al., 2006; Addadi & Weiner, 2014). The theory has been extended to imply that calcium leakage into the cell results in its transport into the endoplasmic reticulum and the production of vesicles of amorphous calcium carbonate. This route may also release exosomes from cell debris (Zang et al., 2012). (3) Facilitated diffusion of calcium through the epithelium based on the interpretation of avian shell formation (Jonchere et al., 2012). Calcium ions bound onto calbindin are delivered to calcium pumps and ion exchangers. (4) Molluscan calcium cells form intracellular deposits that may subsequently release calcium and bicarbonate ions through membrane pores (Sminia et al., 1977; Watabe & Blackwelder, 1980).

2.4.1   THEORY 1: ELECTROPHYSIOLOGY AND PROTONS

As explained in Section 2.3.2, this approach has utilized some of the best techniques to track the movement of ions across the bivalve mantle (Istin & Kirschner, 1968; Coimbra et al., 1988). The short circuit data indicate, however, that there is an acidification of the extrapallial fluid that bathes the shell (Machado et al., 1990). This is balanced across the outer mantle epithelium by a Cl^-/HCO_3^- exchange but it is a counter intuitive result in that the shell would be expected to dissolve in the presence of increasing levels of protons. The ability to deposit calcium carbonate on the shell is also complicated as no calcium pump can be detected, and it has been suggested that calcium moves across the mantle through the paracellular route (Bleher & Machado, 2004). Such a system usually invokes diffusion. A possible explanation of shell formation in the freshwater mussel has been related to a study throughout a year since there are seasonal variations in the composition of both the hemolymph and the extrapallial fluids that may explain some of these anomalies. Environmental acidosis may also have considerable effects on the calcareous deposits that are formed and resorbed by the mussel during an annual cycle (Moura et al., 2003).

2.4.2   THEORY 2: FRAMEWORKS, AMORPHOUS MINERALS, AND VESICLES

In the early 1980s, Bevelander and Nakahara (1969, 1980) observed organic compartments that they interpreted as preformed structures for nacre formation. Shortly afterward, Wiener et al. (1983) proposed an organic framework with silk fibroid and acidic protein linings. These two concepts were to become a new approach for an understanding of the molluscan matrix and its relation to aragonite nacre.

According to the theory of Addadi et al. (2006) and Addadi and Weiner (2014), the formation of the nacreous layer of the molluscan shell involves the creation of a structural framework of chitin onto which hydrophobic silk gel and acid-rich proteins are adsorbed to form a compartment for the growth of aragonite tablets. There has been some discussion as to whether these compartments precede the mineralization process or whether the growth of the aragonite crystals contributes to the organic phase and its dehydration. A detailed model of the components of this structural framework and their relation to shell formation has been discussed by Furuhashi et al. (2009). It is suggested that once the structural framework has been developed, an amorphous form of calcium carbonate is transported in vesicles from the epithelial cells of the mantle into these organic structures to nucleate the formation of aragonite crystals. This raises the question as to whether or not the compartments are influenced as a result of mineral formation, but once the process has started, there may be further interactions between the organic proteins and the growth of the inorganic crystals. It is an extensive theory in that it encompasses a large database of experimental work including Weiner's demonstration of amorphous calcium carbonates at the sites of biomineralization in many different invertebrate shells (Weiss et al., 1991).

An intracellular pathway of vesicles carrying ions in an amorphous state to the sites of biomineralization in molluscs was originally proposed by Abolins-Krogis (1965, 1970) and reviewed for a variety of invertebrates by Simkiss (1976). Such vesicular routes are involved in a whole range of functions but mineralization is a particularly challenging one as the concentration of calcium ions in the blood is so much higher than its level in the cytoplasm of the cell.

The properties of vesicles that contain amorphous calcium salts have been studied using both the very soluble calcium carbonate-based granules of *Helix aspersa* (Greaves et al., 1984; Taylor et al., 1986) and the highly insoluble amorphous phosphate/pyrophosphate granules of the same organism (Taylor & Simkiss, 1989). As can be seen (Table 2.3), the amorphous calcium carbonate is very soluble and could probably drive the nucleation and formation of calcareous minerals. The structural and analytical properties of the amorphous minerals have been investigated using a random network model of amorphous calcium carbonate to explore their potential involvement in biomineralization (Simkiss, 1991). There is clear evidence in the abalone *Haliotis discus hannai* for an amorphous calcium carbonate-binding protein that has been isolated by Huang et al. (2009) and this could provide an insight into how such an unstable molecule could be controlled (Su et al., 2013). Further evidence for such a system was provided by Jacob et al. (2011) who discovered an extremely thin layer of amorphous calcium carbonate between the periostracum and the prismatic layers of the shells of *Hyriopsis cumingii* and *Diplodon chilensis*; the implication being perhaps, that this indicates where the first nucleation of calcium carbonate crystals occurs. There are hints that some amorphous materials may already contain "ghosts" of the various minerals into which they transform and the concept has now reached the stage where there is a question of "how many amorphous calcium carbonates are there?" (Cartwright et al., 2012).

TABLE 2.3   Composition and Solubility of Amorphous Calcium Carbonate and Amorphous Calcium Phosphate Granules Obtained from Different Organs of the Snail *Helix aspersa* (Simkiss, 1976).

Type of granule	Composition of granule (molar ratios)				Solubility in Krebs saline (mM)			Solubility in snail saline (mM)			
	Ca^{2+}	Mg^{2-}	PO_4^{3-}	CO_3^{2-}	Ca^{2-}	PO_4^{3-}	pH	Ca^{2+}	Mg^{2+}	PO_4^{3-}	pH
None (saline)			–		2.3	1.3	7.6	4.2	4.8	0.0	7.9
Foot	1	: 0.06	: 0.06	: 1.04	9.4	0.6	7.7	12.5	4.5	0.2	7.8
Hepatopancreas	1	: 0.96	: 1.35	: 0.53	2.1	1.5	7.7	5.3	8.7	0.3	7.8

The theory as presented does not seem to make many suggestions as to the origin of prismatic layers in those molluscs such as *Mytilus edulis* that have an outer calcite layer of this mineral, although Bevelander and Nakahara (1980) have described envelopes and compartments in detail in the prisms of *Pinctada radiatae.* They have also suggested that there may be considerable similarities between the protein components of the two types of compartments that house calcite and aragonite minerals. These possibilities have been advanced by Nudelman et al. (2007) who found clear evidence for acidic proteins in both prismatic and nacreous layers but with no direct evidence for the presence of amorphous minerals at either of those sites.

It is not clear as to whether the carbonate ions are contained in the "delivery vesicles" or whether carbonic anhydrase (Nielsen & Frieden, 1972) or nacrein-related proteins (Norizuki & Samata, 2008) provide the anion during crystallization. This may not be as straightforward as expected as Miyamoto et al. (2005) have found that nacrein can also act as a negative regulator of calcification in *Pinctada jucata*, although binding it to other proteins may help to regulate the formation of the aragonite crystal.

2.4.3   THEORY 3: FACILITATED DIFFUSION

This is an alternative approach to the "framework, gel, vesicle, and amorphous mineral theory" of biomineralization in that it contains none of those components. It is a cell-based approach derived from vertebrate studies rather than invertebrates so it could be ignored here for having little experimental relevance to molluscan shells. The interest in calcium ions as a crucial component of the "second messenger" signaling system has shown that the cytoplasm has to be maintained at the very low concentration of around

1 μM. That concentration would appear to rule out any direct involvement in transporting the ion through the cytoplasm but it has also stimulated a lot of interest in calcium movements across epithelia (Hoenderop et al., 2005).

The theory to explain calcium movements across epithelia was initially developed by Wasserman (Feher et al., 1992) and is referred to as "facilitated diffusion." Calcium enters the cell passively through a specific ion channel in the serosal surface of the cell and is then buffered by the protein calbindin that also facilitates its diffusion through the cytoplasm (Lambers et al., 2006). Free calcium is then released by calcium ATPase into crossing the apical membrane to where it is involved in extracellular shell formation. Carbon dioxide similarly diffuses into the cells and is hydrated via carbonic anhydrase to form bicarbonate ions that are pumped by a variety of ion exchange systems to produce carbonate ions.

A worked example of this system has compared proteomic and transcriptomic analyses of the shell gland of the fowl in laying or resting states. The results have been interpreted using the facilitated diffusion model for intracellular calcium transport (Jonchere et al., 2012). It is abbreviated in Figure 2.2 simply to show that the initial concern that transcellular calcium movements would be potentially lethal to cells were no longer applicable. Calcium ions pass through an open calcium channel that is closed if intracellular calcium ions accumulate near the channel exit. The calcium-binding protein calbindin plays a dual role in that it removes these ions, so as to keep the channel open, and binds them so as to buffer the cytoplasm. The calbindin calcium is then delivered to a calcium ATPase pump at the basolateral membrane and transported into the lumen of the shell gland where it reacts with carbonate to form a calcite shell. A number of features should be emphasized. The calcium pump is a calcium/proton exchanger together with a calcium/sodium exchanger. There is a proton flux away from the shell and a large potassium movement toward the mineralizing surface. This might help to explain the potassium current in the electric potential observed in the molluscan mantle. Overall, there are a total of 37 highly expressed ion transport genes balancing the fluid movements associated with avian shell formation.

At the present time, there is limited data on the molluscan mantle that could be used to test the application of the facilitated diffusion model although the molluscan genome appears to contain a rich selection of interesting genes. Evidence exists for gated calcium channels with calcium dependent inactivation (Kits and Mansveider 1999). There is a plasma membrane calcium ATPase (Lopes-Lima et al., 2008) with calconectin (Duplat et al., 2006), some calmodulin (Fang et al., 2008), and more importantly calbindin (Jackson et al., 2007) as calcium-binding proteins together with inositol triphosphate sensitivity (Fink et al., 1988). Clearly, the data are not sufficient for a definitive study and equally clearly it need not be in conflict with the Weiner/Addadi studies that could be interpreted as an endoplasmic reticulum safeguard removing free calcium ions.

2.4.4   THEORY 4: MOLLUSCAN "CALCIUM CELLS" AND EXOSOMES

Virtually, every study on shell formation in molluscs eventually ends up discovering intracellular deposits of calcium in some isolated cells. The cells are sometimes in connective tissue, sometimes in the mantle and other organs, and occasionally free in the blood. They are variously described as a subgroup of amoebocyte as granulocytes, as interstitial cells, as calcium cells, or as hemocytes (Watabe et al., 1976). Any attempt to quantify the distribution of calcium stores in the molluscs usually identifies a significant quantity variously positioned around the animal (Greenaway 1971; Sminia et al., 1977). Recently, Mount et al. (2004) described these granulocyte cells as having macrophage-like functions and moving through the outer mantle of the "oyster" to deliver crystals of calcium carbonate at sites of shell damage. It was a well-documented example of cell-mediated biomineralization.

A few years later a consortium of roughly 80 researchers published an analysis of the genome of the oyster *Crassostrea gigas* (Zhang et al., 2012) commenting on the stress responses of the organism and the complexity of its shell formation. They identified 259 shell proteins including fibronectin but found no evidence of silk-like material. Genes coding for fibronectins and integrins suggested that hemocytes and other active cells might be the source of exosomes near the sites of biomineralization.

Exosomes were discovered about 30 years ago but have only been treated as a major development in cell biology in the past decade. They are small (c. 100 nm diameter) microvesicles that are shed by many cell lines and are found in virtually all body fluids (blood, seminal fluid, urine, etc.) of many vertebrates and invertebrates. They variously carry packages of enzymes, hormones, and nucleic acids and fuse with other cells around the body (Keller et al., 2006). Currently, exosomes and other microvesicles are seen as organelles that can form intercellular communications systems capable of distributing molecules such as mRNA, DNA and various enzymes so as to modify the activities of other cells (Thery, 2011; Raposo & Stoorvogel, 2013). These activities have attracted a great deal of attention in both the medical and pharmaceutical sciences. The vesicles occur in almost all the body fluids at concentrations of up to 10^{10}/ml and they have recently been classified into two groups by Cocucci and Meldolesi (2015). The vesicles are formed in the endosome systems of cells with the exosomes arising from the multivesicular bodies.

The second group of vesicles arise directly beneath the plasma membranes and are referred to as ectosomes. The discovery, therefore of exosomes around the sites of oyster shell damage could have potentially great significance. Were they involved as part of a defense system or a repair mechanism? Alternatively, suppose that what was proposed as the "electro-physiology theory of shell formation" (Section 2.4.1) is actually intercon-nected by exosome interactions to what was suggested as the "calcium cell theory" (Section 2.4.3). "Dream on?" Perhaps but if the 80-odd researchers of the oyster genome project ever got together some must have had such thoughts. Certainly, 10 of the original participants extended the results of the genome analysis of the oyster to question whether the "chitin, silk and acidic protein" model of shell formation really reflected the activity of the mantle tissue or whether there should be a reconsideration of the model for shell formation (Wang et al., 2013). What this group had discovered was that when the shell was damaged many of the proteins that had previously been identified in the oyster shell as secretory products of the oyster mantle had actually been produced by a wide range of other organs around the body of the oyster. The repair of the damaged shell involved significant activity by other cells such as those in the digestive gland, the gills, and gonads with the transport of their products to the site of usage. Is biomineralization a whole body experience? Exosomes in action by intercellular signaling?

2.5 CONCLUSIONS

The simplistic view of shell formation in molluscs is that it consists of three components. The first is the involvement with the movement of calcium cations (usually out of the cell), the second is the production of carbonate anions (probably via carbonic anhydrase), and the third is the regulation of the mineral products by organic molecules (almost certainly involving proteins). These are all basic properties of aerobic cells so the literature is full of the search for the genetic assembly of a "biomineralization tool kit." That concept would be severely damaged if it turned out not to exist as a single system but to involve convergent evolution (Jackson et al., 2010). The following information provides an exemplar example.

Watabe and Wilbur (1960) demonstrated that the decalcified matrix of a molluscan aragonite shell would deposit calcite crystals if it was remineralized in a calcite-forming snail. It was a clear demonstration that the organic composition of the shell matrix determined the mineral form of the inorganic crystals. Roughly 20 years later, Wheeler and Sikes (1984, 1989) showed that it was possible to separate the matrix components to form crystal-facilitating and crystal-inhibiting fractions. The question that was then raised was "why was there an inhibitory factor in shell matrices" and one of the answers was that "life originated in a pre-Cambrian sea that had a very high calcium content" and the invertebrates used mucins to inhibit being smothered by calcium deposits that crystallized out of the sea water. As the calcium level of the post-Cambrian seas fell the situation changed and the mucins evolved into a shell binding system (Marin et al., 2000). What originated as a calcium-inhibiting product evolved into a calcium-mineralizing system.

There is now a medical approach that considers that the vertebrate bone system is too sensitive, and it has a tendency to produce ectopic pathological calcification (Jahnen-Dechent, 2004). The soft tissues of the body would mineralize unless they are continually exposed to inhibitory proteins such as Fetuin-A (Jahnen-Dechent et al., 2011). In the absence of such inhibitory molecules, ectopic biomineralization occurs in the cardiovascular system and smooth muscle tissues (Kapustin et al., 2015; Leopold, 2015) The implication is that biomineralization is a potentially dangerous system and that we should expect the physiological process to be regulated by an inhibitory counter activity. That would require a major rethinking of biomineralization processes.

The second novel reconsideration of calcification processes is the suggestion that they may involve the cellular derived systems of exosome/ectosome vesicles (Section 2.4.4). A frequently cited theory for bone formation in vertebrates is that the process is initiated by cells that release vesicles and nucleate mineral deposition (Anderson et al., 2005; Golub, 2011; Shapiro et al., 2015). A possible molluscan form of this theory is as follows. Molluscs, as with all other studied species, have cells that release exosomes into their body fluids. The cells contain cytoplasmic endosomal systems that form the multivesicular bodies, and lysosomal structures with the potential for controlling ion fluxes (Scott & Gruenberg, 2010). The endosome system produces exosomes and ectosomes along two somewhat different routes and pass the membrane vesicles into the blood system. There are two studies where the exosomes are found in oysters that are remineralizing damaged shells. The first study found that oyster shell proteins were released from most of the organs in the recovering mollusc. The response appeared to be associated with exosomal activity (Wang et al., 2013) The second study traced hemo-cyte cells, with exosomes containing calcite particles, being transported to the site of shell damage (Johnstone et al., 2014). A possible interpretation of these studies is shown in Figure 2.3.

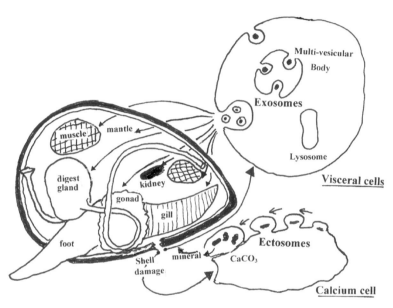

FIGURE 2.3   Illustration of how damage to the shell could be repaired by the exosomes and ectosomes of a bivalve. The cells of the viscera produce exosomes from the multivesicular bodies causing the release of shell proteins from most of the molluscs organs (data derived from Wang et al, 2013). The mineralization of the shell proteins is facilitated by the release of calcium carbonate deposits formed by ectosomes in calcium cells (data derived from Johnstone et al., 2014). Much more data would be necessary to support these concepts although analagous work on bones provides some support.

Finally the history of an absence of good control over the breeding of oysters and the problems of diseases have led to a change in the technology away from the traditional use of oyster farms and mantle implants (Simkiss & Wada, 1980) to the study of isolated cells and tissue culture. The aim is to produce pearls *in vitro* by using cell cultures. It seems quite likely that the responses of isolated mantle cells probably involve different cellular responses from the pathways that were involved in the forming the original shell. Despite this, the general observation on the use of primary cultures of mantle epithelial cells to grow pearls in tissue culture would seem to indicate that the distinction between the biomineralization products of epithelial and isolated cells may not be very large (Barik et al., 2004) and that the *in vivo* and *in vitro* activities will show many new similarities (Awaji & Machhii, 2011; McGinty et al. 2012). On the technology side, however, there is a question as to the extent that organs and cells can be studied in isolation. The study of pearl sacs and isolated cell cultures could involve some very fundamental experiments in biomineralization.

The studies that will benefit from these approaches should stimulate the somewhat isolated areas of enzymology, endocrinology, hormonal specificity, and the implications of secretory signals. This information is necessary in many experiments where much could be learnt if one could block or stimulate one variable and then measure the response.

As far as the aspirations made in Section 2.2 of this review are concerned, one accepts the reality of mineralizing compartments and the revelations of the amorphous phase but the benefits of a "systems approach" are not there yet. There is relatively little in terms of a full testable theory for how shell formation occurs although it doesn't seem to be a particularly difficult system. But, if there are 85,000 species of molluscs that indulge in convergent evolution we will have to decide which few should be considered in detail. Perhaps as Freer et al. (2014) commented on completion of their transcriptome study of mussel mantle tissue, what is now required is "the careful application of bioinformatics with an essential hands-on approach and where researchers use their savvy to derive the best from multiple data sets." To that one would simply add two sets of comments. First, where in the physiology of shell formation does calcium meet the carbonate ion? What conditions determine whether carbonic anhydrase is intracellular or extracellular? Are we dealing with the diffusion of a neutral gas or the pumping of a bicarbonate anion? Where is the proton sink? What is the calcium pathway? Where do the nacreins fit in and what are the properties of proteins that dictate the structure of crystals?" Second, and more briefly, is the molluscan mantle a manufacturing or a packaging station? "Bring on the bioinformatics."

KEYWORDS

- molluscan
- biomineralization
- extrapallial space

- **extrapallial fluid**
- **calcium fluxes**
- **ectosomes**
- **exosomes crystal formation**
- **gene expression**
- **matrix inhibitors**

REFERENCES

Abolins-Krogis, A. Electron Microscopic Observations on Calcium Cells in the Hepatopancreas of the Snail *Helix aspersa. Ark. Zool.* **1965**, *18*, 85–92.

Abolins-Krogis, A. Electron Microscope Studies of the Intracellular Origin and Formation of Calcifying Granules and Calcium Spherites in the Hepatopancreas of the Snail, *Helix pomatia. Z. Zellfors. Mikrosk Anat.* **1970**, *108*, 501–515.

Addadi, L.; Joester, D.; Nudelman, F.; Weiner, S. Mollusc Shell Formation: A Source of New Concepts for Understanding Biomineralization Process. *Chem.: Eur. J.* **2006**, *12*(4), 980–987.

Addadi, L.; Wiener, S. Biomineralization: Mineral Formation by Organisms. *Phys. Scr.* **2014**, *89*, 1–13. DOI:10.1088/0031-8949/89/9/098003.

Anderson, H. C.; Garimella, R.; Tague, S. E. The Role of Matrix Vesicles in Growth Plate Development and Biomineralization. *Front. Biosci.* **2005**, *10*, 822–837.

Awaji, M.; Machhii, A. Fundamental Studies on *In Vivo* and *In Vitro* Pearl Formation–Contribution of Outer Epithelial Cells of Pearl Oyster Mantle and Pearl Sacs. *Aqua—BioSci. Monogr. (ABSM)* **2011**, *4*(1), 1–39. DOI:10.5047/absm.2011.00401.0001.

Barik, S. K.; Jena, J. K.; Ram, K. J. $CaCO_3$ Crystallization in Primary Culture of Mantle Epithelial Cells of Freshwater Pearl Oyster. *Curr. Sci.* **2004**, *86*(5), 730–734.

Belcher, A. M.; Wu, X. H.; Christensen, R. J.; Hansma, P. K.; Stucky, G. D.; Morse, D. E. Control of Crystal phase Switching and Orientation by Soluble Mollusc-shell Proteins. *Nature* **1996**, *381*, 56–58.

Berridge, M. J.; Lipp, P.; Bootman, M. D. The Versatility and Universality of Calcium Signalling. *Nat. Rev.* **2000**, *1*, 11–21.

Bevelander, G.; Nakahara, H. An Electron Microscope Study of the Formation of the Nacreous Layer in the Shell of certain Bivalve Molluscs. *Calcif. Tissue Res.* **1969**, *3*, 84–92.

Bevelander, G.; Nakahara, H. Compartment and Envelope Formation in the Process of Biological Mineralization. In *The Mechanisms of Biomineralization in Animals and Plants*; Omori, M., Watabe, N., Eds.; Tokai University Press: Tokyo, 1980; pp 19–27.

Bleher, R.; Machado, J. Paracellular Pathway in the Shell Epithelium of *Anadonta cygnea. J. Exp. Zool. A: Comp. Exp. Biol.* **2004**, *301*(5), 419–427.

Cartwright, J. H.; Checa, A. G.; Gale, J. D.; Gebauer, D.; Sainz-Diaz, C. I. Calcium Carbonate Polymorphism and its Role in Biomineralization: How many Amorphous

Calcium Carbonates are there? *Angew. Chem. Int. Ed. Engl.* **2012**, *51*(48), 11960–11970. DOI:10.1002/anie.201203125.

Checa, A. G.; Salas, C.; Herper. E. M.; Bueno-Peres, J. de Dios. Early Stage Biomineralization in the Periostracum of the 'Living Fossil' bivalve *Neotrigonia.* PLoS ONE **2014**, *9*(2), e90033. DOI:10.1371/journal.pone.0090033.Clapham, D. E. Calcium Signalling. *Cell* **2007**, *131*, 1047–1058. DOI:10.1016/j.cell.2007.11.026.

Cocucci, E.; Meldolesi, J. Ectosomes and Exosomes: Shedding the Confusion between Extracellular Vesicles. *Trends Cell Biol.* **2015**, *25*(*4*), 364–372.

Coimbra, J.; Machado, J.; Fernandes, P. L; Ferreira, H. G.; Ferreira, K. Electrophysiology of the Mantle of *Anadonta cygnea. J. Exp. Biol.* **1988**, *140*, 65–88.

Crenshaw, M. A. The Inorganic Composition of Molluscan Extrapallial Fluid. *Biol. Bull.* **1972**, *143*, 506–512.

Crenshaw, M. A.; Neff, J. M. Decalcification at the Mantle-shell Interface in Molluscs. *Am. Zool. (Integr. Compar. Biol.)* **1969**, *9*, 881–885.

Currey, J. D. Shell form and Strength. In *The Mollusca*; Trueman, E. R., Clarke, M. R., Eds.; Academic Press Inc.: San Diego, CA, 1988; Vol. 11, pp 183–210.

Currey, J. D. The Design of Mineralized Hard Tissues for their Mechanical Functions. *J. Exp. Biol.* **1999**, *202*, 3285–3292.

da Silva, J. J. R. F., Williams, R. J. P. *The Biological Chemistry of the Elements*; Clarendon Press: Oxford, 1991; p 561.

Duplat, D.; Puissegur, M.; Beduuet, L.; Rousseau, M.; Boulzaguet, H.; Milet, C.; Selios, D.; van W. A.; Lopez, E. Identification of Calconectin: A Calcium-binding Protein Specifically Expressed by the Mantle of *Pinctada margaritifera. FEBS Lett.* **2006**, *580*(10), 2435–2441.

Fang, Z.; Yan, Z.; Li, S.; Wang, Q.; Cao, W.; Xu, G.; Xiong, X.; Xie, L.; Zhang, R. Localization of Calmodulin and Calmodulin-like Protein and their Functions in Biomineralization in *P. jucata. Prog. Nat. Sci.* **2008**, *18*, 405–412.

Feher, J. J.; Fulmer, C. S.; Wasserman, R. H. Role of Facilitated Diffusion by Calbindin in Intestinal Calcium Absorption. *Am. J. Physiol.* **1992**, *262*, C517–C526.

Fink, L. A.; Connor, J. A.; Kaczmarek, L. K. Insitol Trisphosphate Releases Intracellularly Stored Calcium and Modulates Ion Channels in Molluscan Neurons. *J. Neurosci.* **1988**, *8*(7), 2544–2555.

Freer, A.; Bidgett, S.; Jiang, Y.; Cusack, M. Biomineral Proteins from *Mytilus edulis* Tissue Transcriptome. *Mar. Biotechnol.* **2014**, *16*, 34–45.

Furuhashi, T.; Schwarzinger, C.; Miksik, I.; Smrz, M.; Beran, A. Molluscan Shell Evolution with Review of Calcification Hypothesis. *Comp. Biochem. Physiol., B* **2009**, *154*, 351–371.

Gardener, L. D.; Mills, D.; Wiegand, A.; Leavesley, D.; Elizur, A. Spatial Analysis of Biomineralisation Associated Gene Expression from the Mantle Organ of the Pearl Oyster *Pinctada maxima. BMC Genomics* **2011**, *12*, 455–470. DOI:10.1186/1471-2164-12-455.

Golub, E. E. Biomineralization and Matrix Vesicles in Biology and Pathology. *Semin. Immunopathol.* **2011**, *33*(5), 409–417.

Gordon, J.; Carriker, M. R. Sclerotized Protein in the Shell Matrix of a Bivalve Mollusc. *Mar. Biol. (Berl.)* **1980**, *57*, 251–260.

Greaves, G. N.; Simkiss, K.; Taylor, M. G.; Binsted, N. The Local Environment of Metal Sites in Intracellular Granules Investigated by using X-ray-absorption Spectroscopy. *Biochem. J.* **1984**, *221*, 855–868.

Greenhalgh, T. *How to Read a Paper: The Basics of Evidence Based Medicine*; Wiley-Blackwell: London, 2010; p 238.

Greenaway, P. Calcium Regulation in the Freshwater Mollusc, *Limnaea stagnalis* (L.) (Gastropod: Pulmanata) II. Calcium Movements between Internal Calcium Compartments. *J. Exp. Biol.* **1971**, *54*, 609–620.

Hattan, S. J.; Laue, T. M.; Chasteen, N. D. Purification and Characterization of a Novel Calcium-Binding Protein from the Extrapallial Fluid of the Mollusc, *Mytilus edulis. J. Biol. Chem.* **2001**, *276*, 4461–4468.

Hoenderop, J. G. J.; Nilius, B.; Bindels, R. J. M. Calcium Absorption Across Epithelia. *Physiol. Rev.* **2005**, *85*, 373–422. DOI:10.1152/physrev.00003.204.

Huang, J.; Wang, H.; Cui, Y.; Zhang, G.; Zheng, G.; Liu, S.; Xie, L.; Zhang, R. Identification and Comparison of Amorphous Calcium-binding Protein and Acetylcholine-binding Protein in the Abalone, *Haliotus discus hanna. Mar. Biotechnol.* **2009**, *11*(5), 596–607.

Hurwitz, S.; Cohen, I.; Bar. A. The Transmembrane Electrical Potential Difference in the Uterus (Shell Gland) of Birds. *Comp. Biochem. Physiol.* **1970**, *35*(4), 873–878.

Istin, M.; Kirschner, L. B. On the Origin of the Bioelectrical Potential Generated by the Freshwater Clam Mantle. *J. Gen. Physiol.* **1968**, *51*, 478–496.

Jackson, D. J.; McDougall, C.; Green, K.; Simpson, F.; Worheide, G.; Degnan, B. M. A Rapidly Evolving Secretome Builds and Patterns a Sea Shell. *BMC Biol.* **2006**, *4*, 40. DOI:10.1186/1741-7007-4-40.

Jackson, D. J.; Worheide, G.; Degnan, B. M. Dynamic Expression of Ancient and Novel Molluscan Shell Genes During Ecological Transitions. *BMC Evol. Biol.* **2007**, *7*, 160. DOI:10.1186/1471-2148/7.160.

Jackson, D. J.; McDougall, C.; Woodcroft, B.; Moase, P.; Rose, R.; Kube, M.; Reinhardt, R.; Rokhsar, D. S.; Montagnani, C.; Joubert, C.; Piquemal, D.; Degnan B. M. Parallel Evolution of Nacre Building Gene Sets in Molluscs. *Mol. Biol. Evol.* **2010**, *27*(3), 591–608. DOI:10.1093/molbev/msp278.

Jacob, D. E.; Wirth, R.; Soldati, A. L.; Wehrmeister, U.; Schreiber, A. Amorphous Calcium Carbonate in the Shells of Adult Unionoida. *J. Struct. Biol.* **2011**, *173*, 241–249.

Jahnen-Dechent, W. Lot's Wife's Problem Revisited: How We Prevent Pathological Calcification. *Biomineralization* **2004**, *15*, 243–268.

Jahnen-Dechent, W.; Heiss, A.; Schafer, C.; Kettler, M. Fetuin-A Regulation of Calcified Matrix Metabolism. *Circ. Res.* **2011**, *108*, 1494–509.

Johnstone, M. B.; Gohad, N. V.; Falwell, E. P.; Hansen, D. C.; Hansen, K. M.; Mount, A. S. Cellular Orchestrated Biomineralization of Crystalline Composites on Implant Surfaces by the Eastern Oyster *Crassostrea virginica. J. Exp. Mar. Biol Ecol* **2014**, *463*, 8–16.

Jonchere, V.; Brionne, A.; Gautron, J.; Nys,Y. Identification of Uterine Ion Transporters for Mineralisation Precursors of the Avian Eggshell. *BMC Physiol.* **2012**, *12*, 10. DOI:10.1186/1472-6793.

Joubert, C.; Piquemal, D.; Marie, B.; Manchon, L.; Pierrat, F.; Zanella-Cleon, I.; Cochennec-Laureau, N.; Gueguen, Y.; Montagnant, T. Transcriptome and Proteome Analysis of *Pinctada margaritifera* Calcifying Mantle and Shell: Focus on Biomineralization. *BMC Genomics* **2010**, *11*, 613. DOI:10.1186/1471-2164-11-613.

Kapustin, A. N.; Chatrou, M. L. L.; Drozdov, I.; Zheng, Y.; Davidson, S. M.; Soong, D.; Furmanik, M.; Sanchis, P.; de Rosales, R. T. M.; et al. Vascular Smooth Muscle Cell Calcification is Mediated by Regulated Exosome Secretion. *Circulation Res* **2015**, *116*, 1312–1323.

Keller, S.; Sanderson, M. P.; Stoeck, M. A.; Altevogt, P. Exosomes: From Biogenesis and Secretion to Biological Function. *Immunol. Lett.* **2006**, *107*, 102–108.

Khanal, R. C.; Nemere, I. Regulation of Intestinal Calcium Transport. *Ann. Rev. Nut.* **2008**, *28*, 179–196.

Kits, K. S. Mansvelder, H. D. Voltage Gated Calcium Channels in Molluscs: Classification, Ca^{2+} Dependent Inactivation, Modulation and Functional Roles. *Invert. Neurosci.* **1999,** *2*(1), 9–34.

Krebs, H. A. The August Krogh Principle 'For Many Problems There is an Animal on Which It Can be Most Conveniently Studied'. *J. Exp. Zool.* **1975,** *194,* 221–226.

Lambers, T. T.; Mahieu, F.; Oancea, E.; Hoold, L.; deLange, F.; Mensenkamp, A. R.; Voets, T.; Nilius, B.; Clapham, D. F.; Hoenderop, J. G.; Bindels, R. J. Calbindin-D_{28K} dynamically Controls TRPV5-mediated Ca^{2+} Transport. *EMBO* **2006,** *25*(13), 2978–2988.

Leopold, J. A. Vascular Calcification: Mechanisms of Vascular Smooth Muscle Cell Calcification. *Trends Cardiovasc. Med.* **2015,** *25,* 267–274.

Lopes-Lima, M.; Bleher, R.; Forg, T.; Hafner, M.; Machado, J. Studies on a PMCA-like Protein in the Outer Mantle Epithelium of *Anodonta cygnea*: Insights on Calcium Transcellular Dynamics. *J. Comp. Physiol. B* **2008,** *178*(1) 17–25.

Machado, J.; Ferreira, K. G.; Ferreira, H. G.; Fernandes, P. L. The Acid–base Balance of the Outer Mantle Epithelium of *Anadonta cygnea. J. Exp. Biol.* **1990,** *150,* 159–169.

McDougall, C.; Green, K.; Jackson, D. J.; Degnan, B. M. Ultrastructure of the Mantle of the Gastropod *Haliotis asinine* and Mechanisms of Shell Regionalization. *Cells Tissues Organ.* **2011,** *194,* 103–107.

McGinty, E. L.; Zenger, K. R.; Jones, D. B.; Jerry, D. R. Transcriptome Analysis of Biomineralisation-related Genes within Pearl Sac: Host and Donor Oyster Contribution. *Mar. Genomics* **2012,** *5,* 27–33.

Mann, K.; Edsinger-Gonzales, E.; Mann, M. In Depth Proteomic Analysis of a Mollusc Shell: Acid-soluble and Acid-insoluble Matrix of the Limpet *Lottia gigantean. Proteome Sci.* **2012,** *10,* 28. www.protermsci.com/content/10/1/28.

Mann, S. Biomineralization in Lower Plants and Animals—Chemical Perspectives. In *Systematics Association*; Leadbeater, B. S. C., Riding, R., Eds.; 1986; Vol. 10, p 39–54.

Mann, S. Crystallochemical Strategies inBiomineralization . In *Biomineralization. Chemical and Biochemical Perspectives*; Mann, S., Webb, J., Williams, R. J. P., Eds.; VCH: Wienheim, 1989, pp 35–62.

Mann, S. *Biomineralization: Principles and Concepts in Bioinorganic Material Chemistry*; Oxford University Press: Oxford, 2001; p 198.

Marie, B.; Joubert, C.; Tayale, A.; Zanella-Cleon, I.; Belliard, C.; Piquemal, D.; Cochenn-Laureau, N.; Marin, F.; Gueguen, Y.; Montagnani, C. Different Secretory Repertoires Control the Biomineralization Processes of Prism and Nacre Deposition of the Pearl Oyster Shell. *Proc. Natl. Acad. Sci.* **2012,** *109*(51), 20986–20991.

Marin, F.; Corstjens, P.; de Gaulejac, B.; Vrind de Jong, E.; Westbroek, P. Mucins and Molluscan Calcification. *J. Biol. Chem.* **2000,** *275*(27), 20667–20675.

Marin, F.; Luquet, G.; Marie, B.; Medakovic, D. Molluscan Shell Proteins, Primary Structure, Origin and Evolution. *Curr. Topics Dev. Biol.* **2008,** *80,* 209–276.

McElwain, A.; Bullard, S. A. Histological Atlas of Freshwater Mussels (Bivalvia Unionidae). *Villosa nebulosa* (*Ambleminae lampsilini*), *Fusconaia cerina* (*Ambleminae pleurobemini*), and *Strophitus connasaugaensis* (*Unioninae Anodontini*). *Malacologia* **2014,** *57*(1), 99–239.

McGinty, E. L.; Zenger, K. R.; Jones, D. B.; Jerry, D. R. Transcriptome Analysis of Biomineralisation-related Genes within the Pearl sac: Host and Donor Oyster Contribution. *Mar. Genomics* **2012,** *5,* 27–33.

Misogianes, M.; Chasteen, N. D. A Chemical and Spectral Characterization of the Extrapallial Fluid of *Mytilus edulis. Anal. Biochem.* **1979,** *100,* 324–334.

Miyamoto, H.; Miyoshi, F.; Kohno, J. The Carbonic Anhydrase Domain Protein Nacrein is Expressed in the Epithelial cells of the Mantle and Acts as a Negative Regulator in Calcification in the Mollusc *Pinctada jucata. Zool. Sci.* **2005,** *22*(3) 311–315.

Mount, A. S.; Wheeler, A. P.; Paradkar, P.; Snider, D. Hemocyte-mediated Shell Mineralization in the Eastern Oyster. *Science* **2004,** *304*, 297–300.

Moura, G.; Almeida, M. J.; Machado, M. J.; Vilarinho, L.; Machado, J. The Action of Environmental Acidosis on the Calcification Process of *Anadonta cygnea* (L.). In *Biomineralization, Formation, Diversity, Evolution and Application*; Kobayashi, I., Ozawa, H., Eds.; Tokai University Press: Tokyo, Japan, 2003; pp 178–182.

Nakahara, H. Nacre Formation in Bivalve and Gastropod Molluscs. In *Mechanisms and Phylogeny of Mineralization in Biological Systems.* Suga, S., Nakahara, H., Eds.; Springer-Verlag, 1991; pp 343–350.

Neff, J. M. Ultrastructure of the Outer epithelium of the mantle in the Clam *Mercenaria mercenaria* in Relation to Calcification of the Shell. *Tissue Cell* **1972,** *4*(4), 591–600.

Nielsen, S. A.; Frieden, E. Carbonic anhydrase Activity in Molluscs. *Comp. Biochem. Physiol. B: Comp. Biochem.* **1972,** *41*(3), 461–468.

Norizuki, M.; Samata. T. Distribution and Function of the Nacrein-related Proteins Inferred from Structural Analysis. *Mar. Biotechnol. (NY)* **2008,** *10*(3), 234–241.

Nudelman, F. Nacre biomineralisation: A Review on the Mechanisms of Crystal Nucleation. *Semin Cell Dev. Biol.* **2015,** *46*, 2–10.

Nudelman, F.; Chen, H. H.; Goldberg, H. A.; Weiner, S.; Addadi, L. Lessons from Biomineralization: Comparing the Growth Strategies of Mollusc Shell Prismatic and Nacreous Layers in *Atrina rigida. Faraday Dis.* **2007,** *136*, 9–25.

Raposo, G.; Stoorvogel, W. Extracellular Vesicles: Exosomes, Microvesicles and Friends. *J. Cell Biol.* **2013,** *200*(4), 373. DOI:10.1083/jcb201211138.

Saleuddin, A. S. M.; Petit, H. P. The Mode of Formation and Structure of the Periostracum. In *The Mollusca*; Wilbur, K. M., Ed.; Academic Press: San Diego, CA, 1983; Vol. 4(1), pp 199–234.

Scott, C. C.; Gruenberg, J. Ion Flux and the Function of Endosomes and Lysosomes: pH is Just the Start. *Bioessays* **2010,** *33*, 103–110.

Shapiro, I. M.; Landis, W. J.; Risbud, M. V. Matrix Vesicles: Are They Anchored Exosomes? *Bone* **2015,** *79*, 29–36.

Simkiss, K. Intracellular and Extracellular Routes in Biomineralization. In *Calcium in Biological Systems*; Duncan, C. J., Ed.; *Symp. Soc. Exp. Biol.* **1976,** *30*, 423–444.

Simkiss, K. Molluscan Skin (excluding cephalopods). In *The Mollusca*; Wilbur, K. M., Ed.; Academic Press: San Diego, CA, 1988; Vol. 11, pp 11–35.

Simkiss, K. The Processes of Biomineralization in Lower Plants and Animals—An Overview. In *Biomineralization in lower plants and animals*. Leadbeater, B. S. C.; Riding, R. Eds.; Systematics Association. 1986, 30, 19–37.

Simkiss, K. Amorphous Minerals and Theories of Biomineralization In *Mechanisms and Phylogeny of Mineralization* in *Biological Systems.* Suga, S.; Nakahara, H. Eds.; Springer-Verlag: Tokyo 1991, 375–382.

Simkiss, K. Calcium Transport Across Calcium-regulated Cells. *Physiol. Zool.* **1996,** *69*, 343–350.

Simkiss, K. Extracellular Vesicles and Biomineralization. *J.J. Physiol.* **2015,** *1*(2), 009.

Simkiss, K.; Wada, K. Cultured Pearls-commercialised Biomineralisation. *Endeav., New Ser.* **1980,** *4*(1), 32–37.

Sminia, T.; de With, N. D.; Bos, J. L.; van Nieuwmegen, M. E.; Witter, M. P.; Wondergem, J. Structure and Function of the Calcium Cells of the Freshwater Pulmonate Snail *Lymnaea stagnalis. Netherlands J. Zool.* **1977**, *27*(2), 195–208.

Sorenson, A. L.; Wood, D. S.; Kirschner, L. B. Electrophysiological Properties of Resting Secretory Membranes of lamellibranch mantle. Interactions between Calcium and Potassium. *J. Gen. Physiol.* **1980**, *75*(1), 21–37.

Sowerby, J. *The Mineral Conchology of Great Britain*. Printed by Meredith, B.; Arding, W., 1812; Vols. 1–6.

Su, J.; Liang, X.; Zhou, Q.; Zhang, G.; Wang, H.; Xie, L.; Zhang, T. R. Structural Characterization of Amorphous Calcium Carbonate-binding Protein: An Insight into the Mechanism of Amorphous Calcium Carbonate Formation. *Biochem. J.* **2013**, *453*(2), 179–188.

Taylor, M. G.; Simkiss, K. Structural and Analytical studies on Metal Ion-containing Granules. In *Biomineralization, Chemical and Biochemical Perspectives*; Mann, S.; Webb, J.; Williams, R. J. P., Eds.; VCH: Wienheim, 1989; pp 427–460.

Taylor, M. G.; Simkiss, K.; Greaves, G. N. Amorphous Structure of Intracellular Mineral Granules. *Biochem. Soc. Trans.* **1986**, *14*, 549–552.

Thery, C. Exosomes: Secreted Vesicles and Intercellular Communications. *F1000 Biol. Rep.* **2011**, *3*, 15. DOI:10.3410/B3-15.

Vincent, J. F. V. *Structural Biomaterials*; MacMillan Press: London, 1982; p 206.

Waite, J. H. Quinone-tanned Scleroproteins. In *The Mollusca*. Wilbur, K. W., Ed.; Academic Press: San Diego, CA, 1983; Vol. 4, pp 467–504.

Wang, N.; Li, Li.; Zhu, Y.; Du, Y.; Song, M.; Chen, Y.; Huang, R.; Que, H.; Fang, X.; Zang, G. Oyster Shell Proteins Originate from Multiple Organs and their Probable Transport Pathway to the Shell Formation Front. *PLoS ONE* **2013**, *8*(6), e66522. DOI:10.1371/journal.pone0066522.

Watabe, N. Shell Structure. In *The Mollusca*; Trueman, E. R., Clark, M. R., Eds.; Academic Press: San Diego, CA, 1988; Vol. 1, pp 69–104.

Watabe, N.; Meenakshi, V.; Blackwelder, P.; Kurtz, E. M.; Dunkelberger, D. G. Calcareous Spherules in the Gastropod *Pomacea paludosa*. In *The Mechanisms of Mineralization in the Invertebrates and Plants*; 'Watabe, N., Wilbur, K. M., Eds.'; Univ. South Carolina Press: Columbia, SC, 1976; pp 283–308.

Watabe, N.; Blackwelder, P. L. Ultrastructure and Calcium Localization in the Mantle of the Freshwater Gastropod *Pomacea paludosa* During Shell Regeneration. In *The Mechanisms of Biomineralization in Animals and Plants*; Omori, M., 'Watabe, N., Eds.'; Tokai University Press: Tokyo, 1980; pp 131–144.

Watabe, N.; Wilbur, K. M. Influence of the Organic Matrix on Crystal Type in Molluscs. *Nat. Lond.* **1960**, *188*, 344.

Weiner, S.; Traub, W.; Lowenstam, H. A. Organic Matrix in Calcified exoskeletons. In *Biomineralization and Biological Metal Accumulation*; Westbroek, P., de Jong, E. W., Eds.; D. Reidel Publishing Company, 1983, pp 205–224.

Weiss, I. M.; Tuross, L.; Lahau, M.; Leiserowitz, L. Mollusc Larval Shell Formation: Amorphous Calcium Carbonate is a Precursor for Aragonite. *J. Exp. Zool.* **1991**, *293*, 478–491.

Wheeler, A. P.; Sikes, C. S. Regulation of Carbonate Calcification by Organic Matrix. *Am. Zool.* **1984**, *24*, 933–944.

Wheeler, A. P.; Sikes, C. S. Matrix–Crystal Interactions in $CaCO_3$ Biomineralization. In *Biomineralization, Chemical and Biochemical Perspectives*; Mann, S.; Webb, J.; Williams, R. J. P., Eds.; VCH: Wienheim, 1989, pp 95–131.

Wilbur, K. M. Cells, Crystals and Skeletons. In *The Mechanisms of Biomineralization in Animals and Plants*; Omori, M.; Watabe, N., Eds.; Tokai University Press: Tokyo, Japan, 1980, pp 3–11.

Wilbur, K. M.; Saleuddin, A. S. M. *Shell Formation*. In *The Mollusca*; Wilbur, K. M., Ed.; Academic Press: New York, 1983, *4*(1), pp 235–287.

Williams, R. J. P. Calcium Chemistry and its Relation to Biological Function. In *Calcium in Biological Systems*; Duncan, C. J., Ed.; Symposia of the Society for Experimental Biology. Cambridge University Press: Cambridge. 1976, *30*, pp 1–17.

Williams, R. J. P. The Functional Forms of Biominerals. In *Biomineralization, Chemical and Biochemical Perspectives*; Mann, S., Webb, J., Williams, R. J. P., Eds.; VCH: Wienheim, 1989; pp 1–34.

Yin, Y.; Huang, J.; Paine, M. L.; Reinhold, V. N.; Chasteen, N. D. Structural Characterization of the Major Extrapallial Fluid of the Mollusc *Mytilus edulis*. *Biochemistry* **2005**, *44*, 10720–10732.

Zang, G.; Fang, X.; Guo, X.; Li, li.; Luo, R.; Xu, F.; Yang, P.; Zhang, L.; Wang, X.; Qi, H.; et al. The Oyster Genome Reveals Stress Adaptation and Complexity of Shell Formation. *Nature* **2012**, *490*, 49–54. DOI:10.1038/nature11413.

Zhang, C.; Zhang, R. Matrix Proteins in the Outer Shells of Mollusca. *Mar. Biotechnol.* **2006**, *8* (6), 572–586.

Zylstra, U.; Boer, H.; H.; Sminia, T. Ultrastructure, Histology, and Innervation of the Mantle Edge of the Freshwater Pulmonate Snails *Lymnaea stagnalis* and *Biompharalaria pfenferi*. *Calcif. Tissue Res.* **1978**, *26*(1), 271–282.

CHAPTER 3

DRILLING INTO HARD SUBSTRATE BY NATICID AND MURICID GASTROPODS: A CHEMO-MECHANICAL PROCESS INVOLVED IN FEEDING

ERIC S. CLELLAND[1] and NICOLE B. WEBSTER[2]

[1]Bamfield Marine Sciences Center 100 Pachena Rd., Bamfield, BC, Canada V0R 1B0.E-mail: research@bamfieldmsc.com

[2]Department of Biological Sciences, University of Alberta, Edmonton, AB, Canada T6G 2E9. E-mail: nwebster@ualberta.ca

CONTENTS

ABSTRACT

Evidence of the drilling muricid and naticid gastropods, shells with precise holes, are evident on marine coastlines the world over. Although there are other groups that bore into hard substrates substrates, including other gastropods, snails of the Naticidae (moon snails) and Muricidae (whelks and drills) are the most widespread and diverse, as well the most well studied groups. Drilling in these two families is convergent, the process involves alternating phases of mechanical rasping with a radula, and chemical dissolution with acid-produced by the accessory boring organ. Although some of the specifics vary, this gives them acces to their prey directly through the shell. The ABO is a complex organ that uses carbonic anhydrase and V-ATPase pumps to control acid production, in a process similar to mitochondria rich cells in other groups. The boreholes produced are diagnostic, allowing for a rare case of direct evidence of behavior and ecological interactions both in recent and fossil populations. Modern techniques have opened the door to newly feasible studies on the drilling process at a cellular and molecular level, and much is still not known about how drilling works, especially in taxa other than muricids and naticids.

3.1   INTRODUCTION

Traverse a marine coastline virtually anywhere in the world, and one is certain to find evidence that predatory gastropods have been at work. Most obvious on the shells of bivalves or snails, but also in other groups, is often a precisely drilled hole. This indicates that these animals were consumed by a drilling predator. The most familiar of these "boring" predators are the whelks and drills (Muricidae) and the moon snails (Naticidae), groups who almost exclusively drill the shells of their prey (Carriker, 1961). Boring in these groups is now known to result from a chemo-mechanical process: alternating mechanical rasping using the radula with chemical dissolution using acidic secretions from the accessory boring organ (ABO). Eventually piercing the shell, the proboscis is then inserted into the hole, and feeding begins. Naticids prey mostly on infaunal molluscs, usually boring in with a characteristic countersunk round hole, while muricids feed on more diverse prey, leaving a roughly circular hole.

3.2   DRILLING IN OTHER GROUPS

A number of diverse groups are capable of dissolving carbonates, generally as epibionts or endolithic burrowers: bacteria, algae, and fungi; sponges, bryozoans, turbellarians, phoronids, polychaetes, sipunculids, barnacles, and bivalves (Carriker, 1981; Carriker & Gruber, 1999; Carriker & Smith, 1969; Carriker & Yochelson, 1968; Katz et al.,2010). This is substrate boring, where the shell or other substrate is the target, rather than the tissue within (Carriker & Yochelson, 1968). Shell boring is a relatively rare predatory strategy; it has appeared only a few times in Gastropoda, Octopoda, Nematoda, and Turbellaria (Bromley, 1981; Kowalewski, 2002; Matsukuma, 1978; Sohl, 1969).

Within the Gastropoda, drilling is most well studied in the Naticidae and the Muricidae. There are several other gastropod families that include boring members, but these are generally poorly documented, with our understanding of their drilling mechanism, prey specificity, and borehole morphology varying widely (Kowalewski, 2002). The only known opisthobranch borer is *Vayssierea elegans*, a dorid nudibranch (family Okaidaiidae) that bores into spirorbid polychaetes by simultaneously applying the radula for mechanical abrasion and stomodeal secretions for chemical dissolustion in periodic bouts (Young, 1969). Within the pulmonates, three families are known to bore into other land snails. The oleacinid genus *Poiretia* scrape away the sides of the shell, leaving irregular holes (Helwerda & Schilthuizen, 2014; Schilthuizen et al., 1994; Wächtler, 1927). Similar large irregular holes are produced by zonitid snails such as *Aegopinella nitens* (Barker & Efford, 2004; Mordan, 1977). Rathouissid slugs in the genus *Atopos* have been observed drilling small circular holes into their microsnail prey (Schilthuizen & Liew, 2008; Schilthuizen et al., 2006).

There are a number of marine snails that drill. Some species of capulids (cap snails), such as *Capulus danieli*, are kleptoparasties. Some bore holes in their scallop hosts, while others simply notch the shell, allowing them to steal food, while not damaging their host (Matsukuma, 1978; Orr, 1962). These holes are oval or tear shaped, and may be surrounded by a telltale "home scar" or growth deformations. In the Nassaridae, newly settled *Nassarius festivus* bored cannibalistically into conspecifics when starved, although adults show no evidence of boring behavior (Chiu et al., 2010). The resulting borings left evidence of both mechanical and chemical action (Morton & Chan, 1997). There are other claims of nassarid drilling, but they are unsubstantiated (Morton & Chan, 1997). Two species of *Austroginella* are the only reported marginellids to bore bivalve prey. The shape of the crystals in the borehole suggests chemical dissolution, with no sign of radula marks. The holes produced are wide and circular at the outset, with a very small, irregularly shaped inner edge; this inner hole is smaller than the proboscis and is probably used for toxin injection rather than feeding (Ponder & Taylor, 1992). In the Buccinidae, two species of *Cominella* were reported to drill thin bivalves. Boreholes were described as indistinguishable from muricid drill holes, but no figures were presented (Peterson & Black, 1995).

A few families bore echinoderms, whether as parasites or predators (Kowalewski & Nebelsick, 2003). There is evidence that some platyceratids in the Paleozoic were boring parasites on blastoid and crinoids echinoderms, possibly using the drill hole to steal food from its host (Baumiller, 1990, 1993, 1996). Some Eulimidae, though all may be parasites and predators of echinoderms, actively penetrate the test of their echinoderm hosts (Crossland et al., 1991). Snails in the genus *Hypermastus* produce pit-shaped holes with a small terminal hole penetrating to the tissue. *Thyca* are obligate parasites on starfish of the genus *Linckia*, and leave a distinct trace made up of a circular groove surrounding a small hole (Neumann & Wisshak, 2009). Due to the shape of the hole, and the fact that they lack a radula, they are thought to chemically dissolve the holes (Warén, 1983). Several species of Cassidae drill into echinoderms. Drilling is done with a taenioglossate radula and sulfuric acid in buccal secretions (Hughes & Hughes, 1971, 1981). Rather than drilling a countersunk borehole, a circular groove is cut, and then punched out, leaving a circular to ragged outline on the test, with parallel walls (Nebelsick & Kowalewski, 1999). Some cymatid and tonnid gastropods also reportedly drill prey, but no details have been provided (Day, 1969; Morton & Miller, 1973).

3.3   THE DRILLING OF MECHANISM OF MURICID AND NATICID GASTROPODS

3.3.1   *HISTORY OF RESOLVING THE DRILLING MECHANISM*

Some 2300 years ago, Aristotle made the first surviving record that predatory marine snails were able to drill holes in the shells of bivalves to feed (Carriker, 1981; Jensen, 1951). Since boreholes made by muricids and naticids are characteristically cylindrical and smooth inside, Réaumur (1709) first concluded that their drilling was perhaps a chemical process, while in naticids, Schiemenz (1891) provided the first serious evidence for chemical drilling. He argued that the radula was too soft, and the proboscis insufficiently mobile, to bore holes through hard substrate. When he noticed that the secretion from the "boring gland," a hemispherical boss that underlies the ventral lip of the proboscis, had reddened litmus paper, he hypothesized that an acid secretion was responsible for producing the hole (Fretter & Graham, 1962; Schiemenz, 1891). Additional support for a chemical drilling mechanism came from observations that the diameter of an engorged boring gland and that of the borehole were the same (Hirsch, 1915), and a demonstration that secretions obtained from the boring gland of *Natica* could remove the gloss from a polished shell (Ankel, 1937). In later studies, observing the etching effect of ABO on shell was hit or miss; Ankel (1938) proposed that "calcase," a compound active under only certain conditions, was responsible for the chemical dissolution the shell. Other investigators around this time argued that only the rasping action of the radula on the substrate (a mechanical model) was necessary to drill a hole. The naticid boring gland was first described in the mid-nineteenth century by Troschel (1854), who believed it was simply a muscular sucker used to hold the prey during the drilling process, but did remark on the acidic secretions produced by salivary glands (Carriker & Gruber, 1999). The muscular nature of the structure and a failure to color litmus paper (Fischer, 1922) did not support a chemical role for the boring gland. This conclusion was supported by the investigations of Pelseneer (1925) and Loppens (1926) and later studies (Jensen, 1951; Ziegelmeier, 1954), who concluded that drilling must be solely mechanical in nature. This view was held until the 1960s, when ultrastructural analysis of the naticid boring gland (now termed the ABO) showed similarities to the ABO of muricids (Bernard & Bagshaw, 1969).

The mechanism underlying muricid boring was also contested. As with naticids, a mechanical process that involved only the rasping action of the radula on the substrate was favored by a number of researchers (Graham, 1941; Jensen, 1951; Pelseneer, 1925). The muricid ABO, first described in 1941 (Fretter), was also initially believed to act as a sucker to hold the snail in place during drilling (Fretter & Graham, 1962). This makes sense as, during the drilling process, the ABO everts and swells into a fungiform structure with a diameter about equal to that of the proboscis (Carriker, 1981; Carriker & Gruber, 1999; Fretter & Graham, 1962; Webb & Saleuddin, 1977). The role of the ABO in the drilling process was demonstrated through amputation experiments in *Urosalpinx* and *Eupleura*, where only those animals possessing both an ABO and proboscis were able to bore, and furthermore, both the ABO and the proboscis of these muricids regenerated rapidly following amputation (Carriker, 1959). Subsequent studies in 1972 further confirmed that both the proboscis and ABO are essential when drilling hard substrate (Carriker et al., 1972; Carriker & Van Zandt, 1972a). Carriker was a proponent of the chemo-mechanical model of boring, whereby secretions from the ABO containing acid, enzymes, and chelators are used to soften the calcareous substrate, which is then rasped away by the radula. Thus, the borehole is drilled by an alternation of chemical dissolution and mechanical rasping until the hole is complete (Carriker et al., 1963; Fretter & Graham, 1962). Following the removal of the periostracum by radular rasping, Carriker and colleagues recorded periods of apparent inactivity lasting from a few minutes in naticids to approaching an hour in muricids, which were followed by brief periods of rasping. Periods of inactivity and rasping were alternated until a hole was either completed or abandoned (Carriker et al., 1963). Carriker interpreted these extensive periods of inactivity as when chemical dissolution was occur-ring and subsequent studies confirmed that secretions from the ABO contain enzymes, chelators, and acid (reviewed in Carriker, 1978; Carriker, 1981; Carriker & Gruber, 1999). The acidic nature of the secretions from the ABO were clearly demonstrated by Carriker et al., (1967) using a glass-shell model and pH-sensitive glass electrodes (a new development at the time). They determined that the secretions from the ABO of *Urosalpinx* could fall as low as pH 3.8. Moreover, etchings in shells and artificial substrates produced by ABO secretions were similar to those produced by exposure to HCl and EDTA (Carriker & Williams, 1978). Enzymatic secretions from the ABO have also been documented to include chelators (Bernard & Bagshaw, 1969; Person et al., 1967). (See reviews by Carriker, 1981; Carriker & Gruber, 1999; Carriker & Williams, 1978; Kabat, 1990.)

The most recent advance in our understanding comes from the identifica-tion of V-ATPases in the secretory epithelium of the muricid ABO (Clelland & Saleuddin, 2000). This, in addition to various anatomical and biochemical features of the ABO, including the long-established presence of carbonic anhydrase (CA) in the ABO of both muricid and naticid gastropods (Chétail & Fournié, 1969; Smarsh, 1969; Webb & Saleuddin, 1977), established a means of HCl production. This chemo-mechanical model of drilling is now widely accepted.

3.3.2 SHAPE OF BOREHOLES

Boreholes can have a characteristic size and shape based on the properties of the boring mechanism and the size of the borer. Both the chemical weak-ening and radular abrasion of the prey shell affect borehole characteristics

(Kabat, 1990). The diameter of the borehole is generally indicative of the size of the snail; large holes having been drilled by large predators, small holes by small predators (Kabat, 1990). Boreholes left by muricids tend to be uniformly cylindrical, with virtually straight edges; the diameter of the hole matching the size of the ABO of the drilling animal. Boreholes produced by naticids are shallower and appear to be countersunk due to the noticeably beveled edges. Moreover, the shape of the hole conforms to that of the ABO, to the extent that damage to an ABO (due to surgical manipulation or natural causes) can be identified in the borehole (Carriker & van Zandt, 1972b).

Should the thickness of the shell exceed the drilling limits of the predator, or something interrupts the predator, the borehole will be unsuccessful. Incomplete boreholes of muricids have a smooth bottom, whereas those of naticids have a shallower bowl and usually have a raised boss in the center (Fretter & Graham, 1962; Kabat, 1990). These characteristics can be used to differentiate the boreholes of muricids and naticids (and indeed those of other borers, such as octopus) from each other, for both recent and fossilized events (Sohl, 1969). The shell thickness that can be drilled by a muricid is effectively a function of the depth to which the ABO can be extended. For naticids, it has long been held that in order for the diameter of the inner hole to be sufficiently large to permit passage of the proboscis, the ratio of the diameter at the bottom of the borehole to that of the top cannot be less than 0.5 (Grey et al., 2005; Kitchell et al., 1981). Since the top diameter is a function of the diameter of the ABO, larger naticids, with correspondingly larger ABOs, can drill deeper holes than smaller ones. This metric has been used in paleobiological studies to make inferences about the putative size of naticid predators and was assumed to be relevant for all species. However, a species-specific component to borehole geometry, including the ratio of inner to outer borehole diameters, has been demonstrated for both extant and extinct naticid species. The inner-to-outer borehole ratios varied from 0.78 (*Euspira heros*) to 0.53 (*Euspira lewisii*) (Grey et al., 2005). Thus, a given borehole could be produced by either a smaller *E. lewisii* or a larger *E. heros*, demonstrating that caution must be taken when assigning the identity of a predator, based solely on the metrics of a particular borehole. Interestingly, in the case of muricids, a larger animal could potentially enlarge an existing borehole made by a smaller animal in an unsuccessful attempt to breach the shell of the prey, since incomplete boreholes may be detected and used as points of entry (Hughes & de B. Dunkin, 1984). In the case of naticids, boreholes are almost always started fresh, regardless of the presence of incomplete holes, and any reoccupation of holes appears to occur only by chance, due to stereotypic prey handling behavior (Kitchell et al., 1981).

3.3.3   THE MECHANICAL PROCESS

Muricid boreholes are produced mainly by the chemical dissolution of the shell by the secretions of the ABO, combined with relatively infrequent rasping of the radula. Removal of chemically softened and dissolved shell is completed by the rasping action of the radula. During the rasping process, the ABO is retracted and the proboscis is inserted into the borehole at the anterior edge of the foot, through a channel that is formed by an upward folding of the propodium. The posterior part of the foot remains firmly attached to the shell, holding the predator in position (Carriker, 1981). It is the central rachite teeth of the radula that is responsible for the rasping of the shell, as evidenced by the wear they suffer during the boring process. Rasping is completed in a posterior direction, resulting in the front teeth wearing away fastest. Teeth are replaced at the posterior of the odontophore and move anteriorly in a continuous conveyor like fashion. In muricids, the odontophore is able to rotate through at least 180° to the left or to the right, such that the bottom of the hole is rasped in two complete hemispheres, which results in the smooth-bottomed borehole, characteristic of incomplete muricid boreholes. The marginal teeth of the muricids are sickle shaped and are believed to be used for the tearing of flesh rather than the rasping of shell (Carriker, 1969, 1981). Shell material that has been removed by rasping is swallowed and eliminated via the feces (Carriker, 1977; Carriker et al., 1963). The construction of a shell-glass model permitted Carriker et al. (1967) to view these processes, which otherwise are hidden from sight by the foot.

The mechanical boring process of the naticids was thoroughly described by Ziegelmeier (1954). This was made much more difficult due to the infaunal nature of the naticids; drilling is normally completed within the sediment out of sight, and the fact that they often envelop their prey with their large foot, further obscuring the boring process (Fretter & Graham, 1962). As with muricids, drilling is initiated by removal of periostracum, or other organic layers, on the shell of the prey by radular rasping. The ABO, which in naticids is found on the proboscis, is then applied to begin softening of the shell using acidic secretions. The borehole cavity is sealed by mucous secreted from glandular cells which surround the central stroma of the naticid ABO (Bernard & Bagshaw, 1969). After a period of softening, shell fragments are rasped away and swallowed. Unlike muricids, the naticid odontophore can only rotate 90° in either direction, with rasping progressing in a posterior fashion. Rasping is always initiated from the centre of the borehole, and once a 90° sector is rasped, the proboscis is lifted. The ABO is reapplied to soften more shell, after which the proboscis is reinserted, but twisted to rasp a different 90° sector (often the opposite direction). Mechanical rasping is alternated with chemical softening in the described fashion, until the hole is completed or abandoned. The boss that is characteristic of incomplete naticid boreholes results from the fact that rasping is always initiated from the inside to the outside, leaving more unrasped material in the center compared to the periphery. Drilling is completed as a series of overlapping quadrant-sized sweeps of the odontophore, alternating left and right inside the larger circle formed by the central disk area of the ABO, which continue in a progressive fashion (Fretter & Graham, 1962).

3.3.4 ANATOMY OF THE ABO

The ABO of muricid gastropods is located in the anterior mid-ventral region of the foot whereas that of naticids is located on the underside of the proboscis, at the anterior-ventral lip (Fig. 3.1). In males, the muricid ABO is located in a vestibule (crypt) situated behind the transverse furrow (Fig.3.1A), while in females, it is located atop the ventral pedal gland (Fig.3.1B) or in a vestibule situated between the transverse furrow and the ventral pedal gland (Fig.3.1C; Carriker, 1981). The ABO in these two families are remarkably similar, but for their location on the foot and proboscis, respectively. They must have arisen independently, representing a "striking case of convergent evolution" (Carriker & Gruber, 1999; Kabat, 1990). Viewed in cross section, the muricid ABO is mushroom shaped, with a cap and a long stalk (Fig. 3.2). The ABO is everted through muscular contraction of foot around the vestibule, which squeezes blood from the contiguous blood sinus of the stalk into the cap, extruding the ABO from the sole of the foot. When the foot relaxes, the ABO is retracted via longitudinal muscles in the stalk, pulling it back into the crypt and out of sight. The cap of the ABO is composed of a single layer of long epithelial cells with a prominent brush border. Mucous glands are present in the short epithelial cells of the foot tissue surrounding the ABO, rather than being in the ABO proper. Secretions from these mucous glands form a seal around the borehole when the ABO is everted, preventing seawater from diluting the secretions of the gland (Clelland & Saleuddin, 2000).

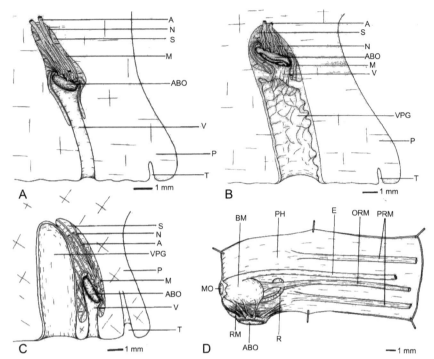

FIGURE 3.1 Muricid and naticid ABO anatomy. (A–C) Drawing of sagittal section of the anterior part of muricid ABO. (A) Foot of a male *Rapana thomasiana*. (B) Foot of female *R. thomsiana*. The ABO is located atop the ventral pedal gland and passes through the lumen of the gland when everted. (C) Foot of a female *Urosalpinx cinerea follyensis*. Arteries, A; nerves, N; muscles, M; propodium, P; transverse furrow, T; S, ABO sinus containing arteries (A), nerves (N) and muscles (M) passing to the back of the ABO; V, ABO vestibule through which ABO is extended to the borehole; ventral pedal gland, VPG. (D) Drawing of the left side of proboscis of the naticid *Polinices duplicatus*, opened laterally. Buccal mass, BM; esophagus, E; mouth, MO; odontophoral retractor mussel, ORM; proboscidial hemocoel, PH; radular sac, R; retractor muscle, RM. (Modified from Carriker, M.R. *Malacologia* **1981**,20, 405–406 with permission.)

In contrast, the naticid ABO appears as a shapeless mass of tissue attached to the lower lip when flaccid, but reveals a fungiform pad with a diameter equal that of the proboscis when engorged with blood (Fig.3.1D). Viewed in cross section (Fig.3.3), the gland is connected to the ventral side of the distal tip of the proboscis via a short stalk. A blood sinus contiguous with the proboscis connects to the ABO. The central epithelium of the naticid ABO is composed of long epithelial cells similar to those of the muricid ABO, but this central disk is surrounded by a ring of short mucous secreting cells that are separated from the central disk by a narrow band of lateral epithelium.

Secretions from the mucous cells of the ABO serve the same function as those in the foot of the muricids; they seal the central region of the ABO against the borehole, preventing dilution of secretions (Bernard & Bagshaw, 1969). The difference in the anatomy of the two ABOs is a major cause of the differences in the resulting borehole shape (Bernard & Bagshaw, 1969; Clelland & Saleuddin, 2000; Fretter & Graham, 1962; Kabat, 1990). The diameter of the ABO and the mean width of the radula (at the drilling position) are loosely related to the size of the individual and to the species, although the correlation is not strong. In a comprehensive listing of boring gastropods (Carriker & Gruber, 1999), the size of the ABOs ranged from 0.9 to 4.4 mm (mean diameter) and the radula varied from 0.12 to 1.4 mm (mean width), while shell height (as an index of size) varied from 12.1 to 115.0 mm in mean diameter.

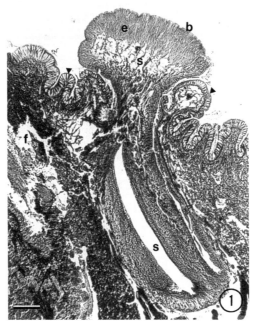

FIGURE 3.2 Section of an everted ABO of *Nucella lamellosa*, stained with Mallory–Heidenhain quick stain. Note the mushroom shape of the organ. The cap is composed of a single layer of epithelial cells (e) surrounding a central sinus (s), which is contiguous with the sinus in the stalk. The stalk invaginates into the foot (f). The epithelial cells of the ABO cap are long (200–300µm), with a prominent brush border (b). By contrast, the epithelial cells of the surrounding foot tissue (arrow heads) are short (40µm). Muscle is seen in the left-hand side of the stalk and is presumably used to retract the ABO into its crypt (vestibule). Scale bar: 200µm. (From Clelland, E.S.;Saleuddin, A.S.M. *Biol. Bull.***2000,***198*, 276,with permission.)

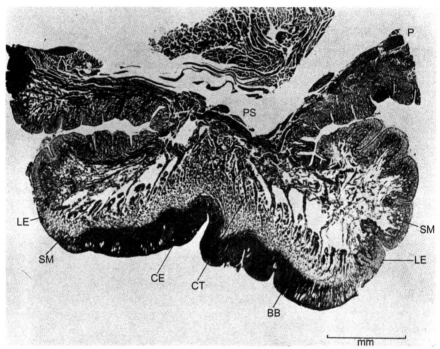

FIGURE 3.3 Median sagittal section of the ABO of *Polinices lewisi*. Stained by Lillie's Allochrome. Brush border, BB; central epithelium, CE; connective tissue, CT; lateral epithelium, LE; proboscis, P; proboscidal sinus, PS; subdermal mucocytes, SM. Scale bar = 1 mm. (From Bernard, F.R.; Bagshaw, J.W. J. *Fish. Res. Bd. Canada* **1969**, *26*, plate 1, © Canadian Science Publishing or its licensors. (Used with permission.)

3.3.5 FINE STRUCTURE OF THE ABO

To our knowledge, only one study has been conducted on the fine structure of a naticid ABO, *Polinices lewisi* (Bernard & Bagshaw, 1969); thus, details are primarily from investigations of muricid ABOs. The cap of the muricid ABO and the central disc of the naticid ABO are formed by a single layer of tall secretory cells (Fig. 3.4). These cells have three regions; apically they possess prominent microvilli and numerous mitochondria, the intermediate zone contains relatively few organelles, and the basal zone contains the nucleus, Golgi complex, and endoplasmic reticulum (Bernard & Bagshaw, 1969; Carriker & Gruber, 1999; Derer, 1975; Nylen et al., 1969; Webb & Saleuddin, 1977). The cells are arranged in compact groups that form a continuum over the surface, but are separated basally by blood-filled interstitial spaces. An elaborate network of blood vessels, nerves, and muscle

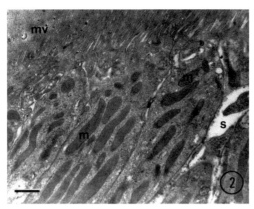

FIGURE 3.4   TEM of the apical region of the ABO cap epithelium. Note the long microvilli (mv) of the brush border, the numerous mitochondria (m), and the interstitial spaces (s) between the cells. Dense granules seen in the cells may contain degradative enzymes. Scale bar = 1.5 mm. (From Clelland, E.S.; Saleuddin, A.S.M. *Biol. Bull.* **2000,** *198,* 281,with permission.)

pass through the ABO sinus to the apex of the epithelium. Blood is drained across the basolateral membranes into the interstitial spaces then into the open sinus. The basal membranes of the cells are highly infolded, increasing surface area, and ensuring efficient transport of gas, nutrients, and metabolites. Pools of glycogen are present in the basal regions of the cells and accumulate during inactive periods, providing energy for cellular functions when the ABO is active (Carriker & Gruber, 1999; Webb & Saleuddin, 1977). Apically, the cells are joined together by various junction complexes, including gap junctions, which serve to coordinate the action of the cells (Clelland & Saleuddin, 2000; Nylen et al., 1969). Secretion granules are produced in the basal region and are believed to travel apically via star-shaped interstitial ducts to the base of the microvilli (Carriker & Gruber, 1999), where they discharge from the ABO (Nylen et al., 1969). Synthesis and discharge of secretory products is highest during periods of active boring (Carriker & Gruber, 1999). Uniform 3 nm particles, lining the inner side of the microvillar membranes, noted in *Nucella lapillus*, were believed to indicate the presence of proton pumps (Derer, 1975), and subsequent identification of V-ATPase in the microvilli of *Nucella lamellosa* (Clelland & Saleuddin, 2000; Fig. 3.5) substantiates this claim. Active ABOs are conspicuously different in appearance than inactive ABOs, with glandular cells that are taller, have longer microvilli, and contain more secretory granules, endoplasmic reticulum, and lysosomes than those of inactive ABOs (Carriker

& Gruber, 1999). Active ABOs also contain relatively little glycogen when compared to inactive ABOs (Chétail et al., 1968) and the interstitial sinuses of active ABOs also have higher concentrations of hemocyanin (Carriker & Gruber, 1999; Provenza et al., 1966).

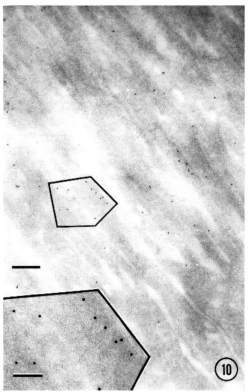

FIGURE 3.5   TEM of the microvilli (mv) that form the brush border of the ABO of *Nucella lamellosa*. Black spots are gold nanoparticles immunoconjugated to V-ATPase (39-kDa d subunit) showing that the immunoreactive sites for V-ATPases are located in the microvillar membrane. Scale bar = 200 nm. Inset: Gold particles lie close to the plasma membranes of the microvilli. Scale bar = 50 nm. (From Clelland, E.S.; Saleuddin, A.S.M. *Biol. Bull.* **2000,** 198, 281, with permission.)

3.3.6 PHYSIOLOGY OF THE ABO

Although the physiology of the ABO has been investigated in only a small number of primarily muricids (Carriker & Gruber, 1999; Clelland & Saleuddin, 2000), it is believed to be similar for all ABOs. The physiology of the ABO varies depending on whether the animals are fed or unfed, and if the ABO is active and everted (Carriker & Gruber, 1999; Franchini et al., 1983). The ABO produces a number of enzymes such as cytochrome oxidase, succinate dehydrogenase, lactate dehydrogenase, lipase alkaline, and acid phosphatases. The ABO is also rich in CA (Carriker & Gruber, 1999 and references therein). ABOs of starved individuals are poor in various secretory granules and vesicles, relative to those of satiated controls, but this situation is quickly remedied after feeding. It is believed that ABO activity and protein synthesis are under nervous control, as part of the overall regulation of drilling behavior (Carriker, 1981; Franchini et al., 1983). Changes in synthesis rates of various enzymes associated with activation of the ABO are difficult to quantify objectively due to differences in ABO isolation techniques, treatment and fixation protocols, assay procedures, and inconsistencies in the literature regarding the state of the ABO at the time of examination (Carriker & Gruber, 1999). However, it is generally accepted that synthesis increases in active ABOs, as aerobic processes are enhanced through increased blood flow into the ABO and by consumption of glycogen reserves built up during inactive periods (Nylen et al., 1969; Person et al., 1967).

The most prominent enzyme of the ABO is CA (Carriker & Gruber, 1999), where it is localized in the secretory cells of the ABO in concentration levels much higher that surrounding pedal tissues (Smarsh, 1969). Smarsh (1969) also found CA in the brush border, whereas Webb and Saleuddin (1977) did not. Similarly, many investigators have noted that CA is about equally abundant in active (boring) and inactive (resting) ABOs, while others have noted an increase in active ABOs (Carriker & Gruber, 1999). CA is associated with secretory cells known for active transport of hydrogen ions, bicarbonate ions, and carbon dioxide (Webb & Saleuddin, 1977) and is a hallmark of the mitochondria rich (MR) cells known to utilize V-ATPase proton pumps (Brown & Breton, 1996; Clelland & Saleuddin, 2000; Harvey et al., 1998; Wieczorek et al., 1999). Here, they generate free protons and bicarbonate ions by catalyzing the reaction of water and carbon dioxide through a carbonic acid intermediary (Fig.3.6); the free protons being extruded from the cytoplasm by V-ATPases at the expense of ATP. While it has long been held (Carriker & Chauncey, 1973) that CA does not act directly as a demineralization agent, it is known to be vitally important in shell dissolution during boring, since application of specific CA inhibitors such as Diamox have been shown to impair the drilling process (Carriker & Gruber, 1999).

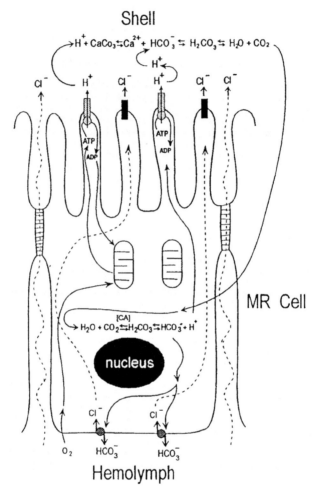

FIGURE 3.6 Model of the mechanism of proton transport in the muricid ABO. Carbonic anhydrase (CA) catalyzes the production of H^+ and HCO_3^- via the carbonic acid pathway. Bicarbonate is removed from the mitochondria-rich (MR) epithelial cell via basal HCO_3^-/ Cl^- antiporters, while protons are extruded from the cell into the bore hole by V-ATPase pumps located in the microvilli. Mitochondria generate ATP to power the extrusion process, generate metabolic CO_2 for the carbonic acid reaction, and provide a reducing environment to stabilize V-ATPase molecules. Chloride ions exit the cell via apical ion channels, and possibly by paracellular routes. The protons and chloride ions (HCl) act to dissolve the mineralized component ($CaCO_3$) of the shell, while degradative enzymes also present in the secretions of the ABO break down the organic matrix. Carbon dioxide liberated from the dissolving shell may diffuse into the cell to enhance the carbonic acid reaction. The presence of HCO_3^-/Cl^- antiporters and, as speculated here, of chloride ion channels is based on comparable studies of MR cells in other animal epithelia. (From Clelland, E.S.;Saleuddin, A.S.M. *Biol. Bull.* 2000, *198*, 281,with permission.)

The secretion of the active ABO that is released onto the surface is granular and viscous, and generally insoluble in sea water. It contains membrane-bound vesicles and granules in a thick mucus, limiting its dispersion during boring (Carriker, 1981; Carriker & Gruber, 1999). It accumulates on the surface of the ABO during the rasping phase and is introduced into the borehole on the next dissolution phase (Carriker et al., 1978). Application of excised live ABOs to polished mollusc shells etches the surface, and is inhibited by co-application with papain, or heat treatment of the ABOs to 80°C prior to application (Carriker et al., 1978), suggesting the inactivation of an enzymatic component (Carriker & Gruber, 1999). The presence of protein in the ABO secretion has been also been demonstrated (Evans, 1980). Blots of the secretions obtained from active, everted ABOs using a "valve model for secretion collection" have been found to contain 1–2 µg of water-absorbent material after drying, but as much as an estimated one-third of the initial secretion is composed of volatile components which evaporate (Carriker et al., 1978). Furthermore, using this model, the authors were able to confirm the acidic nature of the ABO secretion (shown previously by Carriker et al. (1967), with measures ranging from pH 3.8 to pH 4.0, and to demonstrate that while the ABO was everted into the bore hole, the Cl^- concentration increased in a stepwise fashion, with maximal levels ranging from 0.79 to 1.71M; much higher than the 0.5M Cl^- content of the surrounding seawater.

When the ABO is engorged, increased blood flow infuses oxygen into the secretory cells, which stimulates aerobic processes within the cells. Aerobic synthesis is upregulated and acid production is initiated as mitochondrial activity and overall mitochondrial number increases, leading to the generation of additional ATP to drive various catalytic and transport processes (Carriker & Gruber, 1999; Carriker & Williams, 1978; Webb & Saleuddin, 1977). It is well established that the V_0 (transmembrane) and V_1 (catalytic) domains of V-ATPases can reversibly assemble and that the catalytic domain can activate, depending upon on the redox potential inside a cell or organelle (Merzendorfer et al., 1997). V-ATPases in the apical brush border of the ABO secretory cells are activated by changes in the redox potential when the organ is engorged with blood. Energy liberated by the catalytic conversion of ATP to ADP is utilized by these pumps to expel protons across the apical membrane and onto the substrate (Clelland & Saleuddin, 2000). As modeled by Webb and Saleuddin (1977) and Clelland and Saleuddin (2000), CA within the secretory cells catalyzes the production of $NaHCO_3$ and free protons. The free protons are pumped into the cavity of the borehole by the apical proton pumps, while the HCO_3^- is exchanged for Cl^- at the basolateral membrane. The chloride ions pass through apical membrane channels and perhaps via paracellular routes into the borehole, thus maintaining the acid–base balance of the secretory cells, and upon binding with the protons in the borehole cavity, produce HCl (Fig.3.4).

3.3.7   CONVERGENCE OF THE ABO, MR CELLS, AND V-ATPASES

In a 1999 review, Carriker and Gruber describe the ABO of drilling gastropods as "unique among the organs of invertebrates." This compact fungiform organ has twice evolved physiological and biochemical mechanisms that permit penetration of the $CaCO_3$ external shells of their prey. As an organ, the gastropod ABO is unquestionably unique, albeit having evolved separately in muricids and naticids. However, the localization of V-ATPase proton pumps in the ABO of *Nucella lamellosa*, as well as a number of other characteristics common to the ABO cells of all gastropods thus far examined, suggests that the mechanism for decalcification of calcareous substrates is actually conserved (Clelland & Saleuddin, 2000).

A similar process has been shown to occur in bivalve mantle epithelium, where free protons and Cl^- are expelled by mitochondrial-rich (MR) cells against the shell surface to alleviate acidosis when the shells are closed (da Costa et al., 1999; Hudson, 1992; Hudson, 1993) and by MR osteoclast cells of vertebrate bone, during the processes of bone resorption and remodeling (Brown & Breton, 1996). Interestingly, as reviewed by Ehrlich et al. (2008), aquatic vertebrates such as turtles (shell), crocodilians (osteoderms), and amphibian (femur) also utilize the buffering capacity of bone to alleviate systemic acidosis by extruding HCl from MR cells against boney substrate. More recently, a bone resorption process has also been described for the polychaete boneworm, *Osedax*, where the MR cells of their dorsal root are involved in the dissolution of whale bone in which they reside (Tresguerres et al., 2013). In some epithelia, V-ATPases are known to change polarity (i.e., move between the apical and basolateral membranes of the cell), depending upon the requirements at a given time. For example, extrusion of HCl against the cuticle from apically positioned V-ATPases can aid moulting in the terrestrial isopod *Porcellio scaber* through the demineralization of calcium carbonate, whereas assembly in the basolateral membrane can drive processes leading to remineralization of the cuticle through uptake and accumulation of calcium (Ziegler, 2004). V-ATPase containing MR cells are found in the transport epithelia of fish gills (Sullivan et al., 1995), frog skin (Ehrenfeld & Klein, 1997), toad and turtle bladder (Brown et al., 1987), insect Malpighian tubules (al-Fifi et al., 1998; Maddrell & O'Donnell, 1992), kidneys (Brown et al., 1988; Sallman et al., 1986), seminiferous tubules (Brown et al., 1997), etc. V-ATPase pumps are utilized to extrude protons at the expense of ATP, either to directly acidify specific compartments or to utilize the ensuing electrochemical gradient to drive the movement of ions or other molecules across cell membranes (Ehrlich et al., 2008; Ehrlich et al., 2009; Harvey et al., 1998; Merzendorfer et al., 1997; Wieczorek et al., 1999). In the tissues of developing pond snails, *Lymnaea stagnalis*, proton extrusion is linked to calcium uptake for shell deposition (Ebanks et al., 2010), while in insect Malpighian tubules, protons extruded from the apical membrane of the epithelial cells into the lumen of the tubule, return to the cell via cation antiporters in exchange for potassium or sodium, leading to the alkalinization of the tubule (Maddrell & O'Donnell, 1992). Thus, while the nature of these various epithelia (ABO, mantle, gill, Malpighian, root, osteoclast, etc.) are quite different, the fundamental machinery and function of the MR cells is similar; all possess abundant levels of CA to catalyze production of protons and bicarbonate, all have numerous mitochondria to produce ATP as fuel, and all possess electrogenic V-ATPase proton pumps. Subsequent studies in decapods (Ziegler, 2004), pulmonates (Ebanks et al., 2010), and polychaetes (Katz et al., 2010; Tresguerres et al., 2013), among others have provided further evidence in support of a conserved mechanism for decalcification *in vivo*. Reviews on the principles of demineralization (Ehrlich et al., 2008; Ehrlich et al., 2009) have included observations of the gastropod boring mechanism and the structure and function of the gastropod ABO amongst some of the most significant events in the history of demin- eralization in the past 500 years. These authors have also remarked upon the conservative nature of decalcification *in vivo*, and the involvement of V-ATPases in MR cells.

3.4   FEEDING BEHAVIOR AND ENVIRONMENTAL EFFECTS

Drilling gastropods are known to consume a variety of prey, although the majority of muricids eat bivalves, gastropods (including smaller individuals of the same species), and barnacles (Carriker, 1981; Kabat, 1990). Much of what is known about feeding behavior comes from observations of muricid snails, which are common intertidally and thus are relatively easy to study. For example, *Nucella lapillus* feeds primarily on barnacles and mussels (and occasionally other molluscs) in post juvenile stages (Hughes & de B. Dunkin, 1984a), and *Thais melones* feeds on a variety of bivalves, limpets, and polychaetes (West, 1988). Muricids are also known to eat carrion, bryo-zoans, crabs, and although less so in the present, brachiopods (Kelley & Hansen, 2003; Leighton, 2003). Some juvenile muricids feed on ostracods (Reyment & Elewa, 2003). Strangely, there is almost no evidence of muri-cids drilling chitons; there is a single fossil drill hole attributed to muricids found on a chiton in the Late Pleistocene of Uruguay (Rojas et al., 2014), with almost no recent descriptions (Taylor & Morton, 1996).

Studies of naticids are more difficult because much of the activity occurs out of sight beneath the substrate (Kabat, 1990). Naticids are generally thought to be more restricted in their diets than muricids, eating primarily live bivalves (Carriker, 1981), in particular, infaunal species. Naticids are known to feed on other, mainly soft-substrate gastropods, and occasionally scaphopods. *Neverita duplicata* is the sole naticid that has been observed feeding on polychaetes (*Owenia fusiformis*) (Paine, 1963), although another naticid is implicated in the drill holes of *Ditrupa arietina*, another poly-chaete (Morton and Salvador, 2009). One species, *Conuber sordidus* has been observed feeding on Blue soldier crabs (*Mictyris*) and hermit crabs (Huelsken, 2011). There are also reports based on stereotypical drill holes of naticids drilling of egg capsules, of other gastropods and elasmobranchs, as well as brachiopods (Ansell, 1961; Cox et al., 1999; Kabat, 1990; Leighton, 2003). Incidents of scavenging and feeding on fish that seem to be limited to aquarium situations, but field observations have been made (Kabat, 1990; Kelley & Hansen, 2003). Juvenile naticids are even thought to prey on ostra-cods (Reyment & Elewa, 2003) and foraminiferans (Arnold et al., 1985; Culver & Lipps, 2003).

From a distance, prey is located principally by the osphradium (chemo-reception), where chemical odors carried on the current direct the predators toward their source (Carriker, 1981; Morgan, 1972; Rittschof et al., 1983). These odors are primarily composed of peptides and amino acids produced by metabolic processes such as digestion (Rittschof, 1980a; Rittschof, 1980b; Rittschof et al., 1983). Upon reaching the prey, tactile cues allow the predator to further assess the quality of the prey (Carriker & Gruber, 1999; Hughes & de B. Dunkin, 1984a; Kabat, 1990). Predators are able to discrim-inate amongst these cues to identify specific types of prey and can often develop preferences for particular foods (Carriker, 1981; Wood, 1968); with time and experience, they can learn to select the optimum sized prey and attack methodology (Hughes & de B. Dunkin, 1984a; Hughes & Drewett, 1985; Rovero et al., 1999). *Urosalpinx* can discriminate between starved and satiated oysters, preferentially feeding on the latter (Carriker, 1981). Prefer-ence for particular prey is not genetically fixed, and laboratory experiments demonstrate that predators prefer effluents from prey species they have recently fed upon (ingestive conditioning) over those of other prey species (Carriker, 1981). Field studies have also demonstrated that there may be wide intraspecific variation in prey selection (West, 1986). *Nucella lapillus*, for example, may develop preferences for either mussels or barnacles (Hughes & de B. Dunkin, 1984b; Hughes & Drewett, 1985) and learns to how to best manipulate its prey to minimize handling time (Rocha-Barreira et al., 2004; Rovero et al., 1999). Similarly, *N. emarginata* are selective in the size and nature of their bivalve and barnacle prey, depending on their size and experience (Palmer, 1990).

In addition to chemoreception, Naticids also likely detect vibrations produced by their prey, a method that is better suited to their infaunal lifestyle than the hard substrate preferred by many muricids (Kitching & Pearson, 1981). Naticids have also been observed foraging with the siphon extended to the surface, where directionality and strength of chemical cues are less likely to be perturbed or diffused by the substrate. They display stereotypical feeding behaviors involving detection of prey, followed by evaluation and seizure. Prey is enveloped in thick pedal mucous, which may contain an anesthetic, wrapped in the foot, dragged some distance away, and then carried deeper into the sand where boring begins (Carriker, 1981; Kabat, 1990). Naticids are usually absent from high-energy, wave-disturbed beaches where olfactory and vibrational cues tend to be diffused. An interesting exception, is the moon snail *Polinices incei*, native to exposed beaches of Queensland, Australia, which has evolved a surfing behavior. They float upside down via an inflated foot to get caught in the wash and swept up the beach. There they attack juvenile surf clams (*Donax deltoides*). *P. incei* hunts infaunally or by galloping on the surface of the sand to catch their leaping clam prey (Morton, 2008). Once seized, prey is enveloped by the foot and the snail rolls down the beach in the surf, presumably to less turbulent areas for drilling and consumption (Morton, 2008).

Drilling is not always limited to calcareous exoskeletons, which helped confound the debate on the nature of the drilling process (Bernard & Bagshaw, 1969). Penetration of soft tissues of egg cases of gastropods (Jensen, 1951) and elasmobranchs (Ansell, 1961) was likely completed by mechanical rasping alone, and no evidence of this would remain even if there was participation from the ABO. Furthermore, drilling is often unnecessary in instances where there is ready access to soft tissues. For example, the proboscis can be inserted between the operculum plates of barnacles, through the valve openings (gape) of bivalves, or via the aperture of gastropods, to feed without drilling. Field and laboratory studies have shown that with age (i.e., size) and experience, predators are more likely to minimize or forego drilling whenever possible (Kabat, 1990; Morgan, 1972; Palmer, 1990; Rocha-Barreira et al., 2004; Rovero et al., 1999). For example, under laboratory conditions, *Thais haemastoma floridana*, consumed about 31% of feed oysters by forcing the proboscis through the valves, and of those drilled, virtually all holes were found along of adjacent to the shell margins (Rocha-Barreira et al., 2004). Some muricids can penetrate a barnacle or mussel by ramming the margin with a labial tooth, a downward protrusion from the aperture (MacGinitie & MacGinitie, 1968; Perry, 1985; Spight & Lyons, 1974). Some tropical naticid species are also known to preferentially drill bivalve prey at the edge of the valves, where the shell is thinner and easier to bore (and holes are often overlooked) and they may forego drilling entirely if there is sufficient gape for insertion of the proboscis. Still other species have been found to smother their prey, although recent work suggests this may be an artifact of lab conditions (Visaggi et al., 2013). In the case of gastropod prey, it is possible to pry open the operculum of the prey to permit access to the soft tissues via the aperture (Kabat, 1990). The presence of narcotizing or toxic agents in the saliva to paralyze the prey are also in the repertoire of some muricid and naticids (Andrews et al., 1991; Gordillo, 2013; Rovero et al., 1999) and can remove the need for drilling.

The foraging and feeding behaviors of predatory gastropods are complex, involving factors ranging from ingestive conditioning and prey experience to abiotic factors such as temperature or salinity. Especially for intertidal species, variation in the mortality risks associated with feeding and foraging can alter feeding patterns, and when environmental conditions are severe, these may almost entirely determine the timing of foraging bouts and resting periods (Burrows & Hughes, 1991). Furthermore, extended periods of unfavorable foraging conditions tend to synchronize foraging and resting periods in a population, since every individual begins the first subsequent foraging bout with an empty stomach (Burrows & Hughes, 1991). Selection of drilling site is another confounding factor. *Thais haemastoma* can immediately perceive the presence of prey oysters under laboratory conditions and preferentially drills near the shell border to minimize drilling time and forgoes drilling when possible (Rocha-Barreira et al., 2004). Pharmacologically active compounds present in the saliva of predatory gastropods (e.g., *Nucella lapillus*, *Thais floridanum*, etc.) favor edge drilling or gape access via valves. Only small openings are required to introduce paralytic compounds into the tissues, to relaxing adductor muscles, and allowing for rapid access to the tissues (Andrews et al., 1991). The hypobranchial and accessory salivary glands of many muricids contain active compounds, including the potent paralytic urocanylcholine and various other choline esters (Andrews et al., 1991; Keyl et al., 1957; Whittaker, 1960). These are the compounds responsible for the intense color of Tyrian purple, long harvested from Mediterranean muricids for dye (Keyl et al., 1957). Despite the advantages of drilling in areas of the shell where it is the thinnest, the location of the borehole (if present) is often simply a consequence of which part of the shell is accessible for drilling during prey manipulation (Kabat, 1990). Environmental factors also impinge on feeding behavior. Temperature has a major effect on temperate and boreal species such as the naticids *Neverita duplicata* and *Euspira heros* which do not feed at low temperatures (2 and 5°C, respectively), nor at salinities below 10‰ (Kabat, 1990). Adult *Nucella lapillus* are subject to torpor below 5°C, but have their highest growth efficiencies at moderate temperatures (between 10 and 15°C). Temperatures above 20°C and below 10°C have a negative impact on growth (Stickle & Bayne, 1987). For juvenile *N. lapillus*, the drilling time is roughly constant between 10 and 20°C, while the ingestion time drops from approximately 45 to 15 min (i.e., by ~2/3) across this range (Miller, 2013). Warm water species such as the Southern oyster drill, *Stramonita haemastoma*, are also affected by temperature and salinity, with optimal growth rates in the 25–35°C range, suboptimal growth occurring at 20°C, and 50% lethality at 37.5°C. Temperatures in excess of 40°C are 100% fatal. These animals limited to salinities greater than 15‰ (Brown & Stickle, 2001). Aerial exposure also becomes an issue for these snails at high temperatures, most likely due to the elevated metabolic requirements at higher temperatures.

Drilling rates have been published over the years for various muricid and naticid species, for example 0.3–0.5 mm per day for *Urosalpinx cinerea* (Carriker & Williams, 1978), 0.36 mm per day for *Nucella lapillus* (Hughes & de B. Dunkin, 1984a) and for naticids, 0.6 mm per day for *Euspira nitida* (Ziegelmeier, 1954), ~0.54 mm per day for *Neverita duplicata* (Kitchell et al., 1981), and so on (Fretter & Graham, 1962; Kabat, 1990). Environmental effects such as temperature and salinity are now known to affect the feeding rate; prey investigation time, drilling time, and consumption time (Kabat, 1990), so published rates should be viewed only as approximations. They are useful, however, for estimating the quantity of prey consumed under various environmental conditions (Kabat, 1990; Miller, 2013).

Once the shell of a prey animal is pierced, additional predators and scavengers are quickly attracted to the site due to the increased release of odors via enzymatic digestion of the prey tissue (Rittschof, 1980b). These include hermit crabs, which may be attracted as scavengers or as opportunists waiting for an empty shell. These competitors may compete with the original borer for possession of the prey or in the case of larger bivalve prey, may simply wait until the prey is agape, then take advantage of the opportunity to feed at a different site feed alongside at little cost to themselves.

Another interesting facet of the behavior of predatory gastropods involves notions of cannibalism. Cannibalism is relatively common in the animal world (Kelley & Hansen, 2007) and examples of intraspecific predation (true cannibalism) and interspecific predation of gastropods can be found in the literature for both extinct and extant species (Gordillo, 2013). Muricid cannibalism likely occurs in situations where there are too many predators and insufficient prey, a view supported by observations of the fossil record for Halocene and living specimens of *Trophan geversianus* from South Africa (Gordillo, 2013). Muricid cannibalism has been recorded in the field for *Murex trunculus* and in the laboratory for *Triplofusus giganteus* (Kelley & Hansen, 2007). Evidence of naticid cannibalism dates back to the Cretaceous Period (Gordillo, 2013; Kelley & Hansen, 2007); examples from the Miocene St. Marys Formation in Maryland including cannibalism in two common species *E.heros* and *N. duplicata* (Kelley & Hansen, 2007) and in Pliocene naticids from southern France and Antarctica (Gordillo, 2013). Energetically, there is a benefit to cannibalism in naticids; high success rates of penetration due to thin shells and the opportunity for prising, coupled with the high energy richness of the prey's flesh, yield a high energy gain per foraging unit in cost–benefit analyses (Gordillo, 2013; Kelley & Hansen, 2007; Kitchell et al., 1981).

3.5 PALEONTOLOGICAL HISTORY OF DRILLING PREDATION

The drilling predation of gastropods is of great interest to paleontologists. This behavior is one of the few that leaves a direct record of predation through trace fossils (ichnofossils). Paleontological studies are limited in their ability to determine actual ecological interactions. In the case of predation, most evidence is indirect and can be difficult to interpret. Direct evidence comes from traces of predation, such as drill holes or breakage scars. Other types of direct evidence are rare: coprolites, stomach contents, or exceptional preservation events showing an actual interaction preserved (Kowalewski, 2002). There are some clear limitations, however, and steps must be taken to ensure holes are infact due to drilling by a predator (Carriker & Yochelson, 1968; Kelley & Hansen, 2003). Some marks can be made abiotically, or could be made biotically for other purposes such as parasites or epibionts. Drilling offers further advantages over other evidence such as breakage scars as they are easier to recognize and differentiate from abiotic damage (Harper et al., 1998). Numerous authors have listed possible criteria to distinguish drilling from other holes: holes should be circular to oval, perpendicular, drilled from outside to in, regular placement or size, and usually not more than one complete hole (Kelley and Hansen, 2003), and references therein). Furthermore, there are a number of assumptions that must be made about predator and prey behavior and taphonomy that can bias the data, and must be taken into account (Kelley & Hansen, 2003; Kowalewski, 2004; Leighton, 2001).

Fossil traces of predation, including drill holes, offer a number of benefits. These traces are common in a broad range of environments, taxa, and times, and are easily preserved in the hard prey skeletons. This allows for quantification of predation events across time and space on a variety of scales (Kowalewski, 2002; Kowalewski, 2004; Kowalewski et al., 1998). A primary direction of inquiry relates to how predation affects evolution, through coevolution and/or escalation, and how the magnitude of these interactions compares to abiotic selective pressures (Vermeij, 1987; Vermeij, 1994).

Fossil drilling data are commonly used to determine predation frequency, but can also be used to look into other aspects. Predation efficiency can be estimated by comparing complete and incomplete drill holes, although this is more problematic for muricids who can restart previous holes, eat without completing a drill hole, and some drill cooperatively (Kowalewski, 2004; Taylor & Morton, 1996). Size relationships can be determined as the size of the boring is related to the size of the predator, although as previously mentioned, care must be taken if assigning the identity of specific predators, due to uncertainty in the borehole ratio parameters (Grey et al., 2005). Selectivity of predators can be examined by determining the distribution of drill holes in prey, and prey choice is demonstrated by which species show similar drillings (Kelley & Hansen, 2003). Kelley and Hansen (2003) compiled a summary of this body of research, and concluded that the situation is complicated and requires more work; however predator–prey interactions do have important effects on evolution on a large scale, and support hypotheses of escalation. There is evidence for increased selectivity on drilling position, increased size selectivity, increased conforming to optimal foraging strategy, and increased prey defensive capabilities over time.

The first drill holes appear before the Cambrian, in the tubes of *Claudina* (550 mA), and predatory boring is found throughout the fossil record to the present day (Bengtson & Zhao, 1992; Hua et al., 2003; Kelley & Hansen, 2003; Kowalewski et al., 1998). Except during the Cenozoic, most drill holes do not have a hypothesized driller; unknown gastropods are most likely, although octopods, nematodes, or flatworms are also possible (Kabat, 1990; Sohl, 1969). There is a reasonable record of Paleozoic drillers with a peak in the Devonian, where most holes were found in brachiopods and echinoderms (Kelley & Hansen, 2003; Kowalewski et al., 1998; Leighton, 2003). Platyceratid gastropods have been found *in situ* parasitizing echinoderms, identifying at least one Paleozoic driller, though they go extinct before the Mesozoic (Baumiller, 1990; Baumiller, 1993; Baumiller, 1996). Evidence of drilling is quite weak throughout most of the Mesozoic. A sharp increase in drilling occurs in the Late Cretaceous, probably corresponding to the evolution and diversification of modern drillers (Kowalewski et al., 1998).

The origin of the Naticidae and their drilling is not well understood; the evidence of possible naticid boring is from the Late Triassic. This is followed by a large 120 million year gap in the record of drill holes, where there are few reported drillings. Some of these Triassic borings are associated with the Ampullospirinae, whose relationship with the naticids is in unclear (Kabat, 1990; Sohl, 1969). This group has even been suggested to be herbivorous, but this interpretation is not well supported (Aronowsky & Leighton, 2003; Kase & Ishikawa, 2003a; Kase & Ishikawa, 2003b). If these earlier holes were not made by naticids, it leaves in doubt the driller, even after naticids have clearly evolved (Kabat, 1990). Naticid shells and drill holes are generally accepted as present starting with a few records in the late Cretaceous, and then diversifying into the modern fauna (Aronowsky & Leighton, 2003; Harper et al., 1998; Kabat, 1990; Kowalewski et al., 1998; Sohl, 1969).

The origin of the Muricidae is less muddy, there is at least one site with undisputed muricid shells seen in the Late Cretaceous, and these are accompanied by muricid-type boreholes (Harper et al., 1998; Merle et al., 2011). There is evidence of muricid-like boreholes earlier in the Mesozoic, but these are generally attributed to some unknown, convergent, culprit (Harper et al., 1998). The family radiated from there to the current diversity of extant muricids (Kowalewski et al., 1998).

3.6 FUTURE DIRECTIONS

Research conducted over the past 100 years, primarily from the 1940s onward, as led to a solid understanding of the mechanism of drilling utilized by muricids and naticids. The fundamental physiology of the ABO is well understood and the mechanism for production of acidic secretions has been established. This said, there are still some gaps that need to be filled. Almost nothing is known about the other gastropod groups that drill prey, and observations and experiments similar to those previously done on muricids and naticids would broaden our understanding of drilling predation and its evolution. Considering the behavior of so many gastropods is still unknown, it seems likely other groups will continue to be added to this list, whether as confirmation of previous claims or as new groups (Sohl, 1969).

The presence of V-ATPase in the apical epithelia of the ABO secretory cells should be confirmed in naticids, as well as additional muricid species; the comprehensive listing of boring gastropod species provided by Carriker and Gruber (1999) could serve as a guide for these studies. Technology has advanced such that these studies would be much simpler. Whereas Clelland and Saleuddin (2000) utilized noncommercial V-ATPase antibodies, commercially produced antibodies are now available for virtually all of the ~13 V-ATPase subunits. Alternatively, utilization of molecular biological techniques (mRNA isolation, ssDNA synthesis, sequence alignment, etc.) to determine conserved epitopes within target subunits for the production of antibodies (Tresguerres et al., 2013) targeted against V-ATPase peptides of predatory gastropods, should now be a relatively straightforward undertaking. Such antibodies could be employed in microscopic and peptide expression analyzes (Clelland & Saleuddin, 2000) using more modern techniques, such as immunofluorescence imaging and chemiluminescence. In situ hybridization analyzes of mRNA expression might also be informative. Along the same vein, although the cationic pathway is well established, questions still need to be answered regarding the chloride ion component. Identification of basolateral and apical anion exchange molecules (exchangers, channels, etc.) would help clarify the chloride pathway(s) in the ABO. Judicious application of antibodies to fixed specimens of inactive and active ABOs may also shed light on the activation of V-ATPases. If activation results from the assembly of V_0 and V_1 complexes as the ABO is engorged with blood, it should be possible to probe for V_1 specific subunits and compare their localization in active and inactive ABOs.

From a physiological perspective, the advent of relatively inexpensive versions of Bafilomycin, a specific inhibitor of V-ATPases (Clelland & Saleuddin, 2000) presents opportunities to conduct activity assays for V-ATPase in a grander scale. One can also envision use of this inhibitor to directly monitor the effects of its application on proton and chloride fluxes from live ABOs in Carriker's shell-glass model (discussed above). Scanning Ion-selective Electrode Technique (SIET) has been utilized to study such fluxes in most recent studies (Ebanks et al., 2010; Jonusaite et al., 2011; Nguyen & Donini, 2010) and could be adapted to the shell-glass arrangement. SIET provides exquisitely precise measures of ion flux. Effects of other pharmacologicals (CA inhibitors, channel blockers, etc.) on ion flux could also be determined using this technique. Additional study is needed to identify other components of the ABO secretions, most specifically, the as yet unknown calcium chelator (Carriker & Gruber, 1999).

The recent paper by Miller (2013) poses some interesting questions regarding the effect of climate change on the activity of predatory gastropods. Temperature changes could have the effect of prolonging the length of the feeding period in cold water species, but may well extirpate warm water species should temperatures rise to lethal threshold levels (Brown & Stickle, 2001). Effects of increasing ocean acidification may also play an important role. Increased temperature and lower pH are likely to place stress on larval and juvenile snails (and those of their prey) which may impinge on their viability. In adults, shells of both predators and prey may become thinner, which will have an impact on drilling rates, the ability to survive in active surge zones, and an increase overall metabolic rate. From the perspective of drilling, increased temperature and lower pH would be expected to speed up the drilling process.

3.7 CONCLUSIONS

The chemo-mechanical process utilized by naticid and muricid gastropods to drill into hard substrate is a fascinating process. The development of specialized glands, the ABOs, for the production of acids, enzymes, and chelators used to dissolve calcareous substrate is unique, as is the drilling process, where chemical softening of substrate is alternated with mechanical rasping, until a hole is bored. The cells of the ABO themselves, however, fall within a broader grouping of similar MR cells characterized by high levels of CA, V-ATPase proton pumps, chloride transport molecules, etc. which are used by animals of a number of phyla for the dissolution calcified substrates. Whereas the drilling process is well understood in the naticids and muricids, there remains numerous aspects of the process that await further investigation (e.g., the elucidation of the anionic components of ABO physiology) and confirmation of the process as modeled, as well as its conservation or diversity in other groups. We look forward to reading such reports in the future.

KEYWORDS

- **moon snails**
- **molluscs**
- **muricids**
- **naticids**
- **accessory boring organ**
- **drilling predation**

REFERENCES

al-Fifi, Z.I.; Marshall, S.L.; Hyde, D.; Anstee, J.H.; Bowler, K. Characterization of ATPases of Apical Membrane Fractions from Locusta Migratoria Malpighian Tubules. *Insect Biochem. Mol. Biol.* **1998**, *28*, 201–211.

Andrews, E.B.; Elphick, M.R.; Thorndyke, M.C. Pharmacologically Active Constituents of the Accessory Salivary and Hypobranchial Glands of *Nucella lapillus*.*J. Mollus. Stud.* **1991**, *57*, 136–138.

Ankel, W.E. Wie bohrt Natica?*Biol. Zetra* **1937**, *57*, 75–82.

Ankel, W.E. Erwerb und Aufnahme der Nahrung bei den Gastropoden. *Zool. Anz. Suppl.* **1938**, *11*, 223–295.

Ansell, A.D. Egg Capsules of the Dogfish (*Scylliorhinus canicula* L.) Bored by Natica (Gastropoda, Prosobranchia).*Proc. Malacol. Soc. Lond.* **1961**, *34*, 248–249.

Arnold, A.J.; d'Escrivan, F.; Parker, W.C. Predation and Avoidance Responses in the Foraminifera of the Galapagos Hydrothermal Mounds.*J. Foramin. Res.* **1985**, *15*, 38–42.

Aronowsky, A.; Leighton, L.R. Mystery of Naticid Predation History Solved: Evidence from a "living fossil" Species: Comment and Reply COMMENT. *Geology* **2003**, *31*, e34–e35.

Barker, G.M.; Efford, M. Predatory Gastropods as Natural Enemies of Terrestrial Gastropods and Other Invertebrates.In *Natural Enemies of Terrestrial Molluscs*; Barker, G.M., Ed.; CABI Publishers, 2004; p 279.

Baumiller, T.K. Non-predatory Drilling of Mississippian Crinoids by Platyceratid Gastropods. *Palaeontology* **1990**, *33*, 743–748.

Baumiller, T.K. Boreholes in Devonian Blastoids and their Implications for Boring by Platyceratids. *Lethaia* **1993**, *26*, 41–47.

Baumiller, T.K. Boreholes in the Middle Devonian Blastoid Heteroschisma and their Implications for Gastropod Drilling. *Palaeogeogr., Palaeoclimatol., Palaeoecol.* **1996**, *123*, 343–351.

Bengtson, S.; Zhao, Y. Predatorial Borings in Late Precambrian Mineralized Exoskeletons. *Science* **1992**, *257*, 367–369.

Bernard, F.R.; Bagshaw, J.W. Histology and Fine Structure of the Accessory Boring Organ of Polinices Lewisi (Gastropoda, Prosobranchia). *J. Fish. Res. Board Can.* **1969**, *26*, 1451–1457.

Bromley, R.G. Concepts in Ichnotaxonomy Illustrated by Small Round Holes in Shells. *Acta Geol. Hisp.* **1981**, *16*, 55–64.

Brown, D.; Breton, S. Mitochondria-rich, Proton-secreting Epithelial Cells. *J. Exp. Biol.* **1996**, *199*, 2345–2358.

Brown, D.; Gluck, S.; Hartwig, J. Structure of the Novel Membrane-coating Material in Proton-secreting Epithelial Cells and Identification as an H⁺ATPase.*J. Cell Biol.* **1987**, *105*, 1637–1648.

Brown, D.; Hirsch, S.; Gluck, S. Localization of a Proton-pumping ATPase in Rat Kidney. *J. Clin. Invest.* **1988**, *82*, 2114–2126.

Brown, D.; Smith, P.J.; Breton, S. Role of V-ATPase-rich Cells in Acidification of the Male Reproductive Tract. *J. Exp. Biol.* **1997**, *200*, 257–262.

Brown, K.M.; Stickle, W.B. Physical Constraints on the Foraging Ecology of a Predatory Snail. *Mar. Fresh. Behav. Physiol.* **2001**, *35*, 157–166.

Burrows, M. T.; Hughes, R.N. Variation in Foraging Behaviour Among Individuals and Populations of Dogwhelks, *Nucella Lapillus*: Natural Constraints on Energy Uptake. *J. Anim. Ecol.* **1991**, *60*, 497–514.

Carriker, M. R.; Person, P.; Libbin, R.; Van Zandt, D. Regeneration of the Proboscis of Muricid Gastropods after Amputation, with Emphasis on the Radula and Cartilages. *Biol. Bull.* **1972**, *143*, 317–331.

Carriker, M. R.; Van Zandt, D. Regeneration of the Accessory Boring Organ of Muricid Gastropods after Excision Trans. *Am. Microsc. Soc.* **1972a**, *91*, 455–466.

Carriker, M. R.; Van Zandt, D.; Charlton, G. Gastropod *Urosalpinx*: pH of Accessory Boring Organ while Boring. Science **1967**, *158*, 920–922.

Carriker, M.R. Comparative Functional Morphology of the Drilling Mechanism in Urosalpinx and Eupleura (Muricid Gastropods). Proceedings of the XVIH International Congress on Zoology, London, 1959, pp 373–376.

Carriker, M.R. Comparative Functional Morphology of Boring Mechanisms in Gastropods. *Am. Zool.* **1961**, *1*, 263–266.

Carriker, M.R. Excavation of Boreholes by the Gastropod, Urosalpinx: An Analysis by Light and Scanning Electron Microscopy. *Am. Zool.* **1969**, *9*, 917–933.

Carriker, M.R. Ultrastructural Evidence that Gastropods Swallow Shell Rasped During Hole Boring. *Biol. Bull.* **1977**, *152*, 325–336.

Carriker, M.R. Ultrastructural Analysis of Dissolution of Shell of the Bivalve *Mytilus edulis* by the Accessory Boring Organ of the Gastropod *Urosalpinx cinerea. Mar. Biol.* **1978**, *48*, 105–134.

Carriker, M.R. Shell Penetration and Feeding by the naticacean and muricacean Predatory Gastropods: A Synthesis. *Malacologia* **1981**, *20*, 403–422.

Carriker, M.R.; Chauncey, H.H. Effect of Carbonic Anhydrase Inhibition on Shell Penetration by the muricid Gastropod *Urosalpinx cinerea. Malacologia* **1973**, *12*, 247–263.

Carriker, M.R.; Gruber, G.L. Uniqueness of the Gastropod Accessory Boring Organ (ABO): Comparative Biology, an Update. *J. Shellfish Res.* **1999**, *18*, 579–595.

Carriker, M.R.; Scott, D.B.; Martin, G.N. Demineralization Mechanism of the Boring Gastropods. In *Mechanisms of Hard Tissue Destruction*; Sognannes, R.F., Ed.; American Association for the Advancement of Science: Washington, DC, 1963; pp 55–89.

Carriker, M.R.; Smith, E.H. Comparative Calcibiocavitology: Summary and Conclusions. *Am. Zool.* **1969**, *9*, 1011–1020.

Carriker, M.R.; van Zandt, D. Regeneration of the Accessory Boring Organ of Muricid Gastropods after Excision. *Trans. Am. Microsc. Soc.* **1972b**, *91*, 455–466.

Carriker, M.R.; Williams, L.G.The Chemical Mechanism of Shell Dissolution by Predatory Boring Gastropods: A Review and an Hypothesis. *Malacologia* **1978**, *17*, 143–156.

Carriker, M.R.; Williams, L.G.; Van Zandt, D. Preliminary Characterization of the Secretion of the Accessory Boring Organ of the Shell-penetrating Muricid Gastropod *Urosalpinx cinerea*. *Malacologia* **1978**, *17*, 125–142.

Carriker, M.R.; Yochelson, E.L.Recent Gastropod Boreholes and Ordovician Cylindrical Borings, U.S. Government Printing Office, 1968.

Chétail, M.; Binot, D.; Bensalem, M. Organe de perforation de *Purpura lapillus* (L.) (Muricidae): histochemie et histoenzymologie. *Cah. Biol. Mar.* **1968**, *9*, 13–22.

Chétail, M.; Fournié, J. Shell-boring Mechanism of the Gastropod, *Purpura* (Thaïs) *lapillus*: A Physiological Demonstration of the Role of Carbonic Anhydrase in the Dissolution of CaCO$_3$. *Am. Zool.* **1969**, *9*, 983–990.

Chiu, J.M.; Shin, P.K.; Wong, K.-P.; Cheung, S.-G. Sibling Cannibalism in Juveniles of the Marine Gastropod *Nassarius festivus* (Powys, 1835).*Malacologia* **2010**, *52*, 157–161.

Clelland, E.S.; Saleuddin, A.S.M. Vacuolar-type ATPase in the Accessory Boring Organ of *Nucella lamellosa* (Gmelin) (Mollusca: Gastropoda): Role in Shell Penetration. *Biol. Bull.* **2000**, *198*, 272–283.

Cox, D.L.; Walker, P.; Koob, T.J. Predation on Eggs of the Thorny Skate.*Trans. Am. Fish. Soc.* **1999**, *128*, 380–384.

Crossland, M.; Alford, R.; Collins, J. Population Dynamics of an Ectoparasitic Gastropod, *Hypermastus* sp. (Eulimidae), on the Sand Dollar, Arachnoides placenta (Echinoidea).*Mar Freshwater Res.* **1991**, *42*, 69–76.

Culver, S.J.; Lipps, J.H. Predation on and by Foraminifera. In Predator—Prey Interactions in the Fossil Record; Kelley, P.H., Kowalewski, M., Hansen, T.A., Eds.; Springer US, 2003; p 7–32.

da Costa, A.R.; Oliveira, P.F.; Barrias, C.; Ferreira, H.G. Identification of a V-type Proton Pump in the Outer Mantle Epithelium of *Anodonta cygnea*. Comp. *Biochem. Physiol., A: Mol. Integr. Physiol.* **1999**, *123*, 337–342.

Day, J.A. Feeding of the Cymatiid Gastropod, *Argobuccinum argus*, in Relation to the Structure of the Proboscis and Secretions of the Proboscis Gland. *Am. Zool.* **1969**, *9*, 909–916.

Derer, M. The Perforation Organ of *Thais lapillus* L. (Gasteropodes, Prosobranches). Optical and Electron Microscopic Study.*Arch. Anat. Microsc. Morphol. Exp.* **1975**, *64*, 1–26.

Ebanks, S.C.; O'Donnell, M.J.; Grosell, M. Characterization of Mechanisms for Ca^{2+} and HCO$_3{}^-$/CO$_3{}^{2-}$ acquisition for Shell Formation in Embryos of the Freshwater Common Pond Snail *Lymnaea stagnalis*. *J. Exp. Biol.* **2010**, *213*, 4092–4098.

Ehrenfeld, J.; Klein, U.The Key Role of the H$^+$ V-ATPase in Acid–Base Balance and Na$^+$ transport Processes in Frog Skin. *J. Exp. Biol.* **1997**, *200*, 247–256.

Ehrlich, H.; Koutsoukos, P.G.; Demadis, K.D.; Pokrovsky, O.S. Principles of Demineralization: Modern Strategies for the Isolation of Organic Frameworks. Part I. Common Definitions and History.*Micron* **2008**, *39*, 1062–1091.

Ehrlich, H.; Koutsoukos, P.G.; Demadis, K.D.; Pokrovsky, O.S. Principles of Demineralization: Modern Strategies for the Isolation of Organic Frameworks. Part II. Decalcification. *Micron* **2009**, *40*, 169–193.

Evans, T.B. Optical and Ionic Characterization of the Secretion of the Accessory Boring Organ of the Predatory Gastropod *Urosalpinx cinerea* (Say). *Am. Zool.* **1980**, *20*, 769.

Fischer, P.H. Sur les Gasteropodes Perceurs. *J. Conchyliol.* **1922**, *67*, 3–56.

Franchini, A.; Fantin, M.B.; Caselli, P. Fine Structure of the Accessory Boring Organ of Starvd and Satiated Specimens of *Ocinebrina edwardsi* (Payr.). *J. Exp. Mar. Biol. Ecol.* **1983**, *72*, 59–66.

Fretter, V.The Genital Ducts of some British Stenoglossan Prosobranchs.*J. Mar. Biol. Assoc. U.K.* **1941**, *25*, 173–211.

Fretter, V.; Graham, A. British Prosobranch Molluscs: Their Functional Anatomy and Ecology, Bartholomew Press: Dorking, 1962.

Gordillo, S. Cannibalism in Holocene Muricid Snails in the Beagle Channel, at the Extreme Southern Tip of South America:An Opportunistic Response? *Palaeontol.Electron.* **2013**, *16*, 13p.

Graham, A.The Oesophagus of the Stenoglossan Prosobranchs. *Proc. R. Soc. Edinb. B.* **1941**, *61*, 1–23.

Grey, M.; Boulding, E.G.; Brookfield, M.E. Shape Difference among Boreholes Drilled By Three Species of Naticid Gastropods.*J. Mollus. Stud.* **2005**, *71*, 253–256.

Harper, E.M.; Forsythe, G.T.W.; Palmer, T. Taphonomy and the Mesozoic Marine Revolution; Preservation State Masks the Importance of Boring Predators. *Palaios* **1998**, *13*, 352–360.

Harvey, W.R.; Maddrell, S.H.P.; Telfer, W.H.; Wieczorek, H. H+ V-ATPases Energize Animal Plasma Membranes for Secretion and Absorption of Ions and Fluids. *Am. Zool.* **1998**, *38*, 426–441.

Helwerda, R.A.; Schilthuizen, M. Predation on Greek Albinaria (Pulmonata: Clausiliidae) by Poiretia (Pulmonata: Oleacinidae) and by an Unknown Organism making Circular Holes: Possible Drivers of Shell Evolution. *J. Mollus. Stud.* **2014**, *80*, 272–279.

Hirsch, G.C. Die Ernährungsbiologie fleischfressender Gastropoden: (Murex, Natica, Pterotrachea, Pleurobranchaea, Tritonium); T. 1, Makroskopischer Bau, Nahrung, Nahrungsaufnahme, Verdauung, Sekretion, Lippert, 1915.

Hua, H.; Pratt, B.R.; Zhang, L.-Y. Borings in Cloudina Shells: Complex Predator–Prey Dynamics in the Terminal Neoproterozoic. *Palaios* **2003**, *18*, 454–459.

Hudson, R.L. Ion Transport by the Isolated Mantle Epithelium of the Freshwater Clam, Unio Complanatus. *Am. J. Physiol.* **1992**, *263*, R76–83.

Hudson, R.L. Bafilomycin-sensitive Acid Secretion by Mantle Epithelium of the Freshwater Clam, Unio Complanatus.*Am. J. Physiol.—Reg. I* **1993**, *264*, R946–R951.

Huelsken, T. First Evidence of Drilling Predation by *Conuber sordidus* (Swainson, 1821) (Gastropoda: Naticidae) on Soldier Crabs (Crustacea: Mictyridae). *Molluscan Res.* **2011**, *31*, 125–132.

Hughes, R.N.; de B. Dunkin, S. Behavioural Components of Prey Selection by Dogwhelks, *Nucella lapillus* (L.), Feeding on Mussels, *Mytilus edulis* L., in the Laboratory. *J. Exp. Mar. Biol. Ecol.* **1984a**, *77*, 45–68.

Hughes, R.N.; de B. Dunkin, S. Effect of Dietary History on Selection of Prey, and Foraging Behaviour among Patches of Prey, by the Dogwhelk, *Nucella lapillus* (L.). *J. Exp. Mar. Biol. Ecol.* **1984b**, *79*, 159–172.

Hughes, R.N.; Drewett, D.A Comparison of the Foraging Behaviour of Dogwhelks, *Nucella lapillus* (L.), Feeding on Barnacles or Mussels on the Shore.*J. Mollus. Stud.* **1985**, *51*, 73–77.

Hughes, R.N.; de B. Dunkin, S. Behavioural Components of Prey Selection by Dogwhelks, *Nucella Lapillus* (L.), Feeding on Mussels, *Mytilus edulis* L., in the Laboratory. *J. Exp. Mar. Biol. Ecol.* **1984**, *77*, 45–68.

Hughes, R.N.; Hughes, H.P.I.A Study of the Gastropod *Cassis tuberosa* (L.)Preying Upon Sea Urchins.*J. Exp. Mar. Biol. Ecol.* **1971**, *7*, 305–314.

Hughes, R.N.; Hughes, H.P.I. Morphological and Behavioural Aspects of Feeding in the Cassidea (Tonnacea, Mesogastropoda).*Malacologia* **1981,** *20,* 385–402.

Jensen, A.S. Do the Naticidæ Drill by Mechanical or by Chemical Means? Nature **1951,** *167,* 901–902.

Jonusaite, S.; Kelly, S.P.; Donini, A.The Physiological Response of Larval *Chironomus riparius* (Meigen) to Abrupt Brackish Water Exposure.*J. Comp. Physiol. B.* **2011,** *181,* 343–352.

Kabat, A.R. Predatory Ecology of Naticid Gastropods with a Review of Shell Boring Predation. *Malacologia* **1990,** *32,* 155–193.

Kase, T.; Ishikawa, M. Mystery of Naticid Predation History Solved: Evidence from a "Living Fossil" Species. *Geology* **2003a,** *31,* 403–406.

Kase, T.; Ishikawa, M. Mystery of Naticid Predation History Solved: Evidence from a "Living Fossil" Species: Comment and Reply. *Geology* **2003b,** *31,* e35–e35.

Katz, S.; Klepal, W.; Bright, M. The Skin of Osedax (Siboglinidae, Annelida): An Ultrastructural Investigation of its Epidermis. *J. Morphol.* **2010,** *271,* 1272–1280.

Kelley, P.H.; Hansen, T.A.The Fossil Record of Drilling Predation on Bivalves and Gastropods. In *Predator—Prey Interactions in the Fossil Record*; Kelley, P.H.; Kowalewski, M.; Hansen, T.A., Eds.; Springer US: 2003; pp 113–139.

Kelley, P.H.; Hansen, T.A.A Case for Cannibalism: Confamilial and Conspecific Predation by Naticid Gastropods, Cretaceous through Pleistocene of the United States Coastal Plain. In *Predation in Organisms*; Springer, 2007; pp 151–170.

Keyl, M.J.; Michaelson, I.A.; Whittaker, V.P. Physiologically Active Choline Esters in Certain Marine Gastropods and Other Invertebrates.*J. Physiol.* **1957,** *139,* 434–454.

Kitchell, J.A.; Boggs, C.H.; Kitchell, J.F.; Rice, J.A. Prey Selection by Naticid Gastropods: Experimental Tests and Application to Application to the Fossil Record. *Paleobiology* **1981,** *7,* 533–552.

Kitching, R.L.; Pearson, J. Prey Localization by Sound in a Predatory Intertidal Gastropod. *Mar.Biol.Lett.* **1981,** *2,* 313–321.

Kowalewski, M. The Fossil Record of Predation: An Overview of Analytical Methods. *Paleontol.Soc. Papers* **2002,** *8,* 3–42.

Kowalewski, M. Drill Holes Produced by the Predatory Gastropod *Nucella lamellosa* (Muricidae): Palaeobiological and Ecological Implications. *J. Mollus. Stud.* 2004,*70,* 359–370.

Kowalewski, M.; Dulai, A.; Fürsich, F.T.A Fossil Record Full of Holes: The Phanerozoic History of Drilling Predation. *Geology* **1998,** *26,* 1091–1094.

Kowalewski, M.; Nebelsick, J.H. Predation on Recent and Fossil Echinoids. In *Predator—Prey Interactions in the Fossil Record*; Kelley, P.H.; Kowalewski, M.; Hansen, T.A., Eds.; Springer, 2003; pp 279–302.

Leighton, L. Evaluating the Accuracy of Drilling Frequency as an Estimate of Prey Preference and Predation Intensity. *PaleoBios* **2001,** *21,* 83.

Leighton, L. Predation on Brachiopods. In *Predator—Prey Interactions in the Fossil Record*; Kelley, P.H., Kowalewski, M., Hansen, T.A., Eds.; Springer US, 2003; pp 215–237.

Loppens, K. La perforation des coquilles des mollusques par les gastropodes et les eponges. *Ann. Soc. R. Zool. Belg.* **1926,** *57,* 14–18.

MacGinitie, G.E.; MacGinitie, N. Natural History of Marine Animals, McGraw Hill Text: US, 1968.

Maddrell, S.H.; O'Donnell, M.J. Insect Malpighian Tubules: V-ATPase Action in Ion and Fluid Transport. *J. Exp. Biol.* **1992,** *172,* 417–429.

Matsukuma, A. Fossil Boreholes Made by Shell-boring Predators or Commensals, Part I: Boreholes of Capulid Gastropods. *Venus* **1978**, *27*, 29–45.

Merle, D.; Garrigues, B.; Pointier, J.-P. Fossil and Recent Muricidae of the World: Part Muricinae, ConchBooks: Hackenheim, Germany, 2011.

Merzendorfer, H.; Graf, R.; Huss, M.; Harvey, W.R.; Wieczorek, H. Regulation of Proton-translocating V-ATPases.*J. Exp. Biol.* **1997**, *200*, 225–235.

Miller, L. P. The Effect of Water Temperature on Drilling and Ingestion Rates of the Dogwhelk *Nucella lapillus* Feeding on *Mytilus edulis* Mussels in the Laboratory. *Mar. Biol.* **2013**, *160*, 1489–1496.

Mordan, P.B. Factors Affecting the Distribution and Abundance of *Aegopinella* and *Nesovitrea* (Pulmonata: Zonitidae) at Monks Wood National Nature Reserve, Huntingdonshire. *Biol. J. Linn. Soc.* **1977**, *9*, 59–72.

Morgan, P. R.*Nucella lapillus* (L.) as a Predator of Edible Cockles.*J. Exp. Mar. Biol. Ecol.* **1972**, *8*, 45–52.

Morton, B. Biology of the Swash-riding Moon Snail *Polinices incei* (Gastropoda: Naticidae) Predating the Pipi, *Donax deltoides* (Bivalvia: Donacidae), on Wave-exposed Sandy Beaches of North Stradbroke Island, Queensland, Australia. In *Proceedings of the Thirteenth International Marine Biology Workshop, The Marine Fauna and Flora of Moreton Bay, Queensland*; Davie, P.F., Philips, J.A., Eds.; Memoirs of the Queensland Museum—Nature: Brisbane, 2008; Vol. 54, pp 303–322.

Morton, B.; Chan, K.First Report of Shell Boring Predation by a Member of the *Nassariidae* (Gastropoda).*J. Mollus. Stud.* **1997**, *63*, 476–478.

Morton, B.; Salvador, A.The Biology of the Zoning Subtidal Polychaete *Ditrupa arietina* (Serpulidae) in the Açores, Portugal, with a Description of the Life History of its Tube. *Açoreana (Suppl.)* **2009**, *6*, 146–155.

Morton, J.E.; Miller, M.C.*The New Zealand Sea Shore*, Collins, 1973.

Nebelsick, J.H.; Kowalewski, M. Drilling Predation on Recent Clypeasteroid Echinoids from the Red Sea. *Palaios* **1999**, *14*, 127.

Neumann, C.; Wisshak, M. Gastropod parasitism on Late Cretaceous to Early Paleocene holasteroid echinoids—Evidence from *Oichnus halo*isp. n. *Palaeogeogr., Palaeoclimatol., Palaeoecol.* **2009**, *284*, 115–119.

Nguyen, H.; Donini, A. Larvae of the Midge *Chironomus riparius* Possess Two Distinct Mechanisms for Ionoregulation in Response to Ion-poor Conditions.*Am. J. Physiol. Regul. Integr.Comp. Physiol.* **2010**, *299*, R762–R773.

Nylen, M.U.; Provenza, D.V.; Carriker, M.R. Fine Structure of the Accessory Boring Organ of the Gastropod, Urosalpinx.*Am. Zool.* **1969**, *9*, 935–965.

Orr, V. The Drilling Habit of *Capulus danieli* (Crosse) (Mollusca Gastropoda). *Veliger* **1962**, *5*, 63–67.

Paine, R.T. Trophic Relationships of 8 Sympatric Predatory Gastropods.*Ecology* **1963**, 63–73.

Palmer, A.R. Predator Size, Prey Size and the Scaling of Vulnerability: Hatchling Gastropods vs. Barnacles. *Ecology* **1990**, *71*, 759–775.

Pelseneer, P. Gastropodes Marine Carnivores Natica et Purpura. *Ann. Soc. Zool. Belg.* **1925**, *55*, 37–39.

Perry, D.M. Function of the Shell Spine in the Predaceous Rocky Intertidal Snail *Acanthina spirata* (Prosobranchia: Muricacea). *Mar. Biol.* **1985**, *88*, 51–58.

Person, P.; Smarsh, A.; Lipson, S. J.; Carriker, M. R. Enzymes of the Accessory Boring Organ of the Muricid Gastropod *Urosalpinx cinerea follyensis*. I. Aerobic and Related Oxidative Systems. *Biol. Bull.* **1967**, *133*, 401–410.

Peterson, C.H.; Black, R. Drilling by Buccinid Gastropods of the Genus *Cominella* in Australia.*Veliger* **1995**, *38*, 37.

Ponder, W.F.; Taylor, J.D. Predatory Shell Drilling by Two Species of Austroginella (Gastropoda: Marginellidae). *J. Zool.* **1992**, *228*, 317–328.

Provenza, D.C.; Nylen, M.U.; Carriker, M.R.Some Cytologic Observations of the Secretory Epithelium of the Accessory Boring Organ of the Gatropods Urosalpinx and Eupleura.*Am. Zool.* **1966**, *6*, 322.

Réaumur, R. De la formation et de l'accroissment des coquilles des animaux tant terrestres qu'aquatiques, soit de mer soit de rivière. *Mem. Hist. Acad. Sci. Annie* **1709**, 364–400.

Reyment, R.A.; Elewa, A.M.T. Predation by Drills on Ostracoda. In *Predator—Prey Interactions in the Fossil Record*; Kelley, P.H., Kowalewski, M., Hansen, T.A., Eds.; Springer US, 2003; pp 93–111.

Rittschof, D. Chemical Attraction of Hermit Crabs and Other Attendants to Simulated Gastropod Predation Sites. *J. Chem. Ecol.* **1980a,** *6*, 103–118.

Rittschof, D. Enzymatic Production of Small Molecules Attracting Hermit Crabs to Simulated Gastropod Predation Sites. *J. Chem. Ecol.* **1980b,** *6*, 665–675.

Rittschof, D.; Williams, L.G.; Brown, B.; Carriker, M.R. Chemical Attraction of Newly Hatched Oyster Drills.*Biol. Bull.* **1983**, *164*, 493–505.

Rocha-Barreira, C.; Santana, I.C.H.; Franklin-Junior, W. Predatory Behavior of *Thais haemastoma floridana* (Conrad 1837) in Laboratory.*Thalassas* **2004**, *20*, 55–60.

Rojas, A.; Verde, M.; Urteaga, D.; Scarabino, F.; Martínez, S.The First Predatory Drillhole on a Fossil Chiton Plate: An Occasional Prey Item or an Erroneous Attack? *Palaios* **2014,** *29*, 414–419.

Rovero, F.; Hughes, R.N.; Chelazzi, G. Effect of Experience on Predatory Behaviour of Dogwhelks.*Anim. Behav.* **1999**, *57*, 1241–1249.

Sallman, A.L.; Lubansky, H.J.; Talor, Z.; Arruda, J.A. Plasma Membrane Proton ATPase from Human Kidney. *Eur. J. Biochem.* **1986**, *157*, 547–551.

Schiemenz, P. Wie bohrt Natica die Muscheln an? *Mitt. Zool. Stat. Neapel.* **1891**, *10*, 153–169.

Schilthuizen, M.; Kemperman, T.C.M.; Gittenberger, E. Parasites and Predators in Albinaria (Gastropoda Pulmonata: Clausiliidae). *Bios (Macedonia, Greece)* **1994**, *2*, 177–186.

Schilthuizen, M.; Liew, T.-S. The slugs and semislugs of Sabah, Malaysian Borneo (Gastropoda, Pulmonata: Veronicellidae, Rathouisiidae, Ariophantidae, Limacidae, Philomycidae). *Basteria* **2008**, *72*, 287–306.

Schilthuizen, M.; van Til, A.; Salverda, M.; Liew, T.-S.; James, S.S.; Elahan, B.b.; Vermeulen, J.J.; O'Foighil, D. Microgeographic Evolution of Snail Shell Shape and Predator Behavior. *Evolution* **2006**, *60*, 1851–1858.

Smarsh, A. Carbonic Anhydrase in the Accessory Boring Organ of the Gastropod, Urosalpinx. *Am. Zool.* **1969**, *9*, 967–982.

Sohl, N.F. The Fossil Record of Shell Boring by Snails.*Am. Zool.* **1969**, *9*, 725–734.

Spight, T.M.; Lyons, A. Development and Functions of the Shell Sculpture of the Marine Snail *Ceratostoma foliatum*.*Mar. Biol.* **1974**, *24*, 77–83.

Stickle, W.B.; Bayne, B.L. Energetics of the muricid gastropod *Thais* (*Nucella*) *lapillus* (L.). *J. Exp. Mar. Biol. Ecol.* **1987**, *107*, 263–278.

Sullivan, G.V.; Fryer, J.N.; Perry, S.F. Immunolocalization of Proton Pumps (H$^+$–ATPase) in Pavement Cells of Rainbow Trout Gill. *J. Exp. Biol.* **1995**, *198*, 2619–2629.

Taylor, J.D.; Morton, B.The Diets of Predatory Gastropods in the Cape d'Aguilar Marine Reserve, Hong Kong.*Asian Mar. Biol.* **1996**, *13*, 141–166.

Tresguerres, M.; Katz, S.; Rouse, G.W. How to Get into Bones: Proton Pump and Carbonic Anhydrase in Osedax boneworms. *Proc. R. Soc. B*.**2013**, *280*, 20130625.

Troschel, F.H. Ueber die Spiechel von Dolium galea. Journal prakt.*Chemie* **1854**, *63*, 173–179.

Vermeij, G.J. *Evolution and Escalation: An Ecological History of Life*, Princeton University Press, 1987.

Vermeij, G.J. The Evolutionary Interaction among Species: Selection, Escalation, and Coevolution. *Annu. Rev. Ecol. Syst.* **1994**, 219–236.

Visaggi, C.C.; Dietl, G.P.; Kelley, P.H. Testing the Influence of Sediment Depth on Drilling Behaviour of *Neverita duplicata* (Gastropoda: Naticidae), with a Review of Alternative Modes of Predation by Naticids. *J. Mollus. Stud.* **2013**, *79*, 310–322.

Wächtler, V.W. Zur biologie der Raubling euschnecke Poiretia (Glandina) algira Brug.*Zool. Anz.* **1927**, *72*, 191–197.

Warén, A.A Generic Revision of the Family Eulimidae (Gastropoda, Prosobranchia).*J. Mollus. Stud.* **1983**, *49*, 1–96.

Webb, R.S.; Saleuddin, A.S.M. Role of Enzymes in the Mechanism of Shell Penetration by the Muricid Gastropod, *Thais lapillus* (L.).*Can. J. Zool.* **1977**, *55*, 1846–1857.

West, L. Intertidal variation in Prey Selection by the snail *Nucella* (=Thais) *emarginata*. *Ecology* **1986**, *67*, 798–809.

West, L. Prey Selection by the Tropical Snail *Thais melones*: A Study of Interindividual Variation. *Ecology* **1988**, *69*, 1839–1854.

Whittaker, V.P. Pharmacologically Active Choline Esters in Marine Gastropods.*Ann. N. Y. Acad. Sci.* **1960**, *90*, 695–705.

Wieczorek, H.; Brown, D.; Grinstein, S.; Ehrenfeld, J.; Harvey, W.R. Animal Plasma Membrane Energization by Proton-motive V-ATPases.*Bioessays* **1999**, *21*, 637–648.

Wood, L. Physiological and Ecological Aspects of Prey Selection by the Marine Gastropod Urosalpinx Cinerea (Prosobranchia: Muricidae). *Malacologia* **1968**, *6*, 267–320.

Young, D.K. *Okadaia elegans*, a Tube-boring Nudibranch Mollusc from the Central and West Pacific.*Am. Zool.* **1969**, *9*, 903–907.

Ziegelmeier, E. Beobachtungen uber den nahrungerwerb bei der naticide Lunatia nitida Donovan (gastropoda prosobranchia).*Helgol.Wiss.Meeresunters.* **1954**, *5*, 1–33.

Ziegler, A. Expression and Polarity Reversal of V-type H^+-ATPase During the Mineralization–Demineralization Cycle in Porcellio Scaber Sternal Epithelial Cells. *J. Exp. Biol.* **2004**, *207*, 1749–1756.

CHAPTER 4

THE ROLE OF METAL IONS IN THE MUSSEL BYSSUS

ANTJE REINECKE and MATTHEW J. HARRINGTON*

Department of Biomaterials, Max Planck Institute of Colloids and Interfaces, Potsdam 14424, Germany. E-mail: matt.harrington@mpikg. mpg.de

Corresponding author.

CONTENTS

ABSTRACT

Marine mussels (Mytilidae) are sessile bivalve mollusks that populate the highest wave-impact regions of rocky seashore habitats. A key to their evolutionary success in the intertidal environment is the byssus—a protein-based fibrous anchor that is fabricated by the mussel. The byssus is made up of numerous byssal threads, which are without hyperbole, the lifeline of the organism. Each byssal thread can be subdivided into three distinct functional elements—the core, the cuticle, and the plaque, which function as a self-healing shock-absorbing tether, an abrasion resistant coating, and a versatile underwater glue, respectively. A critical feature of each of these three elements is the presence of protein–metal coordination interactions that influence to a large degree the material properties. In particular, interactions between histidine and Zn/Cu contribute significantly to the deformation and self-healing behavior of the thread core, while interactions between 3, 4-dihydroxyphenylalanine and Fe enhance the mechanical function of the cuticle, as well as the formation and adhesive properties of the plaque. In the present chapter, we will provide a comprehensive overview of the state-of-the-art understanding of the role of protein–metal interactions in the material properties and formation of the byssus, while highlighting open questions.

4.1 INTRODUCTION

4.1.1   THE MUSSEL BYSSUS

With the right equipment, survival is possible under even the most extreme circumstances. Just as mountain climbers scaling Everest must rely on the proper gear, organisms inhabiting extreme environments have evolved distinctive adaptations to not only survive, but to thrive under extremely unfavorable conditions. While humans are not generally accustomed to thinking of the seaside intertidal zone as an extreme environment, the animals inhabiting this nutrient-rich niche face tremendous daily challenges to survival. These include exposure to temperature extremes, solar radiation, predation, sandblasting, and most notably, intense and incessant forces from crashing waves (Denny et al., 1998; Denny and Gaylord, 2010; Harrington & Waite, 2008a) (Fig. 4.1A). Waves provide a primary selective pressure in the marine intertidal zone, capable of producing enormous lift and drag forces (Denny and Gaylord, 2002) (Fig. 4.1B). Not coincidentally, the organisms residing in the high-intertidal zone often possess remarkable strategies and materials for producing a secure attachment to the rocky substrates characteristic of seashore environments. Examples of successful anchoring strategies include the cement of barnacles (Kamino, 2010), the tube feet of sea stars (Hennebert et al., 2014) and the topic of the current chapter, the mussel byssus.

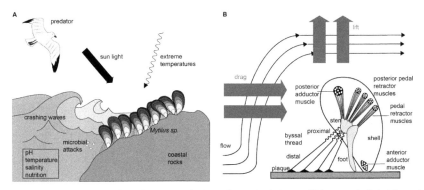

FIGURE 4.1   Habitat of marine mussels (*Mytilus sp.*). (A) Mytilid mussels inhabit rocky seashore environments where they face a number of challenges to survival. (B) Mussels attach to hard substrates with byssal threads, which prevent dislodgement from lift and drag forces caused by crashing waves. The inner anatomy of a mussel is shown in schematic emphasizing the musculature and the foot, which is the organ that forms the byssal threads. Figure adapted from source details (Source: Denny & Garland, 2010; illustration by Antje Reinecke.)

Marine mussels from the species *Mytilus* are bivalve mollusks that typically reside in sprawling beds along rocky coastal regions between high and low tides. The dominance of the organism in this habitat makes it vital part of the complex intertidal ecosystem (Denny and Gaylord, 2010). For the majority of their lives, besides early-larval and postlarval stages, the mytilid mussels maintain an entirely sessile way of life (Yonge, 1962) in which they sustain themselves on ample nutrient sources by filter feeding. Due to their sessile lifestyle, however, mussels cannot escape the often physically challenging conditions defining their environment. In particular, the enormous lift and drag forces associated with incessant wave action carries the constant threat of dislodging mussels from their attachment point or shattering their hard shells against the rocks (Fig. 4.1B) (Denny and Gaylord, 2010). Mussels overcome this dominant challenge to survival by fabricating a protein-based attachment holdfast called the byssus (Harrington & Waite, 2008a; Yonge, 1962).

The byssus is a nearly universal feature of almost all bivalves in the postlarval stage, used for securing attachment to various substrates during settlement. However, only a few bivalve groups retain the byssus into adulthood, including marine mussels from the family Mytilidae (Brazee & Carrington, 2006; Yonge, 1962). The byssus of mussels from *Mytilus* species are composed of numerous extracellular protein-based fibrous byssal threads, each of which is several centimeters long, possessing an ellipsoidal cross-section ranging from approximately 100–200 μm along the long axis (Fig. 4.2A) (Harrington & Waite, 2008a). In the present chapter, the byssal threads of *Mytilus* species *Mytilus californianus*, *Mytilus galloprovincialis*, and *Mytilus edulis* will provide the main focus, as they are the most thoroughly characterized.

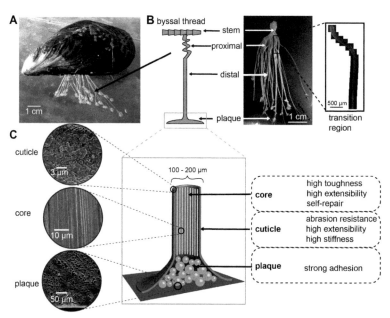

FIGURE 4.2   Byssal thread morphology. (A) Image of a mussel with its byssus. (B) Byssus removed from the mussel with a schematic showing important morphological features. SEM image of the transition region highlights the gradient between the distal and proximal regions of the thread. (C) Schematic of the core, cuticle, and plaque regions of the thread. SEM images highlight the micron-scale structural features.

Mussel byssal threads not only provide a means of anchoring mussels to the hard substratum, but they also function collectively as a robust, shock-absorbing holdfast (Qin & Buehler, 2013). Each individual thread possesses remarkable mechanical features, such as high extensibility, high stiffness, high toughness, abrasion-resistance, self-healing behavior, and underwater adhesion (Carrington & Gosline, 2004). Along the length of a single thread, it can be divided into distinct regions (Fig. 4.2B, Table 4.1). At the most distal point (i.e., at the byssus-surface interface), each thread is glued to the hard substratum by an adhesive plaque, which functions as a strong anchoring point to a range of different surface types. The ability of the plaque to create strong adhesive interfaces under harsh seawater conditions has made it a primary target of biomimetic research in recent years (Lee et al., 2011) (Fig. 4.2B and C, Table 4.2). The other end of the thread, a region known as the stem, is inserted into the living mussel tissue anchoring the thread to the organism (Fig. 4.2A).

TABLE 4.1 Overview of Metal-dependent Performance in the Byssus.

	Thread core	Cuticle	Plaque
Function	Holdfast fiber	Protective coating	Attachment
Metal-dependent material properties	Mechanical energy dissipation Self-healing Extensibility Stiffness	Abrastion resistance Hardness Self-healing Extensibility	Underwater adhesion
Metal–ligand	His–Zn (Broomell et al., 2008) His–Cu (Broomell et al., 2008) His–Ni?	DOPA–Fe (Sun & Waite, 2005)	DOPA–Fe (Hwang et al., 2010)
Protein with metal–ligand	PreCol, PTMP-1, TMP-1	Mfp-1	Mfp-2–6
Ligand content (mol % per AA)	2% His total, 20% in His-rich domains (Harrington & Waite, 2007; Waite et al., 1998)	10–15% DOPA (Lin et al., 2007; Sun & Waite, 2005)	2% (Mfp-4) (Zhao & Waite, 2006b)–28% (Mfp-5) (Waite & Qin, 2011) DOPA 11 Cys in Mfp-6 (Yu et al., 2011b; Zhao & Waite, 2006a)

Between the plaque and stem, the fiber-like thread, which can reach 3–4 cm in length, can be subdivided into the distal and the proximal regions, which exhibit distinctive morphological and mechanical properties (Harrington & Waite, 2008a). The proximal region of the thread is relatively thick and corrugated with a crimped surface and comprises between 20% and 44% of the total thread length of *M. californianus* and *M. edulis*, respectively (Carrington, 2002; Mascolo & Waite, 1986; Qin & Buehler, 2013). It is

TABLE 4.2 Overview of Byssal Thread Proteins and Their Properties.

Protein	Location	Mass (kDa)	pI	Amino acids with mechanical function	Interactions with metal ions	Potential function	
Mfp-1	Cuticle	90–108	10.5	Hyp 15 mol% DOPA	Cross-linking of mfp-1 by DOPA–Fe^{3+}	Abrasion resistance & extensibility of cuticle	Lee et al. (2011)
Mfp-2	Interior of plaque	45	9.5	Cross-linked Cys 5 mol% DOPA	Ca^{2+}-binding, DOPA–Fe^{3+} Cross-linking of mfp-2 by DOPA–Fe^{3+} and with Ca^{2+}	Stability of foam-like structure of plaque	Hwang et al. (2010), Inoue et al. (1995), Rzepecki et al. (1992)
Mfp-3	Plaque–substratum interface	5–7		10–20 mol% DOPA		Adhesion	Carrington et al. (2015), Papov et al. (1995), Wei et al. (2013)
Mfp-4	Plaque–thread core interface	93		2 mol% DOPA	Ca^{2+}-binding motif might bind to plaque proteins, His-rich N-terminal might bind to his-rich domains of preCols (by Cu^{2+}-binding)	Interconnection of preCols and plaque proteins (especially mfp-2)	Lee et al. (2011), Zhao and Waite (2006b)
Mfp-5	Plaque–substratum interface	9	9	28 mol% DOPA	Ca^{2+}/Mg^{2+}-binding	Adhesion	Lee et al. (2011), Waite and Qin (2011)
Mfp-6	Plaque–substratum interface	11	9.5	3 mol% DOPA High amount of Cys		Anti-oxidant for DOPA in mfp-3 and mfp-5 Crosslinking with other mfps by cysteinyl–DOPA crosslinks	Yu et al. (2011b), Zhao and Waite (2006a)

TABLE 4.2 (*Continued*)

Protein	Location	Mass (kDa)	pI	Amino acids with mechanical function	Interactions with metal ions	Potential function	
PreCol-D	Thread core (more distal)	240	10.5	< 1 mol% DOPA Hyp	His–Zn^{2+} His–Cu^{2+}	Toughness, self-healing behavior, extensibility of thread core	Qin et al. (1997)
PreCol-P	Thread core (more proximal)	250		~20 mol% His in his-rich domains	His–Ni^{2+}		Carrington et al. (2015), Coyne et al. (1997)
PreCol-NG	Thread core	230	7.5				Qin and Waite (1998)
TMP-1		57	9.5	3–5 mol% DOPA			Carrington et al. (2015), Sagert and Waite (2009)
PTMP-1	Proximal region of thread core	250		3 mol% DOPA Disulfide bonds	Zn^{2+}/Cu^{2+}	Integrity of collagen/matrix Interconnection of preCols and other byssal proteins by metal coordination	Lucas et al. (2002), Suhre et al. (2014), Sun et al. (2002)

The protein in the light gray row is in the cuticle, proteins in the normal gray and dark gray rows are located in the plaque and in the thread core, respectively.

highly extensible ($\varepsilon_{ult} > 200\%$) and exhibits a relatively low stiffness ($E \sim 20$ MPa) (Carrington & Gosline, 2004). In contrast, the distal region is more fibrous and exhibits a much higher stiffness ($E \sim 500$–800 MPa depending on the species), high toughness, and the ability to self-heal following damage during cyclic loading (Carrington & Gosline, 2004; Mascolo & Waite, 1986).

Surrounding the core of the *Mytilus* byssal thread is a thin outer cuticle (thickness = 5–10 μm) that possesses a characteristic granular morphology (Fig. 4.2C, Table 4.1). Based on a combination of high hardness and high extensibility, the cuticle is proposed to function as a protective coating for the softer thread core (Holten-Andersen et al., 2007). Mussels fabricate byssal threads one-by-one as an orchestrated secretion of an assortment of at least 11 distinct proteins into the byssus-forming organ, known as the foot (Hagenau et al., 2014; Silverman & Roberto, 2010) (Fig. 4.1B). During secretion, the proteins self-assemble and cross-link into a fully functional byssal thread that is attached to the existing byssus (Silverman & Roberto, 2010). Despite the wealth of descriptive data and molecular information of the comprising protein components, presently the details of the assembly process are poorly understood. The final section of this chapter will be dedicated to this topic.

Considering that the byssus is an acellular proteinaceous material that functions extracorporeally (i.e., outside the living organism), its material properties must arise from specific features of the material itself, programed through intrinsic physicochemical properties of the biomolecular protein building blocks and their self-assembly. Thus, insights into the origin of the remarkable (and industrially attractive) material properties of the byssus (e.g., self-healing, abrasion resistance, underwater adhesion—Fig. 4.2C, Table 4.1) can be ascertained by elucidating the underlying material design principles. Over the last 30 years, intensive biochemical and structural–mechanical investigations of the mussel byssal threads from Mytilid species have led to an impressive understanding of the structure–function relationships that define this material (Carrington et al., 2015). A particularly exciting aspect of byssal thread design that has emerged in recent years is the role of protein–metal interactions as mechanical cross-links that contribute to the material performance of the fibrous core, cuticle, and adhesive plaque (Degtyar et al., 2014). In fact, based on these findings, in the last 5 years, there has been a surge in the production of "mussel-inspired" metallopolymers that harness protein–metal chemistry of the byssus for applications in technical and biomedical settings (Li et al., 2015). Thus, the mussel byssus is attracting a broad interest in diverse fields ranging from evolution and biochemistry to ecology and materials science.

In the present chapter, we will discuss the structure–function relationships that define the material function of the mussel byssus, with a particular focus on the essential role of metal coordination cross-links in the self-healing fibrous core, the abrasion-resistant extensible cuticle, and the versatile adhesive plaque. The next section will provide a general overview of the role of metal-coordination in biological systems and the growing understanding of protein–metal cross-links in biological materials, specifically the byssus.

4.1.2 METAL IONS AS MECHANICAL CROSS-LINKING AGENTS

Metal acquisition and metal homeostasis are vital functions of nearly every living organism (Outten et al., 2007). In excessive amounts metals can be dangerous cellular toxins; however, in small doses, particular metals are essential for proper cellular function. For example, countless proteins require metal ions such as Zn, Cu, Fe, Ca, and Mg as cofactors in physiologically important functions and essential biochemical pathways including gas transport, respiration, nitrogen fixation, and photosynthesis (Outten et al., 2007). On top of these roles, metal coordination bonds in proteins also serve to increase protein stability (Glusker, 1991; Holm et al., 1996). Protein-bound metal ions act as a bridge between multiple amino acid side chains to form a coordination complex in which each amino acid side chain donates a pair of electrons to empty orbitals in the outer electron shell of the metal ion. Amino acid side chains such as histidine, cysteine, aspartate, and glutamate are common ligands for a range of metal ions found bound to proteins (Glusker, 1991).

From a mechanical perspective, metal coordination bonds are intermediate to covalent and non-covalent bonds in bond strength and lability. This simply means that they exhibit breaking forces significantly higher than hydrogen bonds, while maintaining the transient ability to reform on biologically relevant timescales once ruptured (Lee et al., 2006). The mechanical properties of biologically relevant protein–metal coordination bonds have been tested using single-molecule force spectroscopy techniques such as atomic force microscopy (AFM), which confirm, for example, that histidine coordination interactions with Zn, Cu, and Ni are both strong and reversible (Schmitt et al., 2000). Furthermore, it has also been demonstrated that bioengineering histidine–metal-binding sites into proteins can lead to an increase in both the thermodynamic (Kellis et al., 1991) and mechanical (Cao et al., 2008) stability of the folded protein.

Based on some of the same features that make metal ions attractive as cofactors for physiologically relevant functions (e.g., redox activity, low kinetic lability, high bond strength), certain protein-based biological materials have evolved the capacity to harness metal ions as essential cross-linking agents that enhance material performance. (See recent review by Degtyar et al., 2014.) For example, the hard biting parts and stingers of arachnids such as spiders and scorpions, the mandibles of insects such as termites, and the piercing jaws of particular marine worms (*Nereis* sp. and *Glycera* sp.) utilize histidine-rich proteins coordinated to metal ions such as Cu^{2+} and Zn^{2+} in order to harden and stiffen the protein-based scaffolds as a lightweight alternative to mineralization (Broomell et al., 2006; Degtyar et al., 2014; Politi et al., 2008; Schofield et al., 2002). Protein–metal interactions in materials, however, are quite versatile and can be used for more than just hardening and stiffening organic structures. As will become clear in this chapter, metal coordination chemistry can be harnessed and tailored to achieve a broad range of mechanical functions from self-healing, adhesion, abrasion resistance, and triggered self-assembly—all of which are exploited in the mussel byssus (Degtyar et al., 2014) (Fig. 4.2, Table 4.1).

4.1.3   METAL IONS IN THE MUSSEL BYSSUS

The presence of metals in the mussel byssus was first highlighted in the 1970s as researchers searched for a reliable method for determining pollution levels in ocean waters around the world (Coombs & Keller, 1981; Phillips, 1976a, b). As filter feeders, mussels are constantly sampling the local water column, and it was found that they incorporate various metals from the surrounding environment into their shells, soft tissues, and their byssus. Due to the mussels ability to concentrate an expansive range of trace elements existing in the seawater into their byssus (e.g., Ag, Cd, Cr, Pb, Ti, Fe, Zn, Cu, and Ni) (Coombs & Keller, 1981), byssal threads have become a standard and reliable environmental biomarker for harmful heavy metals in polluted coastal and estuarine habitats (Szefer et al., 2006). For example, based on their propensity for concentrating V and Ni ions, mussels can be used to assess pollution levels following oil spills (Amiard et al., 2004), as well as radioactive pollution levels in sources of nuclear effluents through the uptake of radionuclides such as uranium, plutonium, and polonium (Hamilton, 1980; Hodge et al., 1979). As a consequence, large efforts have been dedicated to elucidating the complex mechanisms underlying metal uptake, bioaccumulation, transport, and release of metal ions in mussels. Here, we provide only a brief overview of these processes, as this topic has been exhaustively covered in several earlier reviews (Marigomez et al., 2002; Viarengo, 1989; Viarengo & Nott, 1993) (Fig. 4.3).

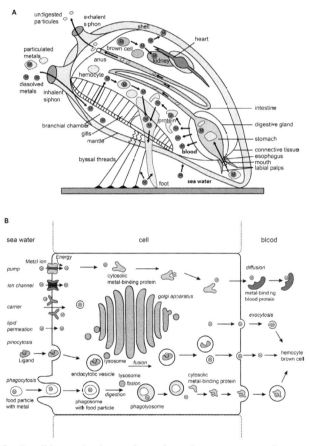

FIGURE 4.3 Possible mechanisms of metal uptake, transport, and accumulation in the byssus. (A) Schematic illustrating proposed movement of dissolved and particle-bound metal ions into and through the mussel tissue. Dissolved metal ions and metals bound to food particles are inhaled through the ingoing siphon, transported via the mouth to the esophagus and stomach where digestion begins. Metals are taken up by cells of the digestive glands and released into the blood. Dissolved metals can also enter the cells of the gills, to be released into the blood. Metals are transported by proteins of the blood plasma or in blood cells (hemocytes/brown cells) to target tissues, for example, to detoxification organs or to the foot for byssus formation. (B) Possible cellular pathways of metal ions from the inhaled sea water into the blood (Simkiss & Taylor, 1989). Dissolved metal ions can be taken up by ion pumps, ion channels, carriers, or lipid permeation and bound to metal-binding proteins inside the cell. Particulated metals can be engulfed by endocytotic pathways. The resulting vesicle can fuse with lysosomes for hydrolysis. On the other side of the cell, metal ions are released into the blood plasma or are incorporated into blood cells after exocytosis.

The bioavailability of metals in seawater strongly depends on several biotic and abiotic factors. For example, metal uptake by mussels is influenced by the age of the organism, as well as by annual growth cycles (Langston & Spence, 1995; Wang & Fisher, 1997). Abiotic factors, on the other hand, such as seawater salinity, temperature, pH, oxygen concentration, redox conditions, and the presence of other trace elements or binding partners may also affect an organisms ability to uptake metals (Abbe & Sanders, 1990; Bjerregaard & Depledge, 1994). Furthermore, metals in seawater exist in many different forms such as dissolved free ions, complexes with various ligands or bound to particles (Abbe & Sanders, 1990), each of which requires different strategies for acquisition.

Mussels, as well as other bivalves, regulate metal uptake, trafficking, accumulation, and removal through complex cellular processes and organ systems (Marigomez et al., 2002) (Fig. 4.3). Specific pathways, for example, exist in the mussel for the acquisition of dissolved vis-à-vis particle-bound metal ions. In both cases, metal uptake commences when seawater containing suspended food particles is inhaled through the incurrent siphon, transported into the branchial chamber and released through the exhalent siphon (Fig. 4.3A). While resident within the confines of the mussel shell, dissolved metal ions in the inhaled seawater are taken up primarily through the gills via passive diffusion across the cell membrane, through ion-channels or via pinocytosis (Carpene & George, 1981; Marigomez et al., 2002) (Fig. 4.3A and B). Once they have entered the epithelial cells of the gills, metal ions can be bound to metal-binding proteins such as metallothioneins, engulfed into lysosomes and transported in vesicles across the cell membrane to be released into the blood plasma or incorporated into blood cells for transport to other tissues and destinations (Fig. 4.3A and B) (Marigomez et al., 2002; Mason, 1983). Metallothioneins are a class of low molecular weight cysteine-rich proteins that possess a high affinity for essential metals such as Zn and Cu, as well as for toxic metal such as Cd, Ag, and Hg (Viarengo & Nott, 1993). In the former case, Zn and Cu binding by metallothioneins is thought to function as a means of storing these metals for later use in enzymes or possibly, in the byssus. Notably, it was shown that lysosomes of marine mussels contain byssus-relevant metal ions including Cu, Zn, and Fe (Mason et al., 1984).

In contrast to dissolved metal ions which pass through the gills, particle-bound metal ions are primarily taken up by the digestive system of the mussel (Viarengo, 1989) (Fig. 4.3). In this case, the labial palps, located near the mouth, move particulate food via movement of cilia toward the esophageal opening and into the stomach where digestion begins (Fig. 4.3A). Surrounding the stomach is the digestive gland, which functions, among other things, to facilitate the acquisition of metal ions from the digested food particles (Marigomez et al., 2002). To achieve this, one possible way is that digestive cells first take up metal-containing particles by phagocytosis into vesicles, which fuse with lysosomes where the particle is further hydrolyzed (Marigomez et al., 2002) (Fig. 4.3B). Similar to acquisition of dissolved ions through the gill epithelial cells, the specialized epithelial cells of the digestive gland (i.e., the digestive and basophilic cells) possess a range of metal-binding proteins, including metallothioneins, which present ligands that have affinities for a range of metal ions. Metal selectivity is primarily based on whether the ions prefer to complex with ligands possessing oxygen donor atoms (e.g., carbonate, phosphate, and sulfate ligands) or non-oxygen donor atoms (e.g., sulfur, nitrogen ligands) (Marigomez et al., 2002). However, there are some biomolecules, such as ferritin, that specifically bind only certain metal ions, in this case iron (Taylor, 1995). This differentiation in metal specificity leads to the non-homogenous distribution of metal ions between various cell types (Marigomez et al., 2002), which are then available for use in various physiological functions or alternatively, are destined for removal from the organism.

Essential metal ions (i.e., those destined for specific physiological roles) are transported to target tissues either bound to blood plasma-binding proteins or by cell-mediated means in hemocytes or brown cells, migratory phagocytic cells specialized for metal transport (Haszprunar, 1996) (Fig. 4.3A and B). Presently, however, the pathway by which the huge range of metals found in the byssus are specifically incorporated into byssal threads is not well understood. The best-characterized case is that of iron incorporation. It was previously demonstrated by raising mussels in seawater spiked with radiolabeled Fe ions that the iron present in the byssus is largely acquired through filter feeding, rather than passive diffusion into the threads (George et al., 1976). In this study, radiolabeled Fe was shown to first accumulate in the visceral tissue and then only during a period of approximately a week were radiolabeled Fe ions incorporated into byssal threads. Autoradiography studies on mussel foot sections have suggested that Fe might accumulate in localized micron-sized "hot spots" in the foot tissue; however, this was never further corroborated (Pentreath, 1973). In addition to this evidence for the "active" incorporation of metal ions into threads, metal ions can also clearly be integrated passively into the threads under certain conditions (Harrington et al., 2010). Thus, at this point, the mechanism of incorporation of metal ions into the byssus remains an open question.

While the puzzle of how metal ions are integrated into the byssus is still not entirely obvious, the question of what they are doing there is beginning to become more clear. Based on the sheer diversity of metals found in the byssus, it was originally assumed that the byssus serves as a waste depository for excess and poisonous metal ions that had accumulated in the mussel soft tissue (Coombs & Keller, 1981; George et al., 1976). However, it was later proposed that at least some of the accumulated metal ions might instead be fulfilling a functional role in the material based on the high concentration of specific amino acid residues prevalent in byssal thread proteins known for their metal-binding prowess (Taylor et al., 1994; Waite et al., 1998). This includes histidine residues that were found to be concentrated in highly conserved patches at the end of proteins in the thread core (Harrington & Waite, 2007) (Fig. 4.4) and an elevated presence of a relatively rare post-translational modification of tyrosine called 3, 4-dihydroxyphenylalanine (DOPA), which is known for forming very stable complexes with a number of metal ions (Sever & Wilker, 2004) (Figs. 4.5 and 4.6).

A growing body of evidence now supports a critical role of protein–metal coordination bonds in byssal thread assembly and material performance between histidine–Zn/Cu and DOPA–Fe as summarized in Table 4.1. The next three sections will describe the state-of-the-art understanding of the relationship between material behavior and structure of byssal threads with the focus on the various proposed functions of metal coordination bonds, in the material performance of the self-healing core, abrasion-resistant cuticle, and adhesive plaque of byssal threads. The final section will provide an overview of the current understanding of byssal thread assembly, emphasizing the proposed role of protein metal interactions.

4.2   SACRIFICIAL METAL COORDINATION BONDS IN THE BYSSAL THREAD CORE

Living organisms invest ample resources and energy into building materials such as skeletal elements, soft tissues, and the extracellular matrix. Thus, inherent in the design of many biological materials are features to enhance the durability, toughness, and damage tolerance (Chen et al., 2012). Along these lines, a defining feature of many living organisms is the ability to heal damage. Normally, this is achieved by cellular-dependent processes as observed in bone mending (Fratzl & Weinkamer, 2007), wound healing (Werner & Grose, 2003), and even full limb regeneration (Nacu & Tanaka, 2011). Less common, however, is the ability of acellular materials to exhibit self-healing behaviors, as exhibited by the mussel byssus (Carrington & Gosline, 2004). In the absence of cellular intervention, damage tolerance and self-healing in byssal threads must be programmed into the material itself through the biochemical features of the protein building blocks and their hierarchical organization.

The functionality and efficacy of the byssal attachment system is due in part to the ability of threads to combine the features of high stiffness, high strength, and high extensibility (Coyne et al., 1997). As previously mentioned, the byssal thread is divided into two regions, the proximal and distal region, each of which possesses different mechanical properties (Fig. 4.4A). The proximal region possesses a stiffness of ~20 MPa (Gosline et al., 2002) and is extensible up to 200% of its initial length, which is comparable to elastin and resilin (Bell & Gosline, 1996). In contrast, the distal region is less extensible (~100% strain), but stiffer (up to 800 MPa) and stronger than the proximal region (Fig. 4.4A). In particular, mechanical tests performed on the distal region exhibit a stress–strain curve with three phases during loading: (1) a linear region at low strain, (2) a yield point with a post-yield plateau, and (3) a post-yield stiffening before breakage (Fig. 4.4A). The combination of high extensibility and high strength provides the material with a high toughness. At low cyclic extensions (<10% strain), the distal portion of the thread deforms elastically and returns to its initial length when the applied stress is removed (Carrington & Gosline, 2004); however, when strained beyond the yield point, the material exhibits significant mechanical hysteresis during cyclic loading (up to 70%) (Fig. 4.4B).

While material yield provides an important means of dissipating energy from crashing waves, it also results in an apparent damage to the material that can be observed during subsequent loading cycles (Carrington & Gosline, 2004) (Fig. 4.4B). For example, when the distal thread is cyclically loaded to 35% strain, a second cycle following the first shows a reduction of ~65% in the stiffness and the energy dissipated (Harrington et al., 2009). Notably, however, the byssal thread possesses the remarkable ability to self-heal. Here, self-healing is simply defined as the time-dependent recovery of initial material properties following yield-induced pseudo-damage (Carrington & Gosline, 2004) (Fig. 4.4B). Times required to recover a significant proportion of initial properties are on the order of several hours, which is consistent with the low-tide periods in which mussels may be given a brief reprieve from the onslaught of crashing waves. However, at this point, there have been no studies specifically examining the role of self-healing in the natural environment. Nonetheless, it is noteworthy that byssal thread self-healing behavior occurs in the absence of an active metabolism and that threads regain functionality after damage in a completely acellular manner.

Therefore, the origin of this behavior must be intrinsic to the composition and structure of byssal threads.

The thread core consists of more than 95% protein by dry weight (Waite et al., 2002) as well as a small amount of metal ions (<1% by dry weight) (Coombs & Keller, 1981), particularly Zn^{2+} and Cu^{2+} (Fig. 4.4D, Table 4.2). X-ray diffraction studies performed as early as the 1950s suggested that the fibrous core of byssal threads is a fibrillar tendon-like collagen-based material (Mercer, 1952; Rudall, 1955). Biochemical confirmation of this hypothesis was hindered for quite some time due to the intractability of protein extraction; however, Waite and colleagues were eventually able to extract partial sequence of a collagen-like protein by pepsin digestion of threads (Benedict & Waite, 1986; Waite et al., 1998). Subsequent acquisition of the full-length sequences from cDNA revealed three variants of collagen-like proteins that were named preCol (for "pre-pepsized" collagen) (Waite et al., 1998) (Fig. 4.4D, Table 4.2). All three variants, denoted preCol-D, -NG, and -P, possess a dominant central collagen domain with a typical [Gly-X-Y]$_n$ repeat sequence. However, at the N- and C-terminal ends, of all three proteins are non-collagenous domains, that unlike the pro-domains of type I collagen are not cleaved off and remain in the mature form of the protein (Waite et al., 1998). The non-collagen domains consist of the so-called flanking domains and the histidine-rich domains, which are described in the next two paragraphs (Fig. 4.4D).

The preCol flanking domains surround the central collagen domain at both ends and contain sequences that vary between the three variants. PreCol-D flanking domains exhibit multiple runs of polyalanine reminiscent of beta-sheet forming sequence in spider silk, whereas preCol-P flanking domains are proline-rich and have been shown to have sequence similarity to elastic proteins such as elastin and flagelliform silk protein (Harrington & Waite, 2007). PreCol-NG possesses flanking domains with sequences that combine features of both preCol-D and -P flanking domains (Harrington & Waite, 2007; Qin & Waite, 1998). Recent NMR- and FT-IR-based studies on byssal threads support the hypothesis that the flanking domain of both preCol-D and -NG form beta-sheet secondary structures (Arnold et al., 2013; Hagenau et al., 2011). PreCol-NG (*N*on-*G*raded) exists uniformly along the length of the byssal thread, while preCol-D (*D*istal) and -P (*P*roximal) exist in a complementary gradient with preCol-P more prevalent at the extensible proximal end of the thread and preCol-D more prevalent at the distal end. It is believed that this molecular gradient plays a key role in the mechanical gradient existing along the thread axis (Harrington & Waite, 2009; Waite et al., 2004) (Fig. 4.2D).

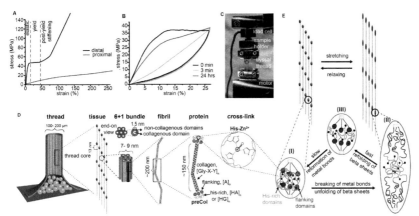

FIGURE 4.4 Structural hierarchy of the byssal thread core. (A) Comparative tensile stress–strain curves of the distal and proximal region of mussel byssal threads. (B) Self-healing behavior of the distal region during cyclic loading. (C) Mechanical tester used for tensile loading experiments. (D) The core of the byssal thread is composed primarily of a family of proteins known as preCols. PreCols possess several domains including a large central collagen domain, beta-sheet forming flanking domains and histidine-rich domains. PreCols form triple helices, which further assemble into 6 + 1 hexagonal bundles. PreCol bundles are arranged end-to-end in series with a 13 nm stagger between adjacent bundles, forming a highly organized semicrystalline framework. PreCol ends interact via the His-rich domains through metal coordination cross-links. (E) One proposed molecular level model of stretching, relaxing, and healing in mussel byssal threads. Stretching beyond the yield point results in the rupture of His–Zn^{2+} cross-links and unfolding of flanking domain beta-sheet structure. When the thread is relaxed, the beta-sheet domains are expected to refold immediately; however, reformation of broken metal coordination cross-links into a stable topology likely requires more time and provides a rate-limiting step in mechanical healing.

At the N- and C-terminal end of all preCols are the His-rich domains (Fig. 4.4D). The His-rich domains contain variable sequences, which are between 21 and 126 amino acids in length and invariably contain at least 20 mol% of histidine, which is ~10-fold higher than found in average protein sequences (Harrington & Waite, 2007; Lucas et al., 2002; Waite et al., 1998). The spacer residues between histidines are typically Ala and Gly, but also occasionally other residues such as Ser and Val (Harrington & Waite, 2007). The preCol sequences of three species are now known, and while interspecies variation exists in the spacer residues and surrounding sequences of the His-rich domains, the His residues themselves are almost completely conserved—strongly suggesting an important functional role (Harrington & Waite, 2007). In particular, the well-known metal-binding affinity of His for metal ions such as Zn and Cu, known to be in the thread, led to the hypothesis that they might contribute as cross-linking ligands (Fig. 4.4D and E) (Waite et al., 1998).

Based on the presence of a large collagen domain, the preCol mono-mers assemble into preCol triple helices, which further self-organize to hexagonal 6 + 1 bundles (Krauss et al., 2013) (Fig. 4.4D). X-ray diffraction (XRD) (Harrington et al., 2009; Krauss et al., 2013) and AFM (Hassenkam et al., 2004) investigations on the thread core have revealed that preCols are arranged in a highly organized semicrystalline framework in which preCol bundles are aligned end-to-end in series and organized laterally in a quasi-hexagonal array (Fig. 4.4D). End-to-end alignment of preCols indi-cates that His-rich domains from consecutive preCol bundles are in contact, and it has been proposed that the formation of coordination bonds between His-residues and transition metal ions bridge the ends of adjacent preCols (Harrington & Waite, 2007) (Fig. 4.4D).

In addition to the preCols, two other proteins have been localized to the byssal thread core—namely, thread matrix protein-1 (TMP-1) and proximal thread matrix protein-1 (PTMP-1) (Sun et al., 2002) (Table 4.2). As the names imply, these proteins are believed to be present in the matrix material surrounding the preCols, with PTMP-1 present primarily in the proximal region of the thread (Sun et al., 2002) and TMP-1 present in both the distal and proximal region (Sagert & Waite, 2009). The proximal region contains a higher content of matrix proteins, whereas the distal region exhibits much lower matrix content (2% or less by dry weight) (Sagert & Waite, 2009). TMP-1 has the distinctive property of undergoing spontaneous deamidation of numerous asparagine residues over time to form aspartate and isoaspar-tate residues, effectively lowering the isoelectric point of the protein (Sagert & Waite, 2009); however, it is unclear what the functional role of this might be. Recent work by the group of Scheibel has demonstrated an effect of PTMP-1 on collagen assembly *in vitro*, supporting an important role of the matrix proteins on assembly and function (Suhre et al., 2014). Furthermore, it was shown that PTMP-1 contains motifs consistent with metal-binding geometries found in other proteins and that the collagen binding of PTMP-1 is slightly enhanced in the presence of Zn^{2+} ions (Suhre et al., 2014). In general, the thread matrix proteins are proposed to separate and lubricate preCol fibrils; however, their specific role is still under discussion (Sagert & Waite, 2009; Suhre et al., 2014).

The understanding of the structure–function relationships defining the byssal thread core has been advanced in recent years by the use of spec-troscopic and X-ray diffraction techniques in combination with *in situ* mechanical testing (Fig. 4.4E). For example, wide angle X-ray diffraction studies combined with *in situ* tensile tests have revealed that the collagenous domain strains by a mere 2%, even when the thread is strained to more than 70% of its initial length (Harrington et al., 2009; Krauss et al., 2013). Because the preCols are aligned in series, this implies that the high extensibility of the thread must arise from the unfolding of folded protein structure (i.e., hidden length) of the non-collagenous domains of preCols (Fig. 4.4E) (Harrington et al., 2009; Krauss et al., 2013). In particular, the unfolding of the predicted beta-sheet forming flanking domains of preCol-D and possibly -NG is a prime candidate to provide extensibility as suggested by FT-IR and NMR-based studies (Arnold et al., 2013; Hagenau et al., 2011). Notably, however, small-angle X-ray diffraction studies suggest that the unfolded protein structure refolds without delay when the thread is relaxed (Fig. 4.4E) (Krauss et al., 2013). Considering that the mechanical properties recover only on a much longer timescale (Carrington & Gosline, 2004; Harrington et al., 2009), it was proposed that byssal thread healing proceeds by a two-step process consisting of a fast recovery of the folded protein length (presumably by refolding of unfolded beta-sheet structure) and a slower recovery of mechanical properties by reformation of a sacrificial bonding network (Krauss et al., 2013).

Sacrificial bonding is a strategy for enhancing the toughness that has been identified in a broad range of biological materials including wood, bone, and silk (Becker et al., 2003; Fantner et al., 2004; Keckes et al., 2003; Smith et al., 1999). The basic concept underlying sacrificial bonding is that weaker non-covalent bonds are strategically positioned to rupture prior to covalent bonds, dissipating applied mechanical energy and avoiding catastrophic rupture of the biomolecular backbone. Often, rupture of sacrificial bonds in protein-based biological materials results in the unfolding of folded protein length—so-called hidden length—which provides an effective means of increasing the extensibility and toughness of the material (Smith et al., 1999). Furthermore, if the "hidden length" can refold when unloaded and the sacrificial bonds are reversible (i.e., can reform on biologically relevant timescales), this can also lead to self-healing behavior.

Waite and colleagues first proposed that metal coordination bonds between histidine residues in the His-rich domains might be functioning as sacrificial bonds in the byssal thread core (Coyne et al., 1997). Since then, several pieces of important evidence have emerged that support this hypothesis (Harrington & Waite, 2007; Holten-Andersen et al., 2009a; Vaccaro & Waite, 2001). First, removal of Zn and Cu ions from byssal threads following their incubation with ethylenediamine tetracetic acid (EDTA), a metal-chelation agent, resulted in a significant decrease in thread stiffness and the loss of the yield point (Vaccaro & Waite, 2001). Second, byssal thread stiffness can be reduced by treatment at low pH, where histidine is positively charged and unable to bind metal ions (Harrington & Waite, 2007) (pKa ~6.5). Furthermore, plotting thread stiffness as a function of pH reveals a sigmoidal curve that almost perfectly overlaps with the titration curve of histidine (Harrington & Waite, 2007). Third, it was demonstrated that the ability of byssal threads to heal is completely eradicated when threads were incubated at pH 4 (Harrington et al., 2009). Finally, synthetic peptides based on sequences of the histidine-rich domains of preCols were shown to exhibit reversible metal-dependent interactions *in vitro* (Schmidt et al., 2014), adding further support to potential use of histidine–metal cross-links as sacrificial bonds.

Based on these and other findings, a molecular-level model has been developed to explain the complex deformation and self-healing properties of byssal threads (Fig. 4.4E). His–metal bonds in byssal threads are believed to break at the yield point functioning as reversible sacrificial bonds. The rupture of sacrificial bonds leads to the unfolding of the flanking domain beta-sheet structure, providing the thread with increased extensibility and energy dissipation. Refolding of unfolded protein length is believed to occur immediately upon unloading (Krauss et al., 2013); however, as already mentioned, mechanical recovery requires longer rest periods. Therefore, based on the evidence provided in the previous paragraph, the reformation of a network of stable protein–metal coordination bonds seems a likely candidate for the slow step in the self-healing process. However, further research is undoubtedly needed to investigate the mechanical roles of the beta-sheet forming flanking domains and the metal-cross linking His-rich domains.

It is worth noting that the compositional and structural properties and thus, the mechanical behavior of byssus fibers vary between *Mytilus* species (Harrington & Waite, 2007). For example, *M. californianus* produces threads which are about twice as stiff as threads from *M. galloprovincialis* (Bell & Gosline, 1996). Furthermore, the mechanical properties of byssal threads produced by individual mussels can be influenced by external factors including water quality, predation, temperature, nutrition, sea current, density of the mussel bed, and quality of substratum (Bell & Gosline, 1997; Carrington, 2002; Carrington et al., 2008; Côté, 1995; Garner & Litvaitis, 2013; Lachance et al., 2008; Moeser & Carrington, 2006). For example, one study demonstrated that during fall and summer months *M. edulis* mussels on the northeastern seaboard of North America synthesized threads that were half as strong and extensible as those formed in the winter and spring (Moeser & Carrington, 2006). This reduced mechanical performance has been proposed to contribute to higher incidence of dislodgement events during these seasons (Carrington et al., 2015; Denny and Gaylord, 2010). Additionally, it was found that threads grown under conditions mimicking forecasted ocean acidification exhibited weaker adhesive attachment than native threads (O'Donnell et al., 2013), while warmer waters were found to weaken the proximal region of the thread (Carrington et al., 2015). Presently, the molecular-level mechanisms underlying these mechanical modifications have not yet been elucidated; however, it is tempting to posit that they may reflect fundamental changes in the composition and structure of the byssal thread.

4.3   A STRETCHY AND ABRASION-RESISTANT COATING REINFORCED BY DOPA–FE^{3+}

Abrasion resistant coatings in technical applications are typically very hard materials with low strain limits. In contrast, the byssal thread cuticle, which is proposed to protect the thread against abrasion, degradation due to solar radiation and microbial attacks, is able to exhibit high stiffness and hardness values comparable to modern engineering epoxies, while still remaining surprisingly flexible and extensible (Holten-Andersen et al., 2007). These materials properties—that is, high hardness and high extensibility—are traditionally diametrically opposed and are almost never found in the same material in man-made polymers or composites. Thus, the byssal thread cuticle has emerged as an exciting biological archetype for the design of man-made coatings for a variety of possible biomedical and technical applications (Holten-Andersen & Waite, 2008).

The byssal thread cuticle is a thin (5–10 μm) coating that completely surrounds the fibrous core (Fig. 4.2C and Fig. 4.5A, Table 4.1). It has been shown that the cuticle of *Mytilus* species exhibits hardness and stiffness values that are about fivefold higher than that of the fibrous core, but still stays intact up to tensile strains of 100% in certain species (Holten-Andersen et al., 2007, 2009b; Lee et al., 2011) (Fig. 4.5A). The extraordinary combination of mechanical properties exhibited by the byssus cuticle is believed to arise mainly as a result of its composite-like structure in which micron-sized granular inclusions are embedded in an amorphous homogenous matrix (Fig. 4.2C and Fig. 4.5A and D). While the granules were observed to deform at low strain (<30%), they behave as stiff reinforcing elements at higher strains and microcracking is observed in the matrix between granules (Holten-Andersen et al., 2007) (Fig. 4.5D). Microcracking has been proposed as an effective toughening mechanism that distributes damage over a larger volume, preventing the propagation of larger catastrophic cracks through the material (Holten-Andersen et al., 2007). In support of this hypothesis, it was shown that species with smaller granules, and thus, higher granular surface area, exhibit higher extensibilities than those with larger granules (Holten-Andersen et al., 2009b). Mussels living in subtidal habitats, such as *Perna canaliculus*, on the other hand possess cuticles without granular microarchitecture and exhibit fracture strains of ~30% (vs. up to 100% strain for *M. californianus*), suggesting that extensible granular cuticles might be an adaptation to high wave-exposure of mytilids (Holten-Andersen et al., 2007).

Presently, the only protein confirmed to be present in the cuticle is a DOPA-rich protein known as *m*ussel *f*oot *p*rotein-1 (mfp-1) (Table 4.2). Additionally, a small amount of Fe^{3+} and Ca^{2+} have been co-localized in the cuticle (Holten-Andersen et al., 2009a). As already mentioned, DOPA is known to form extremely stable coordination complexes with Fe^{3+} (Fig. 4.5C), leading the proposal that such interactions might be present in the cuticle stabilizing mfp-1 (Taylor et al., 1994). Recently, confocal Raman spectroscopic imaging has confirmed the presence of tris–DOPA–Fe^{3+} cross-links in byssal threads from *M. galloprovincialis* and *M. californianus*, which are specifically localized in the cuticle (Harrington et al., 2010) (Fig. 4.5B).

The mechanical importance of the DOPA–Fe^{3+} cross-links in the cuticle was shown by the fact that EDTA-mediated removal of Fe^{3+} ions resulted in a reduction of the DOPA–Fe Raman signal and in a 85% loss in hardness (Harrington et al., 2010; Holten-Andersen et al., 2009a; Schmitt et al., 2015). DOPA–metal cross-links have the advantage that they are at the same time strong and reversible, as demonstrated by the work of Messersmith and colleagues in AFM-based single molecule force experiments of a single DOPA side chain and a TiO_2 surface (Lee et al., 2006). In this study, it was demonstrated that a single DOPA–Ti interaction possesses a breaking force of approximately half that of a typical covalent bond, while being able to break and reform reversibly over hundreds of times (Lee et al., 2006). Furthermore, surface force apparatus (SFA) experiments have enabled the measurement of the adhesion forces between films of mfp-1 proteins in the presence and absence of Fe^{3+}. In this technique, the attractive and repulsive forces are measured between two mica surfaces coated with mfp-1, which are first brought together and then pulled apart (Israelachvili et al., 2010). Using SFA, the adhesion energy of the interaction is determined based on interaction forces and contact areas measured. In the absence of Fe^{3+}, layers coated with mfp-1 were not bridged (e.g., did not interact); however, in the presence of Fe^{3+}, adhesion energies of up to 5 mJ/m² were calculated indicating very stable, but reversibly breakable cohesive interactions between proteins (Holten-Andersen & Waite, 2008; Lin et al., 2007; Zeng et al., 2010).

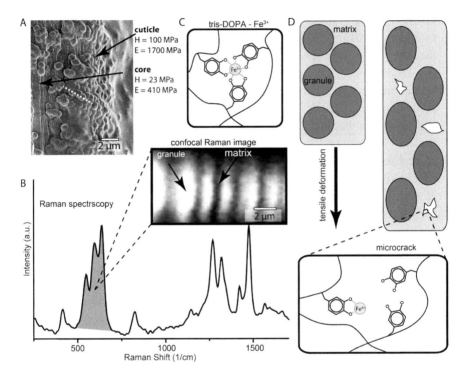

FIGURE 4.5 DOPA–Fe^{3+} cross-links in the byssal thread cuticle. (A) SEM image showing the granular morphology of the cuticle and the fibrous morphology of the core. Hardness (H) and stiffness (E) values are included for both. (B) Raman spectrum and confocal Raman image showing the presence and distribution of tris–DOPA–Fe cross-links (C) in the cuticle. Granules contain elevated cross-link density compared to the matrix. (D) Schematic model of microcrack formation in the cuticle. When stretched, the more heavily cross-linked granules behave stiffly. In the less cross-linked matrix, DOPA–Fe^{3+} complexes break sacrificially, allowing the formation of microcracks between granules.

The link between the unusual DOPA–Fe^{3+} cross-linking strategy and the hard, yet extensible behavior of the byssal thread cuticle was further elucidated by confocal Raman spectroscopic imaging of thin sections of the cuticle (Harrington et al., 2010) (Fig. 4.5B). This investigation demonstrated that the tris–DOPA–Fe^{3+} cross-links are more concentrated in the granular regions of the cuticle than in the surrounding matrix (Harrington et al., 2010;

Taylor et al., 1996). This is consistent with the more stiff behavior of the granules under tensile loading (Holten-Andersen et al., 2007) and suggests that the granules and matrix fulfill different mechanical functions (Fig. 4.5D). More specifically, the granules, owing to their elevated DOPA–Fe^{3+} cross-link density were proposed to provide hardness under compressive loading, as the granules are forced together during abrasion in marine environments (Harrington et al., 2010). The less cross-linked matrix material on the other hand, is proposed to provide extensibility, via rupture of sacrificial tris–DOPA–Fe^{3+} complexes at high strains (i.e., >30%), leading to formation of microcracks between the granules (Harrington et al., 2010; Holten-Andersen et al., 2007). Based on the ability of DOPA–metal bonds to break and reform reversibly, it was proposed that microcracks may self-heal once the cuticle is relaxed following extension; however, this is yet to be demonstrated experimentally. As already mentioned, elevated concentrations of Ca^{2+} are also co-localized with mfp-1 and Fe in the cuticle (Holten-Andersen et al., 2009a) yet presently, they have no clear function. DOPA is not known for its tendency to bind Ca^{2+} and furthermore, mfp-1 carries a strong net positive charge at seawater pH due to its high pI value (~10.5), making electrostatic interactions equally unlikely (Lee et al., 2011) (Table 4.2). It has been proposed that negatively charged fatty acids might also be present in the cuticle, which interact with Ca^{2+}; however, this has not yet been corroborated (Holten-Andersen et al., 2009a). It is, of course, possible that there is an as-of-yet unidentified cuticle component that might interact with Ca^{2+}, but presently no suitable candidates have been revealed.

4.4 ROLE OF PROTEIN–METAL INTERACTIONS IN THE ADHESIVE PLAQUE

Strong adhesion between two surfaces is dependent on the ability to establish strong physical interactions at the molecular interface (Stewart et al., 2011). On a dry clean surface, this is relatively straightforward; however, in seawater, this can become exceedingly tricky. The major impediment to establishing adhesion in marine environments is infiltrating the layer of ions and organic macromolecules that is fixed firmly to exposed surfaces in the ocean (Stewart et al., 2011). While man-made adhesives still struggle to overcome this fundamental challenge, the mussel rapidly and reliably fabricates a versatile glue, the byssal thread plaque, that is, compatible with an astonishing range of surface chemistries in marine environments. Based on its impressive performance, a substantial amount of work has been invested in the last 30 years in order to unlock the secrets of mussel byssal thread plaque.

Biochemical investigations have identified at least five proteins that are localized in the byssus adhesive plaque—namely, mfp-2, -3, -4, -5, and -6 (Table 4.2). Like the cuticle protein mfp-1, these five proteins are characterized by a high isoelectric point and the presence of DOPA (Lee et al., 2011) (Table 4.2). Many years of work have localized each of the mfp proteins to specific regions of the byssal thread plaque as summarized in Figure 4.6. More recently, surface force apparatus (SFA—described in the previous section) experiments performed on extracted and purified mfp proteins by the groups of Waite and Israelachvili have begun to shed light on the functional roles of various mfp protein in the plaque. The following section provides an overview of the current understanding; however, we direct the reader to a recent and more thorough review specifically focused on the byssal adhesive plaque (Lee et al., 2011).

MALDI–TOF mass spectrometry performed on the adhesive residue remaining after plaque removal was able to provide strong evidence that the adhesive interface of the plaque consists primarily of mfp-3 and -5 (Zhao et al., 2006; Zhao & Waite, 2006a) (Fig. 4.6D). The adhesive function of mfp-3 and -5 was further supported by SFA measurements demonstrating that these two proteins display the highest adhesion energies of all the mfps. The fact that mfp-3 and -5 also have the highest DOPA content led to the hypothesis that DOPA is the primary adhesive agent. Contrary to initial speculations that DOPA–metal interactions, specifically, were the primary bonds controlling plaque adhesion, very high adhesive energies were recorded even in the absence of metal ions on surfaces such as mica where hydrogen bonding likely dominates (Lee et al., 2006; Lin et al., 2007; Yu et al., 2013a) (Fig. 4.6E). Nevertheless, mfp-3 was shown to create strong adhesion with TiO_2-coated mica surfaces via coordination with Ti at neutral to basic pH (Yu et al., 2013b), suggesting that plaque adhesion in nature likely occurs through an assortment of different physical interactions (Fig. 4.6E).

While these investigations indicate that DOPA residues in mfp-3 and -5 are able to mediate strong adhesion to a variety of surface chemistries, DOPA is nonetheless highly susceptible to oxidation into the quinone form of DOPA, which has a significantly reduced affinity for adhering to surfaces and a propensity for further cross-linking reactions (Lee et al., 2006; Yu et al., 2011a). Interestingly, however, the mussel has apparently evolved a mechanism for counteracting the oxidation of DOPA in mfp-3 and -5 at the plaque adhesive interface (Nicklisch et al., 2013; Yu et al., 2011b). In this case, mfp-6, a cysteine-rich protein that is also localized near the plaque–surface interface (Fig. 4.6D) is proposed to function as a reducing agent that counteracts DOPA oxidation and enables more robust adhesion by mfp-3 and -5 (Yu et al., 2011b). In the proposed mechanism, free thiolates in mfp-6 are oxidized at the plaque–surface interface in a reaction coupled with the reduction of DOPA–quinone back to DOPA.

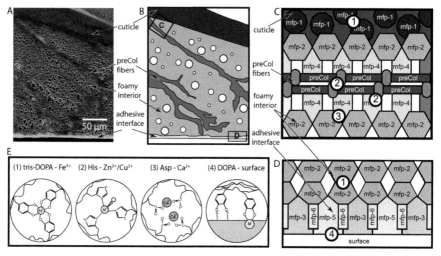

FIGURE 4.6 Metal interactions in the byssal thread adhesive plaque. (A) An SEM image of the adhesive plaque and (B) a schematic representation, which highlights specific morphological features (e.g., cuticle, preCol fibers, foamy interior, and adhesive interface). (C) and (D) Schematic models indicating the different protein variants found in the plaque in the region of the foamy interior near the cuticle (C) and at the adhesive interface (D) as indicated in panel (B). (E) Overview of different proposed metal-based interactions between the various protein components in the plaque. Numbers correspond to specific interactions identified in panels (C) and (D).

In contrast to mfp-3 and -5, the adhesive energy measured with SFA for mfp-2 (~5 mol% DOPA) on mica is almost negligible (Hwang et al., 2010). However, the cohesive forces between two symmetric layers of mfp-2 in the presence of Fe^{3+} ions were shown to be quite high, suggesting a nonadhesive role for this protein in the byssal plaque. *In situ* Raman microscopic imaging of the plaque confirmed that mfp-2, makes up the foamy interior and furthermore, that it forms tris–DOPA–Fe^{3+} interactions that are proposed to provide mechanical stability to the structure (Hwang et al., 2010) (Fig. 4.6C–E). Interestingly, it was also shown that Ca^{2+} could mediate weak adhesion between layers of mfp-2 in SFA measurements, possibly by electrostatic interactions with a putative calcium-binding consensus sequence in the protein (Hwang et al., 2010) (Fig. 4.6D and E).

The protein mfp-4 is believed to be localized at the interface between the plaque and the distal region of the thread core and is therefore proposed to mediate the interaction between the preCols and the plaque proteins (Zhao & Waite, 2006b) (Fig. 4.6C). Unlike the other mfp proteins, mfp-4 is enriched in histidine residues (~20 mol%), which are arranged in numerous tandem decapeptide repeats and has been shown to have affinity for binding Cu^{2+}, similar to His-rich domains of the preCols (Zhao & Waite, 2006b). Additionally, mfp-4 also contains tandem repeats of aspartate-rich motifs that possess an affinity for binding Ca^{2+} similar to the calcium-binding domains of mfp-2 (Zhao & Waite, 2006b). Thus, it was proposed that mfp-4 might be a molecular bridge at the interface between the preCol fibers that interpenetrate into the foamy core of the plaque, mediated primarily through metal binding (Zhao & Waite, 2006b) (Fig. 4.6C and E).

To sum up, the byssal thread plaque is a complex underwater adhesive composed of at least five different protein variants localized to particular regions of the plaque where they are proposed to serve specific functions (Fig. 4.6, Table 4.2). Interactions between the different protein variants are vital to their synergistic function and are primarily mediated by molecular bridging and cross-linking via metal ions (Fig. 4.6E). In this respect, the example of the plaque truly underscores the extent to which the structure and biochemistry of the mussel byssus has been tuned and refined through evolution, and especially highlights the fact that not all protein metal bonds are created equal. It is of course, not completely understood why particular combinations of amino acid and metal ions are found in specific regions of the byssal thread and not others; however, it is tempting to speculate that the specific physicochemical properties of the different interactions (i.e., His-based vs. DOPA-based cross-linking) offer specific advantages during the function or formation of the material. Although there is much that still needs to be understood about the byssal thread plaque, the biochemical design principles extracted through their study have already led to the development of bioinspired materials for medical and technical applications. In particular, DOPA has been engineered into polymer chains to generate surgical glues, underwater adhesives, as well as, antifouling coatings in aquatic environments (Lee et al., 2011).

4.5 THE ROLE OF METAL IONS IN MUSSEL BYSSAL THREAD ASSEMBLY

It should be clear at this point that the mussel byssus is a multifaceted and structurally complex material. Thus, it is all the more surprising that in contrast to similarly complex biological materials, such as vertebrate tendon that forms over extended time periods, each byssal thread is formed in only a few minutes time. Byssal thread formation involves a choreographed secretion of the byssal protein precursor molecules, which then rapidly self-assemble into the specific nano-architectured structures that define the core, cuticle, and plaque (Pujol, 1967) (Fig. 4.2C and Fig. 4.7). As has been shown, quite a lot is now understood about the structure–function relationships that define these materials; however, comparatively little is understood about how byssal threads are fabricated by the mussel. This section will review the most current understanding of byssal thread assembly, highlighting the importance of protein–metal interactions.

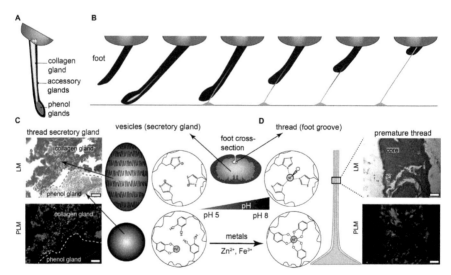

FIGURE 4.7   Byssal thread formation process and the role of metal ions. (A) Schematic view of the mussel foot indicating locations of the various secretory glands. (B) Schematic indicating the thread formation process. The mussel foot reaches out of the shell to find a spot on the surface to form a thread. Proteins are secreted into a groove in the foot, where they self-assemble and cross-link in a matter of minutes. (C) Thread proteins are stored in secretory vesicles as indicated in the light microscopy (LM) image, showing stained histological sections of the collagen gland (core) and phenol gland (plaque) in the a mussel foot. The polarized light microscopy (PLM) image indicates that the contents of the collagen gland are birefringent, suggesting liquid crystalline storage of preCols prior to thread formation.

Proteins are stored at low pH (pH < 5) in the vesicles, which prevents premature formation of metal coordination complexes by DOPA and histidine. (D) Thread formation involves the secretion of the proteins into the foot groove, which has a pH ~ 8, allowing DOPA and histidine to form complexes with available metal ions. LM and PLM images show early stages of thread core formation and alignment. All scale bars represent 10 μm.

The organ responsible for byssal thread formation is known as the foot. The foot is an extendable tongue-like appendage, which the mussel can extend outside of its shell (Fig. 4.7). Within the mussel foot are specialized secretory glands in which all the protein precursor molecules that comprise the thread are synthesized, packaged into secretory vesicles and stored (Fig. 4.7A and C). Extensive TEM studies of the mussel foot secretory glands were able to identify three different glands that produce three different vesicle types (Tamarin & Keller, 1972; Zuccarello, 1980, 1981)—namely, the phenol gland, the enzyme gland, and the collagen gland, which produce the plaque, cuticle, and thread core, respectively (Fig. 4.7A and C). The seemingly random gland nomenclature predates the discovery of the individual proteins that compose the threads.

Thread formation begins when the mussel extends its foot out of the protective confines of the shell and onto the surrounding surface (Fig. 4.7B). The mussel first searches the surface briefly until it settles on an appropriate spot to form a thread. At this point, the contents of the secretory vesicles are released into a groove running along one side of the mussel foot, which is sealed to the seawater (Fig. 4.7B and C). The end of the groove furthest from the mussel, known as the distal depression, is the site of plaque formation and is pressed tightly against the surface creating a water-tight groove into which the mfp-2, -3, -4, -5, and -6 are secreted (Hwang et al., 2010; Tamarin et al., 1976). Along the length of the foot groove, the mussel first secretes the collagen gland contents, forming the core, which is then followed by secretion of the cuticle-forming vesicles containing mfp-1. When the secretion process is over, the foot groove opens its seal and pulls away, leaving a white-golden byssal thread that will immediately be loaded by the next set of crashing waves (Fig. 4.7B).

The intricate details of the secretion process occurring within the foot are hidden from the prying eyes of researchers; thus, most of our knowledge comes from invasive techniques such as TEM or *in vitro* experiments on purified byssal thread proteins. For example, the ellipsoidal secretory vesicles of the collagen gland, which encapsulate the preCol molecules that will form the byssal thread core, exhibit a characteristic fibrous banding pattern in TEM imaging (Fig. 4.7C). Based on similarities between the spacing of these fibrillar contents and those in the native thread (Hassenkam et al., 2004; Krauss et al., 2013), it was proposed that the preCols are pre-organized in the vesicles into a liquid crystalline-like phase with smectic structure (i.e., the preCols are aligned in the same direction with the collagen domains packed like books on a shelf) (Hassenkam et al., 2004) (Fig. 4.7C). A liquid crystalline phase would be advantageous during formation by allowing rapid assembly of the preCols into a highly ordered state reminiscent of a solid crystal, while still allowing the vesicle contents to flow like a fluid. Once the structure is set, however, the order must be locked in rapidly before the thread can assume its load-bearing role.

Several studies have suggested that pH-triggered His–metal cross-linking may contribute to the locking mechanism during byssal thread formation (Fig. 4.7C). The terminal ends of adjacent preCol bundles along the length of the thread core interact via their His-rich domains. As already mentioned, histidine, which possesses a typical pKa value of ~6.5, is highly sensitive to the pH of the local environment. The thread secretion in the region of the plaque has been measured to be quite acidic (pH < 5), suggesting that the preCols may also be stored under acidic conditions (Fullenkamp et al., 2014; Yu et al., 2011b). Assuming an average pKa of 6.5, the overwhelming majority of histidine residues would be positively charged under these acidic conditions and unable to bind metal ions (Fig. 4.7C). Within the secretory vesicles, this would provide a reliable means of preventing premature assembly of preCols, while still permitting their liquid crystalline alignment. Upon secretion into seawater (pH 8.2), the vast majority of histidine residues would be deprotonated, removing the charge repulsion and allowing coordination of divalent metal ions such as Zn^{2+} and Cu^{2+} (Fig. 4.7D).

Support for this hypothesis of thread core assembly was provided from *in vitro* assembly studies on purified preCol proteins, in which it was shown that preCol fibers were only formed at pH values above the pKa of histidine (Harrington & Waite, 2008b). Furthermore, it was possible to reduce the metal-dependent interaction energy between two layers of peptides based on preCol His-rich domain sequences by a factor of six when the pH of the surrounding medium was reduced from 8.0 to 4.8 (Schmidt et al., 2014). In spite of the growing support for this hypothesis, it may only be one contributing factor to the fast locking of preCol structure. For example, highly conserved tyrosine residues in the preCol His-rich domains are believed to be posttranslationally converted to DOPA, which could also contribute to cross-linking between neighboring preCol bundles during thread formation (Harrington & Waite, 2007).

After a new byssal core is assembled in the foot groove, the cuticle-forming vesicles are released into the groove where they are thought to merge and coalesce around the thread to form the protective cuticle (Holten-Andersen et al., 2011; Holten-Andersen & Waite, 2008; Zuccarello, 1980). As described above, the hard, yet extensible performance of the cuticle is highly dependent on the formation of a network of DOPA–Fe^{3+} cross-links with mfp-1, which are localized into dense cross-link centers (granules) in a less densely cross-linked milieu (matrix) (Harrington et al., 2010) (Fig. 4.5). It has been proposed that, similar to the formation of His–metal interactions in the thread core, the formation of the DOPA–Fe complexes in the cuticle is also triggered by a pH jump from acidic to basic conditions (Holten-Andersen et al., 2011) (Fig. 4.7C and D). This hypothesis is supported by *in vitro* experiments on DOPA–Fe^{3+} cross-linked polymer gels, which undergo a transition from the mono (1 DOPA:1 Fe^{3+}) to bis (2 DOPA:1 Fe^{3+}) to tris (3 DOPA:1 Fe^{3+}) forms of the DOPA–Fe complex as the pH was increased stepwise (Holten-Andersen et al., 2011). Transitions in DOPA coordination are accompanied by a concomitant transition in mechanical performance, going from a fluid-like behavior at low pH to a viscoelastic hydrogel at high pH (Holten-Andersen et al., 2011).

A key structural feature of the cuticle is its granular morphology; however, it is still an open question as to how this forms. Based on the fact that higher amounts of DOPA–Fe^{3+} cross-links are present in the granules than in the matrix (Harrington et al., 2010), two alternative hypotheses were proposed to account for this: (1) the granular protein is more highly condensed than the matrix or (2) the protein in the granules contains a higher content of DOPA. In support of the former hypothesis, the cuticle forming secretory vesicles undergo a maturation process in which the contents apparently undergo a condensation process to form the granule and matrix material within a single vesicle (Zuccarello, 1981). In support of the latter hypothesis, there have been two variants of mfp-1 identified, which different only in the degree to which the Tyr residues are converted to DOPA (Harrington et al., 2010; Sun & Waite, 2005). Presently, however, these hypotheses await further testing. Interestingly, a recent study suggests that subtle sequence differences in mfp-1 between *M. californianus* and *M. edulis* might contribute to the observed differences in granule morphology and mechanical performance of cuticles between the species (Das et al., 2015).

Like the plaque structure itself (Fig. 4.6), the plaque formation process appears to be quite complex. MALDI–TOF analysis of the forming plaque at various time points revealed a progression of different proteins being secreted successively into the distal depression of the foot where the adhe-sive forms (Yu et al., 2011b). It was observed that variants of the adhesive protein mfp-3 are first secreted, followed shortly thereafter by the Cys-rich protein, mfp-6. As already mentioned, the authors of this study proposed that the adhesion of mfp-3 is enhanced on surfaces because oxidation to DOPA–quinone in mfp-3 is counteracted by the reducing thiolate groups of mfp-6. A recent investigation in which plaques were formed by mussels on surfaces coated with pH-sensitive dyes indicated that the initial adhesive secretion may have a pH as low as 2–3. These highly acidic conditions would be favorable for creating adhesion because they additionally hinder DOPA oxidation (Martinez Rodriguez et al., 2015). Furthermore, another recent study has suggested that under pH and ionic strength conditions similar to those expected during plaque secretion, mfp-3 will self-coacervate *in vitro*, which means that it forms a dense fluid phase of the protein separated from bulk water (Wei et al., 2014). The mfp-3 coacervate was found to have low interfacial energy, which may help it to spread over surfaces, enhancing adhesion mediated by DOPA. Based on the importance of DOPA–Fe^{3+} as stabilizing cross-links in the mfp-2 proteins comprising the foamy interior of the plaque (Hwang et al., 2010), it is not a stretch of the imagination to posit that a similar pH-triggered metal-dependent cross-linking mechanism might occur as proposed for the thread cuticle (Holten-Andersen et al., 2011) (Fig. 4.7C and D). However, at this point, it is still not clear how the complex foamy architecture is formed. Interestingly, recent *in vitro* results suggest that it would not be a clever strategy to store Fe and DOPA together at low pH. In this study, mixing purified mfp-1 and Fe^{3+} at low pH was shown to result in the oxidation of DOPA residues leading to formation of covalent diDOPA cross-links, which would be a highly unfavorable result prior to secretion (Fullenkamp et al., 2014).

To summarize, the rapid formation of byssal threads appears to be a highly orchestrated process, about which very little is known at present. Preliminary insights into this process reveal several common themes, which are worth repeating. First, the precursor molecules are stored in secretory vesicles in which they are pre-organized to facilitate rapid acquisition of the desired structure upon secretion (Fig. 4.7C). Second, the pH jump going from the acidic secretory vesicles (pH < 5) to the basic seawater (pH ~ 8) environment may play an important role in quickly locking the thread struc-ture immediately after secretion of the precursors (Fig. 4.7C and D). Based on the sensitivity of both histidine and DOPA to pH in this physiologically relevant range (pH 5–8), protein–metal cross-linking has been highlighted as a likely candidate for the initial locking mechanism. While these insights are exciting and already leading to the development of mussel inspired hydro-gels (Fullenkamp et al., 2013; Holten-Andersen et al., 2011), there is still a great deal to be understood about byssal thread assembly.

4.6 SUMMARY

Byssal threads are remarkable protein-based fibers crucial to the evolu-tionary success of mussels. This is in no small part due to their impres-sive material properties—for example, self-healing, abrasion resistance, and wet adhesion. A key design feature contributing to these properties is the clever use of metal–protein interactions as critical cross-links. Enormous efforts over the last 30 years have advanced our understanding of the struc-ture–function relationships that define this material to the point that mussel-inspired synthetic polymers are being developed and even applied in clinical settings (Haller et al., 2011). In spite of the rapid progress in our under-standing of the mussel byssus, however, there are many exciting questions left to be answered that will provide even further insights into the formation and behavior of this extraordinary material.

ACKNOWLEDGMENTS

We acknowledge funding from the DFG priority program SPP-1568 (HA6369/4-1) and the Max Planck Society. We thank E. Degtyar and T. Priemel for helpful discussion and for providing light microscopy images and J. H. Waite for providing an SEM image.

KEYWORDS

- marine mussels
- byssal threads
- histidine
- DOPA
- metal coordination

REFERENCES

Abbe, G. R.; Sanders, J. G. Pathways of Silver Uptake and Accumulation by the American Oyster (*Crassostrea virginica*) in Chesapeake Bay *Estuarine Coastal and Shelf Science* **1990,** *31*(2), 113–123.

Amiard, J.-C.; Bacheley, H.; Barillé, A.-L.; Barillé, L.; Geffard, A.; Himery, N. Temporal Changes in Nickel and Vanadium Concentrations and in Condition Index and Metallo-thionein Levels in Three Species of Molluscs Following the "Erika" Oil Spill. *Aquat. Liv. Resour.* **2004,** *17*(03), 281–288.

Arnold, A. A.; Byette, F.; Seguin-Heine, M.-O.; LeBlanc, A.; Sleno, L.; Tremblay, R.; Pellerin, C.; Marcotte, I. Solid-State NMR Structure Determination of Whole Anchoring Threads from the Blue Mussel *Mytilus edulis. Biomacromolecules* **2013,** *14*(1), 132–141.

Becker, N.; Oroudjev, E.; Mutz, S.; Cleveland, J. P.; Hansma, P. K.; Hayashi, C. Y.; Makarov, D. E.; Hansma, H. G. Molecular Nanosprings in Spider Capture-silk Threads. *Nat. Mater.* **2003,** *2*(4), 278–283.

Bell, E. C.; Gosline, J. M. Mechanical Design of Mussel Byssus: Material Yield Enhances Attachment Strength. *J. Exp. Biol.* **1996,** 1999, 1005–1017.

Bell, E. C.; Gosline, J. M. Strategies for Life in Flow: Tenacity, Morphometry, and Prob-ability of Dislodgment of Two *Mytilus* Species. *Mar. Ecol. Progr. Ser.* **1997,** 159, 197–208.

Benedict, C. V.; Waite, J. H. Location and Analysis of Byssal Structural Proteins of *Mytilus edulis. J. Morphol.* **1986,** *189*(2), 171–181.

Bjerregaard, P.; Depledge, M. H. Cadmium Accumulation in *Littorina littorea, Mytilus edulis* and *Carcinus maenes*—The Influence of Salinity and Calcium-ion Concentrations. *Mar. Biol.* **1994,** *119*(3), 385–395.

Brazee, S. L.; Carrington, E. Interspecific Comparison of the Mechanical Properties of Mussel Byssus. *Biol. Bull.* **2006,** *211*(3), 263–274.

Broomell, C. C.; Mattoni, M. A.; Zok, F. W.; Waite, J. H. Critical Role of Zinc in Hardening of Nereis Jaws. *J. Exp. Biol.* **2006,** *209*(16), 3219–3225.

Broomell, C. C.; Zok, F. W.; Waite, J. H. Role of Transition Metals in Sclerotization of Biological Tissue. *Acta Biomater.* **2008,** *4*(6), 2045–2051.

Cao, Y.; Yoo, T.; Li, H. Single Molecule Force Spectroscopy Reveals Engineered Metal Chelation is a General Approach to Enhance Mechanical Stability of Proteins. *Proc. Natl. Acad. Sci.* **2008,** *105*(32), 11152–11157.

Carpene, E.; George, S. G. Absorption of Cadmium by Gills of *Mytilus* edulis (L.). *Mol. Physiol.* **1981,** *1*(1), 23–34.

Carrington, E. The Economics of Mussel Attachment: From Molecules to Ecosystems. *Integr. Compar. Biol.* **2002,** 42, 846–852.

Carrington, E.; Gosline, J. M. Mechanical Design of Mussel Byssus: Load Cycle and Strain Rate Dependence. *Am. Macal. Bull.* **2004,** 18.

Carrington, E.; Moeser, G. M.; Thompson, S. B.; Coutts, L. C.; Craig, C. A. Mussel Attach-ment on Rocky Shores: The Effect of Flow on Byssus Production. *Integr. Compar. Biol.* **2008,** *48*(6), 801–807.

Carrington, E.; Waite, J. H.; Sara, G.; Sebens, K. P. Mussels as a Model System for Integrative Ecomechanics. *Annu. Rev. Mar. Sci.* **2015,** 7, 443–469.

Chen, P.-Y.; McKittrick, J.; Meyers, M. A. Biological Materials: Functional Adaptations and Bioinspired Designs. *Prog. Mater. Sci.* **2012,** *57*(8), 1492–1704.

Coombs, T. L.; Keller, P. J. *Mytilus* Byssal Threads as an Environmental Marker for Metals. *Aquat. Toxicol.* **1981,** *1*(5–6), 291–300.

Côté, I. M. Effects of Predatory Crab Effluent on Byssus Production in Mussels. *J. Exp. Mar. Biol. Ecol.* **1995,** *188*(2), 233–241.

Coyne, K. J.; Qin, X.-X.; Waite, J. H. Extensible Collagen in Mussel Byssus: A Natural Block Copolymer. *Science (Washington, D. C.)* **1997,** *277*(5333), 1830–1832.

Das, S.; Miller, D. R.; Kaufman, Y.; Martinez Rodriguez, N. R.; Pallaoro, A.; Harrington, M. J.; Gylys, M.; Israelachvili, J. N.; Waite, J. H. Tough Coating Proteins: Subtle Sequence Variation Modulates Cohesion. *Biomacromolecules* **2015,** *16*(3), 1002–1008.

Degtyar, E.; Harrington, M. J.; Politi, Y.; Fratzl, P. The Mechanical Role of Metal Ions in Biogenic Protein-based Materials. *Angew. Chem. Int. Ed.* **2014,** *53*(45), 12026–12044.

Denny, M. W.; Gaylord, B. Marine Ecomechanics. In *Annual Review of Marine Science*; Carlson, C. A.; Giovannoni, S. J., Eds.; 2010; Vol. 2, pp 89–114.

Denny, M.; Gaylord, B. The Mechanics of Wave-swept Algae. *J. Exp. Biol.* **2002,** *205*(10), 1355–1362.

Denny, M.; Gaylord, B.; Helmuth, B.; Daniel, T. The Menace of Momentum: Dynamic Forces on Flexible Organisms. *Limnol. Oceanogr.* **1998,** *43*(5), 955–968.

Fantner, G. E.; Birkedal, H.; Kindt, J. H.; Hassenkam, T.; Weaver, J. C.; Cutroni, J. A.; Bosma, B. L.; Bawazer, L.; Finch, M. M.; Cidade, G. A. G.; Morse, D. E.; Stucky, G. D.; Hansma, P. K. Influence of the Degradation of the Organic Matrix on the Microscopic Fracture Behavior of Trabecular Bone. *Bone* **2004,** *35*(5), 1013–1022.

Fratzl, P.; Weinkamer, R. Nature's Hierarchical Materials. *Prog. Mater. Sci.* **2007,** *52*(8), 1263–1334.

Fullenkamp, D. E.; Barrett, D. G.; Miller, D. R.; Kurutz, J. W.; Messersmith, P. B. pH-Dependent Cross-linking of Catechols Through Oxidation via Fe^{3+} and Potential Implications for Mussel Adhesion. *RSC Adv.* **2014,** *4*(48), 25127–25134.

Fullenkamp, D. E.; He, L.; Barrett, D. G.; Burghardt, W. R.; Messersmith, P. B. Mussel-inspired Histidine-based Transient Network Metal Coordination Hydrogels. *Macromolecules* **2013,** *46*(3), 1167–1174.

Garner, Y. L.; Litvaitis, M. K. Effects of Injured Conspecifics and Predators on Byssogenesis, Attachment Strength and Movement in the Blue Mussel, *Mytilus edulis. J. Exp. Mar. Biol. Ecol.* **2013,** 448, 136–140.

George, S. G.; Pirie, B. J. S.; Coombs, T. L. Kinetics of Accumulation and Excretion of Ferric Hydroxide in *Mytilus edulis* (L.) and its Distribution in Tissues. *J. Exp. Mar. Biol. Ecol.* **1976,** *23*(1), 71–84.

Glusker, J. P. Structural Aspects of Metal Liganding to Functional-groups in Proteins. *Adv. Protein Chem.* **1991,** *42*, 1–76.

Gosline, J.; Lillie, M.; Carrington, E.; Guerette, P.; Ortlepp, C.; Savage, K. Elastic Proteins: Biological Roles and Mechanical Properties. *Philos. Trans. R. Soc. Lond., Ser. B—Biol. Sci.* **2002,** *357*(1418), 121–132.

Hagenau, A.; Papadopoulos, P.; Kremer, F.; Scheibel, T. Mussel Collagen Molecules with Silk-like Domains as Load-bearing Elements in Distal Byssal Threads. *J. Struct. Biol.* **2011,** *175*(3), 339–347.

Hagenau, A.; Suhre, M. H.; Scheibel, T. R. Nature as a Blueprint for Polymer Material Concepts: Protein Fiber-reinforced Composites as Holdfasts of Mussels. *Prog. Polym. Sci.* **2014,** *39*(8), 1564–1583.

Haller, C. M.; Buerzle, W.; Brubaker, C. E.; Messersmith, P. B.; Mazza, E.; Ochsenbein-Koelble, N.; Zimmermann, R.; Ehrbar, M. Mussel-mimetic Tissue Adhesive for Fetal Membrane Repair: A Standardized Ex Vivo Evaluation Using Elastomeric Membranes. *Prenat. Diagn.* **2011,** *31*(7), 654–660.

Hamilton, E. I. Concentration and Distribution of Uranium in *Mytilus edulis* and Associated Materials. *Mar. Ecol. Progr. Ser.* **1980,** *2*(1), 61–73.

Harrington, M. J.; Gupta, H. S.; Fratzl, P.; Waite, J. H. Collagen Insulated from Tensile Damage by Domains that Unfold Reversibly: In Situ X-ray Investigation of Mechanical Yield and Damage Repair in the Mussel Byssus. *J. Struct. Biol.* **2009,** *167*(1), 47–54.

Harrington, M. J.; Masic, A.; Holten-Andersen, N.; Waite, J. H.; Fratzl, P. Iron-Clad Fibers: A Metal-based Biological Strategy for Hard Flexible Coatings. *Science (Washington, DC, U.S.)* **2010,** *328*(5975), 216–220.

Harrington, M. J.; Waite, J. H. Holdfast Heroics: Comparing the Molecular and Mechanical Properties of *Mytilus californianus* Byssal Threads. *J. Exp. Biol.* **2007,** *210*(24), 4307–4318.

Harrington, M. J.; Waite, J. H. How Nature Modulates a Fiber's Mechanical Properties: Mechanically Distinct Fibers Drawn from Natural Mesogenic Block Copolymer Variants. *Adv. Mater. (Weinheim, Ger.)* **2009,** *21*(4), 440–444.

Harrington, M. J.; Waite, J. H. pH-Dependent Locking of Giant Mesogens in Fibers Drawn from Mussel Byssal Collagens. *Biomacromolecules* **2008,** *9*(5), 1480–1486.

Harrington, M. J.; Waite, J. H. Short-order Tendons: Liquid Crystal Mesophases, Metal-complexation and Protein Gradients in the Externalized Collagens of Mussel Byssal Threads; In *Fibrous Proteins*; Scheibel, T., Ed.; Landes Bioscience, Austin, TX, 2008; pp 30–45.

Hassenkam, T.; Gutsmann, T.; Hansma, P.; Sagert, J.; Waite, J. H. Giant Bent-Core Mesogens in the Thread Forming Process of Marine Mussels. *Biomacromolecules* **2004,** *5*(4), 1351–1355.

Haszprunar, G. The Molluscan Rhogocyte (pore-cell, Blasenzelle, cellule nucale), and Its Significance for Ideas on Nephridial Evolution. *J. Mollusc. Stud.* **1996,** *62*, 185–211.

Hennebert, E.; Wattiez, R.; Demeuldre, M.; Ladurner, P.; Hwang, D. S.; Waite, J. H.; Flammang, P. Sea Star Tenacity Mediated by a Protein that Fragments, Then Aggregates. *Proc. Natl. Acad. Sci. U.S.A.* **2014,** *111*(17), 6317–6322.

Hodge, V. F.; Koide, M.; Goldberg, E. D. Particulate Uranium, Plutonium and Polonium in the Bio-geo-chemistries of the Coastel Zone. *Nature* **1979,** *277*(5693), 206–209.

Holm, R. H.; Kennepohl, P.; Solomon, E. I. Structural and Functional Aspects of Metal Sites in Biology. *Chem. Rev.* **1996,** *96*(7), 2239–2314.

Holten-Andersen, N.; Fantner, G. E.; Hohlbauch, S.; Waite, J. H.; Zok, F. W. Protective Coatings on Extensible Biofibres. *Nat. Mater.* **2007,** *6*(9), 669–672.

Holten-Andersen, N.; Harrington, M. J.; Birkedal, H.; Lee, B. P.; Messersmith, P. B.; Lee, K. Y. C.; Waite, J. H. pH-induced Metal–Ligand Cross-links Inspired by Mussel Yield Self-healing Polymer Networks with Near-covalent Elastic Moduli. *Proc. Natl. Acad. Sci. U.S.A.* **2011,** *108*(7), 2651–2655.

Holten-Andersen, N.; Mates, T. E.; Toprak, M. S.; Stucky, G. D.; Zok, F. W.; Waite, J. H. Metals and the Integrity of a Biological Coating: The Cuticle of Mussel Byssus. *Langmuir* **2009,** *25*(6), 3323–3326.

Holten-Andersen, N.; Waite, J. H. Mussel-designed Protective Coatings for Compliant Substrates. *J. Dent. Res.* **2008,** *87*(8), 701–709.

Holten-Andersen, N.; Zhao, H.; Waite, J. H. Stiff Coatings on Compliant Biofibers: The Cuticle of *Mytilus californianus* Byssal Threads. *Biochemistry* **2009,** *48*(12), 2752–2759.

Hwang, D. S.; Zeng, H.; Masic, A.; Harrington, M. J.; Israelachvili, J. N.; Waite, J. H. Protein-and Metal-dependent Interactions of a Prominent Protein in Mussel Adhesive Plaques. *J. Biol. Chem.* **2010,** *285*(33), 25850–25858.

Inoue, K.; Takeuchi, Y.; Miki, D.; Odo, S. Mussel Adhesive Plaque Protein Gene is a Novel Member of Epidermal Growth Factor-like Gene Family. *J. Biol. Chem.* **1995,** *270*(12), 6698–6701.

Israelachvili, J.; Min, Y.; Akbulut, M.; Alig, A.; Carver, G.; Greene, W.; Kristiansen, K.; Meyer, E.; Pesika, N.; Rosenberg, K.; Zeng, H. Recent Advances in the Surface Forces Apparatus (SFA) Technique. *Rep. Prog. Phys.* **2010,** *73*(3).

Kamino, K. Molecular Design of Barnacle Cement in Comparison with Those of Mussel and Tubeworm. *J. Adhes.* **2010,** *86*(1), 96–110.

Keckes, J.; Burgert, I.; Fruhmann, K.; Muller, M.; Kolln, K.; Hamilton, M.; Burghammer, M.; Roth, S. V.; Stanzl-Tschegg, S.; Fratzl, P. Cell-wall Recovery After Irreversible Deformation of Wood. *Nat. Mater.* **2003,** *2*(12), 810–814.

Kellis, J. T.; Todd, R. J.; Arnold, F. H. Protein Stabilization by Engineered Metal Chelation. *Bio-Technology* **1991,** *9*(10), 994–995.

Krauss, S.; Metzger, T. H.; Fratzl, P.; Harrington, M. J. Self-repair of a Biological Fiber Guided by an Ordered Elastic Framework. *Biomacromolecules* **2013,** *14*(5), 1520–1528.

Lachance, A. A.; Myrand, B.; Tremblay, R.; Koutitonsky, V.; Carrington, E. Biotic and Abiotic Factors Influencing Attachment Strength of Blue Mussels *Mytilus edulis* in Suspended Culture. *Aquatic Biology* **2008,** *2*(2), 119–129.

Langston, W. J.; Spence, S. K. Biological Factors Involved in Metal Concentrations Observed in Aquatic Organisms. In *Metal Speciation and Bioavailability*; Tessier, A.; Turner, D. R., Eds.; Wiley: Chichester/New York, 1995; pp 407–478.

Lee, B. P.; Messersmith, P. B.; Israelachvili, J. N.; Waite, J. H. Mussel-inspired Adhesives and Coatings. In *Annual Review of Materials Research*; Clarke, D. R.; Fratzl, P., Eds.; 2011; Vol. 41, pp 99–132.

Lee, H.; Scherer, N. F.; Messersmith, P. B. Single-molecule Mechanics of Mussel Adhesion. *Proc. Natl. Acad. Sci. U.S.A.* **2006,** *103*(35), 12999–13003.

Li, L.; Smitthipong, W.; Zeng, H. Mussel-inspired Hydrogels for Biomedical and Environmental Applications. *Polym. Chem.* **2015,** *6*(3), 353–358.

Lin, Q.; Gourdon, D.; Sun, C.; Holten-Andersen, N.; Anderson, T. H.; Waite, J. H.; Israelachvili, J. N. Adhesion Mechanisms of the Mussel Foot Proteins mfp-1 and mfp-3. *Proc. Natl. Acad. Sci. U.S.A.* **2007,** *104*(10), 3782–3786.

Lucas, J. M.; Vaccaro, E.; Waite, J. H. A Molecular, Morphometric and Mechanical Comparison of the Structural Elements of Byssus from *Mytilus edulis* and *Mytilus galloprovincialis. J. Exp. Biol.* **2002,** *205*(12), 1807–1817.

Marigomez, I.; Soto, M.; Cajaraville, M. P.; Angulo, E.; Giamberini, L. Cellular and Subcellular Distribution of Metals in Molluscs. *Microsc. Res. Technol.* **2002,** *56*(5), 358–392.

Martinez Rodriguez, N. R.; Das, S.; Kaufman, Y.; Israelachvili, J. N.; Waite, J. H. Interfacial pH During Mussel Adhesive Plaque Formation. *Biofouling* **2015,** *31*(2), 221–227. DOI: 10.1080/08927014.2015.1026337.

Mascolo, J. M.; Waite, J. H. Protein Gradients in Byssal Threads of Some Marine Bivalve Mollusks. *J. Exp. Zool.* **1986,** *240*(1), 1–7.

Harrington, M. J.; Masic, A.; Holten-Andersen, N.; Waite, J. H.; Fratzl, P. Iron-Clad Fibers: A Metal-based Biological Strategy for Hard Flexible Coatings. *Science (Washington, DC, U. S.)* **2010,** *328*(5975), 216–220.

Mason, A. Z. *The Uptake, Accumulation and Excretion of Metals by the Marine Prosobranch Gastropod Mollusc Littorina littorea (L.)*. University of Wales: Bangor, 1983.

Mason, A. Z.; Simkiss, K.; Ryan, K. P. The Ultrastructural-localization of Metals in Specimens of *Littorina littorea* Collected from Clean and Polluted Sites. *J. Mar. Biol. Assoc. U.K.* **1984**, *64*(3), 699–720.

Mercer, E. H. Observations on the Molecular Structure of Byssus Fibres. *Austr. J. Mar. Freshwater Res.* **1952**, *3*(2), 199–204.

Moeser, G. M.; Carrington, E. Seasonal Variation in Mussel Byssal Thread Mechanics. *J. Exp. Biol.* **2006**, *209*(10), 1996–2003.

Nacu, E.; Tanaka, E. M. Limb Regeneration: A New Development? *Annu. Rev. Cell Dev. Biol.* **2011**, *27*(1), 409–440.

Nicklisch, S. C. T.; Das, S.; Rodriguez, N. R. M.; Waite, J. H.; Israelachvili, J. N. Antioxidant Efficacy and Adhesion Rescue by a Recombinant Mussel Foot Protein-6. *Biotechnol. Progr.* **2013**, *29*(6), 1587–1593.

O'Donnell, M. J.; George, M. N.; Carrington, E. Mussel Byssus Attachment Weakened by Ocean Acidification. *Nat. Clim. Chan.* **2013**, *3*(6), 587–590.

Outten, F. W.; Twining, B. S.; Begley, T. P. Metal Homeostasis. In *Wiley Encyclopedia of Chemical Biology*; John Wiley & Sons, Inc.: Hoboken, NJ, 2007.

Papov, V. V.; Diamond, T. V.; Biemann, K.; Waite, J. H. Hydroxyarginine-containing Polyphenolic Proteins in the Adhesive Plaques of the Marine Mussel *Mytilus edulis*. *J. Biol. Chem.* **1995**, *270*(34), 20183–20192.

Pentreath, R. J. The Accumulation from Water of ^{65}Zn, ^{54}Mn, ^{58}Co and ^{59}Fe by the Mussel, *Mytilus edulis*. *J. Mar. Biol. Assoc. U.K.* **1973**, *53*(01), 127–143.

Phillips, D. J. H. Common Mussel *Mytilus edulis* as an Indicator of Pollution by Zinc, Cadnium, Lead and Copper. 1. Effects of Environmental Variable on Uptake of Metals *Mar. Biol.* **1976**, *38*(1), 59–69.

Phillips, D. J. H. Common mussel *Mytilus edulis* as an Indicator of Pollution by Zinc, Cadnium, Lead and Copper. 2. Relationship of Metals in Mussel to Those Discharged by Industry. *Mar. Biol.* **1976**, *38*(1), 71–80.

Politi, Y.; Priewasser, M.; Pippel, E.; Zaslansky, P.; Hartmann, J.; Siegel, S.; Li, C.; Barth, F. G.; Fratzl, P. A Spider's Fang: How to Design an Injection Needle Using Chitin-based Composite Material. *Adv. Funct. Mater.* **2012**, *22*(12), 2519–2528.

Pujol, J. P. Formation of Byssus in Common Mussel (*Mytilus edulis* L). *Nature* **1967**, *214*(5084), 204–205.

Qin, X. X.; Coyne, K. J.; Waite, J. H. Tough Tendons: Mussel Byssus has Collagen with Silk-like Domains. *J. Biol. Chem.* **1997**, *272*, 32623–32627.

Qin, X.; Waite, J. H. A Potential Mediator of Collagenous Block Copolymer Gradients in Mussel Byssal Threads. *Proc. Natl. Acad. Sci. U.S.A.* **1998**, 95, 10517–10522.

Qin, Z.; Buehler, M. J. Impact Tolerance in Mussel Thread Networks by Heterogeneous Material Distribution. *Nat. Commun.* **2013**, 4.

Rudall, K. The Distribution of Collagen and Chitin. *Symp. Soc. Exp. Biol.* **1955**, *9*, 49–71.

Rzepecki, L. M.; Hansen, K. M.; Waite, J. H. Characterization of a Cystin-rich Polyphenolic Protein Family from the Blue Mussel *Mytilus edulis* L. *Biol. Bull.* **1992**, *183*(1), 123–137.

Sagert, J.; Waite, J. H. Hyperunstable Matrix Proteins in the Byssus of *Mytilus galloprovincialis*. *J. Exp. Biol.* **2009**, *212*(14), 2224–2236.

Schmidt, S.; Reinecke, A.; Wojcik, F.; Pussak, D.; Hartmann, L.; Harrington, M. J. Metal-mediated Molecular Self-healing in Histidine-rich Mussel Peptides. *Biomacromolecules* **2014**, *15*(5), 1644–1652.

Schmitt, L.; Ludwig, M.; Gaub, H. E.; Tampe, R. A Metal-chelating Microscopy Tip as a New Toolbox for Single-molecule Experiments by Atomic Force Microscopy. *Biophys. J.* **2000**, *78*(6), 3275–3285.

Schmitt, C. N. Z.; Winter, A.; Bertinetti, L.; Masic, A.; Strauch, P.; Harrington, M. J. Mechanical homeostasis of a DOPA-enriched biological coating from mussels in response to metal variation. *J. R. Soc. Interface* **2015**, *12*, 20150466.

Schofield, R. M. S.; Nesson, M. H.; Richardson, K. A. Tooth Hardness Increases with Zinc-content in Mandibles of Young Adult Leaf-cutter Ants. *Naturwissenschaften* **2002**, *89*(12), 579–583.

Sever, M. J.; Wilker, J. J. Visible Absorption Spectra of Metal-catecholate and Metal-tironate Complexes. *Dalton Trans.* **2004** (7), 1061–1072.

Silverman, H. G.; Roberto, F. F. *Byssus Formation in Mytilus.* 2010; pp 273–283.

Simkiss, K.; Taylor, M. G. Metal Fluxes Across the Membranes of Aquatic Organisms. *Rev. Aquat. Sci.* **1989**, *1*(1), 173–188.

Smith, B. L.; Schaffer, T. E.; Viani, M.; Thompson, J. B.; Frederick, N. A.; Kindt, J.; Belcher, A.; Stucky, G. D.; Morse, D. E.; Hansma, P. K. Molecular Mechanistic Origin of the Toughness of Natural Adhesives, Fibres and Composites. *Nature* **1999**, *399*(6738), 761–763.

Stewart, R. J.; Ransom, T. C.; Hlady, V. Natural Underwater Adhesives. *J. Polym. Sci., B—Polym. Phys.* **2011**, *49*(11), 757–771.

Suhre, M. H.; Gertz, M.; Steegborn, C.; Scheibel, T. Structural and Functional Features of a Collagen-binding Matrix Protein from the Mussel Byssus. *Nat. Commun.* **2014**, 5.

Sun, C. J.; Lucas, J. M.; Waite, J. H. Collagen-binding Matrix Proteins from Elastomeric Extraorganismic Byssal Fibers. *Biomacromolecules* **2002**, *3*(6), 1240–1248.

Sun, C. J.; Waite, J. H. Mapping Chemical Gradients within and Along a Fibrous Structural Tissue, Mussel Byssal Threads. *J. Biol. Chem.* **2005**, *280*(47), 39332–39336.

Szefer, P.; Fowler, S. W.; Ikuta, K.; Osuna, F. P.; Ali, A. A.; Kim, B. S.; Fernandes, H. M.; Belzunce, M. J.; Guterstam, B.; Kunzendorf, H.; Wolowicz, M.; Hummel, H.; Deslous-Paoli, M. A Comparative Assessment of Heavy Metal Accumulation in Soft Parts and Byssus of Mussels from Subarctic, Temperate, Subtropical and Tropical Marine Environments. *Environ. Pollut.* **2006**, *139*(1), 70–78.

Tamarin, A.; Keller, P. J. An Ultrastructutal Study of the Byssal Thread Forming System in *Mytilus. J. Ultrastruct. Res.* **1972**, *40*(3–4), 401–416.

Tamarin, A.; Lewis, P.; Askey, J. The Structure and Formation of the Byssus Attachment Plaque in *Mytilus. J. Morphol.* **1976**, *149*(2), 199–221.

Taylor, M. G. Mechanisms of Metal Immobilization and Transport in Cells. In *Cell Biology in Environmental Toxicology*; Cajaraville, M. P., Ed.; University of the Basque Country Press: Bilbao, Spain, 1995; pp 155–170.

Taylor, S. W.; Chase, D. B.; Emptage, M. H.; Nelson, M. J.; Waite, J. H. Ferric Ion Complexes of a DOPA-containing Adhesive Protein from *Mytilus edulis. Inorg. Chem.* **1996**, *35*(26), 7572–7577.

Taylor, S. W.; Luther, G. W.; Waite, J. H. Polarographic and Spectrophotometric Investigation of Iron(III) Complexation to 3, 4-Dihydroxyphenylalanine-containing Peptides and Proteins from *Mytilus edulis. Inorg. Chem.* **1994**, *33*(25), 5819–5824.

Vaccaro, E.; Waite, J. H. Yield and Post-yield Behavior of Mussel Byssal Thread: A Self-healing Biomolecular Material. *Biomacromolecules* **2001**, *2*(3), 906–911.

Viarengo, A. Heavy Metals in Marine Invertebrates Mechanisms of Regulation and Toxicity at the Cellular Level. *Rev. Aquat. Sci.* **1989**, *1*(2), 295–317.

Viarengo, A.; Nott, J. A. Mechanisms of Heavy Metal Cation Homeostasis in Marine Invertebrates. *Compar. Biochem. Physiol., C: Compar. Pharmacol.* **1993,** *104*(3), 355–372.

Waite, J. H.; Lichtenegger, H. C.; Stucky, G. D.; Hansma, P. Exploring Molecular and Mechanical Gradients in Structural Bioscaffolds. *Biochemistry* **2004,** *43*(24), 7653–7662.

Waite, J. H.; Qin, X. X. Polyphenolic Phosphoprotein from the Adhesive Pads of *Mytilus edulis. Biochemistry* **2001,** *40*(9), 2887–2893.

Waite, J. H.; Qin, X. X.; Coyne, K. J. The Peculiar Collagens of Mussel Byssus. *Matrix Biol.* **1998,** *17*(2), 93–106.

Waite, J. H.; Vaccaro, E.; Sun, C.; Lucas, J. M. Elastomeric Gradients: A Hedge against Stress Concentration in Marine Holdfasts? *R. Soc. Philos. Trans. Biol. Sci.* **2002,** *357*(1418), 143–153.

Wang, W.-X.; Fisher, N. S. Modeling Metal Bioavailability for Marine Mussels. *Rev. Environ. Contam. Toxicol.* **1997,** *151*, 39–65.

Wei, W.; Tan, Y.; Rodriguez, N. R. M.; Yu, J.; Israelachvili, J. N.; Waite, J. H. A Mussel-derived One Component Adhesive Coacervate. *Acta Biomater.* **2014,** *10*(4), 1663–1670.

Wei, W.; Yu, J.; Broomell, C.; Israelachvili, J. N.; Waite, J. H. Hydrophobic Enhancement of DOPA-mediated Adhesion in a Mussel Foot Protein. *J. Am. Chem. Soc.* **2013,** *135*(1), 377–383.

Werner, S.; Grose, R. Regulation of Wound Healing by Growth Factors and Cytokines. *Physiol. Rev.* **2003,** *83*(3), 835–870.

Yonge, C. M. On Primitive Significance of Byssus in Bivalvia and Its Effects in Evolution. *J. Mar. Biol. Assoc. U.K.* **1962,** *42*(1), 113–125.

Yu, J.; Kan, Y.; Rapp, M.; Danner, E.; Wei, W.; Das, S.; Miller, D. R.; Chen, Y.; Waite, J. H.; Israelachvili, J. N. Adaptive Hydrophobic and Hydrophilic Interactions of Mussel Foot Proteins with Organic Thin Films. *Proc. Natl. Acad. Sci. U.S.A.* **2013,** *110*(39), 15680–15685.

Yu, J.; Wei, W.; Danner, E.; Ashley, R. K.; Israelachvili, J. N.; Waite, J. H. Mussel Protein Adhesion Depends on Interprotein Thiol-mediated Redox Modulation. *Nat. Chem. Biol.* **2011,** *7*(9), 588–590.

Yu, J.; Wei, W.; Danner, E.; Israelachvili, J. N.; Waite, J. H. Effects of Interfacial Redox in Mussel Adhesive Protein Films on Mica. *Adv. Mater.* **2011,** *23*(20), 2362–2366.

Yu, J.; Wei, W.; Menyo, M. S.; Masic, A.; Waite, J. H.; Israelachvili, J. N. Adhesion of Mussel Foot Protein-3 to TiO$_2$ Surfaces: The Effect of pH. *Biomacromolecules* **2013,** *14*(4), 1072–1077.

Zeng, H.; Hwang, D. S.; Israelachvili, J. N.; Waite, J. H. Strong Reversible Fe^{3+}-mediated Bridging Between DOPA-containing Protein Films in Water. *Proc. Natl. Acad. Sci. U.S.A.* **2010,** *107*(29), 12850–12853.

Zhao, H.; Robertson, N. B.; Jewhurst, S. A.; Waite, J. H. Probing the Adhesive Footprints of *Mytilus californianus* Byssus. *J. Biol. Chem.* **2006,** *281*(16), 11090–11096.

Zhao, H.; Waite, J. H. Linking Adhesive and Structural Proteins in the Attachment Plaque of *Mytilus californianus. J. Biol. Chem.* **2006,** *281*(36), 26150–26158.

Zhao, H.; Waite, J. H. Proteins in Load-bearing Junctions: The Histidine-rich Metal-binding Protein of Mussel Byssus. *Biochemistry* **2006,** *45*(47), 14223–14231.

Zuccarello, L. V. The Collagen Gland of *Mytilus galloprovincialis*: An Ultrastructural and Cytochemical Study on Secretory Granules. *J. Ultrastruct. Res.* **1980,** 73, 135–147.

Zuccarello, L. V. Ultrastructural and Cytochemical Study on the Enzyme Gland of the Foot of a Mollusc. *Tissue Cell* **1981,** *13*(4), 701–713.

CHAPTER 5

PHYSIOLOGY OF ENVENOMATION BY CONOIDEAN GASTROPODS

BALDOMERO M. OLIVERA[1], ALEXANDER FEDOSOV[2], JULITA S. IMPERIAL[1], and YURI KANTOR[2]

[1]Department of Biology, University of Utah, Salt Lake City, UT, USA

[2]A. N. Severtzov Institute of Ecology and Evolution, Russian Academy of Science, Leninsky Prospect, 33, Moscow 119071, Russia

CONTENTS

5.1   INTRODUCTION

This chapter focuses on the physiology of envenomation by gastropods in the superfamily Conoidea (Puillandre et al., 2011; Tucker & Tenorio, 2009); the best known of these are the cone snails (Röckel et al., 1995). This may be the most species-rich superfamily in the phylum Mollusca (it is estimated that there are over 12, 000 extant conoidean species) (Bouchet et al., 2009; Olivera et al., 2014). The vast majority of all conoideans are venomous, although a small minority of lineages have secondarily lost their venom apparatus (Holford et al., 2009). Significant advances have been made in elucidating physiological mechanisms that underlie the prey capture strategy of a few fish-hunting cone snail species.

In contrast to other predatory lineages of molluscs, except for their venom apparatus, conoidean snails are notably lacking in specialized morphological adaptations that facilitate prey capture. Among venomous animals, conoidean gastropods are at one extreme end of a continuous spectrum, where venom is used in conjunction with other types of weaponry for prey capture. Molluscs such as cephalopods use a combination of speed and anatomical adaptations such as tentacles for prey capture, but the slow moving conoidean snails are generally totally dependent on venom for capturing their prey. Not surprisingly, their evolution has resulted in highly complex and sophisticated venoms (Olivera, 2002).

The superfamily Conoidea is one of the major groups of predatory marine snails that have undergone a major adaptive radiation since the cretaceous extinction in the Order Neogastropoda and are prominent components of present-day shallow-water tropical marine communities. As a group, neogastropods have undergone an enormous anatomical diversification of foregut structures, with many neogastropod lineages evolving specialized glands for the biosynthesis of secretions that aid these predators in capturing their prey. The investigation of the chemical strategies of neogastropods is in its infancy, and the analyses of their biochemistry and molecular biology has been carried out on very few species. A recent pioneering study has been on the colubrarid snails, which are the vampires of the ocean that suck blood from fish (see Modica et al., 2015). However, in this chapter, the focus will be restricted to the superfamily Conoidea, which secrete the gene products for interacting with their prey, predators, and competitors in a venom synthesized characteristic and defining anatomical structure, the venom duct (also called the venom gland). At present, there are more biochemical/molecular data available for the superfamily Conoidea than for all other neogastropod lineages combined.

Conoideans have been the subject of several phylogenetic studies (see Puillandre et al., 2011); classically, three large divisions within the superfamily, cone snails, terebrids (or auger snails), and turrids were recognized, but it has become clear that the last group, the turrids is not monophyletic (see Taylor et al., 1993). In the taxonomic treatment for Conoidea by Puillandre et al. (2011), 14 family groups are proposed (instead of the 3 classically recognized in earlier work). Some of the larger groups comprise hundreds and even thousands of species; these appear to fall into two major clades; one that includes the family Conidae, with the Terebridae and Turridae (sensu stricto), as family groups in the other major clade.

Traditionally, all ~800 species of cone snails were assigned to a single genus, *Conus*. The molecular phylogeny that has recently been carried out (Puillandre et al., 2014) has led to the recognition that three groups of cone snails are quite distant phylogenetically from most *Conus* species, and these have been assigned to other genera. In this scheme, the family Conidae now comprises four genera, *Conus*, *Conasprella*, *Profundiconus*, and *Californiconus*. The redefined genus *Conus* encompasses the vast majority of species (>500). Two of the other lineages, *Profundiconus* and *Conasprella* are primarily deep-water groups (except in the Western Atlantic and Panamic regions where the latter is represented by a handful of shallow-water forms). Surprisingly, little is known regarding the physiology of envenomation of these groups. The available evidence from aquarium observations and gut-content analysis is that *Conasprella* species prey on polychaetes (Costa, 1994; Kohn, 2014). The genus *Californiconus* is a monospecific lineage, with *Californiconus californicus* being the only known species (Duda et al., 2001; Espiritu et al., 2001; Puillandre et al., 2015).

Within the impressive biodiversity of venomous conoideans, the documented knowledge regarding envenomation physiology is highly skewed. For a few species, physiological mechanisms that underlie prey capture are understood in exquisite mechanistic detail, to the point where detailed molecular interactions can be defined. In contrast, however, for most conoidean gastropods, virtually nothing is known about envenomation, and in some cases, even the major prey of entire major lineages in the superfamily have not been identified. Consequently, this chapter is somewhat schizoid: although case studies are presented where understanding of envenomation physiology is highly sophisticated, for most species in Conoidea, only fragmentary descriptive observations relevant to envenomation are available.

We begin by describing envenomation for two species of conoidean molluscs for which an intensive investigation has been carried out of both the pharmacology of the venom components and the physiology of prey capture. These two species are both cone snails (genus *Conus*). *Conus geographus*, the geography cone, is the first case study; this conoidean species is generally regarded as the most dangerous to man, having caused over a dozen human fatalities (Fegan & Andresen, 1997; Kohn, 1958) (which is why it has been intensively investigated). A second species, the purple cone *Conus purpurascens* has an entirely different strategy for prey capture. The contrast between the two species will illustrate how even though both have the same prey (fish), strikingly different physiological mechanisms have evolved in these congeners, and there is compelling evidence for distinct evolutionary pathways to fish hunting for these cone snails (Olivera et al., 2014, 2015).

5.2 PHYSIOLOGY OF ENVENOMATION: FISH-HUNTING *CONUS*

5.2.1 OVERVIEW OF ENVENOMATION

What emerges from the best-studied examples of conoidean envenomation is that multiple components of a conoidean venom acting coordinately are required for prey capture (Olivera, 1997, 2002). Each individual venom component functions as a potent pharmacological agent that specifically targets a particular molecular site in potential prey; when a venom component acts at its pharmacologically relevant site, it clearly elicits a downstream physiological consequence. The typical scenario is for multiple venom components to act together on a targeted circuitry, thus efficiently achieving a desired physiological endpoint relevant to prey capture.

Because the target sites can be functionally related to each other (and often are on signaling proteins that act sequentially within a physiological circuit), by targeting multiple sites simultaneously, the venom potently alters circuit function. One specific example is the circuitry responsible for neuromuscular transmission. The group of toxins that target distinct pharmacological sites on the same physiological circuitry is called a "cabal," with reference to the secret societies that are out to overthrow existing authority. Each cabal comprises groups of peptides that act in synergy to achieve a specific end point. The set of individual molecular targets that are functionally linked are referred to as a "constellation" of signaling components, required for the proper functioning of that circuit. In effect, conoidean gastropods have evolved a highly sophisticated form of combination drug therapy that we shall refer to as "constellation pharmacology;" their strategy is to target functionally linked constellations coordinately, not just one individual molecular target at a time. The detailed physiology of prey capture (and how constellation pharmacology is applied) will be discussed using the two specific case studies that follow.

5.2.2 CASE STUDY 1: CONUS GEOGRAPHUS

C. geographus, the geography cone is the most deadly of all *Conus* species, responsible for most documented human fatalities from cone snail stings. In the absence of medical intervention, the fatality rate is ~70% (Yoshiba, 1984). Because of the considerable interest in the venom components from this species, it is possible to reconstruct in unprecedented detail the physiology of envenomation that this cone snail uses to capture its prey. There are probably more mechanistic insights for venom components from *C. geographus* than for any other species in the superfamily Conoidea.

The geography cone is found in the Indo-Pacific region, from the Western Pacific through the Indian Ocean to East Africa. *C. geographus* appears to be a highly specialized predator that can potentially capture a whole school of small fish hiding in reef crevices at night (presumably to avoid predation by sharks). For prey capture this snail has two distinctive suites of toxins, with each resulting in a different physiological end point.

One group of toxins, the "nirvana cabal, " is released into the water as the snail approaches a school of fish. The physiological endpoint for these venom components is to disorient the fish, largely by jamming sensory circuitry and to deter their escape, primarily by making the fish severely hypoglycemic and placid. The peptides of the "nirvana cabal" would render all of the fish in a school easier for the snail to engulf using its large distensible false mouth (rostrum) (Fig. 5.1 shows how the snail approaches prey with its distended rostrum).

Among the components of the nirvana cabal is a specialized, posttranslationally modified insulin (Safavi-Hemami et al., 2015), smaller than any other insulin peptide identified so far. When taken up through the gills, this triggers a fall in blood-glucose levels, leading to hypoglycemic shock, thereby deterring any attempt by the fish to escape. Other venom components of the nirvana cabal include a subtype-selective NMDA receptor antagonist (Donevan & McCabe, 2000), a neurotensin agonist (Craig et al., 1999b), a 5HT3 receptor antagonist (England et al., 1998) and an $\alpha 9$ $\alpha 10$

nicotinic receptor antagonist (S. Christensen et al., in preparation); together, these are thought to inhibit the sensory circuitry of the fish, particularly the lateral line system. As a result, an entire school of fish can be rendered both hypoglycemic and sensory deprived, and when the snail opens its cavernous false mouth (rostrum), the school can be engulfed. In effect, through the nirvana cabal, the school is primed for capture using the snail's "net." This prey capture strategy has been referred to as the "net" or "net engulfment" strategy (Olivera et al., 2015).

FIGURE 5.1 *Conus geographus*, foraging. As the snail approaches a fish, it opens its rostrum (false mouth) releasing the nirvana cabal components of the venom (see text).

The snail then injects venom into each captured fish, uses a hollow, needle-like radular tooth (see below); this delivers a group of paralytic toxins referred to as the "motor cabal." One inhibits the presynaptic Ca channels that control neurotransmitter release (Adams et al., 1993), another antagonizes the postsynaptic nicotinic acetylcholine receptor that is essential for depolarizing the muscle end plate (Gray et al., 1981), and finally, a third toxin blocks the voltage-gated Na channels (Nav1.4) that underlie the muscle action potential (Cruz et al., 1985). Together, these cause an irreversible paralysis, by antagonizing neuromuscular transmission at multiple pharmacological sites. These individual venom components comprise the "motor cabal;" a cartoon representation of the molecular physiology is shown in Figure 5.2. Thus, by using two combinations of highly specific pharmacological agents, the snail can successfully capture multiple fish at one time.

Cases of human envenomation are clearly due to defensive stings by the snail; it is likely that it is the motor cabal components that cause human fatality. When hunting for prey, it appears that *C. geographus* can control the amount of venom injected. After being engulfed, each fish is injected with only a small amount of venom. However, when delivering a defensive sting to a human, the snail probably injects all available venom in a desperate bid to escape the predator. Thus, one factor likely to contribute to the high rate of human fatality is that when *C. geographus* injects all the venom in the duct, this exceeds a lethal dose for humans; the inhibitor of the nicotinic acetylcholine receptor and of voltage-gated Na channels are particularly potent peptides, and their activity is consistent with the symptomatology in clinical reports of human envenomation cases (i.e., paralysis of the diaphragm, leading to respiratory distress and death).

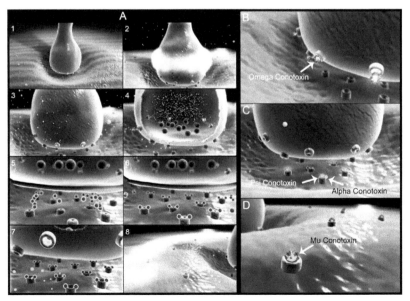

FIGURE 5.2 The Motor Cabal. Panel (A) Normal neurotransmission across the neuromuscular synapse. This panel illustrates the molecular events that occur between the arrival of an action potential at the presynaptic terminus (A1 and A2) to the generation of

FIGURE 5.2   *(Caption continued)*

the action potential on the postsynaptic muscle membrane (A8). Motor axon depolarization results in opening of voltage-gated Ca channels and the entry of calcium into the presynaptic terminus (A3 and A4). The elevated calcium triggers neurotransmitter release; the acetylcholine released then binds and activates the nicotinic acetylcholine receptor (A5 and A6), and channel opening causes depolarization of the postsynaptic terminus, which activates voltage-gated Na channels and triggers the muscle action potential (A7 and A8). Panels (B), (C), and (D) show block by various components of the motor cabal. Panel (B) shows the block by ω-conotoxin, which targets the voltage-gated Ca channels at the presynaptic terminus. Panel (C) shows block of the postsynaptic nicotinic acetylcholine receptor by α-conotoxins and psi-conotoxins; α-conotoxins are competitive antagonists and psi-conotoxins are channel blockers. Panel (D) shows the activity of μ-conotoxins—by blocking voltage-gated Na channels, the action potential on the postsynaptic side is inhibited. Cone snails use all of the inhibitory mechanisms in Panels (B), (C), and (D) in a combination to potently block neuromuscular transmission.

5.2.3   CASE STUDY 2: CONUS PURPURASCENS

C. purpurascens, the purple cone, is found in the Panamic biogeographic province, from the Galapagos to the Sea of Cortez. It is the only fish-hunting *Conus* in the Panamic region; its only close relative is *Conus ermineus*, which is the only fish-hunting species known in the Caribbean and Tropical Atlantic. A comprehensive analysis of the venom of *C. purpurascens* has been carried out (Hopkins et al., 1995; Shon et al., 1995; Terlau et al., 1996). As is the case described above for *C. geographus*, there are two distinctive suites of toxins in the venom that act to achieve different physiological end points that are essential for prey capture (these are referred to as the "lightning-strike cabal" and the motor cabal). The lightning-strike cabal causes an extremely rapid tetanic immobilization of the envenomated fish. In contrast, the motor cabal results in an irreversible block of neuromuscular transmission. This latter group has the same targeted end point as the motor cabal discussed above for *C. geographus*.

C. purpurascens has a highly distensible proboscis, which is colored bright red (see Fig. 5.3). When approaching potential fish prey, the proboscis is extended far out of the rostrum, and once the tip of the proboscis touches the skin of a fish, a disposable radular tooth jets out of the proboscis—this functions both as a hypodermic needle for venom injection, but also as a harpoon to tether the prey. The snail grasps the basal end of the harpoon with a powerful muscular sphincter at the tip of the proboscis. As the radular tooth pierces the skin of the fish, venom is injected, and the snail typically pulls back the proboscis. Since the radular tooth is highly barbed with an accessory process (see Section 5.3.2), the fish becomes tethered through the radular tooth and as the proboscis is reeled back into the rostrum of the snail, the fish is completely engulfed by the rostrum.

FIGURE 5.3 *Conus purpurascens*, envenomation sequence. Upon detecting a fish, the snail extends its bright red proboscis (top). After the fish is struck, the snail retracts its proboscis toward the rostrum, and the immobilized fish is engulfed by the "false mouth" and predigested. The scales and bones of the fish, and the radular tooth used to inject the venom are regurgitated, and the soft parts of the prey move further into the gut of the snail.

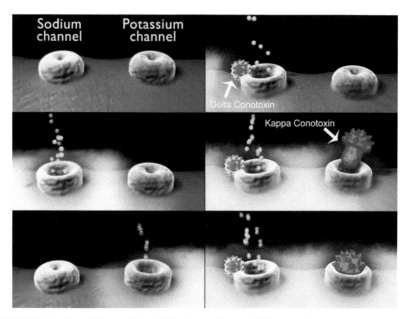

FIGURE 5.4 The effects of the "lightning-strike cabal" on action potentials of axons. Shown on the left panel is the normal progression of an action potential. As the membrane potential is depolarized, sodium channels open and rapidly inactivate (middle and lower left panels). This inactivation of sodium channels and the delayed opening of voltage-gated K channels together repolarize the cell membrane. Thus, these key channel properties, acting coordinately, are responsible for making an action potential transient. On the right panel is an illustration of how a cone snail targets these channels and modulates basic mechanisms of action potential generation. A delta toxin binds to voltage-gated sodium channels (top right) and when the channel opens, the delta toxin inhibits inactivation. This results in the channel remaining open and continuously depolarizing the surrounding membrane. The inhibition of fast inactivation of Na channels, when coupled to the binding of a kappa conotoxin to the voltage-gated K channels (right middle and lower panel) means that both molecular mechanisms for repolarizing the cell membrane are inhibited. As a consequence, the two toxins acting in concert cause a sustained depolarization of affected axonal membranes. In effect, this is the same as applying a powerful taser to a particular site. This is thus a key mechanism of the "taser and tether" strategy for prey capture.

would presumably be rendered largely quiescent, leading to sensory deprivation of the envenomated animal. Thus in many ways, the physiological endpoints are diametrically opposite: hyperstimulation of neuronal circuitry by the lightning-strike cabal, depression of the activity of neuronal circuitry by the nirvana cabal.

Both species of *Conus*, however, have a motor cabal, and, therefore, the broad physiological goals of these venom components are parallel: the complete suppression of neuromuscular transmission, ultimately resulting

The first suite of toxins, the lightning-strike cabal (see Fig. 5.4), has as key components, a δ-conotoxin (in *C. purpurascens*, the specific peptide is δ-conotoxin PVIA) (Terlau et al., 1996) and κ-conotoxins (κ-conotoxin PVIIA) (Shon et al., 1998). δ-Conotoxin PVIA acts by inhibiting fast inactivation of axonal voltage-gated Na channels. Thus, Na channels remain in an open state, and when this is combined with the action of κ-conotoxins, which block voltage-gated K channels on the same axon, all of the mechanisms for terminating an action potential are inhibited. Thus, near the venom injection site, all of the axons become massively depolarized; this is equivalent to applying a powerful electric shock, such as a taser. This generates an electrical storm in the nervous system of the fish, resulting in the tetanic paralysis of the fish within seconds.

This prey capture strategy has been called "taser and tether." The synergy between the two peptides, δ-conotoxin PVIA and κ-conotoxin PVIIA is very striking (Shon et al., 1995). Although δ-conotoxin PVIA will eventually cause a tetanic paralysis of the fish, it has a relatively long onset; κ-conotoxin PVIIA by itself has rather subtle behavioral effects on the fish. However, when the two are injected together, tetanic paralysis occurs in seconds, thereby recapitulating what is observed when the whole venom is injected into the fish by the snail. It is the joint action of the two peptides that is key to the extremely rapid immobilization of the fish prey.

In addition to the lightning-strike cabal of venom components, *C. purpurascens* also has a motor cabal that blocks neuromuscular transmission. The similarities and differences between the motor cabals of *C. purpurascens* and *C. geographus* will be discussed in the next section.

5.2.4 ENVENOMATION BY CONUS PURPURASCENS AND CONUS GEOGRAPHUS: SIMILARITIES AND DIFFERENCES

The most striking contrast between the physiological strategy of *C. geographus* versus *C. purpurascens* are the effects of the lightning-strike cabal of *C. purpurascens* versus the nirvana cabal of *C. geographus*. The *C. purpurascens* lightning-strike cabal causes nervous system circuitry to become hyperstimulated, presumably triggering the generation of trains of action potentials that ultimately cause tetanic paralysis. In contrast, venom components that comprise the *C. geographus* nirvana cabal quiet down the targeted neuronal circuitry, which appears to be largely sensory. Individual components of the *C. geographus* nirvana cabal inhibit molecular targets necessary for activity in the targeted circuitry, and by antagonizing these, the circuitry in paralysis. However, analysis of detailed physiological mechanisms shows both similarities and differences at the molecular level. This is most clearly illustrated by the venom peptides targeted to the postsynaptic receptor. In *C. geographus*, the nicotinic receptor antagonist, α-conotoxin G1 (McManus et al., 1981) is a competitive blocker of acetylcholine binding to its ligand site. Similarly, in *C. purpurascens*, αA-conotoxin PIVA (Hopkins et al., 1995) is a competitive antagonist at the same site. Thus, although biochemically these peptides differ significantly from each other, they are genetically related, and physiologically homologous in their mechanism of activity. What is striking, however, is that *C. purpurascens* has a second peptide that is highly expressed in the venom that inhibits the postsynaptic nicotinic acetylcholine receptor, ψ-conotoxin PIIIE (Shon et al., 1997). This peptide is also an antagonist of the nicotinic acetylcholine receptor, but binds an entirely different pharmacological site on the receptor complex: instead of being a competitive antagonist at the ligand-binding site for acetylcholine, it appears to be a channel blocker. Thus, by having two peptides acting at two different pharmacological sites, presumably, there will be synergy making *C. purpurascens* venom extremely effective in antagonizing this key molecular target that is critical for neuromuscular transmission.

The other components of the motor cabal show both striking similarities and differences between the two species. Both *C. geographus* and *C. purpurascens* have μ-conotoxins that target the muscle subtype of voltage-gated Na channels (Nav1.4) (Mahdavi & Kuyucak, 2014); these peptides, μ-conotoxin GIIIA (Cruz et al., 1989) from *C. geographus* and μ-conotoxin PIIIA (Shon et al., 1998) from *C. purpurascens*, belong to the same gene superfamily (the M-superfamily) and are structurally similar, although diverge considerably in their AA sequences.

In contrast, *C. geographus* venom has, as a major component, conotoxins that block the voltage-gated Ca channels that control neurotransmitter release from the presynaptic terminus of the neuromuscular synapse (i.e., ω-conotoxin GVIA [Olivera et al., 1984] that targets the Cav2.2 [Yarotskyy & Elmslie, 2009] channel). These are not a major component of *C. purpurascens* venom. Thus, there is convergence in the physiological end points achieved, but considerable divergence in the specific molecular mechanisms used.

5.2.5 MOLECULAR PHYLOGENETICS OF FISH-HUNTING CONE SNAILS

There are over 100 species of *Conus* that are believed to primarily hunt fish. The present molecular phylogenetic evidence, summarized in Figure 5.5, demonstrates the relevant branch of the *Conus* phylogenetic tree that has the clades that are primarily fish hunting. The number of species tentatively assigned to each clade is shown in the figure. It should be noted that there is a considerable disparity in the diversity of each lineage, with two (*Phasmoconus* and *Pionoconus*) being significantly more species-rich than the other piscivorous lineages. These branches of the phylogenetic tree were assigned a subgeneric rank (Puillandre et al., 2015). Thus, *C. geographus* belongs to the subgenus *Gastridium*, while *C. purpurascens* is in *Chelyconus*.

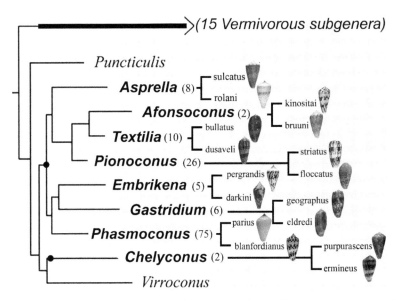

FIGURE 5.5 Molecular phylogeny of fish-hunting *Conus*. Shown are a section of the phylogenetic tree for the Conidae The eight potential piscivorous subgenera, the clades of *Conus* that are believed to be primarily fish hunting are shown in bold, and the number of species that have been assigned to each subgenus is indicated in parentheses (assignments are by Puillandre et al. (2015), except for the subgenus *Asprella*, where newer data (M. Watkins and B. Olivera, unpublished) suggests a larger number of species). Two examples are shown for each fish-hunting clade. The solid dots indicate ancestral nodes where it has been proposed that a fish-hunting ancestor evolved from a worm hunter (Olivera et al., 2015). Note that in this scheme, the subgenus *Chelyconus*, the only clade found in the New World, is postulated to have evolved fish hunting independently of the seven piscivorous Indo-Pacific subgenera.

The observations that have been made on the physiology of prey capture, when combined with the molecular phylogeny shown in Figure 5.5 suggest that fish-hunting arose independently from worm hunting at least twice. The specific molecular evidence that supports this hypothesis has recently been discussed (Aman et al., 2005; Olivera et al., 2015). However, the possibility that some of the present day worm-hunting clades re-evolved from a piscivorous clade cannot be rigorously excluded. This hypothesis is consistent with some earlier studies that have demonstrated that the fish-hunting *Conus* species do not form a single monophyletic group (Puillandre et al., 2014). Although two of the putative subgenera shown in Figure 5.5, *Chelyconus* and *Pionoconus*, have the same general physiological strategies for capturing fish, they diverge in the molecular genetics and biochemistry of the venom components that have evolved in each subgenus for parallel physiological purposes.

The subgenus *Chelyconus*, which is restricted to the new world, has only two species, *C. purpurascens*, the purple cone, which is found in the Panamic region and *C. ermineus*, which is found in the entire tropical Atlantic, from the Gulf of Mexico to West Africa. There are no other known fish-hunting species in these marine biogeographic provinces. It is thought that the two species are very closely related and arose as a result of geographic isolation when the Central American isthmus created a barrier between the Panamic marine province and the Caribbean Sea (Puillandre et al., 2014).

In contrast, the *Pionoconus* clade is strictly Indo-Pacific, and is much more species rich. Both *Pionoconus* and *Chelyconus* (Shon et al., 1998) have a lightning-strike cabal and a motor cabal as described in detail above for *C. purpurascens*. The difference is that the gene superfamilies that have been recruited for similar physiological roles are entirely different in the two lineages. An example are the K-channel blockers that are an essential component of the lightning-strike cabal: these belong to the O-superfamily in *Chelyconus*, but in *Pionoconus*, a family of kunitz-domain containing polypeptides called conkunitzins (Bayrhuber et al., 2005) has been recruited as K-channel blockers in all of the species of *Pionoconus* examined so far.

5.2.6 FISH-HUNTING CONE SNAILS: ECOLOGY AS A DETERMINANT OF PHYSIOLOGY AND A DRIVER OF SPECIATION

For cone snails that primarily prey on fish, a key determinant for the success of an individual species is solving the problem of how to get close enough to a fish to be able to strike or engulf the potential prey. There are multiple solutions, but these are primarily dependent on the ecological setting: different habitats present different opportunities for approaching prey that is clearly equipped to quickly escape, once the predator is detected.

Species such as *C. geographus* that use a net strategy are primarily found in the tropical Indo-Pacific, in habitats where schools of small fish might hide close to where the snails may approach and engulf them. Thus, *C. geographus* has some unusual features; it has an extremely light shell; this allows the snail to be unusually active and agile. Furthermore, *C. geographus* can secrete a thick mucous thread that allows it to levitate itself from the edge of a rock, much like a spider lowering itself using a silk thread. It is a remarkable sight to see a large snail such as *C. geographus* lowering itself from an edge to which it attaches the mucous thread and secreting more of the thread as it lowers itself slowly downward; presumably this is an adaptation to allow it to approach a school of fish hiding in crevices—as it releases the nirvana cabal components of the venom, disorienting the school and making the fish hypoglycemic, it has a much higher probability of feasting on such a school. The other species that is known to use a net strategy, *Conus tulipa* is smaller with a proportionally heavier shell; however, the edges of its rostrum have cilia, the function of which is unknown. It has been speculated by Alan Kohn (personal communication) that the snail can use these to make itself resemble a sea anemone when it extends its rostrum, with the cilia looking like anemone tentacles—this could, therefore, presumably attract clownfish, but there is no field data to provide support for this speculation, and the true function of the unique cilia is unknown.

Probably a far more common approach is to ambush fish as they hide at night and sleep. Many snails have a very long, transparent proboscis that can be many times the length of the shell, and therefore as teleost fish hide from sharks, they are detected by the snails presumably through their potent chemosensory receptor, the osphradium. Thus, the hiding fish can be ambushed by the proboscis that would be almost invisible at night, and after the sting, the lightning-strike cabal of venom components would instantly immobilize the fish. Once tethered, upon retraction of the proboscis, the snail would then deliver the fish to its rostrum for predigestion. Typically, immediately after the snails engulf the tethered fish in their rostrum, they bury themselves, escaping predators of their own.

There appear to be many variations for approaching fish and successfully ambushing them. One species, *Conus obscurus* is reported to be collected under flat coral slabs in about 15–20 m off Hawaii. Presumably, fish hide under these flat coral slabs and the snail can, therefore, ambush fish from on top as they enter the cave-like compartment under the flat coral slab. Other species, such as *Conus stercusmuscarum* are active at night at low tide in pools created by the receding water. Often, many fish are trapped in such tide pools, and the very long proboscis of *Conus stercusmuscarum* allows it to strike at fish that may come close in the shallow water that has become dense with potential fish prey (personal observations). Some other species may actively attract fish toward them. The bright red proboscis of *C. purpurascens* may serve as a lure, since when a fish is present, this species extends its brightly colored proboscis above the sand and writhes it enticingly, presumably making it look like an attractive worm for some unsuspecting fish (Kerstitch, 1979). There have even been reports by divers (P. Poppe, personal communication) that some species such as *Conus striatus* seem to attract certain species of fish toward them—these cone snails may secrete some attractant for particular species of potential fish prey. Another strategy is to remain cryptic: thus, *Conus monachus* has a deep black proboscis that resembles the muddy background in which it typically thrives; when it extends its proboscis, this is completely cryptic and can strike any fish that happens to be close by (personal observations).

5.3   ANATOMY AND BIOCHEMISTRY OF CONOIDEANS

5.3.1   *STRUCTURE AND SYNTHESIS OF CONOTOXINS*

Almost all known toxins produced in conoidean venom glands share some key structural features: these are short peptides (mostly 12–46 amino acids in length that usually have a high frequency of cysteine residues (Olivera, 2006; Norton & Olivera, 2006), which form disulphide cross-links that stabilize the secondary structure of the conopeptide and define its conformation. The number of cysteine residues and their arrangement (the so called Cys pattern or Cys framework) are one determinant of a conopeptide's physiological activity; one classification of the conopeptides is based on their Cys patterns. While the arrangement of cysteine residues in conotoxins is highly conserved, the rest of the venom peptide may vary considerably in amino acid sequence; the accelerated evolution of the genes encoding conopeptides makes them exceptionally diverse and evolutionarily flexible.

A high frequency of posttranslational modification is another unusual feature of many conopeptides (Craig et al., 1999a), which also contributes to their chemical diversity. Some of these modifications are well known (e.g., hydroxyproline, *O*-glycosylated serine or threonine [Craig et al., 1998]), while others are rare and/or unusual (6-bromotryptophan [Jimenez et al., 1997], [gamma]-carboxyglutamate [Bandyopadhyay et al., 2002], sulfotyrosine [Loughnan et al., 1998]). It was also found that some conopeptides contain D-amino acids (Buczek et al., 2008; Jimenez et al., 1996).

The venom components are synthesized in the tubular convoluted venom gland, opening at the proboscis base. The synthesis of all known *Conus* toxins in the venom gland epithelial cells follows a rather conserved scenario. All conopeptide's mRNAs are translated on ribosomes to generate a peptide precursor, its general structure is shared among most known conopeptides. This includes an N-terminal signal sequence (pre-site), followed by the pro-peptide region and a mature toxin region on the C-terminal end of the precursor; the latter is always present in a single copy (in contrast to many neuropeptide precursors that generate multiple mature peptides after posttranslational processing). All conopeptide precursors undergo proteolytic cleavage in the course of maturation with the removal of the signal sequence and pro-peptide region (Puillandre et al., 2010; Terlau & Olivera, 2004).

It was found that the sequence of the signal region is highly conserved across structurally similar conopeptides (Terlau & Olivera, 2004). The study of cDNA clones revealed that even the third codon position is highly conserved in the mRNAs, which encode the signal region of conotoxins (Woodward et al., 1990), although the signal sequences of some conoidean lineages may not be as stringently conserved as in the family Conidae (M. Watkins, unpublished results). The peptides sharing a common signal sequence are encoded by related genes; the conserved signal sequences provide a genetic basis for the classification of conopeptides, defining the gene superfamilies to which individual conopeptides belong.

The study of conopeptide expression levels in different portions of the venom gland revealed notably different expression profiles in distal, medial, and proximal portions (Garrett et al., 2005; Safavi-Hemami et al., 2014). In *Conus textile*, the conotoxins of A-, M-, P-, and T-superfamilies were preferentially expressed in the proximal ½–⅔ of the venom gland, whereas the levels of mRNAs corresponding to these toxins were significantly lower in the distal quarter of the gland. Conversely the conotoxins of the O-superfamily showed highest expression levels in the distal part of the venom gland. The most characteristic peaks, obtained in HPLC of the *C. textile* crude venom, corresponded to the group of μ-conotoxins produced in the proximal half of the venom gland and [delta]-conotoxin TxVIA in the medial part of the gland (Garrett et al., 2005).

Some more recent studies on the spatial differentiation of toxins expression have been carried out (Dutertre et al., 2014). It was suggested that the predatory/defensive functions of *Conus* venoms employed different complements of toxins that are produced in different portions of the venom gland. How the venom cocktail is actually deployed may be more subtle and sophisticated than the two alternatives suggested, since the behavior of species such as *C. geographus* suggests a more complex strategy in capturing free-swimming prey (Olivera et al., 2015) than can be reconstituted under rather rigid and artificial experimental conditions in the laboratory.

5.3.2 BIOMECHANICS OF ENVENOMATION: MORPHOLOGICAL ADAPTATIONS FOR EFFICIENT VENOM DELIVERY

The sophisticated conoidean feeding mechanism based on envenomation of the prey was enabled by several unique morphological adaptations of the anterior foregut. In gastropod molluscs, the radular apparatus generally consists of a radular ribbon with numerous transverse rows of teeth (radula per se) and odontophore—a massive organ, consisting of several subradular cartilages and muscles, providing its movement. The radular apparatus is situated in the buccal cavity in close proximity to the mouth and can be partially everted through the mouth opening. The radula serves as an integrated organ for rasping or gripping food objects. An unusual peculiarity of conoidean foregut anatomy is that the buccal cavity together with radular apparatus is situated at the proboscis base (Fig. 5.6A and C). Consequently, the radula cannot be protruded through the mouth and used for grabbing and rasping the prey.

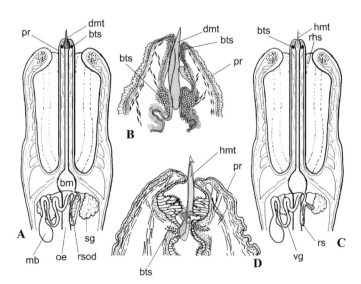

FIGURE 5.6 Diagrammatic sections through the anterior foregut of Conoidea. (A) Anterior foregut of the Conoidea with nonhypodermic marginal radular teeth and odontophore (generalized representative of the clade B in Fig. 5.8). A duplex marginal tooth detached from

FIGURE 5.6 *(Caption continued)*

the subradular membrane is used at the proboscis tip for stabbing and envenomating the prey. (B) Section of the tip of the proboscis with the duplex marginal tooth held by sphincters of the buccal tube (actual specimen of *Aforia kupriyanovi* Sysoev & Kantor, 1988—Cochlespiridae). (B) Anterior foregut of the Conoidea with hypodermic marginal radular teeth and lacking odontophore (generalized representative of clade A in Fig. 5.8). A hypodermic marginal tooth detached from the subradular membrane is used at the proboscis tip. (D) Section of the tip of the proboscis with the hypodermic marginal tooth held by a sphincter of the buccal tube (actual specimen of *Phymorhynchus wareni* Sysoev & Kantor, 1995—Raphitomidae). Abbreviations: bts—buccal tube sphincter, holding the base of the tooth at proboscis tip; dmt—duplex (nonhypodermic) marginal tooth at the proboscis tip; hmt—hypodermic marginal tooth at the proboscis tip; mb—muscular bulb of the venom gland; oe—esophagus; pr—proboscis; rhs—rhynchostome, or false mouth, through which the proboscis is everted; rs—radular sac without odontophore; rsod—radular sac with odontophore; sg—salivary gland; vg—venom gland.

What is essentially unique about the envenomation by conoideans is the use of individual radular teeth at the proboscis tip for stabbing and injecting venom into prey. This was long known for *Conus* spp. that possess elongate, barbed, harpoon-like, hollow marginal teeth (Kohn et al., 1999; Kohn, 1956, 1990; Olivera et al., 1990) (Fig. 5.7A), through which venom is injected into the prey. Recently, it was demonstrated that in the fish-hunting species *Conus catus* the tooth is propelled by a high-speed ballistic mechanism after the proboscis tip makes contact with the fish skin (Schulz et al., 2004) and then gripped by the proboscis tip to retain control of the stung fish prey while the proboscis is retracting. Within 50 ms, the onset of the tetanic immobilization elicited by the lightning-strike cabal toxins in the venom can be observed.

The tubular convoluted venom gland opens at the proboscis base and terminates in a large muscular bulb. The latter propels the venom through proboscis and the tooth cavity by contraction of muscles of the bulb walls. The proboscis of conoideans thus performs different functions: holding the tooth and bringing it in close proximity to the prey; functioning as a channel for venom to flow from the venom gland to the mouth (proboscis tip) and through the tooth (in the case of hypodermic teeth); and finally swallowing the prey. *Conus* species and other Conoidea implement a muscular hydrostat mechanism to enable the rapid (for some *Conus* species, a 7–8-ms long) strike of the proboscis. The muscular hydrostat is based on the fact that liquids are not compressed, and is realized through the coordinated contraction of the cephalic hemocoel muscles, causing the massive flow of the hemolymph into the proboscis. This results in the rapid protraction of the latter. The prey

cannot be rasped or fragmented, since the radula does not function as integral organ, and therefore it is swallowed whole. Observations on mollusc-
hunting cone snails showed that mouth opening can expand more than 50
times in diameter to allow swallowing the prey, which can be comparable in
size to the predator (Kantor, 2007).

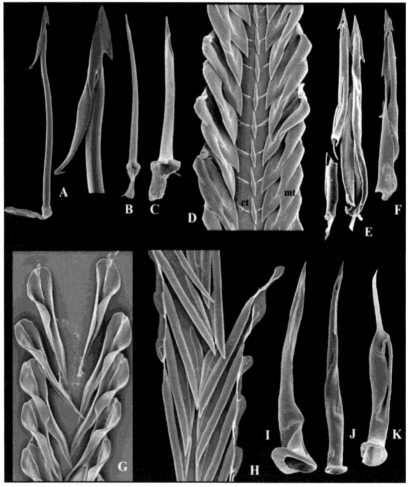

FIGURE 5.7   Variability of radulae of Conoidea. (A)–(C) Hypodermic marginal teeth
of clade A in Fig. 5.8 of Conoidea, including only the species with hypodermic teeth and
without odontophore. (A) Highly barbed tooth of fish-hunting *Conus striatus* (family
Conidae) and enlarged tip of the tooth. (B) *Bathytoma neocaledonica* (family Borsoniidae).
(C) *Toxicochlespira pagoda* (family Mangeliidae). (D)–(K) Radulae of clade B in Fig. 5.8
of Conoidea, which includes species with odontophore. (D) Typical radula with duplex
marginal teeth and central tooth, *Turridrupa jubata* (Turridae). (E) Semienrolled, nearly

FIGURE 5.7 *(Caption continued)*

hypodermic marginal teeth, *Toxiclionella tumida* (family Clavatulidae). (F) Semienrolled, nearly hypodermic marginal tooth, *Cruziturricula arcuata* (family Drilliidae). (G) Radula with semienrolled, trough-shaped marginal teeth, *Ptychobela suturalis* (family Pseudomelatomidae). (H)–(K) Different radulae of Terebridae. (H) Primitive radula with duplex marginal teeth, *Clathroterebra poppei*. Next three figures depict hypodermic marginal teeth in Terebridae, originated independently in three clades, identified by molecular phylogeny: (I) *Terebra cingulifera*; (J) *Hastula lanceata*; (K) *Myurella kilburni*. Abbreviations in Figure 5.7D: ct—central tooth; mt—marginal tooth.

According to the latest molecular phylogeny of Conoidea (Puillandre et al., 2011), the group is split in two major branches (Fig. 5.8, clades A and B). One branch (clade A) includes *Conus* (belonging to family Conidae) and a number of related conoideans (currently assigned to nine families [Kantor et al., 2012b]). In this group the radular apparatus has undergone a profound transformation compared to the typical gastropod design, and the odontophore has completely disappeared (Taylor et al., 1993) (Fig. 5.6C). In conoideans of this branch, the radula comprises only a pair of hollow (hypodermic) marginal teeth in each transverse row and the radular membrane is greatly reduced (Fig. 5.7A–C). Sometimes, the teeth are rather simplified and form a trough rather than a tube.

Conversely, in most conoideans that fall within the second major branch (Fig. 5.8, clade B), the radular apparatus includes a well-developed subradular membrane and a fully functional odontophore with muscles, suggesting that the radula still has some (although maybe limited) function as an integral organ. These conoideans, classified in eight families (Bouchet et al., 2011; Puillandre et al., 2011), also have the radula and odontophore situated at the proboscis base and it normally cannot be protruded through the mouth (Fig. 5.6A). The radula in these conoidean families has very diverse morphology both in number of teeth in the transverse row (from 5 to 2) and in the morphology of the teeth themselves. In molluscs of this branch, the separate marginal tooth was very often (in most preserved specimens examined) found held at the proboscis tip gripped by special sphincter(s) (Kantor & Taylor, 1991; Sysoev & Kantor, 1987, 1989) (Fig. 5.6B). Therefore, it can be supposed that conoideans of this branch also use radular tooth for envenomation, but not via venom injection through the tooth (i.e., a hypodermic needle mechanism), but rather by dagger-style laceration of the prey's integument and releasing the toxic liquid through mouth. Very few direct observations on feeding of the conoideans of this second branch (including the example reported herein, the feeding of *Turridrupa*) are congruent with the anatomical observations made on preserved specimens. In particular, the marginal teeth of *Turridrupa* are non-hypodermic, and belong to so-called duplex type (Kantor & Taylor, 2000) (Fig. 5.7D).

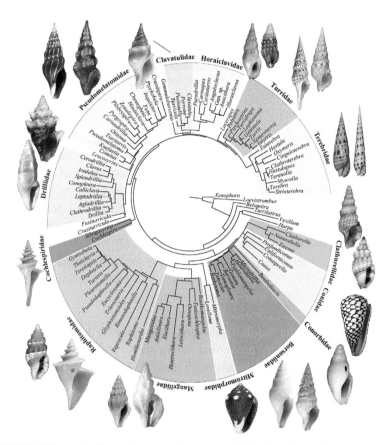

FIGURE 5.8 A phylogenetic tree of the superfamily Conoidea. The phylogenetic tree shows all major families included in the superfamily of Conoidea and is derived from the analysis of Puillandre et al. (2012). The section of the tree with fish-hunting cone snails comprises a very minor subset of the sector of the tree labeled Conidae. In this chapter, most of the physiology of envenomation discussed has focused on a species in the family Conidae, with the diversity of strategies for prey capture in the family Terebridae, a secondary focus. These are the most prominent shallow-water groups with relatively large species accessible by routine collection methods. Two major divisions of Conoidea are marked and referred to in the text as clades A and B.

It is remarkable that within conoideans of the second branch numerous cases of independent origin of hypodermic-like marginal teeth, resembling harpoons of *Conus*, were confirmed based on molecular phylogeny of conoideans (Kantor & Puillandre, 2012; Taylor et al., 1993). In some cases, they remarkably resemble those "true" hypodermic teeth of the first branch (Fig. 5.7E–F, I–K); in other cases they are more simple, trough-like ("semi enrolled" [Kantor & Taylor, 2000]). The semienrolled teeth were found in at least three independent lineages—in the families Drilliidae (genera *Cruziturricula* Marks, 1951 and *Imaclava* Bartsch, 1944), Pseudomelatomidae (genera *Zonulispira* Bartsch, 1950, *Ptychobela* Thiele, 1925, others). The hypodermic teeth were recorded in the families Clavatulidae, Terebridae, and Drilliidae. It was also demonstrated based on extensive molecular phylogeny that within the single family Terebridae, hypodermic teeth originated at least thrice (Castelin et al., 2012) (Fig. 5.6, Radula I–K).

Reconstruction of the major morphological transformations of the radular apparatus in Conoidea was recently conducted based on the molecular phylogeny (Kantor & Puillandre, 2012). It revealed that the use of separate (individual) marginal teeth at proboscis tip was a key synapomorphy of Conoidea that appeared prior to the divergence of the two major branches.

A very remarkable and unexpected tendency in Conoidea is the complete loss of the radula and venom apparatus that is recorded in several unrelated lineages. This is most common in the families Raphitomidae and Terebridae belonging to two different major branches, clades A and B (Fig. 5.8). In both families, it occurred multiple times and is usually associated with the development of rhynchostomal introvert, or labial tube—greatly extended invertible extension of the head, similar to the one in *C. geographus* (Kantor & Sysoev, 1989; Taylor et al., 1993). The introvert is obviously used in the prey capture, as was demonstrated in live observation on feeding of different Terebridae (Miller, 1975). The latter family is one of the best studied of Conoidea in terms of molecular phylogeny, and it was convincingly demonstrated that the venom apparatus was lost eight times within terebrids (Castelin et al., 2012). Its worth mentioning that species without a venom apparatus were among most abundant Tereibridae (Kantor et al., 2012a), suggesting that toxin production is a high energy-consuming process, leaving less resources for reproduction. In addition to the mentioned families, some species lacking the venom apparatus were found also among Horaiclavidae and Borsoniidae (Fedosov & Kantor, 2008).

Data on feeding and diet of Conoidea are still very limited. *Conus* species can be separated into three major groups in relation to their diet: worm-, mollusc-, and fish-hunting. With the exception of *Conus*, information on feeding is available for fewer than 50 species (e.g., Taylor, 1980, 1986), and these reports involved much less direct observation (e.g., Heralde et al., 2010). Most of the conoideans (other than *Conus*) feed on sedentary and errant polychaetes, although feeding on other worms (sipunculans and nemerteans), enteropneusts, and even molluscs has been recorded (see e.g., Miller, 1975, 1979). Both rather specialized and generalist species are known. For example, *Turricula nelliae spurius* in Hong Kong consumes at least 16 species of errant and sedentary polychaetes, while the sympatric species *Lophiotoma leucotropis* appeared to be more specialized, consuming mostly single species of polychaete worm (Taylor, 1980). This information is derived mainly from gut content analysis.

5.4   OTHER CONOIDEA

In the first sections of this chapter, the primary focus was on fish-hunting cone snails. Far more has been elucidated about the underlying physiological mechanisms for this miniscule fraction of all Conoidean species (<1% of the total) than for all of the other conoideans combined. Here, we provide a somewhat sketchy overview for the rest of the superfamily Conoidea, other than the fish-hunting cone snails. These range from other members of the family Conidae to major groups of Conoidea that clearly branched off from Conidae early in the evolution of the superfamily. As a representative of the latter, we will primarily focus our discussion on the auger snails (family Terebridae).

5.4.1   MOLLUSCIVOROUS AND VERMIVOROUS CONE SNAILS

Of the 750 named species of Conidae, over 600 are likely *not* fish hunting, with the vast majority believed to be vermivorous. It is generally assumed that the ancestral forms of Conidae were vermivorous, with polychaetes being the primary prey. Most of the rest of the superfamily Conoidea is widely assumed to be vermivorous, but there are very few documented observations in the literature.

Although there has been significantly less work on molluscivorous *Conus* species than on the fish-hunting clades, at the transcriptome level, there is an accelerating pace of elucidating venom components from several molluscivorous species. Many of these cone snails are easily accessible, larger, shallow-water species (such as *C. textile* and *Conus marmoreus*), so they are relatively straightforward to study at a biochemical and genomic level.

It seems clear that some of the results of the physiological studies carried out on fish-hunting *Conus* apply to molluscivorous *Conus* as well. Thus, a key mechanism used in the envenomation of fish prey is the "lightning-strike cabal" strategy, to overstimulate neuronal circuitry by inhibiting the inactivation of voltage-gated Na channels and concomitantly blocking voltage-gated K channels in the same circuitry. At least some of the molluscivorous *Conus*, such as *C. textile* and *Conus gloriamaris* probably use an analogous strategy when envenomating their snail prey. However, while the purpose of the lightning-strike cabal in fish envenomation is to elicit an almost instant tetanic paralysis, the primary physiological purpose of the homologous toxins in snail-hunting cone snails is probably to guarantee that prey will not retract into its shell after the predator's initial strike.

By hyperstimulating the neuromuscular circuitry of the prey snail, the effect of envenomation is uncoordinated seizure-like motor activity: after the first venom injection, the envenomated snail is typically observed to be alternately contracting and relaxing its musculature, progressively extending further and further out of the shell. This guarantees that the prey does not retreat deep into its shell where it would be inaccessible. As the prey is moving seizure like outside the shell, the predator typically continues to inject additional venom, and it carefully examines where it envenomates the now helpless and uncoordinated prey. These species of molluscivorous *Conus* inject their prey multiple times (in contrast to fish-hunting cone snails that only envenomate once). What is remarkable is that after the predator has made a meal of the envenomated snail, literally nothing is left except for the shell. This must mean that the columellar muscles have been totally relaxed, making it possible for the predator to recover even the hepatopancreas, which is usually tightly coiled deep within the gastropod shell, and easily broken off.

Molluscivorous cone snails appear to be susceptible to their own venom, leading to some unusual facets of envenomation by molluscivorous *Conus*. When a colony of *C. textile* were maintained in an aquarium, and were not fed, the snails were quiescent, but otherwise did not attack each other. However, in one aquarium after an extended starvation period, some of the larger individuals began to envenomate smaller conspecifics and consume them; thus, it appears that some molluscivorous species will practice cannibalism, albeit reluctantly.

Another unusual feature of the physiology of envenomation is the competitive interactions. *C. marmoreus* individuals compete for prey; in one specific case observed by the authors, a smaller individual reached the potential prey snail first and envenomated it. When a larger *C. marmoreus* approached, it stung the small conspecific, apparently not with enough venom to kill it, but the smaller individual appeared to be stunned and was pushed aside and away from the prey. The larger snail then began to consume the prey, and by the next day the smaller individual had recovered completely. Thus, this species can use venom for intraspecific competitive interactions.

Similar competitive interactions have been observed by Dylan Taylor (unpublished observations) for some vermivorous *Conus* species. Two individuals of *Conus lividus* will compete for the same polychaete worm, and both snails may begin to devour the worm from opposite ends. The snails then go into a competitive sucking match, with one individual ultimately being able to suck out the entire worm from the rostrum of the other snail. This was an observed case of competition between conspecifics for a single worm prey. However, cross-species competition was also observed by authors among worm-hunting *Conus*. One interaction in the aquarium is competition that would not occur naturally, since the two relevant *Conus* species do not overlap in their native geographic ranges. When a polychaete worm was introduced into the aquarium, *C. californicus*, a species found off the coast of Southern and Baja California, immediately began to attack the worm, and as is characteristic of this species, multiple individuals were observed feeding on the same worm, which was larger than any of the snails. Ultimately, a specimen of *Conus quercinus* (which does not occur in California, but rather throughout the Indo-Pacific, which is not a locality for *C. californicus*), approached the same worm and injected venom. What occurred was an immediate withdrawal by all of the *C. californicus*, implying that there was a strong deterrent to competitors in the venom of *C. quercinus*.

Some *Conus* species have been observed in the field to envenomate extremely large marine worms; a diver once observed *Conus betulinus* devouring an enormous worm (ca. 1-m long) and was able to watch this for over 20 min, at the end of which the snail had not completely engulfed its prey. These circumstances must attract potential competitors. Thus, making the worm unpalatable to anyone else would provide a clear selective advantage.

5.4.2   OTHER CONIDAE

The species *C. californicus* is believed to have been isolated from all other cone snails since the Miocene (Stanton, 1966) and is ecologically notable in that it lives in a temperate habitat and no other cone snails overlap with it throughout most of its geographic range (from the Baja California Pacific coast, north to Monterey Bay). This is the one species in the family Conidae outside the genus *Conus*, for which multiple observations have been recorded regarding its prey capture strategy. The overall impression gleaned from these is that the divergent phylogeny in fact reflects a corresponding divergent biology of prey envenomation.

Most cone snails are highly specialized with regard to the range of prey that they envenomate; *C. californicus* is a notable exception; it has been observed to attempt to envenomate prey in four different phyla (polychaetes, molluscs, shrimp, and fish) (Biggs et al., 2010; Kohn & Waters, 1966; Saunders & Wolfson, 1961; Stewart & Gilly, 2005). The primary prey is likely to be polychaete worms, but aquarium observations have been reported of this species attacking prey from other phyla, and an analysis of gut contents suggests that they successfully capture other prey. One notable difference between this species and all other *Conus* is that they will hunt as a pack to bring down larger prey. There is no record of any other species in the family Conidae that carries outgroup hunting behavior (although this is consistent with the hunting behavior of some Turridae). The piscivory of *C. californicus* has been intensively investigated and recorded (Stewart & Gilly, 2005). The strategy of *C. californicus* for capturing fish differs notably from those of the specialist fish-hunting *Conus* described above. Not only do multiple individuals prey on a single fish, but *C. californicus* will also routinely sting the prey multiple times in order to capture a large fish; the number of stings is directly correlated with the size of the fish, and up to seven stinging events were observed on a single fish before it was captured. No specialist piscivorous *Conus* has ever been observed to sting fish more than once. In part, it appears that a single envenomation event is insufficient to subdue the fish, so that it can be successfully engulfed by the rostrum of the snail. In effect, the multiple stings are a reflection of the increased dose of venom necessary to completely subjugate a larger fish prey.

There is a corresponding divergence in the conotoxins used for prey capture (Biggs et al., 2010; Gilly et al., 2011). Thus, while fish-hunting *Conus* snails use μ-conotoxins in the M-superfamily to block voltage-gated sodium channels, Gilly et al. (2011) have demonstrated that *C. californicus* uses an unusual family of peptides with four disulfide cross-links, more closely related to the O-superfamily. Similarly, the radular tooth of *C. californicus* is highly distinctive and unusually complex.

5.4.3   ENVENOMATION BY TEREBRIDS AND TURRIDS

All conoideans outside the family Conidae are conventionally referred to as either terebrids or turrids. The terebrids are a relatively small and distinctive group (417 species listed in WoRMS), but turrids comprise the major biodiversity of the superfamily and are now divided into many different families. The nominative genus, *Turris*, is assigned to the family Turridae, which is much reduced compared to its former phylogenetic scope, a reduction that is thoroughly justified by the most recent available molecular phylogenetic data (Bouchet et al., 2011; Puillandre et al., 2011).

There are three terebrid species shown in Figure 5.9, and these are representative of divergent strategies of envenomation within the family Terebridae. A large set of terebrids, such as *Terebra* (*Oxymeris*) *maculata* have secondarily lost their venom apparatus, and do not envenomate their prey. The feeding behavior of three species of *Terebra* that lack venom suggests that they use a long pseudoproboscis for prey capture. The pseudoproboscis can be everted and used to simply suck in the prey without the need for using venom. Species that use this strategy will either feed on capitellid polychaetes that live in loosely compacted sand or on hemichordate worms.

However, two groups of terebrid species envemomate their prey, and these use quite different strategies. One group, including *Hastula strigilata* (shown in Fig. 5.9) might be referred to as the "surfboarding conoideans." The snail apparently detects their worm prey by chemoreception and uses its large foot as a surfboard to propel quickly through the surf zone. These species inject venom into their polychaete prey and typically attack worms that are exposed by wave action. Prey capture is completed between the passage of two successive ocean waves (Bratcher & Cernohorsky, 1987; Miller, 1970).

The third group of auger snails belong to the genus *Terebra* s.s. and are represented in Figure 5.9 by *Terebra triseriata*. These species have a small foot, live in deeper calm areas, and immobilize their prey to prevent it from retracting into its burrow. In contrast to *Hastula strigilata*, which burrows into the sand after having ingested prey, these species feed on small tube-dwelling polychaetes and do not burrow into the sand during feeding.

Thus, although all auger snails are specialized for sandy habitats, it is clear that there are a diversity of strategies used for capturing their prey and that there should be a corresponding diversity in the venom composition of the different types of auger snails.

Any discussion of conoidean physiology needs to address an obvious disparity between the distribution of biodiversity within the superfamily and our present knowledge base. The turrids, broadly defined as conoideans that are not cone snails or terebrids, comprise the vast majority of the diversity in the superfamily, but are also the least investigated group. The major reason for this disparity is that cone snails and terebrids are well represented in shallow-water marine environments, and some of the larger species found in shallow water are abundant and can be collected in great numbers. In

FIGURE 5.9   Diverse branches of Conoidea. The superfamily Conoidea, encompassing all of the known venomous gastropods was traditionally divided into three families: Conidae, Terebridae, and Turridae. Representatives of these three groups are shown in the figure. However, the family Turridae was clearly polyphyletic and the modern phylogeny of the superfamily is shown in Figure 5.8. These species in the figure, however, typify the diversity in the superfamily. On the upper left-hand corner are three species in Conidae, *Conus consors* (far left), a fish-hunting cone snail; *Conus tessulatus* (center), a worm-hunting cone snail; and *C. marmoreus* (right), a mollusc-hunting cone snail. On the right-hand side of the figure are three species in the family Terebridae. It is believed that most Terebridae eat polychaete worms and that their elongate shape is an adaptation for a sand-dwelling lifestyle. However, it is known that the far-right species, *Terebra triseriata* is venomous and lives offshore in deeper water; in contrast, *Terebra maculata* (top left), the largest terebrid species has secondarily lost its venom apparatus and apparently simply sucks up its polychaete prey. In contrast, the species shown on the lower left of the group is *Hastula strigilata*, a group of "surfer conoideans" that live near breaking waves, can use their large foot as a sail to essentially surf from one locality to another, and also allows them to, therefore, attack prey that might be exposed by the breaking waves. On the lower left are three species that represent the old family Turridae, but each is now placed in a separate family group. Far left is the wonder shell, *Thatcheria mirabilis*, which is in the family Raphitomidae. In the center is *Crassispira cerithina*, in the family Pseudomelatonidae and on the far right is *Turris grandis*, in the family Turridae. All of these species are venomous, but are not at all closely related to each other. These different groups are discussed in the text.

contrast, turrids are primarily a deeper water group, with very few species readily accessible in large numbers. This is why in this chapter, though the turrids represent greater than 95% of diversity, they are barely represented in the overview we are presenting. It should be noted that this situation has the potential to change very rapidly; deep-water collection methods, primarily developed by commercial fishermen in the Philippines, such as gill nets or *Lumun lumun* (Seronay et al., 2010) nets have made live specimens of turrids increasingly available. A continuing problem is that many turrid species are

FIGURE 5.10   Envenomation by a *Turridrupa*. Shown is the first recorded observation of a *Turridrupa* species envenomating prey. The *Turridrupa* extended its proboscis and envenomated the prey out of the field of view; the end of the worm that had been stung turned very dark, as if the worm were bleeding internally. This species was formerly identified as *Turridrupa bijubata*, but recent molecular evidence has demonstrated that it is distinct from that species. The genus *Turridrupa* is phylogenetically distant from cone snails and is a monophyletic group in the family Turridae.

extremely small, and therefore applying some of the more conventional methods for elucidating their physiology tend to be a challenge.

An example of how these new methods have made species available is shown in Figure 5.10; a tiny turrid previously unknown to science, *Turridrupa* sp. (still undescribed), has now been filmed envenomating its prey. The figure shows for the first time a turrid attacking its polychaete worm prey. Curiously, once envenomation occurs, there is an abrupt change in the color of the hemolymph of the worm. This raises the possibility that the venom of this species, instead of targeting molecules in the nervous system, which has certainly been the overall picture gained from this study of fish-hunting cone snails, may be targeting the circulatory system of the prey. Thus, the situation may be analogous to venomous snakes—cobra-like snakes have neurotoxins, while rattlesnakes and their relatives have potent factors that affect blood and the circulatory system. It should be possible to elucidate novel mechanisms that have evolved in the different lineages of turrids, now that these can be collected and observed. At present, however, very little is known about the underlying physiological mechanisms that lead to prey capture.

ACKNOWLEDGMENTS

The work of the authors described in this chapter was primarily supported by NIH Grant GM48677. YIK and AF partially were supported by the grants of Russian Foundation of Basic Research (14-04-00481 and 14-04-31048). Some of the work discussed was supported in part by an ICBG Grant 1U01TW008163 from the Fogarty Center, NIH.

KEYWORDS

- **Gastropoda**
- **Mollusca**
- ***Conus***
- **peptides**
- **cone snails**
- **turrids**
- **venom gland**
- **radula**

REFERENCES

Adams, M. E.; Myers, R. A.; Imperial, J. S.; Olivera, B. M. Toxityping Rat Brain Calcium Channels with Omega-toxins from Spider and Cone Snail Venoms. *Biochemistry* **1993**, *32*(47), 12566–12570.

Aman, J. W.; Imperial, J.; Ueberheide, B. M.; Zhang, M. M.; Aguilar, M. B.; Taylor, D.; Watkins, M.; Yoshikami, D.; Showers Corneli, P.; Teichert, R. W.; Olivera, B. M. Insights into the Origins of Fish-hunting in Venomous Cone Snails from Studies on *Conus tessulatus*. *Proc. Natl. Acad. Sci. U.S.A.* **2015**, (PNAS Early Edition), 1–6.

Bandyopadhyay, P. K.; Garrett, J. E.; Shetty, R. P.; Keate, T.; Walker, C. S.; Olivera, B. M. g-Glutamyl Carboxylation: An Extracellular Post-translational Modification that Antedates the Divergence of Molluscs, Arthropods and Chordates. *Proc. Natl. Acad. Sci. U.S.A.* **2002**, *99*, 1264–1269.

Bayrhuber, M.; Vijayan, V.; Ferber, M.; Graf, R.; Korukottu, J.; Imperial, J.; Garrett, J. E.; Olivera, B. M.; Terlau, H.; Zweckstetter, M.; Becker, S. Conkunitzin-S1 is the First Member of a New Kunitz-type Neurotoxin Family. Structural and Functional Characterization. *J. Biol. Chem.* **2005**, *280*(25), 23766–23770.

Biggs, J. S.; Watkins, M.; Puillandre, N.; Ownby, J.-P.; Lopez-Vera, E.; Christensen, S.; Moreno, K. J.; Bernaldez, J.; Navarro, A. L.; Corneli-Showers, P.; Olivera, B. M. Evolution of Conus Peptide Toxins: Analysis of *Conus californicus* Reeve, 1844. *Mol. Phylogenet. Evol.* **2010**, *56*(1), 1–12.

Bouchet, P.; Kantor, Y.; Sysoev, A. V.; Puillandre, N. A New Operational Classification of the Conoidea (Gastropoda). *J. Molluscan Stud.* **2011**, *77*(3), 273–308.

Bouchet, P.; Lozouet, P.; Sysoev, A. V. An Inordinate Fondness for Turrids. *Deep-Sea Res. II Topic. Stud. Oceanogr.* **2009**, *56*(19–20), 1724–1731.

Bratcher, T.; Cernohorsky, W. O. *Living Terebras of the World*. American Malacologists, Inc.: New York, NY, 1987.

Buczek, O.; Jimenez, E. C.; Yoshikami, D.; Imperial, J. S.; Watkins, M.; Morrison, A.; Olivera, B. M. I(1)-Superfamily Conotoxins and Prediction of Single D-Amino Acid Occurrence. *Toxicon* **2008**, *51*(2), 218–229.

Castelin, M.; Puillandre, N.; Kantor, Y.; Modica, M. V.; Terryn, Y.; et al. Macroevolution of Venom Apparatus Innovations in Auger Snails (Gastropoda; Conoidea; Terebridae). *Mol. Phylogenet. Evol.* **2012**, *64*(1), 21–44.

Costa, F. H. A. On the *Conus jaspideus* Complex of the Western Atlantic (Gastropoda: Conidae. *Veliger* **1994**, *37*, 205–213.

Craig, A. G.; Bandyopadhyay, P.; Olivera, B. M. Post-translationally Modified Peptides from *Conus* Venoms. *European Journal of Biochemistry* **1999a**, *264*, 271–275.

Craig, A. G.; Norberg, T.; Griffin, D.; Hoeger, C.; Akhtar, M.; Schmidt, K.; Low, W.; Dykert, J.; Richelson, E.; Navarro, V.; Macella, J.; Watkins, M.; Hillyard, D.; Imperial, J.; Cruz, L. J.; Olivera, B. M. Contulakin-G, an *O*-glycosylated Invertebrate Neurotensin. *J. Biol. Chem.* **1999b**, *274*, 13752–13759.

Craig, A. G.; Zafaralla, G.; Cruz, L. J.; Santos, A. D.; Hillyard, D. R.; Dykert, J.; Rivier, J. E.; Gray, W. R.; Imperial, J.; DelaCruz, R. G.; Sporning, A.; Terlau, H.; West, P. J.; Yoshikami, D.; Olivera, B. M. An *O*-glycosylated Neuroexcitatory *Conus* Peptide. *Biochemistry* **1998**, *37*, 16019–16025.

Cruz, L. J.; Gray, W. R.; Olivera, B. M.; Zeikus, R. D.; Kerr, L.; Yoshikami, D.; Moczydlowski, E. *Conus geographus* Toxins that Discriminate Between Neuronal and Muscle Sodium Channels. *J. Biol. Chem.* **1985**, *260*, 9280–9288.

Cruz, L. J.; Kupryszewski, G.; LeCheminant, G. W.; Gray, W. R.; Olivera, B. M.; Rivier, J. μ-Conotoxin GIIIA, a Peptide Ligand for Muscle Sodium Channels: Chemical Synthesis, Radiolabeling and Receptor Characterization. *Biochemistry* **1989**, *28*, 3437–3442.

Donevan, S. D.; McCabe, R. T. Conantokin-G is an NR2B-selective Competitive Antagonist of *N*-Methyl-D-aspartate Receptors. *Mol. Pharmacol.* **2000**, *58*, 614–623.

Duda, T. F. Jr; Kohn, A. J.; Palumbi, S. R. Origins of Diverse Feeding Ecologies Within *Conus*, a Genus of Venomous Marine Gastropods. *Biol. J. Linn. Soc.* **2001**, *73*, 391–409.

Dutertre, S.; Jin, A. H.; Vetter, I.; Hamilton, B.; Sunagar, K.; Lavergne, V.; Dutertre, V.; Fry, B. G.; Antunes, A.; Venter, D. J.; Alewood, P. F.; Lewis, R. J. Evolution of Separate Predation- and Defence-evoked Venoms in Carnivorous Cone Snails. *Nat. Commun.* **2014**, *5*, 3521.

England, L. J.; Imperial, J.; Jacobsen, R.; Craig, A. G.; Gulyas, J.; Akhtar, M.; Rivier, J.; Julius, D.; Olivera, B. M. Inactivation of a Serotonin-gated Ion Channel by a Polypeptide Toxin from Marine Snails. *Science* **1998**, *281*, 575–578.

Espiritu, D. J. D.; Watkins, M.; Dia-Monje, V.; Cartier, G. E.; Cruz, L. J.; Olivera, B. M. Venomous Cone Snails: Molecular Phylogeny and the Generation of Toxin Diversity. *Toxicon* **2001**, *39*, 1899–1916.

Fedosov, A. E.; Kantor, Y. Toxoglossan Gastropods of the Subfamily Crassispirinae (Turridae) Lacking a Radula, and a Discussion of the Status of the Subfamily Zemaciinae. *J. Molluscan Stud.* **2008**, *74*(1), 27–35.

Fegan, D.; Andresen, D. *Conus geographus* Envenomation. *Lancet* **1997**, *349*, 1672.

Garrett, J. E.; Buczek, O.; Watkins, M.; Olivera, B. M.; Bulaj, G. Biochemical and Gene Expression Analyses of Conotoxins in *Conus textile* Venom Ducts. *Biochem. Biophys. Res. Commun.* **2005**, *328*, 362–367.

Gilly, W. F.; Richmond, T. A.; Duda Jr, T. E.; Elliger, C.; Lebaric, Z.; Schulz, J.; Bingham, J. P.; Sweedler, J. V. A Diverse Family of Novel Peptide Toxins from an Unusual Cone Snail, *Conus californicus*. *J. Exp. Biol.* **2011**, *214*(Pt. 1), 147–161.

Gray, W. R.; Luque, A.; Olivera, B. M.; Barrett, J.; Cruz, L. J. Peptide Toxins from *Conus geographus* Venom. *J. Biol. Chem.* **1981**, *256*, 4734–4740.

Heralde, F. M., 3rd; Kantor, Y.; Astilla, M. A.; Lluisma, A. O.; Geronimo, R.; Alino, P. M.; Watkins, M.; Showers Corneli, P.; Olivera, B. M.; Santos, A. D.; Concepcion, G. P. The Indo-Pacific *Gemmula* Species in the Subfamily Turrinae: Aspects of Field Distribution, Molecular Phylogeny, Radular Anatomy and Feeding Ecology. *Philippine Sci. Lett.* **2010**, *3*(1), 21–34.

Holford, M.; Puillandre, N.; Modica, M. V.; Watkins, M.; Collin, R.; Bermingham, E.; Olivera, B. M. Correlating Molecular Phylogeny with Venom Apparatus Occurrence in *Panamic auger* Snails (Terebridae). *PLoS ONE* **2009**, *4*(11), e7667.

Hopkins, C.; Grilley, M.; Miller, C.; Shon, K.; Cruz, L. J.; Gray, W. R.; Dykert, J.; Rivier, J.; Yoshikami, D.; Olivera, B. M. A New Family of *Conus* Peptides Targeted to the Nicotinic Acetylcholine Receptor. *J. Biol. Chem.* **1995**, *270*, 22361–22367.

Jimenez, E. C.; Craig, A. G.; Watkins, M.; Hillyard, D. R.; Gray, W. R.; Gulyas, J.; Rivier, J.; Cruz, L. J.; Olivera, B. M. Bromocontryphan: Post-translational Bromination of Tryptophan. *Biochemistry* **1997**, *36*, 989–994.

Jimenez, E. C.; Olivera, B. M.; Gray, W. R.; Cruz, L. J. Contryphan is a D-Tryptophan-containing *Conus* Peptide. *J. Biol. Chem.* **1996**, *281*, 28002–28005.

Kantor, Y. How much can *Conus* Swallow? Observations on Molluscivorous Species. *J. Molluscan Stud.* **2007**, *73*(2), 123–127.

Kantor, Y.; Fedosov, A.; Marin, I. An Unusually High Abundance and Diversity of the Terebridae (Gastropods: Conoidea) in Nha Trang Bay, Vietnam. *Zool. Sci.* **2012a**, *51*, 633–670.

Kantor, Y.; Puillandre, N. Evolution of the Radular Apparatus in Conoidea (Gastropoda: Neogastropoda) as Inferred from a Molecular Phylogeny. *Malacologia* **2012**, *55*(1), 55–90.

Kantor, Y.; Strong, E. E.; Puillandre, N. A New Lineage of Conoidea (Gastropoda: Neogastropoda) Revealed by Morphological and Molecular Data. *J. Molluscan Stud.* **2012b**, *78*(3), 246–255.

Kantor, Y.; Sysoev, A. V. On the Morphology of Toxoglossan Gastropods Lacking a Radula, with a Description of New Species and Genus of Turridae. *J. Molluscan Stud.* **1989, 55,** 537–549.

Kantor, Y.; Taylor, J. D. Evolution of the Toxoglossan Feeding Mechanism: New Information of the Use of Radula. *J. Molluscan Stud.* **1991,** *57*(1), 129–134.

Kantor, Y.; Taylor, J. D. Formation of Marginal Radular Teeth in Conoidea (Neogastropoda) and the Evolution of the Hypodermic Envenomation Mechanism. *J. Zool.* **2000,** *252*(2), 251–262.

Kerstitch, A. The Cone with the Come–hither Proboscis. *Hawaiian Shell News* **1979,** *27*(12), 1.

Kohn, A. J. Cone Shell Stings: Recent Cases of Human Injury Due to Venomous Marine Snails of the Genus *Conus. Hawaii Med. J.* **1958,** *17*(6), 528–532.

Kohn, A. J. *Conus of the Southeastern United States and Caribbean.* Princeton University Press: Princeton, 2014.

Kohn, A. J. Piscivorous Gastropods of the Genus *Conus. Proc. Natl. Acad. Sci. U.S.A.* **1956,** *42*, 168–171.

Kohn, A. J. Tempo and Mode of Evolution in Conidae. *Malacologia* **1990,** *32*, 55–67.

Kohn, A. J.; Nishi, M.; Pernet, B. Snail Spears and Scimitars: A Character Analysis of *Conus* Radular Teeth. *J. Molluscan Stud.* **1999, 65,** 461–481.

Kohn, A. J.; Waters, V. Escape Responses to Three Herbivorous Gastropods to the Predatory Gastropod *Conus textile. Anim. Behav.* **1966,** *14*(2), 340–345.

Loughnan, M.; Bond, T.; Atkins, A.; Cuevas, J.; Adams, D. J.; Broxton, N. M.; Livett, B. G.; Down, J. G.; Jones, A.; Alewood, P. F.; Lewis, R. J. a-Conotoxin EpI, A Novel Sulfated Peptide from *Conus episcopatus* that Selectively Targets Neuronal Nicotinic Acetylcholine Receptors. *J. Biol. Chem.* **1998,** *273*, 15667–15674.

Mahdavi, S.; Kuyucak, S. Systematic Study of Binding of Mu-conotoxins to the Sodium Channel NaV1.4. *Toxins (Basel)* **2014,** *6*(12), 3454–3470.

McManus, O. B.; Musick, J. R.; Gonzalez, C. Peptides Isolated from the Venom of *Conus geographus* Block Neuromuscular Transmission. *Neurosci. Lett.* **1981,** *25*(1), 57–62.

Miller, B. A. Studies on the Biology of Indo-Pacific *Terebra*, Ph. D. Dissertation, University of New Hampshire, Durham, NH, 1970.

Miller, B. A. The Biology of *Hastula inconstans* (Hinds, 1844) and a Discussion of Life History Similarities Among Other Hastulas of Similar Proboscis Type. *Pacific Sci.* **1979,** *33*, 289–306.

Miller, B. A. The biology of *Terebra gouldi* Deshayes, 1859, and a Discussion of Life History Similarities Among Other Terebrids of Similar Proboscis Type. *Pac. Sci.* **1975,** *29*, 227–241.

Modica, M. V.; Lombardo, F.; Franchini, P.; Oliverio, M. The Venomous Cocktail of the Vampire Snail *Colubraria reticulata* (Mollusca, Gastropoda). *BMC Genomics* **2015,** *16*(1), 441.

Norton, R. S.; Olivera, B. M. Conotoxins Down Under. *Toxicon* **2006,** *48*(7), 780–798.

Olivera, B. M. *Conus* Peptides: Biodiversity-based Discovery and Exogenomics. *J. Biol. Chem.* **2006,** *281*(42), 31173–31177.

Olivera, B. M. *Conus* Venom Peptides: Reflections from the Biology of Clades and Species. *Annu. Rev. Ecol., Evol. Syst.* **2002**, *33*, 25–42.

Olivera, B. M.; E. E. Just Lecture, 1996. *Conus* Venom Peptides, Receptor and Ion Channel Targets, and Drug Design: 50 Million Years of Neuropharmacology. *Mol. Biol. Cell* **1997**, *8*(11), 2101–2109.

Olivera, B. M.; McIntosh, J. M.; Cruz, L. J.; Luque, F. A.; Gray, W. R. Purification and Sequence of a Presynaptic Peptide Toxin from *Conus geographus* Venom. *Biochemistry* **1984**, *23*, 5087–5090.

Olivera, B. M.; Rivier, J.; Clark, C.; Ramilo, C. A.; Corpuz, G. P.; Abogadie, F. C.; Mena, E. E.; Woodward, S. R.; Hillyard, D. R.; Cruz, L. J. Diversity of *Conus* Neuropeptides. *Science* **1990**, *249*, 257–263.

Olivera, B. M.; Seger, J.; Horvath, M. P.; Fedosov, A. Prey-capture Strategies of Fish-hunting Cone Snails: Behavior, Neurobiology and Evolution. *Brain Behav. Evol.* **2015**, *86*(1), 58–74.

Olivera, B. M.; Showers Corneli, P.; Watkins, M.; Fedosov, A. Biodiversity of Cone Snails and other Venomous Marine Gastropods: Evolutionary Success Through Neuropharmacology. *Annu. Rev. Anim. Biosci.* **2014**, *2*, 487–513.

Puillandre, N.; Bouchet, P.; Duda, T. F., Jr.; Kauferstein, S.; Kohn, A. J.; Olivera, B. M.; Watkins, M.; Meyer, C. Molecular Phylogeny and Evolution of the Cone Snails (Gastropoda, Conoidea). *Mol. Phylogenet. Evol.* **2014**, *78*, 290–303.

Puillandre, N.; Duda, T. F., Jr.; Meyer, C. P.; Olivera, B. M.; Bouchet, P. One, Four or 100 Genera? Classification of the Cone Snails. *J. Molluscan Stud.* **2015**, *81*(1), 1–23.

Puillandre, N.; Kantor, Y.; Sysoev, A. V.; Couloux, A.; Meyer, C. P.; Rawlings, T.; Todd, J. A.; Bouchet, P. The Dragon Tamed? A Molecular Phylogeny of the Conoidea (Mollusca, Gastropoda). *J. Molluscan Stud.* **2011**, *77*, 259–272.

Puillandre, N.; Koua, D.; Favreau, P.; Olivera, B. M.; Stocklin, R. Molecular Phylogeny, Classification and Evolution of Conopeptides. *J. Mol. Evol.* **2012**, *74*(5–6), 297–309.

Puillandre, N.; Watkins, M.; Olivera, B. M. Evolution of Conus Peptide Genes: Duplication and Positive Selection in the a-Superfamily. *J. Mol. Evol.* **2010**, *70*(2), 190–202.

Röckel, D.; Korn, W.; Kohn, A. J. *Manual of the Living Conidae*. Verlag Christa Hemmen: Wiesbaden, Germany, 1995; Vol. I: Indo-Pacific Region, p 517.

Safavi-Hemami, H.; Gajewiak, J.; Karanth, S.; Robinson, S. D.; Ueberheide, B.; Douglass, A. D.; Schlegel, A.; Imperial, J. S.; Watkins, M.; Bandyopadhyay, P. K.; Yandell, M.; Li, Q.; Purcell, A. W.; Norton, R. S.; Ellgaard, L.; Olivera, B. M. Specialized Insulin is Used for Chemical Warfare by Fish-hunting Cone Snails. *Proc. Natl. Acad. Sci. U.S.A.* **2015**, *112*(6), 1743–1748.

Safavi-Hemami, H.; Hu, H.; Gorasia, D. G.; Bandyopadhyay, P. K.; Veith, P. D.; Young, N. D.; Reynolds, E. C.; Yandell, M.; Olivera, B. M.; Purcell, A. W. Combined Proteomic and Transcriptomic Interrogation of the Venom Gland of *Conus geographus* Uncovers Novel Components and Functional Compartmentalization. *Mol. Cell. Proteomics* **2014**, *13*(4), 938–953.

Saunders, P. R.; Wolfson, F. Food and Feeding Behavior in *Conus californicus* Hinds 1844. *Veliger* **1961**, *3*, 73–76.

Schulz, J. R.; Norton, A. G.; Gilly, W. F. The Projectile Tooth of a Fish-hunting Cone Snail: *Conus catus* Injects Venom into Fish Prey Using a High-speed Ballistic Mechanism. *Biol. Bull.* **2004**, *207*(2), 77–79.

Seronay, R. A.; Fedosov, A. E.; Astilla, M. A.; Watkins, M.; Saguil, N.; Heralde, F. M., 3rd; Tagaro, S.; Poppe, G. T.; Alino, P. M.; Oliverio, M.; Kantor, Y. I.; Concepcion, G. P.; Olivera, B. M. Accessing Novel Conoidean Venoms: Biodiverse Lumun–Lumun Marine

Communities, an Untapped Biological and Toxinological Resource. *Toxicon* **2010**, *56*(7), 1257–1266.

Shon, K.; Grilley, M. M.; Marsh, M.; Yoshikami, D.; Hall, A. R.; Kurz, B.; Gray, W. R.; Imperial, J. S.; Hillyard, D. R.; Olivera, B. M. Purification, Characterization and Cloning of the Lockjaw Peptide from *Conus purpurascens* Venom. *Biochemistry* **1995**, *34*, 4913–4918.

Shon, K.; Grilley, M.; Jacobsen, R.; Cartier, G. E.; Hopkins, C.; Gray, W. R.; Watkins, M.; Hillyard, D. R.; Rivier, J.; Torres, J.; Yoshikami, D.; Olivera, B. M. A Noncompetitive Peptide Inhibitor of the Nicotinic Acetylcholine Receptor from *Conus purpurascens* Venom. *Biochemistry* **1997**, *36*, 9581–9587.

Shon, K.; Olivera, B. M.; Watkins, M.; Jacobsen, R. B.; Gray, W. R.; Floresca, C. Z.; Cruz, L. J.; Hillyard, D. R.; Bring, A.; Terlau, H.; Yoshikami, D. µ-Conotoxin PIIIA, a New Peptide for Discriminating among Tetrodotoxin-sensitive Na Channel Subtypes. *J. Neurosci.* **1998**, *18*, 4473–4481.

Shon, K.; Stocker, M.; Terlau, H.; Stühmer, W.; Jacobsen, R.; Walker, C.; Grilley, M.; Watkins, M.; Hillyard, D. R.; Gray, W. R.; Olivera, B. M. k-Conotoxin PVIIA: A Peptide Inhibiting the *Shaker* K+ Channel. *J. Biol. Chem.* **1998**, *273*, 33–38.

Stanton, R. J. Megafauna of the Upper Miocene Castaic Formation, Los Angeles County, California. *J. Paleontol.* **1966**, *40*, 21–40.

Stewart, J.; Gilly, W. F. Piscivorous Behavior of a Temperate Cone Snail, *Conus californicus*. *Biol. Bull.* **2005**, *209*(2), 146–53.

Sysoev, A. V.; Kantor, Y. Anatomy of Molluscs of the Genus *Splendrillia* (Gastropoda: Toxoglossa: Turridae) with Description of Two New Bathyal Species of the Genus from New Zealand. *N. Z. J. Zool.* **1989**, *16*, 205–214.

Sysoev, A. V.; Kantor, Y. Deep-sea Gastropods of the Genus *Aforia* (Turridae) of the Pacific: Species Composition, Systematics, and Functional Morphology of the Digestive System. *Veliger* **1987**, *30*(2), 105–126.

Taylor, J. D. Diets of Sand-living Predatory Gastropods at Piti Bay, Guam. *Asian Mar. Biol.* **1986**, *3*, 47–58.

Taylor, J. D. Diets of Sublittoral Predatory Gastropods of Hong Kong. In *The Marine Flora and Fauna of Hong Kong and Southern China*; Morton, B. S.; Tseng, C. K., Eds.; 1980; pp 907–920.

Taylor, J. D.; Kantor, Y.; Sysoev, A. V. Foregut Anatomy, Feeding Mechanisms, Relationships and Classification of the Conoidea (=Toxoglossa) (Gastropoda). *Bull., Nat. Hist. Mus., Lond. (Zool.)* **1993**, *59*, 125–170.

Terlau, H.; Olivera, B. M. *Conus* Venoms: A Rich Source of Novel Ion Channel-targeted Peptides. *Physiol. Rev.* **2004**, *84*, 41–68.

Terlau, H.; Shon, K.; Grilley, M.; Stocker, M.; Stühmer, W.; Olivera, B. M. Strategy for Rapid Immobilization of Prey by a Fish-hunting Cone Snail. *Nature* **1996**, *381*, 148–151.

Tucker, J. K.; Tenorio, M. J. *Systematic Classification of Recent and Fossil Conoidean Gastropods*. Conchbooks, 2009.

Woodward, S. R.; Cruz, L. J.; Olivera, B. M.; Hillyard, D. R. Constant and Hypervariable Regions in Conotoxin Propeptides. *EMBO J.* **1990**, *9*(4), 1015–1020.

Yarotskyy, V.; Elmslie, K. S. Omega-conotoxin GVIA Alters Gating Charge Movement of N-type (CaV2.2) Calcium Channels. *J. Physiol.* **2009**, *101*(1), 332–340.

Yoshiba, S. An Estimation of the Most Dangerous Species of Cone Shell. *Conus geographus* Venoms Lethal Dose to Humans. *Jpn. J. Hyg.* **1984**, *39*, 565–572.

CHAPTER 6

ESCAPE RESPONSES BY JET PROPULSION IN SCALLOPS[1]

HELGA E. GUDERLEY* and ISABELLE TREMBLAY

Département de biologie, Université Laval, Québec, QC, Canada G1T 2M7

Corresponding author, Email: Helga.Guderley@bio.ulaval.ca

CONTENTS

---

[1]This review is one of a series dealing with trends in the biology of the phylum Mollusca.

"Escape Responses by Jet Propulsion in Scallops" by Helga E. Guderley and Isabelle Tremblay was originally published in the *Canadian Journal of Zoology*, 2013, 91(6): 420-430, 10.1139/cjz-2013-0004. © Canadian Science Publishing or its licensors. Reprinted with permission.

ABSTRACT

The impressive swimming escape response of scallops uses a simple locomotor system that facilitates analysis of the functional relationships between its primary components. One large adductor muscle, two valves, the muscular mantle, and the rubbery hinge ligament are the basic elements allowing swimming by jet propulsion. Although these basic functional elements are shared among scallop species, the exact nature of the escape response varies considerably within and among species. Valve shape and density have opposing influences upon the capacity for swimming and the ease of attack by predators once captured. Patterns of muscle use can partly overcome the constraints imposed by shell characteristics. The depletion of muscle reserves during gametogenesis leads to a trade-off between escape response performance and reproductive investment. However, changes in muscle energetic status influence repeat performance more than initial escape performance. Escape response performance is influenced by habitat temperature and mariculture techniques. During scallop ontogeny, changes in susceptibility to predation and in reproductive investment may influence escape response capacities. These ontogenetic patterns are likely to vary with the longevity and maximal size of each species. Although the basic elements allowing swimming by jet propulsion are common to scallops, their exact use varies considerably among species.

6.1   INTRODUCTION

Many benthic bivalves and gastropods have impressive responses to their predators, ranging from the jumping motions of clams to the elaborate twisting motions of gastropods such as the common whelk (*Buccinum undatum* L., 1758). None of these responses achieves the displacement and, in our opinion, the grace of the swimming response of scallops. This jet-propelled motion raises the animal in the water column and allows them to swim for considerable horizontal distances. The champion swimmers, such as the saucer scallop (*Amusium balloti* [Bernardi, 1861]), have been recorded to swim 30 m in a single swimming bout (Joll, 1989). The intensity of the scallop swimming response finds its closest parallel in vertebrates in the burst flight of pheasants and other gallinaceous birds. As the scallop locomotor system is composed of few functional components, a more complete evaluation of the impact of changes in muscle characteristics upon escape response behavior should be possible than in multifaceted musculoskeletal systems, such as those of vertebrates. In the following review, we will examine the basic components and physiology of the scallop locomotor system and then evaluate the impact of environmental conditions, physiological status, and size upon escape response performance. The latter portion of the review will focus upon studies of the Iceland scallop (*Chlamys islandica* [O. F. Müller, 1776]), sea scallop (*Placopecten magellanicus* [Gmelin, 1791]), and Peruvian scallop (*Argopecten purpuratus* [Lamarck, 1819]).

6.2 BARE BONES LOCOMOTOR SYSTEM

Scallops swim using two valves, one muscle, and a rubbery hinge ligament. The phasic adductor muscle produces the power required for swimming, by rapidly closing the valves and expelling water through small lateral openings that prevent complete valve closure. The muscular mantle controls the size of the jets sent out through the lateral openings. This creates the jet propulsion that moves the scallop forward. When the adductor muscle relaxes, decompression of the hinge ligament opens the valves. The most complete model of this dynamic system focuses upon *P. magellanicus*. The model integrates properties of the hinge ligament, fluid movement around the valves, fluid pressure within the mantle cavity, contraction of the phasic adductor, and valve inertia (Cheng et al., 1996). The authors separated the locomotor system into two parts: a jet producing pressure pump and an oscillator involving the hinge ligament and the outer fluid. Their careful modeling shows that the pressure pump uses most of the mechanical energy produced by muscle contraction. Biomechanical analyses of scallop swimming compare the shells to airfoils and describe scallop swimming as flight through water. For a given scallop species, size, swimming angle, and current speed and direction set lift production and swimming capacity (Gruffydd, 1976; Millward and Whyte, 1992; Thorburn and Gruffydd, 1979). Although repetitive cycles of valve opening and closing allow forward or upward motion, scallops can also use their muscular mantle for fine motor control. This allows them to use adductions to create depressions and to bury themselves, or to move sideways. The precise nature of these motions speaks to integration of information obtained by the many eyes and tentacles on the fringed edge of the mantle.

The capacity for swimming differentiates pectinids from other bivalves. Whereas most bivalves respond to predators by closing their valves for long periods, burying themselves in the substrate through the action of a muscular foot, or using their foot for jumping, scallops can swim to escape their predators. The capacity for jet propulsion is thought to have evolved during colonization of turbid habitats by ancestral scallops. Morphological changes of the valves, together with reduction or loss of the anterior adductor muscle, reduction of byssal attachment, and opening of the mantle (to allow increased flow for filtration) are thought to have permitted ancestral scallops to exploit deeper and more turbid habitats (Yonge, 1936). On the other hand, as the loss of a siphon made it more difficult to evacuate particle loads imposed by turbid waters, two other morphological changes occurred. First, the performance of the ciliary tract was improved. Second, the striated adductor muscle increased markedly in size, enhancing the speed and force of valve closure (Wilkens, 2006; Yonge, 1936). This permits a type of coughing that forcibly evacuates fluid from the mantle cavity (Wilkens, 2006). Furthermore, the infolding of the mantle was enlarged to form openings for the entry and exit of water from the mantle cavity (Wilkens, 2006). In summary, swimming is thought to have arisen in ancestral pectinids that had colonized turbid waters and had gained a capacity for forcible evacuation of water from the mantle cavity. Yonge (1936) suggests that the modifications of the structure of the shell, the mantle, and the adductor muscle are derived adaptations in the monomyairian bivalves.

6.3   SHELL CHARACTERISTICS AND THEIR INFLUENCE UPON SWIMMING CAPACITIES

Their great variety of shell shapes and colors has made scallops favorites of shell collectors around the world. Although the outline of the valves has a characteristic "scalloped" shape, the depth, symmetry, density, and surface characteristics of the valves vary considerably among the more than 300 species of pectinids. A wide range of swimming capacities parallels this structural variety. Minchin (2003) and Alejandrino et al. (2011) classify scallops into 5–6 major groups according to shell morphology, swimming capacities, and life habit. These groups range from the highly active *Amusium* species to the cemented rock scallops, with intermediate groups showing different degrees of byssal attachment. The critical shell characteristics that set swimming capacity include shell density, aspect ratio, and the shape of the shell's leading edge during movement (Dadswell and Weihs, 1990; Gould, 1971; Morton, 1980). Basically, scallops must produce lift to overcome gravity and thrust to overcome drag. Lift and drag are influenced by the angle and speed of swimming, as well as by shell shape. The impact of gravity is set primarily by the mass of the shell. When shell shape and characteristics are unfavorable for swimming, thrust may also be used to produce lift as for species with plano-convex shells such as those in the genus *Pecten* (Millward and Whyte, 1992). Scallops with good swimming abilities generally have shells with a high aspect ratio, an upper valve that is more convex than the lower valve, and light valves with smooth surfaces (Gould, 1971; Soemodihardjo, 1974).

Besides being critical for swimming, shell characteristics also protect against predation. A thick and heavy shell can prevent predation by crabs and tightly sealing valves can hinder predation by starfish. However, heavy shells and reduced openings between the valves reduce swimming capacity. Clearly, characteristics that facilitate swimming are unlikely to protect against predation once a scallop is captured, so a trade-off between swimming capacities and mechanical defense against predation is likely. Furthermore, during growth, shells not only increase their size, but also often increase their thickness (and mass) and may change their surface characteristics as in the extreme case of cementing rock scallops. Overall, shell smoothness, mass, and shape provide considerable information about the probable swimming capacities of a scallop species. A smooth and light shell with a gentle curvature is likely to facilitate swimming, but as it would provide little protection against predation once captured, scallops with such shells are banking upon a strong swimming capacity to avoid predation. The relationship between shell characteristics and escape response capacities is likely to reflect the properties of the scallop's ecosystem. For example, *A. balloti* coexists with a variety of rapid crustacean predators against which a strong swimming capacity is useful (Himmelman et al., 2009). The similarly shaped Antarctic scallop (*Adamussium colbecki* [E. A. Smith, 1902]) remains byssally attached through its adult life and only swims short distances, perhaps due to ice scouring and the lack of crustacean predators in its habitat (Ansell et al., 1998).

Many types of predators attack scallops, with starfish, crustaceans, gastropods, and fish being common predators for most scallop species. Legault and Himmelman (1993) showed that scallops, as other benthic invertebrates, show the strongest escape responses to the predators, in a given guild, that present the greatest predation risk. However, even though crustaceans can present a strong predation risk (Grefsrud et al., 2003; Nadeau et al., 2009), few scallops respond to crustaceans with an escape response. *A. balloti* is an exception to this pattern in that it responds to various crustaceans, including the slipper or flathead lobster (*Thenus orientalis* [Lund, 1793]) and the blue swimming crab (*Portunus pelagicus* [L., 1758]), with intense swimming activity (Himmelman et al., 2009). *A. balloti* does not show a generalized response to crustaceans, as portunid crabs common in its habitat did not elicit a swimming response (Himmelman et al., 2009). *P. magellanicus* generally responds to crustaceans by closing its valves (Barbeau and Scheibling, 1994), but has been reported to move away from crabs (Wong and Barbeau, 2003). Two major starfish predators of *P. magellanicus* differ in their effectiveness, with the common sea star (*Asterias vulgaris* Verrill, 1866 = *Asterias rubens* L., 1758) preying at higher rates than the polar six-rayed star (*Lepasterias polaris* [Müller and Troschel, 1842]) (Nadeau et al., 2009) but both elicit a strong escape response (H. E. Guderley, unpublished data). The giant Atlantic scallop (*Pecten maximus* [L., 1758]) responds differently to the species of predatory starfish it encounters (Thomas and Gruffydd, 1971). Although escape responses can protect against predation, they are evolved responses to predators encountered in the scallops' habitat. Exotic predators may not elicit escape responses by scallops (Hutson et al., 2005), increasing the potential for disruption of ecosystem structure. The specificity of the reactions of scallops to encounters with potential predators suggests that they use precise chemical and visual cues to activate the intense contraction of the phasic adductor muscle required for swimming or jumping.

6.4　STRUCTURE AND ROLES OF THE SCALLOP ADDUCTOR MUSCLE

The phasic and tonic portions of the scallop adductor differ in structure and function (Chantler, 2006). The phasic adductor contracts and fatigues rapidly, with times to peak tension and relaxation similar to those of vertebrate fast fibers (Marsh et al., 1992; Olson and Marsh, 1993; Pérez et al., 2009a; Rall, 1981). The phasic adductor carries out the rapid valve closures needed for scallop swimming (Millman and Bennett, 1976; Nunzi and Franzini-Armstrong, 1981; Olson and Marsh, 1993). In most scallops, the phasic adductor is considerably larger than the tonic adductor (Fig. 6.1); for example, in *P. magellanicus*, the phasic muscle accounts for 80% of the adductor muscle mass (de Zwaan et al., 1980). The tonic adductor muscle is composed of smooth muscle fibers that contract slowly and sustain their contraction for prolonged periods at low metabolic cost (Watabe and Hartshorne, 1990), using a catch mechanism similar to that of the anterior adductor muscle in mytilid bivalves (Nunzi and Franzini-Armstrong, 1981; Chantler, 2006). The tonic muscle is presumably used for maintenance of constant valve openings during filtration by undisturbed scallops, as well as for prolonged valve closure.

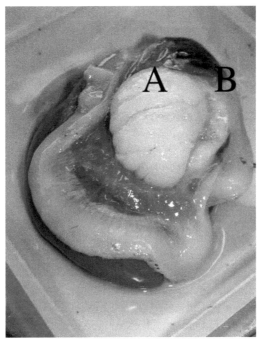

FIGURE 6.1   The two sections of the scallop adductor muscle in a sea scallop (*Placopecten magellanicus*): (A) phasic portion and (B) the tonic portion.

Scallops initiate jet propulsion by a wide gape that increases the volume of water in the mantle cavity. Rapid contraction of the phasic adductor then forces jets of water through focused openings in the mantle, moving the animal forward while closing the valves and compressing the hinge ligament that will open the valves once the muscle relaxes. The moment of force applied to the ligament varies with the distance between the adductor muscle and the ligament (Trueman, 1953; Gould, 1971). In analogy with vertebrate-striated muscles, power production by the phasic adductor should be optimal at intermediate speed and force of contraction. In muscles composed of a single fiber type, force increases with cross-sectional area, whereas speed of contraction rises with fiber length. The large size of the scallop phasic adductor (relative to valve area) clearly enhances force production, whereas the oblique attachment of the phasic adductor increases its length. Effectively, the phasic muscle insertion is closer to the hinge on the right than on the left valve and is generally wider on the left than on the right valve (Thayer, 1972; Soemodihardjo, 1974). This arrangement increases the obliqueness and hence the length of the phasic adductor (Thayer, 1972). In a comparison of several monomyarian bivalves, Thayer (1972) found that the obliqueness of the phasic adductor (in a plane perpendicular to the hinge) increases with swimming ability. In contrast, the tonic adductor is generally attached in a perpendicular position between the two valves, showing much less obliqueness than the phasic adductor (Thayer, 1972; Soemodihardjo, 1974). Thus, the insertions of the tonic and phasic muscles can be separated, particularly on the right valves (Soemodihardjo, 1974).

6.5 METABOLIC SUPPORT OF SWIMMING IN SCALLOPS

As scallops fatigue rapidly during swimming bouts, their swims can be classified as sprints. Vertebrate sprints initially use creatine phosphate followed by anaerobic breakdown of glycogen (Hochachka and Somero, 2002). The limits of vertebrate sprint activity are reached when lactate, the anaerobic breakdown product of glycogen, accumulates to levels that perturb acid–base balance. In pectinids, escape responses also use a phosphagen (arginine phosphate), while anaerobic glycogen breakdown provides a secondary source of ATP. In the species that have been examined most extensively (i.e., *P. magellanicus*, pilgrim's scallop, *Pecten jacobeaeus* [L., 1758]), queen scallop (*Chlamys opercularis* [L., 1758] = *Aequipecten opercularis* [L., 1758]), bay scallop (*Argopecten irradians concentricus* [Say, 1822]), *P. maximus*, and *A. colbecki*), the majority (approximately 70%) of the contractile activity is supported by arginine phosphate breakdown with its rapid generation of 1 mol of ATP per mole (Grieshaber and Gäde, 1977; Grieshaber, 1978; Thompson et al., 1980; de Zwaan et al., 1980; Chih and Ellington, 1983; Bailey et al., 2003). Only the final 30% of rapid valve closures use ATP generated from glycogen breakdown (Grieshaber and Gäde, 1977; Gäde et al., 1978). Instead of accumulating lactate, scallops accumulate octopine, which is the condensation product of pyruvate and arginine. In contrast with vertebrates that primarily produce lactate during exercise, scallops continue to produce octopine after exercise, specifically during the valve closure (Grieshaber, 1978) that typically follows exhaustion. During valve closure produced by tonic contraction, the phasic adductor uses anaerobic glycolysis to partially recuperate adenylate levels (Livingstone et al., 1981; Pérez et al., 2008a). Full metabolic recovery from exhaustive escape responses requires aerobic metabolism (Livingstone et al., 1981), with muscle arginine phosphate levels returning to resting values within 12–24 h of valve opening in *P. magellanicus* (Livingstone et al., 1981; Pérez et al., 2008a), but within 2 h in *A. opercularis* (Grieshaber, 1978). In *A. colbecki, A. opercularis,* and *P.*

maximus, 50% recuperation of arginine phosphate levels occurs within 3–5 h (Bailey et al., 2003).

Although scallop muscles are poorly perfused and their hemolymph does not contain a respiratory pigment, oxygen uptake rises markedly during recovery from exhaustive exercise (Mackay and Shumway, 1980). In juvenile *C. islandica*, oxygen uptake rises to approximately 12-fold resting rates during recovery from exhaustive exercise (Tremblay et al., 2006). Oxygen uptake also rises after swimming in spear scallop (*Chlamys hastata* [G. B. Sowerby II, 1842]), the scallop *Chlamys delicatula* (Hutton, 1873) = *Zygochlamys delicatula* (Hutton, 1873), and *P. magellanicus* (Mackay and Shumway, 1980; Thompson et al., 1980; Donovan et al., 2003; Kraffe et al., 2008). This speaks to the participation of mitochondrial ATP production in the metabolic recovery of the phasic adductor muscle. After opening their valves, scallops recuperate their escape response capacity within minutes (tropical scallops) or hours (temperate-zone scallops). Despite their proximity, no metabolic exchanges appear to occur between the phasic and tonic adductor muscles (Livingstone et al., 1981), such that recovery from exhaustive exercise is accomplished by the mitochondria within phasic adductor fibers.

6.6   THE HINGE LIGAMENT

This critical structure is composed of two portions. The external flexible hinge ligament connects the two valves along the entire hinge. The external ligament contains parallel stratifications (Trueman, 1953). The internal portion of the hinge ligament is made of an elastic material that acts as a compression spring and opens the valves when the adductor muscles relax (Alexander, 1966). This pyramidally shaped section is attached to both valves. The ventral portion of the ligament is curved when the valves are closed. The internal ligament contains a central, noncalcareous rubbery part and two lateral calcified regions that anchor the ligament to the valves. The resilience of the hinge ligament is higher in scallops than in burrowing or sessile bivalves (Kahler et al., 1976). The resilience increases with glycine content and decreases with cysteine and $CaCO_3$ contents (Kahler et al., 1976). The hysteresis loops of the hinge ligaments of the scallops *P. maximus* and *A. opercularis* are considerably tighter than those of sessile or burrowing bivalves (Trueman, 1953), indicating that less energy is lost during cycles of valve opening and closing in scallops than in other bivalves. The gape allowed by the hinge ligament is considerably larger in scallops than in other bivalves (Trueman, 1953). Thus, as with the other major components of the scallop locomotor system, the hinge ligament shows properties that facilitate swimming by jet propulsion.

6.7   STUDIES OF ESCAPE RESPONSE PERFORMANCE

Contraction of the phasic adductor can be easily observed in intact animals, as it leads to dramatic valve closures, often described as "claps" or snaps. Understandably, studies of escape response performance have focused upon these strong valve closures. Although some studies have examined how scallops respond to disturbance or to predators in their natural habitat, the scallop escape response has mainly been studied in the laboratory. Laboratory studies range from observations of unrestrained scallops responding to contact with predators (Brokordt et al., 2000a, 2000b), to high-speed imaging of freely moving scallops (Bailey and Johnston, 2005a, 2005b; Cheng et al., 1996), to measurements of power production during swimming in response to contact with predators by cannulated scallops fitted with piezoelectric transducers (Marsh et al., 1992), and to examination of metabolic parameters within the contracting adductor muscle by magnetic resonance spectroscopy (Bailey et al., 2003). The range of techniques applied to swimming by the lowly scallop is impressive! Although each of these approaches has its strengths and limitations, none of these techniques reveals the integration of contractions of the phasic and tonic adductor muscle.

   To better quantify the activity of the phasic and tonic adductor muscle during escape responses, we developed a simple technique that monitors force development in intact scallops (Fleury et al., 2005). The lower valve of the scallops is clamped to the bottom of an aquarium, while the upper valve is free to move. An extension of the force gauge is inserted under the upper valve at its margin. A test stand allows precise adjustment of the distance between the valves, which we set so that the scallop cannot close its valves more than the distance used during normal ventilation (Fig. 6.2). A computer records any contact of the upper shell with the extension of the force gauge. Rapid peaks reflect contraction of the phasic adductor muscle, whereas sustained force development reflects the contraction of the tonic adductor muscle (Fig. 6.3). In response to having their mantle touched by a predator (e.g., starfish arm), scallops generally increase the gape between the valves and then make a strong phasic contraction. The timing and type of contraction are easily obtained from these recordings (Fig. 6.3). Since the margin of the valve contacts the force gauge, the force recorded is proportional (but not equivalent) to that produced by the muscle, given that the muscle is attached closer to the center of the valve (Pérez et al., 2009a). As this method prevents scallops from swimming or closing their valves, it cannot represent the dynamics of swimming and is best suited to showing the timing of contractions by the phasic and tonic adductor muscles. The relatively noninvasive nature of the force gauge method has allowed us to compare the response of individual scallops before and after a dietary treatment (Guderley et al., 2011), to test individuals during the act of spawning to see if their escape responses change with the effort of spawning (Pérez et al., 2009b), and to examine the metabolic status of individuals sampled in specific activity states (Pérez et al., 2008a, 2008b).

FIGURE 6.2   Force gauge for monitoring muscle activity in intact sea scallops (*Placopecten magellanicus*).

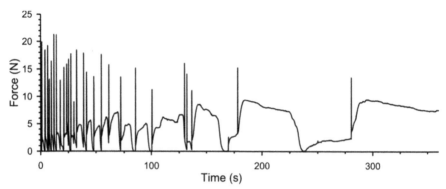

FIGURE 6.3 Typical recording of force production during an escape response by sea scallops (*Placopecten magellanicus*) following Fleury et al. (2005). Sharp peaks indicate phasic contractions, whereas sustained shoulders reflect contraction of the tonic adductor muscle.

Although there is considerable inter- and intraspecific variability in escape response performance, adult scallops typically carry out between 20 and 50 phasic contractions before ceasing to respond to stimulation (Tables 6.1 and 6.2). Generally, the initial response is the most intense, with a flurry of claps occurring at the start of stimulation followed by gradually diminishing numbers of phasic contractions. The frequency of phasic contractions (inverse of minimal interval between claps) varies considerably among species. Among a series of scallops that we compared, the southern scallop (*Pecten fumatus* Reeve, 1852), with its unfavorably shaped valves, made the most rapid succession of phasic contractions (Table 6.2) (Tremblay et al., 2012), followed closely by *A. balloti*. The timing of tonic contractions varies considerably among species, with some species, for example, the doughboy scallop (*Mimachlamys asperrima* [Lamarck, 1819]), starting their response to the predator with a tonic contraction, the strong swimmer, *P. magellanicus*, making many short tonic contractions throughout its escape response, and the best swimmer, *A. balloti*, delaying the onset of tonic contractions until well into its escape response (Table 6.2) (Tremblay et al., 2012). Once scallops cease to respond, most species close their valves and remain shut for several minutes (warm temperate zone scallops) or hours (polar and temperate zone species). The tropical zigzag scallop (*Euvola ziczac* [L., 1758]) (Brokordt et al., 2000a) and some juvenile *P. magellanicus* (H. E. Guderley, personal observation) do not close their valves after exhaustion.

TABLE 6.1 Mean Number of Phasic Contractions (Claps) and Clapping Rate During an Initial Escape Response, as well as the Percent Recuperation of Initial Claps in Adults of Different Scallop Species.

	No. of claps	Clapping rate (no. of claps/min)	Percent recuperation	Reference
Iceland scallop, *Chlamys islandica*	26±1	13±1	95.2±6.3	Brokordt et al. (2000a)
Zigzag scallop, *Euvola ziczac*	52±1	21±2	91.2±2.8	Brokordt et al. (2000b)
Peruvian scallop, *Argopecten purpuratus*				
Cultured	42±2	14±2	78±5	Brokordt et al. (2006)
Wild	33±1	19.5±2	72±3	Brokordt et al. (2006)
Wild sea scallop, *Placopecten magellanicus*	48±3	8.5±3	75.2±9.8	Kraffe et al. (2008)

Note: After visual evaluation of their initial escape response, scallops were given a standard recovery period (4 h for *C. islandica*, 20 min for *E. ziczac*, 20 min for *A. purpuratus*, and 30 min for *P. magellanicus*). These durations were chosen such that recovery was well advanced but not completed. Other details are given in the cited publications. Values are shown as mean ± SE. The data are given for scallops in the gametogenic phase of their reproductive cycle.

TABLE 6.2 Adductor Muscle Performance During Force Recordings of Escape Responses in Scallops with a Wide Range of Shell Morphologies.

	Total phasic contractions	Minimum interval between claps (s)	Time at first tonic (s)	Percent time in tonics
Saucer scallop, *Amusium balloti*	40.9±1.37	0.38±0.036	8.1±2.60	67±3.5
Cultured sea scallop, *Placopecten magellanicus*	27.5±1.65	1.12±0.144	1.1±0.50	82±3.2
Southern scallop, *Pecten fumatus*	32.7±2.86	0.32±0.042	31.7±4.37	65±6.1
Doughboy scallop, *Mimachlamys asperrima*	18.0±1.61	0.65±0.123	0.1±0.14	83±1.6
Giant rock scallop, *Crassadoma gigantea*	1.8±0.84	–	0.0±0.00	93±5.2

Note: Scallops were fixed to the bottom of an aquarium and force development was monitored during their escape responses from their predators, as described in the text. Phasic contractions were identified as sharp peaks in force production, whereas tonic contractions were continuous shoulders of force production. Escape responses were monitored for 355 s (Tremblay et al., 2012).

6.8 PHYSIOLOGICAL STATUS AND ESCAPE RESPONSE CAPACITIES

As the glycogen stores in the adductor muscle are a major energetic reserve, escape response performance could change with the energetic and physiological status of scallops. Seasonal changes in food availability and reproductive investment could modify escape response performance. A reduction in escape response performance by reproductive investment suggests that reproduction could reduce survival in the presence of predators. The impact of temperature and size upon physiological capacities could also modify escape response capacities, as could the stress caused by handling during fishing or culture operations. In the following, we will consider how physiological status modifies escape response performance (as monitored by visual examination of responses to predators) and examine the mechanisms underlying these changes.

Our first inkling that physiological status might modify escape response performance came from studies of mitochondrial physiology in *E. ziczac*. The oxidative capacity of muscle mitochondria changed seasonally, with decreased capacities observed during periods of greatest reproductive investment (Boadas et al., 1997). This suggested a trade-off between reproduction and escape response performance, or more specifically between reproduction and recuperation from exhaustive exercise. We hypothesized that the capacity for repeat escape response performance would be decreased in scallops sampled after gametogenesis and spawning.

K. B. Brokordt validated this hypothesis in two scallop species, *C. islandica* and *E. ziczac*, using visual observations of escape response behaviors of unrestrained scallops. For both species, recuperation from exhaustive escape response performance occurred much more slowly in individuals sampled after gametogenesis and(or) spawning than in individuals with immature gonads (Brokordt et al., 2000a, 2000b). It is well established that gametogenesis depletes macromolecular reserves from muscle and digestive gland in scallops, typically reducing muscle glycogen and even protein levels (Barber and Blake, 1991; Brokordt and Guderley, 2004a). *Chlamys islandica* and *E. ziczac* were no exception to this pattern and demonstrated marked changes in muscle glycogen levels with the reproductive cycle. In parallel, muscle activities of glycolytic and mitochondrial enzymes decreased, potentially due to the decreased availability of glycogen as a matrix for the binding of these enzymes (Brokordt and Guderley, 2004b). Mitochondrial oxidative capacities also decreased as reproductive investment increased.

6.9 REPRODUCTIVE INVESTMENT AND ESCAPE RESPONSE PERFORMANCE

The biochemical changes in muscle physiology during the reproductive cycle suggest that swimming performance could be affected by two, not mutually exclusive, mechanisms. Loss of metabolic and contractile elements in muscle could decrease swimming performance or a reduced aerobic capacity could impede maintenance of muscle status and recuperation from exhaustive exercise. Aerobic capacity could fall through loss of maximal aerobic capacity, or by an increase in routine metabolic requirements, in either case, metabolic recovery would take longer.

Vo_{2max} were reduced or if routine metabolic requirements were increased by reproductive investment, aerobic scope would fall, slowing metabolic recuperation and reducing repeat performance. As reproductive investment requires synthesis of macromolecules and transfer of materials between tissues, routine metabolic requirements are likely to rise during gametogenesis and preparation for spawning. Mobilization of macromolecules from muscle could impair muscle metabolic capacities and reduce maximal rates of oxygen uptake.

To examine this question, we examined the metabolic capacities of *P. magellanicus* from their natural beds at gonadal maturity, after spawning, and during reproductive quiescence (Kraffe et al., 2008). Neither Vo_{2max} nor mitochondrial oxidative capacities changed with reproductive status. In contrast, standard metabolic rate (SMR) was 60% of Vo_{2max} in spawned scallops but only 30–40% of Vo_{2max} in scallops with mature gonads or in reproductive quiescence (Fig. 6.4). Thus, reproductive investment is likely to reduce repeat performance of *P. magellanicus* by increasing maintenance costs and reducing aerobic scope. In other scallops, such as *C. islandica* and *E. ziczac*, as muscle mitochondrial capacities are reduced by reproductive investment (Brokordt et al., 2000a, 2000b). Vo_{2max} could fall with these reduced mitochondrial capacities, slowing recuperation from exhaustive exercise by a second mechanism.

Examination of escape response behavior during the reproductive cycle of several scallop species indicates that the initial response to contact with a predator is quite constant. In *C. islandica*, the number and rate of phasic contractions changed little with gonadal status (Brokordt et al., 2000a). For the tropical scallop *E. ziczac*, reproductive investment did not change the number of phasic contractions, although their rate was slightly higher in spawned than in immature scallops (Brokordt et al., 2000b). The impact of reproductive investment upon escape response performance in *A. purpuratus* differed between wild and cultured scallops, with wild scallops maintaining a constant number and rate of phasic contractions throughout their reproductive cycle, whereas cultured scallops lost performance as reproductive investment increased (Brokordt et al., 2006). Spawned *P. magellanicus* were not able to perform as many phasic contractions as scallops with mature gonads (Kraffe et al., 2008). In these three studies of wild populations, temperature changed little between the sampling times, indicating that changes in performance were due to intrinsic changes in the scallops. Finally, we established that the act of spawning, with its associated valve movements, did not significantly modify the escape response capacity (as measured with the force gauge method) of *A. purpuratus* (Pérez et al., 2009b). Overall, initial escape response performance changed little with reproductive investment, presumably limiting the compromise between survival and reproduction.

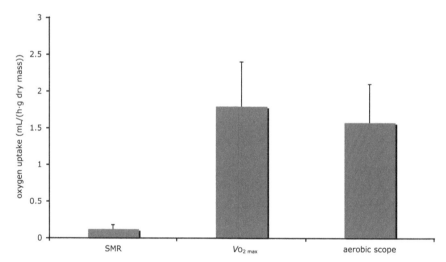

FIGURE 6.4 Changes in rates of oxygen uptake with activity in juvenile Iceland scallops (*Chlamys islandica*). Standard metabolic rates (SMR) were obtained in resting individuals that had fasted for 48 h. To determine the Vo_{2max}, each scallop performed two escape response tests separated by a 45 min recuperation period. The highest oxygen consumption measured for each individual was considered the Vo_{2max} (Tremblay et al., 2006). Aerobic scope was calculated by subtracting the SMR from the Vo_{2max}.

In contrast to initial performance, repeat performance or recuperation from exhaustive exercise is considerably more sensitive to reproductive investment. *Chlamys islandica* with immature gonads can repeat their escape response performance within 4 h, whereas mature, prespawned, and spawned scallops needed 12–18 h for full recuperation (Brokordt et al., 2000a). For *E.*

ziczac, scallops with immature gonads could repeat their initial performance after 20 min of aerobic recuperation, but the time required for recovery increased markedly with reproductive investment (Brokordt et al., 2000b). Reproductive investment also decreased repeat performance in *A. purpuratus* and *P. magellanicus* (Brokordt et al., 2006; Kraffe et al., 2008).

The differing responses of the initial and repeat escape response performance presumably reflect the fact that the initial response is fuelled by anaerobic metabolism, whereas repeat performance requires aerobic recuperation. Effectively, scallops show their maximal rates of oxygen uptake (Vo_{2max}) after exhaustive escape responses (Mackay and Shumway, 1980; Tremblay et al., 2006), with oxygen uptake increasing up to 12-fold (Fig. 6.5).

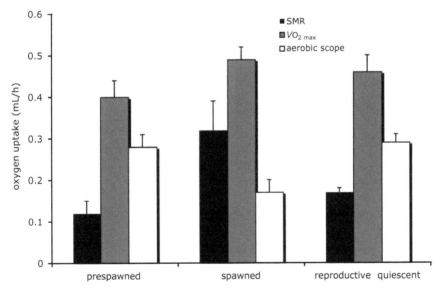

FIGURE 6.5   Oxygen uptake as a function of the reproductive state in adult sea scallops (*Placopecten magellanicus*). Metabolic rates were determined at habitat temperature (6.3–7.2°C) as described in Kraffe et al. (2008). Standard metabolic rate (SMR) was assessed in scallops that had been held in filtered water for 3 days, whereas Vo_{2max} was measured directly after exhaustive escape responses. Aerobic scope was calculated by subtracting SMR from the Vo_{2max}. Oxygen uptake rates are expressed for a standard animal (50 g wet soft tissue mass).

A compromise between reproduction and locomotor capacity may also arise due to the macromolecular requirements for gametogenesis. When reproductive investment occurs during periods of low food availability, muscle glycogen and even protein may fall. Mobilization of muscle glycogen in support of gonadal growth is common in scallops (Barber and Blake, 1981). In *C. islandica*, the loss of intramuscular glycogen reduces binding of glycogen phosphorylase and octopine dehydrogenase, thereby potentially destabilizing the intracellular organization of muscle enzymes (Brokordt and Guderley, 2004b). This loss of binding parallels the decrease in enzyme activity, potentially owing to accelerated turnover of "unbound" enzymes. In keeping with a role of muscle glycogen in maintaining muscle metabolic capacity, reproductive investment by *E. ziczac* leads to greater decreases of muscle metabolic capacity and repeat performance when reproductive cycles occurred during periods of low food availability (Brokordt et al., 2000b).

It is remarkable that initial escape response behavior remains relatively constant during scallop reproductive cycles, despite the marked changes in muscle metabolic status. For *E. ziczac*, *C. islandica*, and *A. purpuratus* that were sampled in their natural habitats, initial escape response performance changed little with reproductive status (Brokordt et al., 2000a, 2000b, 2006). The impact of reproductive investment upon initial escape response performance is much weaker than that upon repeat performance (Brokordt et al., 2006; Kraffe et al., 2008). This pattern emphasizes the selective importance of an effective initial escape from predators (Barbeau and Scheibling, 1994). As repeat escape performance is unlikely to provide much additional protection from predation, the gain in fitness obtained by increasing reproductive investment would seem to easily offset the loss in survival.

6.10   SIZE AND ESCAPE RESPONSE PERFORMANCE

Biomechanical considerations predict that the swimming capacity of scallops changes with size. The power required for swimming increases markedly with size, given the exponential rise in shell mass with size. Changes in shell shape with size will also influence swimming capacity. Although each scallop species has its own ontogenetic changes in shell characteristics, some generalizations are possible. The most extensive literature concerning the impact of size on swimming performance concerns *P. magellanicus* (Gould, 1971; Dadswell and Weihs, 1990; Manuel and Dadswell, 1991, 1993). Diving observations by Caddy (1968, 1972) separated its life cycle into a sessile phase (1–30 mm in shell height), a mobile stage (30–100 mm), and a more sedentary stage (>100 mm). The swimming style of *P. magellanicus* also varies with body size. Small scallops tend to rise in the water column and rarely succeed in swimming horizontally. Scallops of intermediate size rise in a rectilinear motion and can swim steadily for several meters (Caddy,

1968; Dadswell and Weihs, 1990; Manuel and Dadswell, 1993). Larger scallops only swim short distances given the mass of their shells. *Placopecten magellanicus* of intermediate size (40–80 mm) are the fastest swimmers and possess the greatest hydrodynamic efficiency (Dadswell and Weihs, 1990). Ontogenetic changes in swimming capacity are apparent in other scallop species. In contrast to most scallop species, *A. balloti* increases its swimming capacity with size (Joll, 1989). In *A. opercularis*, small scallops (<60 mm) swim more and close their shells more often than larger scallops that tend to jump (Schmidt et al., 2008). The most extreme ontogenetic changes occur in the rock scallop (*Crassadoma gigantea* [J. E. Gray, 1825]), which swims while small (<40 mm) and then cements to the substrate when larger (Yonge, 1951; Lauzier and Bourne, 2006).

In *P. magellanicus*, the impact of size upon muscle metabolic capacities and patterns of muscle use parallels its impact upon swimming performance (Labrecque and Guderley, 2011). Much as swimming performance peaks between 40 and 80 mm shell height, biochemical characteristics increase with size to peak near 60 mm and decrease at greater shell heights. Clearly, these functional characteristics do not follow the simple allometric dependence of metabolic capacity shown in interspecific comparisons of vertebrates (Hochachka and Somero, 2002). The increasing reproductive investment of *P. magellanicus* with size combined with its longevity may underlie this pattern.

As the major role of scallop swimming is predator avoidance (Legault and Himmelman, 1993), the suite of predators in the scallop's habitat must be considered in interpreting the size dependence of escape response performance. Large scallops with thick and heavy shells may have reached a size refuge, relaxing the requirements for high muscle metabolic capacities to power escape response performance. The major predators of *P. magellanicus* are sea stars (*A. vulgaris*), rock crabs (*Cancer irroratus* Say, 1871), American lobsters (*Homarus americanus* H. Milne-Edwards, 1837), and some species of fish (Barbeau and Scheibling, 1994; Elner and Jamieson, 1979; Naidu and Meron, 1986). The preference of sea stars for small *P. magellanicus* may underlie the strong escape responses of small *P. magellanicus* (Barbeau and Scheibling, 1994). Elner and Jamieson (1979) observed that adult rock crabs do not feed on *P. magellanicus* over 72 mm shell height. The stomach contents of American plaice (*Hippoglossoides platessoides* [Fabricius, 1780]) indicate that only small (<25 mm) *P. magellanicus* are consumed (Naidu and Meron, 1986). The simultaneous decline of hydrodynamic efficiency and behavioral and physiological capacities of *P. magellanicus* at shell heights greater than 65 mm suggest that they have reached a size refuge. As *P. magellanicus* is a long-lived species that reaches sizes above 150-mm shell height (Naidu and Robert, 2006), this decline is unlikely to be due to senescence. It seems more likely that changes in predation pressure with size simultaneously affect shell morphology, swimming performance, and the physiological properties of the adductor muscle.

Although more is known about the allometry of escape response performance and its underlying physiology in *P. magellanicus* than in other scallop species, other scallops have similar attributes. Ontogenetic changes in the habitat occupied by *C. islandica* are linked to susceptibility to predation by sea stars and crabs (Arsenault and Himmelman, 1996). Accordingly, small *C. islandica* make more phasic contractions, but remain closed less long after exhaustion than large scallops (Tremblay et al., 2006). *Aequipecten opercularis* changes from swimming to jumping as it increases in size, and shows a marked decline in markers of aerobic metabolism and an increase in indicators of oxidative damage (Philipp et al., 2008). The authors suggest that senescence, with the gradual accumulation of damaged, atrophied tissues, take its toll in older *A. opercularis*.

6.11 TEMPERATURE AND ESCAPE RESPONSE PERFORMANCE

Temperature affects rates of physiological processes in ectotherms, with rates rising to an optimum and then stabilizing and eventually falling at higher temperatures (Hochachka and Somero, 2002). Gill-breathing organisms are particularly tied to environmental temperature, as the simple act of respiring requires intimate contact with water. This is doubly true for filter-feeding animals such as scallops. Thermal effects will arise directly from kinetic effects upon physiological processes and indirectly through changes in oxygen contents of the water and in cardiovascular capacities for oxygen delivery.

Although the thermal sensitivity of scallop escape response performance and of the contractile properties of adductor muscle fibers vary with thermal habitat, generally performance improves with temperature, up to an optimum, and then declines. Valve contraction rate of *P. magellanicus* increases with habitat temperature (Manuel and Dadswell, 1991). Time-related contractile properties (response latency, time to peak tension, and relaxation time) in Atlantic bay scallop (*Argopecten irradians* [Lamarck, 1819]) and *P. magellanicus* decrease with rising temperature, whereas force production changes little (Olson and Marsh, 1993; Pérez et al., 2009a). Perhaps the narrowest thermal optimum for swimming is that of the *A. colbecki*, as it ceases to respond at 2°C (Peck et al., 2004). Interestingly, the abductin in the hinge ligament of *A. colbecki* has a greater resilience than that of temperate zone scallops, partially compensating for the effects of cold temperature on swimming performance (Denny and Miller, 2006). Transferring *P. magellanicus* from 18 to 8°C markedly decreased rates of phasic contraction and slowed recuperation from exhausting exercise for 156 h after the transfer (Lafrance et al., 2002). To examine the thermal sensitivity of swimming performance in *A. opercularis*, without the effects of acute thermal change, Bailey and Johnston (2005b) evaluated performance at acclimation temperatures of 5, 10, and 15°C. Contractile properties were more sensitive to temperature than the total duration of the activity cycle (opening and closing of the valves), presumably because the performance of the hinge ligament is less dependent on temperature than that of the muscle. Thermal acclimation only slightly modifies swimming speed of *A. opercularis*, despite considerable increases in power production with temperature (Bailey and Johnston, 2005a). The thermal sensitivity of peak acceleration is higher in winter- than autumn-acclimatized *A. opercularis* (Bailey and Johnston, 2005b). When escape response performance of *P. magellanicus* was monitored at habitat temperature during spring, summer, and fall, phasic force, the number of phasic contractions, and the minimal interval between phasic contractions remained stable, suggesting that scallops compensate for seasonal changes in temperature (Guderley et al., 2008).

The thermal sensitivity of escape response behavior is influenced not only by the intrinsic thermal sensitivity of the underlying contractile processes, but also by neurosensory integration. The season at which a scallop experiences a given temperature may modify its thermal sensitivities, through direct thermal effects or indirect seasonal effects. Indirect effects include perception of changes in day length and acclimation to prevailing seasonal temperatures and food availabilities. To evaluate the impact of a scallop's thermal history upon the thermal sensitivity of its escape response performance, we compared the responses of scallops sampled in May and September when habitat temperature was 12°C. In May, scallops performed better at 6 than at 12 or 18°C, whereas in September, performance was better at 6 and 12°C than at 19°C (Guderley et al., 2009). Much as with isolated muscle fibers (Olson and Marsh, 1993; Pérez et al., 2009a), mean and maximal force production was not affected by temperature, whereas the duration of phasic contractions fell as temperature rose (Guderley et al., 2009). Clearly, predictions of how temperature will affect scallop escape response performance require knowledge of their thermal history and their typical thermal environment.

6.12 ESCAPE RESPONSE PERFORMANCE AND CULTURE METHODS

As the strength of the scallop escape response is sensitive to their physiological status, we reasoned that escape response performance could be used to assess the impact of handling practices used during scallop culture. Particularly during transfers between the different culture phases, scallops are removed from one site, manipulated, and then brought to the next phase of their culture. Handling is often mechanized and can be quite rough. Virtually all aspects of escape response performance, as well as muscle levels of arginine phosphate and adenylate energy charge, are drastically reduced after juvenile *P. magellanicus* are exposed to a standardized handling stress modeled upon culture techniques (Guderley et al., 2008; Pérez et al., 2008b). Three hours recuperation allows metabolic parameters and escape response performance to return to control values (Pérez et al., 2008b). It is technically much simpler to measure escape response performance than the biochemical status of muscle, suggesting that measurements of escape responses could be used to refine culture practices. Greater reductions in escape performance occurred when the handling stress was applied under summer than fall conditions (Guderleyet al., 2008). The lower temperatures in the fall presumably reduced the impact of handling stress upon physiological status. Thus, when at all possible, culture manipulations should be carried out under cool conditions.

By its very nature, culture selects for rapid growth and may reduce exposure to the predators typically present in the cultured organism's habitat (but see work by J. H. Himmelman and L. Freites). These changing selection pressures may lead cultured organisms to differ from their wild counterparts in their performance characteristics. In one comparison, we found that wild *P. magellanicus* had stronger shells and higher clapping rates but were in weaker energetic condition than equivalently sized cultured scallops (Lafrance et al., 2003). Cultured *A. purpuratus* responded less rapidly to their starfish predator than their wild counterparts (Brokordt et al., 2006). The reduced activity of scallops during culture could affect their swimming capacities. Effectively, as frequent swimming by *P. magellanicus* increases adductor muscle size (Kleinman et al., 1996), culture under confined conditions could lead scallops to become sedentary, reduce muscle size, and swimming performance.

6.13   CONCLUSION

Rapid contractions by the striated adductor muscle allow scallops to use jet propulsion to escape their predators. In scallops with strong swimming capacities, the position and size of the muscle, characteristics of the hinge ligament, and the structure and shape of the valves facilitate swimming. However, species with unfavorable shell morphologies can start escape responses with an intense series of phasic contractions to partially overcome constraints imposed by the shape of their valves. Scallops differ considerably in how they use their adductor muscles, with tonic contractions used much more by some species than others. Even strong swimmers, such as *P. magellanicus*, make extensive use of tonic contractions, although the function of these numerous short tonic contractions is unclear. The behavioral differences suggest that the functional characteristics of the adductor muscle and hinge ligament vary among scallop species. Within scallop species, ontogenetic changes in susceptibility to predation, adjustments to environmental conditions, and reproductive investment can modify escape response performance. However, initial escape responses change less than repeat performance, presumably due to the selective importance of an effective escape response. The functional components required for an escape response are likely to evolve in a coordinated fashion. Modifications in shell morphology that lead to more effective valve closure could reduce the need (and capacity) for effective escape responses, relax requirements for muscle performance, and potentially increase reproductive success. Further analysis of the pectinid locomotor system is likely to reveal such fundamental trade-offs between major fitness functions.

ACKNOWLEDGMENTS

We would be remiss if we did not indicate the major role played by X. Janssoone in the development of the force gauge method that we use for monitoring muscle activity during scallop escape responses. Our research is supported by grants from the Natural Sciences and Engineering Research Council of Canada (NSERC) and Fonds québécois de la recherche sur la nature et les technologies (FQRNT), as well as by the collaboration of many scientists in eastern and western Canada and Australia.

KEYWORDS

- scallop
- Pectinidae
- locomotion
- swimming
- adductor muscle
- escape response capacity

REFERENCES

Alejandrino, A.; Puslednik, L.; Serb, J. M. Convergent and Parallel Evolution in Life Habit in the Scallops (Bivalvia: Pectinidae). *Evol. Biol.* **2011**, *11*, 164. DOI:10.1186/1471-2148-11-164.

Alexander, R. McN. Rubber-like Properties of the Inner Hinge-ligament of Pectinidae. *J. Exp. Biol.* **1966**, *44*, 119–130. PMID:5922731.

Ansell, A. D.; Cattaneo Vietti, R.; Chiantore, M. Swimming in the Antarctic scallop *Adamussium colbecki*: Analysis of In Situ Video Recordings. *Antarct. Sci.* **1998**, *10*, 369–375. DOI:10.1017/S0954102098000455.

Arsenault, D. J.; Himmelman, J. H. Size-related Changes in Vulnerability to Predators and Spatial Refuge Use by Juvenile Iceland Scallops *Chlamys islandica. Mar. Ecol. Prog. Ser.* **1996**, *140*, 115–122. DOI:10.3354/meps140115.

Bailey, D. M.; Johnston, I. A. Scallop Swimming Kinematics and Muscle Performance: Modelling the Effects of "Within-animal" Variation in Temperature Sensitivity. *Mar. Freshw. Behav. Physiol.* **2005a**, *38*, 1–19. DOI:10.1080/10236240500046617.

Bailey, D. M.; Johnston, I. A. Temperature Acclimatisation of Swimming Performance in the European Queen Scallop. *J. Thermal. Biol.* **2005b**, *30*, 119–124. DOI:10.1016/j.jtherbio.2004.08.084.

Bailey, D. M.; Peck, L. S.; Bock, C.; Pörtner, H. O. High Energy Phosphate Metabolism During Exercise and Recovery in Temperate and Antarctic Scallops: An *In Vivo* ^{31}P NMR study. *Physiol. Biochem. Zool.* **2003**, *76*, 622–633. DOI:10.1086/376920. PMID:14671710.

Barbeau, M. A.; Scheibling, R. E. Behavioral Mechanisms of Prey Size Selection by Sea Stars (*Asterias vulgaris* Verrill) and Crabs (*Cancer irroratus* Say) Preying on Juvenile Sea Scallops (*Placopecten magellanicus* (Gmelin)). *J. Exp. Mar. Biol. Ecol.* **1994**, *180*, 103–136. DOI:10.1016/0022-0981(94)90082-5.

Barber, B. J.; Blake, N. J. Energy Storage and Utilization in Relation to Gametogenesis in *Argopecten irradians concentricus* (Say). *J. Exp. Mar. Biol. Ecol.* **1981**, *52*, 121–134. DOI:10.1016/0022-0981(81)90031-9.

Barber, B. J.; Blake, N. J. Reproductive Physiology. In *Scallops: Biology, Ecology and Aquaculture*; Shumway, S. E., Ed.; Elsevier B.V.: Amsterdam, 1991; pp 377–428.

Boadas, M. A.; Nusetti, O.; Mundarain, F.; Lodeiros, C.; Guderley, H. E. Seasonal Variation in the Properties of Muscle Mitochondria from the Tropical Scallop *Euvola* (*Pecten*) *ziczac*. *Mar. Biol. (Berl.)* **1997,** *128*, 247–255. DOI:10.1007/s002270050089.

Brokordt, K. B.; Guderley, H. Energetic Requirements During Gonad Maturation and Spawning in Scallops: Sex Differences in *Chlamys islandica*. *J. Shellfish Res.* **2004a,** *23*, 25–32.

Brokordt, K. B.; Guderley, H. Binding of Glycolytic Enzymes in Adductor Muscle of *Chlamys islandica* is Altered by Reproductive Status. *Mar. Biol. Prog. Ser.* **2004b,** *268*, 141–149. DOI:10.3354/meps268141.

Brokordt, K. B.; Himmelman, J. H.; Guderley, H. E. Effect of Reproduction on Escape Responses and Muscle Metabolic Capacities in the Scallop *Chlamys islandica* Müller 1776. *J. Exp. Mar. Biol. Ecol.* **2000a,** *251*, 205–225. DOI:10.1016/S0022-0981(00)00215-X. PMID:10960615.

Brokordt, K. B.; Himmelman, J. H.; Nusetti, O. A.; Guderley, H. E. Reproductive Investment Reduces Recuperation from Exhaustive Escape Activity in the Tropical Scallop *Euvola ziczac*. *Mar. Biol. (Berl.)* **2000b,** *137*, 857–865. DOI:10.1007/s002270000415.

Brokordt, K. B.; Fernández, M.; Gaymer, C. Domestication Reduces the Capacity to Escape from Predators. *J. Exp. Mar. Biol. Ecol.* **2006,** *329*, 11–19. DOI:10.1016/j.jembe.2005.08.007.

Caddy, J. F. Underwater Observations on Scallop (*Placopecten magellanicus*) Behaviour and Drag Efficiency. *J. Fish. Res. Board Can.* **1968,** *25*(10), 2123–2141. DOI:10.1139/f68-189.

Caddy, J. F. Progressive Loss of Byssus Attachment with Size in the Sea Scallop, *Placopecten magellanicus* (Gmelin). *J. Exp. Mar. Biol. Ecol.* **1972,** *9*, 179–190. DOI:10.1016/0022-0981(72)90047-0.

Chantler, P. D. Scallop Adductor Muscles: Structure and Function. In *Scallops: Biology, Ecology and Aquaculture*; Shumway, S. E., Parsons, G. J., Eds.; Elsevier B.V.: Amsterdam, 2006; pp 229–316.

Cheng, J.-Y.; Davison, I. G.; Demont, M. E. Dynamics and Energetics of Scallop Locomotion. *J. Exp. Biol.* **1996,** *199*, 1931–1946. PMID:9319845.

Chih, P. C.; Ellington, W. S. Energy Metabolism During Contractile Activity and Environmental Hypoxia in the Phasic Adductor Muscle of the Bay Scallop *Argopecten irradians concentricus*. *Physiol. Zool.* **1983,** *56*, 623–631.

Dadswell, M. J.; Weihs, D. Size-related Hydrodynamic Characteristics of the Giant Scallop, *Placopecten magellanicus* (Bivalvia: Pectinidae). *Can. J. Zool.* **1990,** *68*(4), 778–785. DOI:10.1139/z90-112.

Denny, M.; Miller, L. Jet Propulsion in the Cold: Mechanics of Swimming in the Antarctic Scallop, *Adamussium colbecki*. *J. Exp. Biol.* **2006,** *209*, 4503–4514. DOI:10.1242/jeb.02538. PMID:17079720.

de Zwaan, A.; Thompson, R. J.; Livingstone, D. R. Physiological and Biochemical Aspects of the Valve Snap and Valve Closure Responses in the Giant Scallop *Placopecten magellanicus*. II. Biochemistry. *J. Comp. Physiol., B* **1980,** *137*(2), 105–114. DOI:10.1007/BF00689208.

Donovan, D. A.; Bingham, B. L, From, M.; Fleisch, A. F.; Loomis, E. S. Effects of Barnacle Encrustation on the Swimming Behaviour, Energetics, Morphometry and Drag Coefficient of the Scallop, *Chlamys hastata*. *J. Mar. Biol. Assoc. U.K.* **2003,** *83*, 1–7. DOI:10.1017/S0025315403007847h.

Elner, R. W.; Jamieson, G. S. Predation of Sea Scallops, *Placopecten magellanicus*, by the Rock Crab, *Cancer irroratus*, and the American Lobster, *Homarus americanus*. *J. Fish. Res. Board Can.* **1979,** *36*(5), 537–543. DOI:10.1139/f79077.

Fleury, P.-G.; Janssoone, X.; Nadeau, M.; Guderley, H. Force Production During Escape Responses: Sequential Recruitment of Phasic and Tonic Portions of the Adductor Muscle in Juvenile *Placopecten magellanicus* (Gmelin). *J. Shellfish Res.* **2005**, *24*, 905–911.

Gäde, G.; Weeda, E.; Gabbot, P. A. Changes in the Level of Octopine During the Escape Responses of the Scallop, *Pecten maximus* (L). *J. Comp. Physiol.* **1978**, *124*(2), 121–127. DOI:10.1007/BF00689172.

Gould, S. J. Muscular Mechanics and the Ontogeny of Swimming in Scallops. *Palaeontology* **1971**, *14*, 61–94.

Grefsrud, E. S.; Strand, O.; Haugum, G. A. Handling Time and Predation Behaviour by the Crab, *Cancer pagurus*, Preying on Cultured Scallop *Pecten maximus*. *Aquacult. Res.* **2003**, *34*, 1191–1200. DOI:10.1046/j.1365-2109.2003.00927.x.

Grieshaber, M. Breakdown and Formation of High-energy Phosphates and Octopine in the Adductor Muscle of the Scallop, *Chlamys opercularis* (L.), During Escape Swimming and Recovery. *J. Comp. Physiol.* **1978**, *126*, 269–276. DOI:10.1007/BF00688937.

Grieshaber, M.; Gäde, G. Energy Supply and the Formation of Octopine in the Adductor Muscle of the Scallop *Pecten jacobaeus* (Lamarck). *Comp. Biochem. Physiol., B: Compr. Biochem.* **1977**, *58*, 249–252. DOI:10.1016/03050491(77)90198-5.

Gruffydd L.; D. Swimming in *Chlamys islandica* in Relation to Current Speed and an Investigation of Hydrodynamic Lift in This and Other Scallops. *Norw. J. Zool.* **1976**, *24*, 365–378.

Guderley, H.; Janssoone, X.; Nadeau, M.; Bourgeois, M.; Pérez Cortés, H. Force Recordings During Escape Responses by *Placopecten magellanicus* (Gmelin): Seasonal Changes in the Impact of Handling Stress. *J. Exp. Mar. Biol. Ecol.* **2008**, *355*, 85–94. DOI:10.1016/j.jembe.2007.06.037.

Guderley, H.; Labbé-Giguere, S.; Janssoone, X.; Bourgeois, M.; Pérez, H. M.; Tremblay, I. Thermal Sensitivity of Escape Response Performance by the Scallop *Placopecten magellanicus*: Impact of Environmental History. *J. Exp. Mar. Biol. Ecol.* **2009**, *377*, 113–119. DOI:10.1016/j.jembe.2009.07.024.

Guderley, H.; Brokordt, K.; Pérez Cortés, H. M.; Marty, Y.; Kraffe, E. Diet and Performance in the Scallop, *Argopecten purpuratus*: Force Production During Escape Responses and Mitochondrial Oxidative Capacities. *Aquat. Liv. Res.* **2011**, *24*, 261–271. DOI:10.1051/alr/2011116.

Himmelman, J. H.; Guderley, H. E.; Duncan, P. F. Responses of the Saucer Scallop *Amusium balloti* to Potential Predators. *J. Exp. Mar. Biol. Ecol.* **2009** *378*, 58–61. DOI:10.1016/j.jembe.2009.07.029.

Hochachka, P. W.; Somero, G. N. Biochemical Adaptation: Mechanism and Process in Biochemical Evolution. Oxford University Press: New York, 2002.

Hutson, K. S.; Ross, J. D.; Day, R. W.; Ahern, J. J. Australian Scallops Do Not Recognise the Introduced Predatory Seastar *Asterias amurensis*. *Mar. Ecol. Prog. Ser.* **2005**, *298*, 305–309. DOI:10.3354/meps298305.

Joll, L. M. Swimming Behaviour of the Saucer Scallop *Amusium balloti* (Mollusca: Pectinidae). *Mar. Biol. (NY)* **1989**, *102*, 299–305.

Kahler, G. A.; Fisher, F. M., Jr.; Sass, R. L. The Chemical Composition and Mechanical Properties of the Hinge Ligament in Bivalve Molluscs. *Biol. Bull. (Woods Hole)* **1976**, *151*, 161–181. DOI:10.2307/1540712.

Kleinman, S.; Hatcher, B. G.; Scheibling, R. E. Growth and Content of Energy Reserves in Juvenile Sea Scallops, *Placopecten magellanicus*, as a Function of Swimming Frequency and Water Temperature in the Laboratory. *Mar. Biol. (Berl.)* **1996**, *124*, 629–635. DOI:10.1007/BF00351044.

Kraffe, E.; Tremblay, R.; Belvin, S.; LeCoz, J.-R.; Marty, Y.; Guderley, H. Effect of Reproduction on Escape Responses, Metabolic Rates and Muscle Mitochondrial Properties in
the Scallop, *Placopecten magellanicus*. *Mar. Biol. (Berl.)* **2008**, *156*, 25–39. DOI:10.1007/
s00227-008-1062-4.

Labrecque, A. A.; Guderley, H. Size, Muscle Metabolic Capacities and Escape Response
Behaviour in the Giant Scallop. *Aquat. Biol.* **2011**, *13*, 51–64. DOI:10. 3354/ab00342.

Lafrance, M.; Guderley, H.; Cliche, G. Low Temperature, But Not Air Exposure Slows the
Recuperation of Juvenile Scallops *Placopecten magellanicus* from Exhausting Escape
Responses. J. Shellfish Res. **2002**, *21*, 605–618.

Lafrance, M.; Cliche, G.; Haugum, G. A.; Guderley, H. Comparison of Cultured and Wild Sea
Scallops *Placopecten magellanicus*, Using Behavioral Responses and Morphometric and
Biochemical Indices. *Mar. Ecol. Prog. Ser.* **2003**, *250*, 183–195. DOI:10.3354/meps250183.

Lauzier, R. B.; Bourne, N. F. Scallops of the West Coast of North America. In *Scallops:
Biology, Ecology and Aquaculture*; Shumway, S. E., Parsons, G. J., Eds.; Elsevier B.V.:
Amsterdam, 2006; pp 965–989.

Legault, C.; Himmelman, J. H. Relation Between Escape Behaviour of Benthic Marine
Invertebrates and the Risk of Predation. *J. Exp. Mar. Biol. Ecol.* **1993**, *170*, 55–74.
DOI:10.1016/0022-0981(93)90129-C.

Livingstone, D. R.; de Zwaan, A.; Thompson, R. J. Aerobic Metabolism, Octopine Production and Phosphoarginine as Sources of Energy in the Phasic and Catch Adductor Muscles
of the Giant Scallop *Placopecten magellanicus* During Swimming and the Subsequent
Recovery Period. *Comp. Biochem. Physiol., B: Compr. Biochem.* **1981**, *70*(1): 35–44.
DOI:10.1016/0305-0491(81)90120-6.

Mackay, J.; Shumway, S. E. Factors Affecting Oxygen Consumption in the Scallop *Chlamys
delicatula* (Hutton). *Ophelia* **1980**, *19*, 19–26. DOI:10.1080/00785326. 1980.10425503.

Manuel, J. L.; Dadswell, M. J. Swimming Behavior of Juvenile Giant Scallop, *Placopecten
magellanicus*, in Relation to Size and Temperature. *Can. J. Zool.* **1991**, *69*(8), 2250–2254.
DOI:10.1139/z91-315.

Manuel, J. L.; Dadswell, M. J. Swimming of Juvenile Sea Scallops, *Placopecten magellanicus* (Gmelin): A Minimum Size for Effective Swimming? *J. Exp. Mar. Biol. Ecol.* **1993**,
174, 137–175. DOI:10.1016/0022-0981(93)90015-G.

Marsh, R. L.; Olson, J. M.; Guzik, S. K. Mechanical Performance of Scallop Adductor Muscle
During Swimming. *Nature* **1992**, *357*, 411–413. DOI:10.1038/357411a0. PMID:1594046.

Millman, B. M.; Bennett, P. M. Structure of the Cross-striated Adductor Muscle of the Scallop.
J. Mol. Biol. **1976**, *103*, 439–467. DOI:10.1016/0022-2836(76)90212-6. PMID:940156.

Millward, A.; Whyte, M. A. The Hydrodynamic Characteristics of Six Scallops of
the Superfamily Pectinacea, Class Bivalvia. *J. Zool. (Lond.)* **1992**, *227*, 547–566.
DOI:10.1111/j.1469-7998.1992.tb04415.x.

Minchin, D. Introductions: Some Biological and Ecological Characteristics of Scallops.
Aquat. Liv. Res. **2003**, *16*, 521–532. DOI:10.1016/j.aqliv.2003.07.004.

Morton, B. Swimming in *Amusium pleuronectes* (Bivalvia: Pectinidae). *J. Zool. (Lond.)*
1980, *190*, 375–404.

Nadeau, M.; Barbeau, M. A.; Brêthes, J.-C. Behavioural Mechanisms of Sea Stars (*Asterias vulgaris* Verrill and *Leptasterias polaris* Müller) and Crabs (*Cancer irroratus* Say and
Hyas areaneus Linneaus) Preying Upon Juvenile Sea Scallops (*Placopecten magellanicus*
(Gmelin)), and Procedural Effects of Scallop Tethering. *J. Exp. Mar. Biol. Ecol.* **2009**, *374*,
134–143. DOI:10.1016/j.jembe.2009.04.014.

Naidu, K. S.; Meron, S. *Predation of Scallops by American Plaice and Yellowtail Flounder*. Canadian Atlantic Fisheries Scientific Advisory Committee, Research Document 86/62. Fisheries and Ocean Canada: Ottawa, ON, 1986.

Naidu, K. S.; Robert, G. Fisheries Sea Scallop, *Placopecten magellanicus*. In *Scallops: Biology, Ecology and Aquaculture*; Shumway, S. E., Parsons, G. J., Eds.; Elsevier B.V.: Amsterdam, 2006; pp 869–905.

Nunzi, M. G.; Franzini-Armstrong, C. The Structure of Smooth and Striated Portions of the Adductor Muscle of the Valves in a Scallop. *J. Ultrastruc. Res.* **1981**, *76*, 134–148. DOI:10.1016/S0022-5320(81)80012-3.

Olson, J. M.; Marsh, R. L. Contractile Properties of the Striated Adductor Muscle in the Bay Scallop *Argopecten irradians* at Several Temperatures. *J. Exp. Biol.* **1993**, *176*, 175–193. PMID:8478601.

Peck, L. S., Webb, K. S.; Bailey, D. M. Extreme Sensitivity of Biological Function to Temperature in Antarctic Marine Species. *Funct. Ecol.* **2004**, *18*, 625–630. DOI:10.1111/j.0269-8463.2004.00903.x.

Pérez, H. M.; Janssoone, X.; Guderley, H. Tonic Contractions Allow Metabolic Recuperation of the Adductor Muscle During Escape Responses of Giant Scallop *Placopecten magellanicus*. *J. Exp. Mar. Biol. Ecol.* **2008a**, *360*, 78–84. DOI: 10.1016/j.jembe.2008.04.006.

Pérez, H. M.; Janssoone, X.; Nadeau, M.; Guderley, H. Force Production During Escape Responses by *Placopecten magellanicus* is a Sensitive Indicator of Handling Stress: Comparison with Adductor Muscle Adenylate Energy Charge and Phosphoarginine Levels. *Aquaculture* **2008b**, *282*, 142–146. DOI:10.1016/j.aquaculture.2008.07.016.

Pérez, H. M.; Janssoone, X.; Côté, C.; Guderley, H. Comparison Between *In Vivo* Force Recordings During Escape Responses and *In Vitro* Contractile Capacities in the Sea Scallop, *Placopecten magellanicus*. J. Shellfish Res. **2009a**, *28*, 491–495. DOI:10.2983/035.028.0310.

Pérez, H. M.; Brokordt, K. B.; Martinez, G.; Guderley, H. Locomotion Versus Spawning: Escape Responses During and After Spawning in the Scallop *Argopecten purpuratus*. *Mar. Biol. (Berl.)* **2009b**, *156*, 1585–1593. DOI:10.1007/s00227009-1194-1.

Philipp, E. E. R.; Schmidt, M.; Gsottbauer, C.; Sänger, A. M.; Abele, D. Size- and Age-dependent Changes in Adductor Muscle Swimming Physiology of the Scallop *Aequipecten opercularis*. *J. Exp. Biol.* **2008**, *211*, 2492–2501. DOI:10.1242/jeb.015966. PMID:18626084.

Rall, J. A. Mechanics and Energetics of Contraction in Striated Muscle of the Sea Scallop, *Placopecten magellanicus*. *J. Physiol. (Lond.)* **1981**, *321*, 287–295. PMID: 6978395.

Schmidt, M.; Philipp, E. E. R.; Abele, D. Size and Age-dependent Changes of Escape Response to Predator Attack in the Queen Scallop, *Aequipecten opercularis*. *Mar. Biol. Res.* **2008**, *4*, 442–450. DOI:10.1080/17451000802270346.

Soemodihardjo, S. Aspect of the Biology of *Chlamys opercularis* (L.) (Bivalvia) with Comparative Notes on Four Allied Species. Ph.D. Thesis, University of Liverpool: Liverpool, 1974.

Thayer, C. W. Adaptive Features of Swimming Monomyarian Bivalves (Mollusca). *Form Func.* **1972**, *5*, 1–32.

Thomas, G. E.; Gruffydd, L. D. The Types of Escape Reactions Elicited in the Scallop *Pecten maximus* by Selected Sea-star Species. *Mar. Biol. (NY)* **1971**, *10*, 87–93. DOI:10.1007/BF02026771.

Thompson, R. J.; Livingstone, D. R.; de Zwaan, A. Physiological and Biochemical Aspects of Valve Snap and Valve Closure Responses in the Giant Scallop *Placopecten magellanicus*. I. Physiology. *J. Comp. Physiol.* **1980**, *137*(2), 97–104. DOI:10.1007/BF00689207.

Thorburn, I. W.; Gruffydd, L. D. Studies of the Behaviour of the Scallop *Chlamys opercularis* (L.) and Its Shell in Flowing Sea Water. *J. Mar. Biol. Assoc. U.K.* **1979,** *59,* 1003–1023. DOI:10.1017/S0025315400036997.

Tremblay, I.; Guderley, H. E.; Fréchette, M. Swimming Performance, Metabolic Rates, and their Correlates in the Iceland Scallop, *Chlamys islandica*. *Physiol. Biochem. Zool.* **2006,** *79,* 1046–1057. DOI:10.1086/507780. PMID:17041870.

Tremblay, I.; Guderley, H. E.; Himmelman, J. H. Swimming Away or Clamming Up: the Use of Phasic and Tonic Adductor Muscles during Escape Responses Varies with Shell Morphology in Scallops. *J. Exp. Biol.* **2012,** *215,* 4131–4143. DOI:10.1242/jeb.075986. PMID:22972884.

Trueman, E. R. Observations on Certain Mechanical Properties of the Ligament of *Pecten*. *J. Exp. Biol.* **1953,** *30*(4), 453–467.

Watabe, S.; Hartshorne, D. J. Mini-review: Paramyosin and the Catch Mechanism. *Comp. Biochem. Physiol., B* **1990,** *96,* 639–646.

Wilkens, L. A. Neurobiology and Behaviour of the Scallop. In *Scallops: Biology, Ecology and Aquaculture*; Shumway, S. E., Parsons, G. J., Eds.; Elsevier B.V.: Amsterdam, 2006; pp 317–356.

Wong, M. C.; Barbeau, M. A. Effects of Substrate on Interactions Between Juvenile Sea Scallops (*Placopecten magellanicus* Gmelin) and Predatory Sea Stars (*Asterias vulgaris* Verrill) and Rock Crabs (*Cancer irroratus* Say). *J. Exp. Mar. Biol. Ecol.* **2003,** *287,* 155–178. DOI:10.1016/S0022-0981(02)00551-8.

Yonge, C. M. The Evolution of the Swimming Habit in the Lamellibranchia. *Mem. Mus. R. Hist. Nat. Bel. Ser. II* **1936,** *3,* 77–100.

Yonge, C. M. Observations on *Innites multirugosa* (Gale). *Univ. Calif. Publ. Zool.* **1951,** *55,* 409–419.

CHAPTER 7

LOCOMOTION OF COLEOID CEPHALOPODS

JEAN ALUPAY[1*] and JENNIFER MATHER[2]

[1]*Department of Linguistics, University of Southern California, Los Angeles, CA, USA. E-mail: alupay@usc.edu*

[2]*Department of Psychology, University of Lethbridge, Lethbridge, AB, Canada T1K 6T5*

Corresponding author.

CONTENTS

7.1   INTRODUCTION

Locomotion can be broadly defined as movement through one's environment (DeMont et al., 2005; Dickinson et al., 2000). It is important for accomplishing various goals such as finding food and mates, escaping predators, and migrating to find resources. From a physical point of view, locomotion can simply be thought of as a force exerted by an organism on its environment that, in obedience to Newton's laws, produces movement of the organism in the opposite direction. However, we see that the distribution of forces is anything but simple when we compare the diversity of locomotor abilities exhibited by animals (both vertebrates and invertebrates) in a suite of environments (e.g., water, substrate, and air). Locomotor design can be a key driver in the evolutionary history of animals that rely on movement. It can produce changes in morphology and physiology that affect their speed, acceleration, efficiency, endurance, and agility when moving in different environments (Dickinson et al., 2000). This is particularly true for the group of animals that will be discussed in this chapter, the cephalopods, which exhibit a wide array of locomotor types in their mostly water environment (there is one case of air-borne locomotion).

Cephalopods are a bilaterally symmetrical class of invertebrates within the phylum Mollusca that are often considered different from the rest of the group in many ways. They include two extant groups, the small one of nautiloids, which has an external shell, and the larger one of coleoids, which does not. The coleoids, which will be the focus of this chapter, include cuttlefish, squid, octopods (both the cirrates which have fins and the incirrates which do not), and the vampire squid (see Fig. 7.1). Some basic molluscan features that are expressed differently in the cephalopods include the muscular foot, the protective shell, and the mantle cavity (Trueman & Clarke, 1988). These are embodied in molluscan *bauplans* ("a combination of the most significant features of the phylum"; Haszprunar & Wanninger, 2012). The most typical *bauplan*, though others have been proposed (see "urmollusc" in Haszprunar & Wanninger, 2012; see Fig. 2 in Haszprunar, 1992), is the Hypothetical Ancestral Mollusc, or the HAM. The HAM most resembles gastropods with a shell, head, foot, mantle cavity, and visceral mass. Cephalopods have undergone changes in all three features: modifying the foot into funnel and arms, internalizing, reducing, or losing the shell entirely, and co-opting the mantle–funnel for movement (i.e., jet propulsion) as well as respiration of a closed circulatory system (Clarke & Trueman, 1988; Wells, 1994) (see anatomy by group in Figs. 2.1–2.3). These modifications produced fast moving, high metabolic rate animals from a slow moving, low metabolic

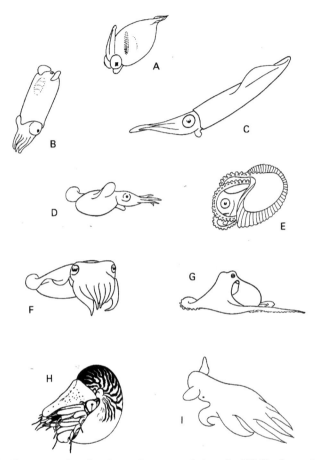

FIGURE 7.1 Representative drawings of extant cephalopods. (A) *Teuthowenia*, a teuthoid squid of the family Cranchiidae, hovering in mid-water with raised arms and tentacles. (Drawn after a photograph of Vecchione and Roper, 1991.) (B) *Spirula*, a sepiolid that can achieve vertical migrations over several hundred meters with a calcified, coiled, and chambered inner shell. (C) A loliginid squid hovering with dynamic lift provided by undulating fin movements and funnel jets. (D) A bobtail squid of the family Sepiolidae with a very small uncalcified internal shell swimming with undulating fins. (E) A female *Argonauta*, a pelagic incirrate octopod, with a calcitic pseudoconch used to brood eggs and maintain buoyancy. (F) *Sepia*, a cuttlefish with a calcified chambered shell and undulating fins for maneuvering. (G) A benthic incirrate octopus of the family Octopodidae crawling. (H) *Nautilus* with a coiled external, heavily calcified chambered shell giving the animal neutral buoyancy. (Drawn after a photograph published by the Japanese research program JECOL, 1977.) (I) A finned cirrate octopod moving with a partly retracted web. (Drawn after a video recording of the French research program CALSUB, 1989.) (From Budelmann, B. U.; Schipp, R.; Boletzky, S. von Cephalopoda. In *Microscopic Anatomy of Invertebrates*; Harrison, F. W., Kohn, A., Eds.; Wiley-Liss: New York, 1997; Vol. 6A, Mollusc II, pp. 119–414. Copyright © (2015) by John Wiley & Sons, Inc. Reprinted by permission of John Wiley & Sons, Inc.)

rate molluscan ancestor. They also had implications for other features such as increased brain and sensory development, appearance and use of fins in all groups but the incirrate octopods, a muscle-based structural support system, and an efficient fuel delivery system, resulting in more effective locomotor activities (Wells, 1994).

Unlike their molluscan relatives that rely primarily on their muscular foot to move in their environment (Trueman, 1983), cephalopods utilize different muscle groups, sometimes in concert, to achieve different modes of loco-motion adapted to habitat differences. Most cephalopods use their mantle–funnel system to swim via jet propulsion in the open ocean. Cuttlefish and squid can also hover with the assistance of undulating fins (Clarke, 1988; Wells & O'Dor, 1991). In some cases, swimming is exclusively achieved by the fins, as in deep sea, low-metabolizing cirrate octopods, like *Cirrothauma* and *Grimpoteuthis* (Aldred et al., 1983; Clarke, 1988; Collins & Villaneuva, 2006; Vecchione, 1995;; Vecchione & Roper, 1991; Vecchione & Young, 1997). Some cirrate octopods like *Opisthoteuthis* and *Stauroteuthis* also use the web between their arms to eject water for medusoid swimming (Collins & Villaneuva, 2006; Vecchione & Roper, 1991; Vecchione & Young, 1997). *Vampyroteuthis infernalis* undergoes a "gait-transition" during development, using more medusoid swimming when it is juvenile and more fin swimming when it is an adult (Seibel et al., 1998). Some benthic cephalopods will also bury themselves using their arms, in the case of octopods, or mantle–funnel jets as seen in cuttlefish and bobtail squids (sepiolids) (Anderson et al., 2004; Boletzky, 1996; Hochberg et al., 2006; Mather, 1986). Octopods in both deep and shallow water have the ability, expressed in their predominantly benthic lifestyle, to crawl using a combination of their arms and suckers to push and pull against a substrate (Huffard, 2006; Mather, 1998; Villanueva et al., 1997a). An important influence on locomotion in the water that all cepha-lopods, nautiloids included, deal with is buoyancy. Although locomotion of nautlioids will not be discussed here, a detailed comprehensive literature is available (Baldwin, 2010; Chamberlain et al., 2010; Chamberlain, 1988, 1991; Jordan et al., 1988; Jacobs & Landman, 1993; Neumeister & Budel-mann, 1997; O'Dor et al., 1990; Packard et al., 1980; Redmond et al., 1978). The trend for fast locomotion in the coleoid cephalopods is hypothesized to be an evolutionary response to the diversification of predatory paired-fin fish with their superior swimming abilities (Packard, 1972). The locomotion and subsequent design of fossil cephalopods, from which present ones evolved, are not covered in this chapter, but some key references are highlighted here (Chamberlain, 1980, 1981; Doguzhaeva et al., 1996; Ebel, 1999; Engeser,

1988; Jacobs et al., 1996; Jacobs, 1992; Klug & Korn, 2004; Kröger et al., 2011; Kröger, 2002; Ritterbush et al., 2014).

Muscles are a major driver of locomotion in Molluscs. Three major systems have been classified: The body wall consisting of circular, medial, diagonal, and longitudinal muscles; a pair of longitudinal muscles on either side of the foot; and the major dorsoventral muscles that move the entire foot (Haszprunar, 1992). Molluscs exhibit two types of skeletal structures, a hard shell or a hydrostatic skeleton. Lacking a hard exoskeleton, coleoids depend on hydrostatic skeletons for both structural support and movement. Hydrostatic skeletons can generate movement either by 1) muscles contracting around a fluid-filled (usually blood) body cavity held at constant volume, resulting in movement of the fluid to another region of the body, or by 2) muscle contractions antagonizing a three-dimensional arrangement of muscle fibers to produce actions (Kier, 1988). Coleoid cephalopods are primary examples of organisms that use muscular hydrostats to achieve movement. Examples of such hydrostats include the mantle–funnel complex in all coleoids to produce jet propulsion and respiration; lateral fins of squid used for swimming; squid arms, tentacles, and octopod arms that allow for extension and mobility, cephalopod suckers, and the tentacles of *Nautilus* (Kier, 1988). It is important to note that although the basic hydrostatic skeletons were thought to be unique generators of movement, many examples in the Mollusca show that muscular hydrostats are widely used. This includes columellar muscles in some gastropods that allow them to extend from the shell, tightly packed muscles in the foot of limpets and chitons that produce crawling waves, the siphons of some bivalves used in deposit feeding, and the tentacles of gastropods, such as abalone (*Haliotis tuberculata*), for sensory perception (Kier, 1988).

Many recent advances in the physiology of cephalopods, including neuronal (Bartol et al., 2008, 2009; Burford et al., 2014; Flash & Hochner, 2005; Hochner, 2012; Laan et al., 2014; Sumbre et al., 2001), developmental (Poirier et al., 2004; Robin et al., 2014; Seibel et al., 1998; Thompson & Kier, 2006; Villanueva et al., 1997b), and genetic (Albertin et al., 2012; Navet et al., 2010), have opened the field for studying many other aspects of locomotion. More information about cephalopod navigation (Alves et al., 2008; Jozet-Alves et al., 2014) and migration, both horizontal (across bodies of water) and vertical (up and down the water column) (Gilly, 2006; Hoving et al., 2014; Rosa & Seibel, 2010b), have led to better understanding of how locomotion affects foraging, reproduction, and other behaviors. Recent field observations of octopods have resulted in more knowledge about flexibility in the muscular hydrostat of the arm (Huffard et al., 2005; Huffard, 2006). In particular, more work in how multiple degrees of freedom give rise to multiple gait patterns is being done (Hochner, 2012; 2013; Huffard et al., 2005; Levy et al., 2015). Better techniques to study cephalopod morphology for taxonomic as well as potential mechanistic purposes are being used (Margheri et al., 2009; Xavier et al., 2014). All these aspects have resulted in a better understanding of cephalopod locomotion and have allowed others to take advantage of that knowledge to develop the growing field of soft robotics. This includes investigations in kinematics (Crimaldi et al., 2002; Kang et al., 2011; Kier & Leeuwen, 1997; Yekutieli, Sagiv-Zohar, Aharonov, Engel, Hochner, & Flash, 2005; Yekutieli, Sagiv-Zohar, Hochner, & Flash, 2005; Zelman et al., 2013), biomaterials (Hou et al., 2011, 2012), sensorimotor control (Flash et al., 2012; Li et al., 2012; McMahan & Jones, 2011; Sfakiotakis et al., 2013b), and dynamical modeling (Calisti et al., 2011, 2012, 2014; Kang et al., 2012; Laschi et al., 2009; Mazzolai et al., 2007; Renda et al., 2014; Sfakiotakis et al., 2013a; Zheng et al., 2013), particularly in octopods though some in cuttlefish (Wang et al., 2008; Willy & Low, 2005).

Cephalopods move differently from most other molluscs, though the muscles that are involved in these movements are essentially the same as those found in other members of the phylum. The goal of this chapter is to provide a comprehensive understanding of the cephalopod muscular hydrostat and its production of different types of locomotion as well as the structures used in the physiological processes of this system.

7.2   FOUNDATIONS

Cephalopods are highly mobile animals that rely on a combination of systems, namely their structural (muscles and connective tissues), nervous, respiratory, and circulatory ones, to adapt to their active lifestyle. The following sections will highlight the basic components of these systems—the muscles, connective tissue, and nervous system—and go over how they work in concert to result in movement. Muscles and muscular structures are the primary effectors of locomotion in cephalopods and will be featured in most of the underlying discussion. There are four major muscular structures of interest, and we will discuss how the physiology of each one is arranged. More details have become available as a result of new technology and the innovative use of tools to visualize many of these structures *in vivo* (King et al., 2005; Margheri et al., 2010).

7.2.2   BASIC STRUCTURES

7.2.2.1   MUSCLES

All forms of locomotion in cephalopods are powered by muscle. The muscle can be studied on multiple levels and the organization within each level can affect the speed and distance of movement. The levels on which this section focuses are the arrangement of muscle-cell components (the ultrastructure), the orientation of muscle fibers (gross arrangement), and the grouping of these fibers into whole muscle structures. The protein filaments that make up the contractile elements of muscle are key to the ultrastructure of muscle cells. These filaments shorten and generate force for movement (DeMont et al., 2005). Most invertebrates have two kinds of filament arrangements, cross-striated or obliquely striated muscles (DeMont et al., 2005). Cross-striated muscles are made up of protein filaments that are highly organized in bands that run transversely. Obliquely striated muscles have protein filaments that are organized in bands that are staggered and at an angle relative to the long axis of the fiber. Like that of most molluscs, cephalopod musculature is dominated by obliquely striated muscles (Kier, 1985). Cross-striated muscles have been identified in the transverse and circular muscles of squid tentacles, the appendages they use for prey capture. The arrangement and dimensions of these protein filaments are key to their differential functions. Protein filaments of cross-striated fibers are much shorter (0.5–0.9 µm) than the protein filaments from obliquely striated fibers (2.8 µm or longer), thus contractions of the cross-striated fibers in the tentacle lead to faster shortening speeds, while obliquely striated muscle fibers in the arm are able to exert more force over a longer distance for bending movements (Kier, 1985, 1991). However, recent studies of obliquely striated fibers in the mantle of the squid indicate that there may be more variation in the function of these muscles depending on the species and even on their structure (Thompson et al., 2014).

At the level of gross arrangement, muscle fibers are tightly packed in bundles and arranged in mutually perpendicular directions (Kier, 1988, 1991; Trueman, 1983). In a basic muscular hydrostat that assumes a cylindrical or tubular shape, the fibers are oriented such that they are circular, transverse, or radial to longitudinal muscle fibers that are parallel to the body axis. Additionally, there are oblique fibers that are oriented in a spiral around the body (Kier & Thompson, 2003). When the muscle fibers of one orientation contract, they are antagonized by fibers oriented in the other directions. Many of the structures to be discussed in cephalopods are more complex shapes than the simple cylinder, but the same muscular hydrostat principles apply.

Bundles of differently oriented muscle fibers are grouped together to form overall muscle structures. These structures include the mantle/funnel (see Fig. 7.2), head/eyes, arms, tentacles, and fins (Boyle & Rodhouse, 2005a). These are surrounded by the skin, which contain structures important for color production (Hanlon & Messenger, 1996). All of these are known to be controlled from centers in the suboesophageal brain. Of the structures listed, the mantle (Ward & Wainwright, 1972), fins of cuttlefish and squid (Kier, 1989), arms and tentacles of squid and octopods, and their associated suckers (Kier, 1985; Kier & Stella, 2007; Van Leeuwen & Kier, 1997) are examples of the muscular hydrostats that are essential to cephalopod locomotion. Most recently, the skin was also identified to be a muscular hydrostat (Allen et al., 2013), but its role is primarily in body patterning and so it will not be discussed further.

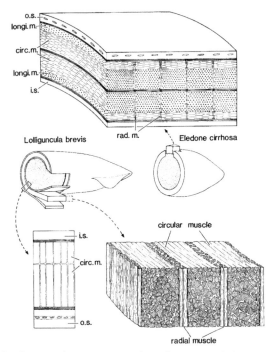

FIGURE 7.2 Mantle musculature cross section of octopus *Eledone cirrhosa* (above) and squid *Lolliguncula brevis* (below). Connective tissue divides layers of circular muscle (circ. m.) and makes up the surrounding tunic. Found in both species: outer skin (o. s.), longitudinal muscles (longi. m.), i. s. (inner skin), rad. m. (radial muscles). (Reprinted from Boyle, P. R. Neural Control of Cephalopod Behavior. In *The Mollusca: Neurobiology and Behavior Part 2*; Wilbur, K. M., Willows, A. O. D., Eds.; Academic Press: London, UK, 1986; Vol 9, pp 1–99. Copyright © (2015), with permission from Elsevier.)

One muscular structure that is physiologically important to locomotion is the heart. To accommodate their fast lifestyle, cephalopods have three hearts, one muscular systemic heart and two branchial hearts, one at the base of each of the two well-vascularized gills in the mantle cavity of dibranchiate coleoids. These hearts drive circulation and pump oxygenated blood through their closed circulatory system. The systemic heart of the octopod *Eledone cirrhosa* is composed of four layers of muscle that are cross-striated (Kier, 1985; Wells, 1983), exhibiting more musculature than the branchial hearts. Contraction of the systemic heart produces high pressure to circulate blood through a network of variously sized vessels. The branchial hearts are composed of only two thin outer layers of striated muscle. The layers run perpendicular to each other with the inside layer parallel to the long axis (Wells, 1983). The branchial hearts on either side of each gill are recruited to maintain circulation of well-oxygenated blood during locomotor activities (Wells, 1983, 1992).

7.2.2.2   CONNECTIVE TISSUES

Connective tissues are important structural components that resist changes in length from muscular contractions. These fibrous tissues also provide an important role in locomotion as they "transmit the force of muscular contraction, control shape change, and store elastic energy to reduce costs of locomotion, movement, or adhesion" (Kier & Thompson, 2003). Shape changes in particular are influenced by the arrangements of cross-linking fibers. These fibers wrap around the body in helices at an angle relative to the long axis, allowing for elongation and shortening (Kier, 2012). Collagen is one of the tensile fibers which resist extension in the direction of the fiber when stretched, but will readily deform when compressed by forces perpendicular to the fibers orientation (Wainwright et al., 1982). Collagen is present in layers of helically arranged fibers around the mantle of cephalopods. This arrangement resists increase in length along the mantle but provides more expansion in width (Wainwright et al., 1982).

Connective tissue fibers can also be embedded within layers of muscle fiber, as seen in the squid mantle and fins (Kier, 1992). Their flexible and resistant properties help restore contracted muscle fibers to their original length. Cartilage is a pliant material made up of flexible amorphous polymers that resist compression and bending forces to act like a rigid structure (Wainwright et al., 1982). It was thought to be vertebrate specific, but similar forms were found in the mouth parts of various molluscs to help control the

radula (Bairati, 1985) as well as in the head of octopods to provide structural support (Bairati et al., 1995). Cephalopods are the only invertebrates
that have cartilage similar to the transparent cartilage found in vertebrates,
known as hyaline (Cole & Hall, 2009). Roper and Lu (1990) systematically
described the presence of cartilage and other connective tissues in oceanic
squids (see Table 1 summary in Roper & Lu, 1990). In coleoids, the function
of cartilage can vary from buoyancy (as in *Cranchia scabra* and *Galiteu-
this glachialis*) to a pseudoskeleton to support muscles (e.g., *Histioteuthis
meleagroteuthis* and *Liocranchia reinhardti*) and reducing drag (like *Tetro-
nychoteuthis massyae*) (Roper & Lu, 1990). Cartilage is also important as
an attachment site for major muscle groups and develops early in cuttlefish
and other coleoid hatchlings with a benthic life history (Cole & Hall, 2009).

7.2.2.3 NERVOUS SYSTEM

The cephalopod nervous system is the most developed and complex one in
the molluscs. This complexity results partly because different parts together
control movement of the cephalopod body. The parts to consider are the
central brain organized in lobes, the peripheral nervous system arranged in
localized ganglia, sensory receptors and organs that gather external information for the central and peripheral systems and neurotransmitters which excite
various nerve endings (Boyle & Rodhouse, 2005a). This section will provide
an overview of these major components, but their detailed mechanisms for
achieving different modes of locomotion will be discussed in subsequent
sections. A summary table of all components of the nervous system based on
the octopus can be found in Table III of Boyle (1986) and detailed diagrams
can be seen in Young (1988) and Budelmann (1995a).

7.2.2.3.1   Central Brain

The coleoid central brain is composed of masses of nerve cells that are organized into lobes. It consists of a cerebral, brachial, pedal, and paired optic
lobes, and it is situated between the eyes and around the esophagus (Boyle,
1986; Hanlon & Messenger, 1996). Sets of lobes above (supraesophageal) and below (subesophageal) the esophagus are central to the brain and
connected by the magnocellular lobes. All of these lobes control different

systemic functions that are discussed in greater detail in the following literature: Boyle (1986); Nixon and Young (2003); in cuttlefish (*Sepia*) Boycott
(1961); in squid (*Loligo*) Young (1974, 1976, 1977, 1979); Messenger
(1979); and in *Octopus* Young (1971). They are not limited to a single function as many connections between multiple lobes allow for coordination.

Several lobes in the sub- and supraesophageal mass are important in
the control of locomotion. The subesophageal mass consists of a pair of
magnocellular lobes between the cerebral and optic lobes dorsally and the
pedal and pallioviseral lobes ventrally. The anterior lobe of the middle
subesophageal area coordinates the arms. Different parts of the pedal lobe
control eye movement, attack initiation, and funnel movement. The pallio-
visceral and magnocellular lobes help control the mantle in its respiratory
and locomotor functions, particularly the fast escape jet. All coleoid cephalopods except the incirrate octopods (those that do not have fins on their
mantle) also have fin lobes connected to the pallioviseral and pedal lobes,
which all together help coordinate fin movement. In octopus, the pedal and
magnocellular lobes are also involved in motor control of the arms (Budelmann, 1995a; Young, 1988).

The supraesophageal mass consists of the basal lobes and the frontal/
vertical lobe system that are involved in numerous interdependent functions. The basal lobes, that include higher motor centers like the peduncle
lobe, receive sensory information and regulate output motor activity. This
includes motor control of swimming, respiration, and muscles in the skin
(Budelmann, 1995a). The peduncle lobe in *Octopus* is part of the highly
interconnected visual-motor system which includes the eyes, optic lobe, and
basal lobes (Messenger, 1967a, 1967b). Visual cues guide many octopus
movement patterns and activities (Gutnick et al., 2011). The peduncle lobe
receives the visual information from the optic lobe and uses it to regulate
motor actions (Messenger, 1967b). The frontal and vertical lobes are particularly important for vision in squid and cuttlefish (Young, 1988). The frontal
and buccal lobes are also concerned with chemotactile information from the
arms of octopods (Young, 1983). Chemoreception also involves feedback
between the olfactory and basal lobes (Messenger, 1971). The suckers and
olfactory pit (an organ close to the eye) of coleoid cephalopods are lined
with ciliated chemoreceptors. Octopus suckers have nearly 100 times more
chemoreceptors than cuttlefish suckers, which is likely a result of the octopuses' benthic lifestyle, exploring the bottom and requiring extraction of
chemical cues from the environment (Budelmann, 1995b). These senses

provide input to the nervous system that can result in motor activity (Budel-mann, 1995a; Young, 1988). The central nervous system lobes and their functional organization are summarized in Table 1 of Wild et al. (2014).

7.2.2.3.2 *Peripheral Nervous System*

The brain and lobes of the central nervous system initiate motor patterns which are carried out locally by the peripheral nervous system (Grasso, 2014). This system is made up of masses of nerve cells called ganglia that provide local control to the organs they are associated with (Boyle & Rodhouse, 2005a). The main ganglia are gastric, cardiac, branchial, buccal, and sub-radial (see Fig. 7.3A). Of particular interest to movement are the stellate ganglia, which control mantle contractions in all coleoid cephalo-pods (Young, 1972). The stellate ganglia are divided into two parts, ventral and dorsal, which control muscles in the mantle important for respiration and locomotion (Young, 1972). In *Octopus*, the interbrachial commissure connecting the eight arms in a ring and the nerve cords (brachial ganglia and sucker ganglia) that extend in each arm (Grasso, 2014) are key to coordina-tion of arm movements. The mechanisms by which these ganglia control movement will be discussed in the following sections on the individual muscular structures.

The giant fiber system of squid is important for fast commands to the mantle musculature. Motor control is provided to the stellate ganglion by the brain through connections with the pallial nerve. The axons of motor cells fuse and create 1-mm diameter fibers (Young, 1938, 1939). The fiber does not need a myelinated sheath because the large diameter of the giant fiber itself results in fast transmission of nerve impulses. The giant fiber system is important for initiation of the escape-jet in squid and cuttle-fish. The giant fiber is not found in octopuses, however, conduction speed to different areas in the body depend on the network of fibers (diameter and number) in the peripheral nervous system. Neurons in the arms are controlled directly from the brain by few but large fibers, whereas there are many small connections to the localized brachial and sucker ganglia (Young, 1965) (see Fig. 7.3B).

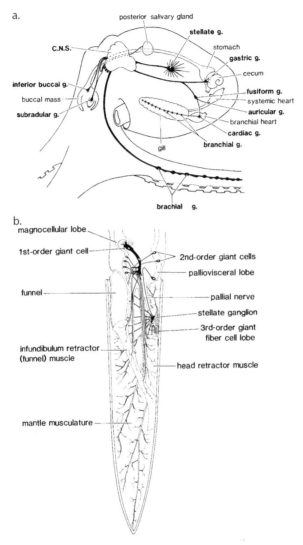

FIGURE 7.3 Illustrations of the central nervous system of two cephalopods. (A) The major ganglionic (g.) masses of the octopus *Eledone cirrhosa* in bold, and the rest of the central nervous system (CNS). Internal structures surrounding the ganglia are also listed in non-bold. (B) The giant nerve fiber system of the squid *Loligo pealei* including the three orders of giant fiber cells and various lobes, nerves, and muscles. (Reprinted from Boyle, P. R. Neural Control of Cephalopod Behavior. In *The Mollusca: Neurobiology and Behavior Part 2*; Wilbur, K. M., Willows, A. O. D., Eds.; Academic Press: London, UK, 1986; Vol 9, pp 1–99. Copyright © (2015), with permission from Elsevier.)

7.2.2.3.3   Sensory Receptors and Organs

A summary of cephalopod sensory systems is beyond the scope of this chapter. However extensive reviews of the different receptors and sense organs can be found in the following sources: Graziadei and Gagne (1976), Williamson (1991), Budelmann (1995b), Hanlon and Messenger (1996). Sensory systems that are of particular importance to locomotion, besides the eyes of these highly visual animals, include mechano-, chemo-, and proprioceptors (those that monitor muscle stretch) in the arms and suckers, especially in *Octopus* (Graziadei, 1962, 1965; Graziadei & Gagne, 1976). These provide tactile, chemical, and positional feedback to the nervous system. The octopus mantle (Boyle, 1976) and the squid and cuttlefish dorsal neck (Budelmann, 1995b; Preuss & Budelmann, 1995) also containss proprioceptors that provide feedback after mantle contractions in the former and control head-to-body movements in the latter. Found in all groups of coleoid cephalopods, paired statocysts serve as the equilibrium receptor system. The organ is composed of polarized hair cells that detect mechanical disturbance in surrounding fluid. These hair cells are divided into two different receptor systems, the macula/statolith/statoconia system which detects gravity and the crista/cupula system which detects angular acceleration (Budelmann, 1990, 1995b; Williamson & Budelmann, 1985; Young, 1984, 1989).

7.2.2.3.4   Neurotransmitters

Communication between neurons in the central and peripheral nervous systems is provided by chemicals known as neurotransmitters. Messenger (1996) and Tansey (1979) published the most comprehensive reviews of neurotransmitters in cephalopods. Initially, four neurotransmitters—acetylcholine, dopamine, noradrenaline, and 5-hydroxytryptamine (or serotonin)—were found in the cephalopod brain (Tansey, 1979). Other pharmacologically active chemicals were also located in different parts of the body of cephalopods (Boyle, 1986). For example, L-glutamate is in the mantle and fins of three squid and cuttlefish species—*Loligo*, *Alloteuthis*, and *Sepia* (Bone & Howarth, 1980; Boyle, 1986) and acetylcholine innervate the nerves in the octopus arm (Nesher et al., 2012). Interestingly, a recent study by Nesher et al. (2014) identified the role chemicals play in the skin in motor control, preventing octopus arms from attaching to one's self. Messenger (1996) added to this many more classes of neuroactive substances, some of which are important to locomotor activity such as L-glutamate, the excitatory neurotransmitter for mantle contraction in the squid giant synapse. Since these reviews, little more has been found about the distribution and functions of these small molecules in the brain. Recently a protocol was published to detect the spatial distribution of different neuroactive substances in the cephalopod nervous system, the best studied so far being acetylcholine, dopamine, serotonin, GABA, and glutamate (Ponte & Fiorito, 2015).

7.2.3 *EFFECTOR UNITS*

Three key muscular hydrostats: the mantle–funnel complex, the fins, and the arms, including the web and suckers, are important for producing movement on a small scale as well as whole-body locomotion. All of these structures will be discussed in depth, including the various physiological processes involved in regulating each. Before that, it is important to understand how muscular hydrostats work, so a brief introduction to principles of hydrostatic skeletons will be given.

Hydrostats are generally composed of three main parts: a fluid, a cavity or container to hold the fluid, and muscles supported by connective tissue to change the shape of the container. Muscular hydrostats differ in that the fluid is replaced by dense muscle fibers oriented in different directions. The key principle that governs all hydrostats, whether based on muscles or on fluid, is that the volume of the container is virtually constant because muscles and fluid are fairly incompressible. When muscle fibers contract, they decrease the extension of the container in one direction. This results in increased pressure within the container and subsequently an increase in extension in another direction. This is the basis for movement in hydrostats (Kier, 1992, 2012; Kier & Smith, 1985).

In a basic muscular hydrostat that assumes a cylindrical shape, there are three muscle fiber orientations to the long axis of the body: perpendicular (arranged in a transverse, radial, or circular pattern), parallel (longitudinal bundles), and helical or oblique. Contraction in these different fiber orientations results in four types of movement. Elongation is achieved by decreasing the cross-sectional area and increasing the length of the body. This is done by contracting transverse, circular, or radial fibers, ones which are oriented perpendicular to the long axis. This activity is used for protrusion, such as in the quick elongation of squid tentacles to capture prey (Kier, 1985). Shortening is done by decreasing the length of the body with a resultant increase in the cross-sectional area. This is done by contracting longitudinal muscle fibers. This shows the antagonistic effects of longitudinal fibers to those perpendicularly oriented to them. The last two movement patterns are due to localized contractions that require resistance in remaining parts of the body. Bending is due to contraction of longitudinal muscles on one side of the body while the animal still maintains a constant diameter by contracting perpendicular muscles. With this mechanism, the body can vary in how much and in what direction it can bend. Lastly, torsion leads to twisting of the body itself along its long axis. This is accomplished by contraction of the helical or oblique muscles. Helical muscles are also involved in shortening and elongation, depending on the angle of arrangement of the fibers with respect to the long axis (Kier & Smith, 1985). Longitudinal and oblique muscle fibers are present in all cephalopod groups, transverse fibers are only found in octopus and squid arms and squid tentacles, and circular muscles are also found in squid tentacles (Kier, 2012). (See Table 1 in Kier, 2012 for more detailed representation of the diversity of muscular hydrostats.)

The following sections will discuss how these different muscle fiber orientations and principles of hydrostats are organized into different muscular structures. For each structure, arrangements of the muscles, connective tissues, and nervous system will be reviewed with respect to the modes of locomotion the structure performs. Table 7.1 summarizes the muscular structures that are most commonly used for locomotion for the major groups of cephalopods.

TABLE 7.1   Locomotor Effector Units used in Cephalopods.

| Group | Octopods | | Squid | Cuttlefish | Vampyromorpha |
	Cirrate	Incirrate			
Arms	xx	xx	x	x	x
Fins	x	–	xx	xx	xx
Tentacles	–	–	x	x	x
Web	xx	x	–	–	xx
Mantle–funnel	x	xx	xx	xx	x

x = occasional xx = common.

7.2.3.1   MANTLE–FUNNEL COMPLEX

The mantle–funnel complex of coleoids is essential for driving many physiological and locomotor activities. It is responsible for respiration and swimming by jet propulsion in all cephalopods, as well as digging or burying in bottom-dwelling cuttlefish (Boletzky, 1996; Mather, 1986). The structure of squid mantles have been studied the most, but variation between the different coleoid groups is minimal.

7.2.3.1.1 Muscle and Connective Tissue

In squid (*Loligo pealei* and *Lolliguncula brevis*), and cuttlefish, the mantle is divided into three layers, a central one of closely packed muscle sandwiched between two layers of connective tissue called the inner and outer tunics (see Fig. 7.2). Muscle fibers in the central layer are oriented circularly around the body as well as radially attaching between the inner and outer tunics. These two types are obliquely striated and arranged in alternating rings along the body. Connective tissues are present not only in the tunics but in the muscle layer also (intramuscular connective fibers). Three intramuscular connective tissue fibers are identified in cuttlefish and squid mantles (Ward & Wainwright, 1972). These are oriented in different directions, with one in the inner and outer tunics that is straight or curved, another that is also in the tunics localized to radial muscle bands, and a third that is parallel to the circular muscle fibers not attached to the tunics (Kier & Thompson, 2003). Collagenous fibers that are tightly cross-linked make up the bulk of the outer tunic (Ward & Wainwright, 1972). The squid mantle also contains a chitinous skeletal structure called the pen. The pen is rigid and resists lateral bending and lengthening of the mantle. The connective tissues, and to some degree the pen, help maintain the squid's shape. The tunics limit length changes and store elastic energy while providing increases in the width or circumference of the body (Ward & Wainwright, 1972).

In the incirrate octopuses (e.g., *Octopus vulgaris* and *E. cirrhosa*), there are two layers of circular muscles, divided by a layer of connective tissue, stellar nerves, and blood vessels. Unlike squid, octopods have two thin layers of continuous longitudinal muscle that bound tightly packed radial and circular muscles (Boyle, 1986; Boyle & Rodhouse, 2005a). A similar organization of mantle musculature is seen in the cirrate octopus species *Cirrothauma murrayi* (Aldred et al., 1983). Some incirrate species (e.g., *Octopus bimaculatus*) do not have the defined inner and outer tunic layers of connective tissue seen in squid. They have collagen fibers arranged in different orientations in a fibrous array surrounding the mantle. They also have connective tissue fibers in the longitudinal muscle layers, oriented parallel to the body axis (Kier & Thompson, 2003). Octopuses have far less mantle musculature than decapods.

A different arrangement in mantle musculature is present in gelatinous and deep water species of squids and octopuses. In these cephalopods, the gelatinous layer is surrounded by two thin layers of circular muscle fibers and a set of radial muscles (Kier & Thompson, 2003). Contraction of radial muscles thins the wall and subsequently expands the mantle (Clarke, 1988). Sheets of connective tissue reinforce the gelatinous layer. Some pelagic squids (Octopoteuthidae, Cycloteuthidae, and Lepidoteuthidae) have one layer of longitudinal muscle fibers, two layers of circular muscle fibers, and radial muscle fibers. These squid do not have well-defined inner and outer tunics of connective tissue. Some of the deep sea squids (Mastigoteuthidae, Chiroteuthidae, Histioteuthidae, and Batoteuthidae) have a similar muscular organization, except they lack longitudinal muscle fibers. These squid have well defined layers of connective tissue (Kier & Thompson, 2003). The reduction of dense musculature exhibited in the mantle of gelatinous species results in the mantle acting like a closed, fluid-filled hydrostatic skeleton rather than a muscular hydrostatic skeleton. The muscle fibers that surround the gelatinous fluid contract and create movement, similar to the hydrostatic skeleton of the gastropod foot. The arrangement of connective tissues in these species also results in the mantle storing less elastic energy than in the loliginid squid.

The funnel is an important component of jet propulsion, as it directs water flow from the mantle cavity. The funnel is highly muscularized with the ability to bend in any direction and to vary the opening of the tip to regulate jetting speed and acceleration. Its muscle fibers are obliquely striated. In squid, the muscle fibers are oriented in in three directions: longitudinal ones on the outer surface, circular ones in the majority of the funnel, and thin radial fibers that run across and attach to the inner and outer funnel walls. This is all enclosed by thin inner and outer connective tissue fiber layers (Kier & Thompson, 2003).

7.2.3.1.2   Nervous Control of Mantle–Funnel

Nervous control of mantle musculature involves the pallioviseral lobe of the brain, which sends messages through the pallial nerve to a pair of stellate ganglia, one on each side of the body on the inner mantle surface, branching into stellar nerves in the muscle (Young, 1972; Boyle, 1986). In squid, the tightly packed circular muscle layer contains a plane of branching stellar nerves that runs horizontal to this muscle layer (Ward & Wainwright, 1972). In octopus, the two layers of circular muscles are divided by branches of the stellar nerve and multipolar nerve cells. The multipolar nerve cells are thought to be receptors in the proprioceptive feedback system of the mantle mediating its contractions (Boyle, 1976).

In squid (*L. pealei*), the giant fiber system is important for control of escape jetting. This is done by an all-or-nothing contraction system, whereby the muscles, particularly the circular fibers, of the mantle and those holding the funnel and head contract at the same time to obtain maximum velocity of water expulsion (Young, 1938). The giant fiber system consists of three orders of giant fibers, increasing in diameter from the brain to the mantle muscles. The first order fibers come from the magnocellular lobe of the brain, the second order fibers innervate the funnel, head retractor muscles, and stellate ganglia, and the third order fibers, which are also the longest and exhibit the largest diameters, connect to mantle musculature (Boyle & Rodhouse, 2005a; Wells, 1988; Young, 1938). The arrangement of the giant fiber system is an important innovation that allows squid to produce a very quick reaction from a simple trigger, based on the gradation of fibers that increase the muscles response amplitude and synchronizes contraction of mantle musculature (Boyle, 1986; Budelmann, 1995b). Cephalopods also have a small fiber system that regulates respiration and slower movements (Young, 1938; Wilson, 1960). Graded contractions of the circular muscles in the mantle are produced by stimulation of smaller fibers in the stellar nerve of *L. pealei* (Young, 1938), *Octopus* (Wilson, 1960), and *Sepia* (Packard & Trueman, 1974) presumed for respiration.

7.2.3.1.3   *Biomechanics of Mantle–Funnel*

During both jet propulsion and respiration, water is drawn into the mantle cavity through a pair of inhalant openings and expelled out of a single exhalant funnel. To produce thrust, a large amount of water passes through the funnel at high speeds due to whole mantle contraction initiated in the posterior. This is different from respiration, where contractions are confined to the anterior half of the mantle so a small amount of water is passed along the gills at slow speeds to maximize oxygen extraction (Packard & Trueman, 1974; Wells, 1988). One way squid deal with these dual incompatible actions of the mantle is through differentiation of their musculature. Their mantle is composed of anaerobic and aerobic muscle types. The anaerobic type, used for fast locomotion, is made of protein filaments with few mitochondria. The aerobic muscle fiber type, used for respiration and steady swimming, is made up of fibers that have many mitochondria and a large supply of oxygenated blood (Bone et al., 1981; Wells, 1988). Also as a consequence of the different functions of the mantle, different groups of coleoids specialize in one of the functions. Squid rely on jet-propelled locomotion and thus their mantle has a more elongated streamlined form and their cavity accommodates larger volumes for greater range in oxygen extraction (Wells, 1990). Octopuses and the Sepioidea use other modes of locomotion more frequently than jet propulsion, and thus their mantle is smaller and more round, to specialize in respiration (Wells & O'Dor, 1991).

In squid (*Loligo opalescens*), three phases of muscular contraction occur during the escape jet (see Fig. 7.4). First, the radial muscles contract, thinning the mantle wall and resulting in hyperinflation of the cavity. Second, the circular muscles contract, decreasing the mantle diameter and increasing the wall thickness to power the jet. Third, the mantle refills and expands,

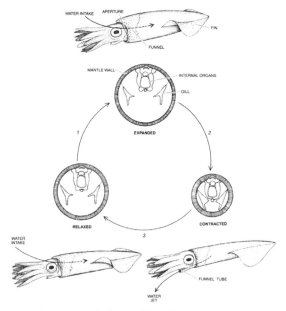

FIGURE 7.4 Escape-jet cycle of the squid *Loligo opalescens*. The cycle starts with hyperinflation of the mantle (1) where the diameter increases as water flows into the mantle cavity. After maximal expansion, the mantle muscles contract (2), pressure inside the mantle cavity increases and water is forced out the funnel, creating a jet. The mantle then refills to its relaxed state (3). (From Gosline, J. M.; DeMont, M. E. Jet-Propelled Swimming in Squids. *Scientific American.* 1985, *256*, pp. 96–103. With permission from illustrator Patricia J. Wynne.)

assisted by the stored elastic energy of the collagen fibers and slight contraction of the radial muscles (Gosline et al., 1983; Shadwick, 1994) (see Fig. 7.4). Two modes of respiration were seen in *L. opalescens*. The first depends on contractions of radial muscles antagonized by the elasticity of connective tissues. The second depends on contractions of the circular muscles, also antagonized by elasticity of the connective tissues. A procedure by which both muscle fibers contract and antagonize each other is possible, but is not observed frequently (Gosline et al., 1983).

Ventilation events in octopus and cuttlefish are similar to muscle actions produced during jet propulsion in squid. Detailed descriptions of water circulation patterns inside the mantle cavity are provided by Wells and Wells (1982) for *Sepia* and *Octopus*. During inhalation in octopuses (e.g., *O. vulgaris*), water is drawn into the mantle cavity as the dorsolateral mantle edge begins to thin out, the funnel tip folds up to close off water, and prebranchial spaces open up. Radial muscles in the rest of the mantle contract to expand the cavity briefly and draw in water which is then passed over the gills into the postbranchial cavity. During exhalation, water is expelled for a longer duration and out the funnel as a result of contraction of circular muscles in the lateral anterior mantle edge and closing of lateral flaps of the funnel over mantle edges (Wells & Smith, 1985; Wells & Wells, 1982). During inhalation in the cuttlefish, *Sepia officinalis*, the locking collar opens and the radial muscles contract to expand the mantle cavity and draw in water. Contraction of the collar flap muscles closes the mantle system and channels water out the funnel during exhalation. Elastic energy stored in the connective tissue network reestablishes the radial fibers to their initial state. Contraction of circular muscle fibers in high-pressure mantles is another mechanism used by some cephalopods during exhalation (Bone et al., 1994).

7.2.3.2   FINS

Cuttlefish and squid use their fins for various forms of swimming and hovering, as well as maneuvering their body in all directions (Bidder & Boycott, 1956; Hoar et al., 1994; Kier & Thompson, 2003; O'Dor & Webber, 1986; Russell & Steven, 1930). The fins of cirrate octopus and *Vampyroteuthis* are also important effectors for swimming movement, as these deep sea species do not rely only on rapid jet propulsion for locomotion (Collins & Villanueva, 2006).

7.2.3.2.1   Muscles and Connective Tissue

Cuttlefish (*S. officinalis*) and some squid (*Loligo forbesi* and *Sepioteuthis sepiodea*) have similar fin musculature arrangements (Kier & Thompson, 2003). The lateral fins of *S. officinalis* and *S. sepiodea* extend the entire length of the mantle, tapering away from the base where the fin is thickest. The fin of *L. forbesi* is similar, except that it does not extend the entire length of the mantle and is triangular in shape (Kier, 1989). The fins are composed of a dorsal and ventral portion of musculature separated by a median connective tissue layer (a fascia). Two other layers of connective tissue, the dorsal and ventral fascias, which are made up of a meshwork of obliquely oriented fibers, are present. Additional connective tissues in the fin include the flattened cartilage at the base of the fin and crossed connective tissue fibers embedded in the musculature. In both the dorsal and ventral portions of the fin, obliquely striated muscle fibers are oriented in three mutually perpendicular directions: transverse, dorsoventral, and longitudinal. Transverse muscle fibers orient parallel to the fin axis, running laterally from the base to the fin margin. Sheets of dorsoventral muscle fibers separate bundles of transverse muscle. These sheets originate at the median fascia and extend either dorsally or ventrally on each side of the median. In the dorsal portion of the fin, dorsoventral muscle fibers connect to the dorsal fascia and vice versa for the ventral portion of the fin. Layers of longitudinal muscles are oriented parallel and adjacent to the median fascia on both the ventral and dorsal sides. Only transverse muscle fibers in both the dorsal and ventral surfaces have a zone with larger mitochondrial cores. These zones are close to the dorsal and ventral fascia. These muscle fibers are similar to the aerobic muscle present in the mantle and suggest that muscle fibers in these zones have higher aerobic capacity than those in the rest of the fin (see Fig. 1 in Kier, 1989 and Fig. 4 in Kier & Thompson, 2003).

Swimming by fin movement is one of the key modes of locomotion among cirrate octopods and *Vampyroteuthis*. Similar to the lateral fins of squid, the fins of cirrate octopods like *Cirrothauma* are tapered from a dense muscular base supporting the fin toward a thinner leading edge that is more gelatinous and transparent. Transparent dorsal cartilage also provides support for the powerful movements of these fins, forming attachment points for fin musculature. A large fin lobe in the brain provides control of the fins (Aldred et al., 1983). Fin shape and size varies between the different groups of cirrate octopods (see Fig. 12.10 in Boyle and Rodhouse (2005b). *Cirroctopus, Cirroteuthis, Cirrothauma* have large fins; *Grimpoteuthis, Luteuthis, Stauroteuthis* have moderately sized fins; and *Cryptoteuthis* and *Opisthoteuthis* have small fins (Collins & Villaneuva, 2006). The cirrate octopod fin is separated into proximal and distal regions, both covered by a thin sheet of muscle. The core of the proximal region is occupied by flattened fin cartilage which provides skeletal support and attachment for muscle fibers. The cartilaginous core is surrounded by bundles of muscle fibers running parallel to the fin. The distal region has dorsal and ventral layers of transversely oriented muscle fibers that are similar to the squid and cuttlefish fins (Vecchione & Young, 1997).

Vampyroteuthis also uses swimming by its broad fins for locomotion. The fins attach to a broad gladius (similar to fin cartilage but chitinous in structure) by a well-developed anterior muscular band (Seibel et al., 1998). Like that of cirrate octopods, the proximal half of the fin contains a carti-laginous core. Early in development, *Vampyroteuthis* develops two pairs of fins. The first juvenile pair becomes resorbed and the second adult fin pair enlarges at the anterior region (Young & Vecchione, 1996).

7.2.3.2.2 Nervous Control of Fins

Little is known about the nervous control of fin movement in cephalo-pods. Electromyographic recordings indicate that gentle fin movements are produced by oxidative muscle fibers and supported by the crossed oblique connective tissues. Short bursts of vigorous fin movements are produced by the anaerobic muscle fibers which make up the bulk of the fin musculature (Kier et al., 1989). Fin nerves with both afferent and efferent connections are located along the plane of the median fascia (Kier et al., 1989). Mechanore-ceptors are also likely found in the fin, which provide position information that help control coordination of the fins (Kier & Thompson, 2003).

7.2.3.2.3 Biomechanics of Fins

Swimming and hovering in cuttlefish and squid is a result of undulatory waves of the fin. Undulatory waves are a result of sequential bending of the dorsal and ventral portions of the fin. Contraction of transverse muscle bundles on one side of the fin results in lateral compression toward that side (either dorsal or ventral). However, bending will only occur if it is resisted on the opposite side of the fin. This can be done by contraction of longitu-dinal muscles supported by fin cartilage to resist length change, contraction of dorsoventral muscles to resist increased thickness, or stored elastic support by intramuscular connective tissues. The mechanism employed depends on the type of movement produced. During hovering and gentle swimming, low-amplitude undulatory waves are produced by contraction of the layer of aerobic transverse muscle bundles and resisted by intramuscular connective tissue. During short bursts of vigorous movements, high-amplitude undulatory waves are produced by contraction of anaerobic transverse muscle bundles and resisted by contraction of dorsoventral muscles and sometimes longitudinal muscles on the opposite side of the fin. These vigorous movements are produced during prey capture, fin beating in agonistic encounters, and maneuvering (Johnsen & Kier, 1993; Kier, 1989; Kier & Thompson, 2003).

During fin swimming, the fins of cirrate octopods move symmetrically to produce backward motion of the body. Their fins move through a cycle of upstrokes and downstrokes, starting with the posterior margins of the fins pointed dorsally. The fins are then pushed ventrally during the downstroke and dorsally during the upstroke. Frequency of the cycle can vary from 4 to 30 strokes per minute (Collins & Villaneuva, 2006).

7.2.3.3   ARMS AND TENTACLES

The arms of octopods are able to do a variety of tasks involving a wide range of movements. Arms are particularly important to incirrate octopuses, which rely on crawling more than jet propulsion to move in their environment. Some species of benthic octopus also use arms for digging into the substrate. In squid and cuttlefish, the arms are used in postures as displays, swimming, and steering. The pair of tentacles in squid is specialized for quick elongation during prey capture, for example *L. pealei* extends its tentacles by 80% within 20–40 ms (Kier & Leeuwen, 1997). Muscular structures associated with these limbs, such as the suckers, their stalks and the interbrachial web between the arms, will also be discussed.

7.2.3.3.1   Muscles and Connective Tissue

All coleoid cephalopod arms consist of dense musculature that surrounds a central axial nerve cord and artery. The musculature is made up of obliquely striated fibers which are oriented in three directions, transverse muscles running perpendicular to the long axis of the arm, bundles of longitudinal muscles running parallel to the long axis, and oblique muscles oriented in helices around the arm. Crossed-fiber connective tissue layers surround both the oral (side facing the mouth, with suckers) and aboral (outward facing, no suckers) sides of the arm and are the insertion points of the oblique muscle fibers. The fiber angle of the connective tissues and the oblique muscles are the same relative to the arm's long axis (Kier & Thompson, 2003). An additional sheet of connective tissue surrounds the octopus axial nerve cord. In octopuses (*O. bimaculoides*, *O. briareus*, and *O. digueti*) arms, the aboral side also has a layer of thin circular muscles wrapping the arm. An additional internal oblique muscle layer is found between transverse and longitudinal muscles of the octopus arms and this is all surrounded by an outer layer of circumferential muscles (Kier & Stella, 2007).

Squid (*L. pealei*) tentacle musculature is similar to that of the arm, with a few differences. One key difference is the cross rather than obliquely striated fiber composition of the transverse and circular muscles of the tentacles that give them their fast contractile properties (Kier, 1985; Kier & Curtin, 2002). Differentiation of the transverse muscle from obliquely to cross-striated fibers has been shown during development of the squid *Sepioteuthis lessoniana* (Kier, 1996). The transverse muscles of squid tentacles are continuous with a layer of circular muscles. Transverse muscle fibers turn and either become part of the circular muscles or insert at the connective tissue layer that encircles the circular muscle layer. Two layers of oblique muscles, oriented in opposing directions, surround the circular muscle layer, which in turn is surrounded by a layer of longitudinal muscle fibers (Kier, 1985; Kier & Thompson, 2003). Despite the presence of an axial nerve cord, presumably supplying efferent commands to the tentacle, little is known about the nervous system dynamics that control its extension (Kier & Leeuwen, 1997).

7.2.3.3.2 Nervous Control of Octopus Arms

The control of octopus arms and their associated structures (web and suckers discussed in the following section) is complex and its investigation is ongoing. It is complicated by the dual centralized and localized control of octopus motor action (Grasso, 2014). Centralized control is executed by their large bilaterally symmetrical brain, which Grasso (2014) refers to as the cerebral ganglia, made up of 35 lobes (Grasso & Basil, 2009; Young, 1971). Results from microstimulation experiments suggest that arm movements are represented by overlapping pathways in higher motor centers of the lobes and that different movement components are not somatotopically represented, as the same movement patterns could be elicited by stimulation of different lobes (Zullo et al., 2009). The large brachial lobes (Maddock & Young, 1987), along with several others in the subesophageal mass (Budelmann & Young, 1985), send information from the brain down the brachial nerves to a series of brachial ganglia in each of the arms.

Localized control of the arms is at the brachial ganglia, where 3.5×10^8 of the total 5×10^8 neurons in the entire nervous system are located (Young, 1971). A series of brachial ganglia extends down each arm, making up the axial nerve cord. The eight nerve cords are connected at the base of the arms by the ring of the interbrachial commissure. This system is referred to as the brachial plexus. Four smaller intramuscular nerve cords run down the periphery of each arm and link to the brachial ganglia. Muscle fibers are innervated by motor neurons in the brachial ganglia and through coordination with the intramuscular nerves, which produces the diversity of movement patterns seen in the arm (Graziadei, 1971; Rowell, 1966). Each individual sucker is also associated with a sucker ganglion that is located at its base, just outside of the muscle. These sucker ganglia are also connected to corresponding brachial ganglia in the arm. Each sucker contains chemo- and mechanosensory receptors on the rim that provide sensory feedback to the sucker ganglion (Graziadei, 1971). Thus an arm can be divided into individual units of neural control called "local brachial modules" which consist of paired brachial and sucker ganglia (see Fig. 7.5). Efforts to understand the coordination of localized sets of suckers have been made (Grasso, 2008) but more information is yet to be learned about how these local brachial modules are coordinated in whole arm movements. The distributed neural control we see in octopus arms is one reason why they have a great deal of flexibility and is a good example of self-organization, which allows for greater adaptability of movement to different environmental and physiological situations (Hochner, 2013).

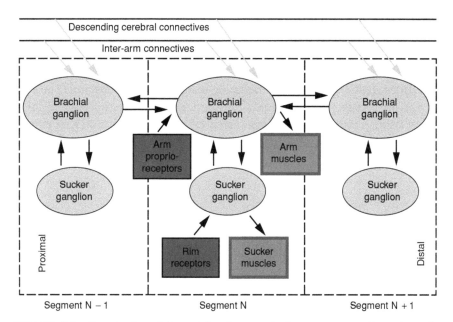

FIGURE 7.5 Organization of the peripheral system in the octopus arm. Repeating units of neural and muscular structures centered around each sucker run down the length of the arm. Segment N in the center illustrates a complete unit including sensory inputs (arm proprioceptors and rim receptors) to the ganglia (brachial and sucker). The brachial ganglion also receives inputs from descending cerebral connectives, connectives from other arms, neighboring brachial ganglia and the sucker ganglion in its segment. The brachial and sucker ganglia process motor commands to arm and sucker muscles in its segment. (From Grasso, F. W. The Octopus with Two Brains: How Are Distributed and Central Representations Integrated in the Octopus Central Nervous System? In *Cephalopod Cognition*; Darmaillacq, A.-S.; Dickel, L., Mather, J., Eds.; Cambridge University Press: Cambridge, UK, 2014; pp 94–124. With permission from Cambridge University Press.)

7.2.3.3.3 Biomechanics of Arms and Tentacles

The arrangement of muscle fibers in these limbs yields a variety of movement patterns based on the principles of muscular hydrostats (Kier, 1992). Elongation of the arm and tentacles is a result of transverse muscle contraction while shortening is a result of longitudinal muscle contraction. Extension of the tentacles occurred in 20–40 ms with maximum extension velocities of the tentacle stalk over 2 m/s (Kier & Leeuwen, 1997). Torsion or twisting of the arm is a result of contraction of the oblique muscle layers. Bending in the arm of squid and octopuses involve antagonistic actions of the longitudinal and transverse muscle fibers. Contraction of the longitudinal muscle bundles on one side of the arm results in shortening. This is resisted by the simultaneous contraction of transverse muscle fibers, resulting in bending toward the side of longitudinal muscle contractions. Concurrent contractions of the transverse and longitudinal muscle increases the flexural stiffness (resistance while undergoing bending) of octopus arms and simultaneous contraction of oblique muscle layers oriented in both right- and left-handed directions results in increased torsional stiffness (Kier & Smith, 2002; Kier & Stella, 2007; Kier & Thompson, 2003) (see Figs. 5–7 in Kier & Thompson, 2003).

7.2.3.4   SUCKERS AND WEB

The arms of coleoid cephalopods contain suckers that are used during locomotion, particularly during crawling in octopuses. Most investigations of morphology have been on octopus suckers, though there are some noticeable differences from squid and cuttlefish ones. Like the arm, the sucker is composed of a dense array of muscle fibers oriented in three directions in the sucker wall: radial muscles that cross the wall, circular muscles that are arranged circumferentially around the wall, and meridional muscles that run perpendicular to the radial and circular muscles. An inner and outer layer of fibrous connective tissue surrounds the sucker and crossed connective tissue fibers are embedded in the muscle (Kier & Smith, 1990, 2002) (see Fig. 3 in Kier & Smith, 1990). The suckers of squid and cuttlefish differ from those of octopuses in that they have a muscular stalk which connects the sucker cup to the arm or tentacle. The stalk is composed of longitudinal and transverse muscle fibers along with connective tissue that result in elongation, shortening, and bending to manipulate placement of the sucker cup. These actions are carried out by similar muscle contraction patterns to those used for arms. Additionally, squid cups are lined with chitin and tooth-like projections. The roof of the sucker also contains muscles arranged like pistons so that when tension is applied to the stalk, it pulls on these muscles, creating more effective attachment to a substrate (Kier & Thompson, 2003).

Suckers attach to a substrate by decreasing pressure within the sucker cavity. When radial muscles contract in a sucker that is sealed to the substrate, the walls thin and the cohesiveness of water resists volume expansion of the sucker. This results in decreased pressure inside the sucker cavity. The circular and meridional muscles antagonize the radial muscles, resulting in a smaller sucker circumference and thicker walls. Elastic energy stored in the connective tissue fibers helps to maintain sucker attachment for a prolonged period (Kier & Smith, 1990, 2002). Pressure differentials of octopus, squid, and cuttlefish suckers are similar at sea level, but at greater depths, squid and cuttlefish exhibit greater sucker strength than octopods (Smith, 1996).

Another component of the arms is the interbrachial membrane or web, which is a muscular fold of skin that extends between each of the arms. This is seen in *Vampyroteuthis* and all species of octopods but is most pronounced among the cirrate octopods. The arrangement of fiber trajectories in the web has been described by Guerin (1908), but since then little is known about the neural mechanisms or pathways that control this muscular structure. Activation of localized muscle fibers in the web can result in localized arm movements and rotations of the base aborally, orally, and laterally (Kier & Stella, 2007). The web can also contract at the base of the arm and produce swimming movements in the cirrate octopods by ejecting interbrachial water but the exact neural mechanism is unknown (Roper and Brundage, 1972).

7.2.4 TYPES OF MOTION

The diversity in physiological control of musculature in coleoid cephalopods underlies the various modes of locomotion that the different groups use. They produce three basic modes of movement: swimming, crawling, and burying. However, several variants of each occur between and among the coleoid cephalopod groups, depending on what muscular structure(s) are the main effectors of motion. Swimming can involve the mantle–funnel complex alone or in combination with fin undulations to produce rapid jet propulsion. It can also involve only the fins or, in the octopods and *Vampyroteuthis*, only the web. Crawling involves the use of one or more limbs and, in the case of coleoids, is almost exclusive to arms in octopods. Benthic cuttlefish sometimes use their arms to walk or crawl (see *Metasepia* in Roper & Hochberg, 1988). Benthic coleoids that hide in their surrounding substrate will bury or dig with a combination of all muscular structures, from the arms in octopus (Boletzky, 1996; Guerra et al., 2006) to arms and mantle–funnel exhalations in cuttlefish (Mather, 1986; Anderson et al., 2004). All types of locomotion are affected by the same principles and depend on what muscular structure and environmental factors are involved. Discussion of these movements will be made according to types—jet propulsion of the mantle or the web and limb-based movement of the arms and fins.

7.2.4.1   JET PROPULSION

Jet propulsion is movement that is achieved by the expulsion of water in one direction, creating an equal and opposite reaction that propels the animal in the opposite direction. The greater the volume of water expelled, the greater the distance that is traveled. A deformable body with elastic capabilities for expansion is required to be able to expel large amounts of fluid to power this mode of locomotion (Trueman, 1980, 1983). Cephalopods are prime examples of how jet-propelled swimming has allowed them to adapt to their environment and compete with fish in the same habitat (Packard, 1972). All cephalopods are able to swim by jet propulsion using their mantle–funnel complex, though squid are the best studied group. Jets from the mantle–funnel complex can also be used by some groups in burying. In addition, cirrate octopods can use another structure, their extensive web, to expel a large volume of water between the arms and propel them posterior first.

7.2.4.1.1   Mantle–Funnel

Jet propulsion by the mantle–funnel complex is well-studied in coleoid ceph-alopods. The mantle contracts to expel a large amount of water as well as expands and takes in water through the funnel to build high hydrostatic pres-sure. Studies have investigated cephalopod, predominantly squid, jet propul-sion from various perspectives: swimming energetics (Bartol, Patterson, & Mann, 2001; O'Dor, 1982, 2002; O'Dor & Webber, 1986, 1991; O'Dor et al., 1994; Wells & O'Dor, 1991), hydrodynamics and mechanics (Anderson & DeMont, 2000; Anderson & Grosenbaugh, 2005; Anderson et al., 2001; Bartol, Patterson, & Mann, 2001; Bartol et al., 2009; O'Dor, 1988), respi-ratory and swimming dynamics (Aitken & O'Dor, 2004; Bartol, Mann, & Patterson, 2001; Payne et al., 2011), kinematics (O'Dor, 1982; Thompson & Kier, 2001, 2002), metabolics (Baldwin, 1982; Finke et al., 1996; O'Dor & Webber, 1986; O'Dor et al., 1994; Pörtner et al., 1993, 1996; Rosa & Seibel, 2010a; Trueblood & Seibel, 2014), swimming dynamics and effi-ciency throughout development (Bartol et al., 2008, 2009; O'Dor & Hoar, 2000; Thompson & Kier, 2001, 2002; York & Bartol, 2016), and physiolog-ical and environmental constraints (O'Dor & Webber, 1986, 1991; O'Dor et al., 2002; Pörtner & Zielinski, 1998). A recent review of jet propulsion and squid swimming and flying highlight current advancements espe-cially for improving technologies in swim tunnel respirometry in the lab and three-dimensional accelerometer chips and radio-acoustic positioning telemetry for tagging and tracking squid in the field.

Compared to other forms of locomotion, jetting is an inefficient way of moving (Alexander, 2003). The energetic costs of jet propulsion in squid are greater than the costs of swimming in fish, based on pressure sensor and acoustic telemetery studies of *L. forbesi* (O'Dor et al., 1994). Fishes undulate their body to push large volumes of water at low speeds for thrust. Cephalopods must accelerate water at high speeds through their funnel to produce the same thrust, but are limited in water volume by the capacity of their mantle cavity (O'Dor, 1988; O'Dor & Webber, 1986, 1991; Webber et al., 2000; Wells & O'Dor, 1991). This response (the rapid escape jet) is best used as an antipredatory response, but is energetically wasteful if used for normal swimming. Normal swimming in *Loligo* occurs with deeper respiratory pulses, where a large volume of water is expelled at low-amplitude jet pulses. During normal swimming, *Loligo vulgaris* generates one-tenth of the maximum pulse pressure in its mantle cavity (Trueman, 1980, 1983). Squid have elastic components (i.e., collagen fibers) in their muscular mantle–funnel propulsion system to help refill the mantle cavity (Gosline & DeMont, 1985). The flexibility of the cephalopod funnel provides maneuverability in controlling movement in any direction and speed (Anderson & DeMont, 2000; Bartol, Man, & Patterson, 2001; O'Dor, 1988). The fins also help compensate for inefficiencies by maintaining control and orientation during locomotion (O'Dor & Webber, 1986).

Jetting speed is influenced by how much musculature is present in the mantle and the diameter of the funnel, as the velocity of the jet is inversely proportional to the funnel's cross-sectional area. Faster and more forceful jetters, such as Loligonidae and Ommastrephidae squid, have muscular mantles that contract to force water out of a small funnel aperture (O'Dor & Webber, 1986). In contrast, the deep-sea squid *Taonis* and the pelagic octopod *Japatella* have wider funnels and slower jet propulsions. *Japatella* produces two additional jets as a result of passing water through the sides of the head via inhalant openings (Clarke, 1988). Larger mantle capacities and powerful mantle muscles are also found in octopods like *Eledone* which rely more on jet propulsion than benthic octopods that have smaller mantle capacities, less developed musculature, which result in short, low-amplitude jet pulses (Trueman & Packard, 1968). Several other factors influence the efficiency of jetting, including the mass of the animal, the mass of water in the cavity, how fast water is expelled, the magnitude of drag force from water motion, and the size of the funnel aperture. Recently, Staaf et al. (2014) generated a theoretical model of the effects of funnel aperture on jet propulsion of squid, covering its range of sizes. Generally, squid of all sizes decrease funnel size during mantle contraction to increase efficiency; however, ecological pressures and the effects of size may lead to fine-tuned changes in contraction speed and funnel diameter.

Different gaits or swimming variants have been identified thought not well investigated, predominantly from swim tunnel studies, for various species of squid. Squid can use a fast escape jet, with hyperinflation of the mantle cavity, or a slow swim, with much less expansion and contraction, and any speed in between (Gosline & DeMont, 1985; O'Dor & Webber, 1986). A combination of slow funnel water pulses with variable use of the fins provides other variants of swimming movement (Hoar et al., 1994; Anderson & Grosenbaugh, 2005). Multiple gait patterns at different speeds, from posterior-first swimming in *L. opalescens* (O'Dor, 1988) and *L. brevis* (Bartol, Patterson, & Mann, 2001) have been identified as depending on variable contractions of several interacting muscle structures. Cruising locomotion, like gliding, soaring, blimping, and climb-and-glide, using jet propulsion and fins have also been observed to take advantage of environmental factors, like upwelling and currents, to reduce energy costs associated with transport and maintaining horizontal position (Gilly et al., 2012; O'Dor, 1988, 2002; O'Dor & Webber, 1986).

Jet propulsion is not only important for swimming, but is also used as the main effector for burying in bottom-living coleoid cephalopods. This action has been referred to as burying, digging, sand-digging, and sand-covering, but all involve an animal covering itself with the substrate, either sand or mud, until it cannot be seen (Hanlon & Messenger, 1996). The behavior pattern is well described in cuttlefish (*S. officinalis*) as consisting of three actions: "blow forward" where the mantle–funnel directs a jet of water anteriorly, "blow backward" which directs the jet of water posteriorly beneath the body, and "wiggle" where the dorsal mantle contracts in a series of small side-to-side movements (Mather, 1986). This stereotyped behavior exhibits some variation, particularly during the "wiggle" phase. Sepiolid squids bury in two phases (Boletzky, 1996). The first phase is similar to the behavioral pattern described for *Sepia*. The funnel directs a series of water jets forward and backward to create a depression for the animal to settle into. Fin undulations help resist the upward component of jetting. The second phase involves the second pair of arms which stretch, enclose, and sweep sediment over the body and head until they are completely covered. The funnel protrudes laterally at the surface between the head and mantle and respiration is maintained in a funnel pouch rather than the mantle. This pouch is a result of pulling back a skirt found at the base of the funnel (Boletzky & Boletzky, 1970). This fixed sequence of behaviors during burying has been seen in *Rossia pacifica* (Anderson et al., 2004), *Euprymna scolopes* (Anderson et al., 1999), and *Sepiola atlantica* (Rodrigues et al., 2010) with variations in behaviors and body patterning due to environmental differences such as substrate type. Benthic octopuses have also been observed digging or burying into mud or sand (Guerra et al., 2006), however mantle–funnel jets are not involved in moving sediment in the process (Boletzky, 1996). The arms are the main effectors in octopus burying, so this action will be considered in the following section concerning arms.

7.2.4.1.2 Web-based

Another way coleoid cephalopods, mainly the deep sea cirrates (Opisthoteuthidae, Cirroteuthidae, Grimpoteuthidae, and Cirroctopodidae), can swim is by using the web to eject water when the arms come together. Seibel et al. (2000) suggested that these animals do not need high-speed locomotion for predator/prey visual detection in their low light environment. These slower modes of movement, based on the web as well as the fins, are more energetically efficient and less costly than jet propulsion (Seibel et al., 1997, 2000).

There are many modes of swimming used by the different cirrate groups. Roper & Brundage (1972) were among the first to describe some of these modes (water ejection, pulsating, and umbrella style) as access for observation of these deep sea species is difficult. With the availability of submersible techniques in more recent years, a suite of swimming patterns involving the web, in some cases assisted by the fins, has been seen *in situ*. "Pumping" produces slow propulsion in Cirroteuthids by alternating expansion and expulsion of water via peristaltic waves in the web (Villanueva et al., 1997a). The "Ballooning-response" was first described by Boletzky (1992) in *Cirroteuthis magna* as similar to pumping during filling, but in this case the water is retained in the web when the distal edges are contracted. All eight sections of the web are expanded simultaneously and filled with water (Villanueva et al., 1997a; Collins & Villaneuva, 2006). This posture may be used as a transition between different modes of locomotion and has also been seen in *Stauroteuthis syrtensis* (Vecchione & Young, 1997; Johnsen et al., 1999) *C. magna* (Villanueva et al., 1997a), and *Opisthoteuthis massyae* (Villanueva, 2000). "Take-off" consists of a single contraction by the web, sometimes proceeded by flapping of the fins. It has been seen as an escape response in *Cirroteuthis* and *Grimpoteuthis* (Villanueva et al., 1997a; Collins

& Villaneuva, 2006). "Umbrella-style drifting" is a passive mode of locomotion that makes use of the animal's neutral buoyancy and bottom currents to maintain position in the water column. As seen in *Cirroteuthis*, the web and arms are spread out with the fins folded in, similar to an open umbrella (Villanueva et al., 1997a). This is also present in *S. syrtensis* (Johnsen et al., 1999; Collins & Villaneuva, 2006). "Medusoid" was first described by Roper & Brundage (1972) in *Vampyroteuthis* to be an escape reaction that consists of rapid movement. The rate of web contractions can be as fast as 1.2/s with 2–3 fin strokes for every pulse of the web (Collins & Villaneuva, 2006). Also known as arm-web contraction, this action consists of pulsations of the round, bell-shape inflated web and is the primary mode of locomotion for *Opisthoteuthis* (Vecchione & Roper, 1991). It has also been seen in *S. syrtensis* and juvenile *Grimpoteuthis* (Vecchione & Young, 1997) which use fin flapping (also referred to as sculling) with medusoid pulsations partly to help steer the body (Johnsen et al., 1999; Collins & Villaneuva, 2006). As has been mentioned in some of the examples, the fins may also play a part in locomotion of these cirrate octopods. The web is inflated in a bell-like shape with the arms separated. Fins on either side of the mantle flap simultaneously for fin swimming (Collins & Villaneuva, 2006; Johnsen et al., 1999; Vecchione & Young, 1997).

Cirrate octopods generally exhibit web-based propulsion, fin swimming, or a combination of the two for fast burst escape responses (Seibel et al., 1998). Opisthoteuthids have oval shaped bodies, small fins, and a thick web with low protein and lipid content in the muscles. These characteristics reflect its predominant use of slow fin swimming. Cirroteuthids, with their long bodies, large fins, and well-developed web, use both the fins and web for swimming. Grimpoteuthidae and Cirroctopodidae have intermediate morphologies to Opisthoteuthidae and Cirroteuthidae and also use both web and fin swimming (Collins & Villaneuva, 2006; Seibel et al., 1998).

7.2.4.2   LIMB MOVEMENT

Both the fins (cirrate octopods, squid, and cuttlefish) and arms (octopods) of cephalopods can be used as effectors for locomotion. For many species of squid and cuttlefish, the fin is used to stabilize body position during jet propulsion. However, in cirrate octopods the fins assist in web-based swimming (discussed in the following section) or may be the primary means of locomotion. Different types of fin movement can result from variation in fin sizes and shape as well as depend on ontogenetic changes of the fins and cephalopod lifestyle. Similarly, different gait patterns can result from variations in flexibility and coordination of arms in octopus. Squid and cuttlefish arms do not exhibit the same diversity of movements or dependence on arms for locomotion as seen in octopuses, with the exception of burying as previously described in Sepiolids and walking ("ambling with arms") exhibited by sequential shuffling of the ventral arms and ambulatory flaps along the bottom in flamboyant cuttlefish, *Metasepia pfefferi* (Roper & Hochberg, 1988).

7.2.4.2.1 Fin

Unlike jet propulsion, locomotion by fins is not well investigated, though there are a few studies (Anderson & DeMont, 2005; Bartol, Patterson, & Mann, 2001; O'Dor, 1988; Hoar et al., 1994; Stewart et al., 2010). Fins are limited by hydrodynamic constraints and scaled to the animal's body during development (Hoar et al., 1994; O'Dor & Hoar, 2000). Clarke (1988) identified nine types of fins that vary in shape and size for different species of finned cephalopods (see Fig. 7.6 for fin and body form examples from Packard, 1972). First, fringing fins, as seen in *Sepia* cuttlefish, produce waves with small amplitude that are used for maneuvering. Second, elongated flapping fins generate large-amplitude waves used for hovering or gentle swimming. Mather et al. (2010) described the different fin positions in *S. sepiodea*. These fins also produce powerful flaps that aid in jet action, which is exemplified in the squid *Thysanoteuthis rhombus*. Dorsoventral stiffness is provided by the squid pen. Hunt et al. (2000) described the locomotor and postural components of behavior for *L. opalescens*. Third, posterior broad triangular fins with a leading edge perpendicular to the sagittal plane of the body are used for fast muscular swimming in negatively buoyant squid. These fins make long undulatory waves that provide stability and produce hovering. There is a parallel variation in the fin structures of batoid fishes (Rosenberger, 2001). Individuals with this fin type usually have narrow pens to help with greater directional control and can interchange between backward and forward hovering. Examples include the Ommastrephids, like shortfin squid (*Illex illecebrosus*) and the Humboldt squid (*Dosidicus gigas*), as well as the Onychoteuthids, hooked squid. Harrop et al. (2014) describe locomotor and postural components of *in situ* behavior of *Illex*. Fourth, medially placed broad rounded fins give cephalopods, like the short-bodied Sepiolidae (bobtail squids), more turning mobility. Fifth, short round fins in species with no buoyancy or a center of buoyancy that is more toward the posterior provide orientation control but do not affect backward or forward movement. Cephalopod examples include the Idiosepiidae, Cranchiidae, and Histioteuthidae. Sixth, oval fins produce large amplitude waves in squids that use ammonia-mediated buoyancy. These fins, which are seen in Architeuthids, Chiroteuthids, and Mastigoteuthids, provide control of orientation and more economical backward movement. Seventh, long circular fins along the mantle produce backward movement by fin beating. Squid with this fin type have robust thick cartilaginous pens. Squid species with this type of fin include *Discoteuthis*, *Ancistrocheirus*, and *Octopoteuthis*. Different postural and locomotor components have been described for *Octopoteuthis deletron* (Bush et al., 2009). Eighth, juvenile members of Grimalditeuthidae and larval Chiroteuthidae have secondary fins which are smaller, develop posterior to the primary fins and whose functions are unknown. Lastly, broad fins with highly muscular attachments to the body generate powerful fin beats for propulsion in Vampyromorpha and the cirrate octopods (Clarke, 1988).

FIGURE 7.6 Different body forms during locomotion, with examples of various forms of cephalopod fins. (A) Adult *Loligo vulgaris*; (B) (i) and (ii) adult *Octopus vulgaris*, (iii) adult *Octopus dofleini*; C. (i) flying ommastrephid squid, (ii) flying fish; (D) adult gonatid squid *Gonatus*; (E) cranchid squid *Leachia*, F. (i) and (ii) different developmental stages of squid *Lepidoteuthis*. (From Packard, A. Cephalopods and Fish: the Limits of Convergence. *Biol. Rev.* 1972, 47, 241–307. Copyright © (2015) by John Wiley & Sons, Inc. Reprinted by permission of John Wiley & Sons, Inc.)

Fins produce two types of movement, either undulations (passing multiple waves down the fin) or fin beating or flapping (oscillatory) (Hoar et al., 1994). Parallel variation in the fin structures of batoid fishes show a continuum between undulatory and oscillatory locomotion (Rosenberger, 2001). Fin flapping predominates in cirrate octopods and Vampyromorpha, which have broad muscular fins. Fin swimming in cirrate octopods consists of beating their pairs of fins synchronously, either in combination with web-based propulsion (discussed in a future section) or alone. This results in mantle-first movement with the arms trailing behind. This is has been observed in *C. murrayi* (Aldred et al., 1983), *Cirrothauma magana* (Villanueva et al., 1997a), and *Grimpoteuthis* (Vecchione & Young, 1997). *V. infernalis* exhibits an ontogenetic change in locomotion based on changes in fin morphology with maturation. Juveniles primarily use jet propulsion, assisted by their small paddle-shaped fins for stability. As they become adults, the fins increase in size and vary in shape such that lift-based propulsion becomes more important and fin swimming is the primary mode of locomotion (Seibel et al., 1998). Other ontogenetic gait changes in locomotion have been seen in *L. brevis* (Bartol et al., 2009) and the effects of scaling of squid fins on locomotion types are discussed in Hoar et al. (1994). The lateral fins of *Sepia* primarily move by undulations to help the animal hover, but several squid patterns of swimming moderated by different fin morphologies fall between the two ends of the continuum. In the long-finned *L. pealei*, the fin is more undulatory at low speeds and more flap-like at high speeds (Anderson & DeMont, 2005). Shallow-water brief squid *L. brevis* also uses a range of fin movements at different speeds (Bartol, Patterson, & Mann, 2001) and recent studies of fin-only hydrodynamics confirm that the fins produce lift and thrust (Stewart et al., 2010). Fins also facilitate "flight" by jet propulsion out of the water into the air, assisted by fin flaps in several squid species (Maciá et al., 2004; O'Dor et al., 2013). These are usually seen in squid with broad fins and spread arm postures.

7.2.4.2.2 Arms

The arms in octopuses are powerful effectors of movement. When the musculature of the mantle and the arms are compared during swimming versus holding, five arms holding onto a side tank can produce enough tension to hold 30–100 times the octopuses' own body weight while tension from swim pulling can only sustain half the body weight (Trueman & Packard, 1968). Arrangements of the muscles also allows arms to have a large number of degrees of freedom: bending, elongating, and twisting in an almost infinite number of directions. Efforts to understand the general principles that organize movement in a single arm, such as "reaching" (Gutfreund et al., 1996) and "fetching" (Sumbre et al., 2006) have been described, along with dynamic modeling of the biomechanics and neural control of the arm "reaching" pattern (Yekutieli, Sagiv-Zohar, Aharonov, Engel, Hochner, & Flash, 2005; Yekutieli, Sagiv-Zohar, Hochner, & Flash, 2005).

Descriptions for crawling are not as frequent as those for jet propulsion in cephalopods. However, benthic octopuses spend most of their time on the bottom using their arms as primary way to move around their environment (Huffard, 2006). Crawling patterns have been described for *Grimpoteuthis* (Villanueva et al., 1997a), *Amphioctopus marginatus* (Huffard & Godfrey-Smith, 2010; Sreeja & Bijukumar, 2013), and *Abdopus aculeatus* (Huffard, 2006) but also mentioned in behavioral and taxonomic accounts of other species (Hanlon & Wolterding, 1989; Hanlon et al., 1999; Mather, 1998; Mather & Mather, 1994; Norman & Finn, 2001; Packard & Sanders, 1971; Roper & Hochberg, 1988; Wells et al., 1983). Observations of arm movement across the substrate in the deep sea octopods *Graneledone, Benthoctopus*, and *Vulcanoctopus* have been made, but with no description of specific gait patterns such as crawling (Voight, 2008). "Crawling" has sometimes been used interchangeably with "walking" in the literature. Crawling can generally be described as actions by the arms and their suckers, with the arms spread on the bottom, pushing and pulling with the more proximal and medial suckers on the substrate (Mather, 1998). Multiple points of contact can be made along the arms with the substrate and in the direction of movement. The posterior and lateral pairs of arms are more often used during crawling, a preference that has been seen *O. vulgaris* with anterior arm preference to explore and forage (Byrne et al., 2006). Levy et al. (2015) recently described crawling in octopus and found that they lack a stereotyped pattern of arm coordination and that crawling direction can be determined by a force vector resulting from a combination of the pushing arms. They also found that instead of rotating the body, choosing to use a certain set of arms and using their vector combination will change the octopus's crawling direction. This suggests that motor control system of crawling consists of relatively stereotyped movements (i.e., a series of arm elongations and sucker adherence) produced by the peripheral system with the coordination of the arms determined by the central brain. Postures may vary within and between species during crawling (see Fig. 1 in Huffard, 2006).

Other gaits of arm-mediated locomotion have been described for a few species of octopus. Bipedal walking in *A. marginatus* and *A. aculeatus* uses two arms (usually the posterior pair) to move backward using a rolling action of the arms (Huffard et al., 2005). The octopus pushes or rolls from the tip of the arm with its suckers against the substrate, alternating arms. Postures used during bipedal walking varied between *A. marginatus*, which typically draws in all six arms closer to the body, and *A. aculeatus*, which raises its two dorsal arms in a "flamboyant display" (Packard & Sanders, 1971). Another variant on arm position that involves multitasking is "stilt walking" and is only seen in *A. marginatus*. It involves bipedal walking in combination with holding an object with the other arms (Finn et al., 2009). "Tiptoe," which is a variant of crawling, involves adhesion, swinging, and release by a subset of suckers on all eight arms on proximal areas resulting in slow gliding movement (Mather, 1998). Locomotion speeds relative to body lengths were compared between different modes in *A. aculeatus* and although crawling (1.94 body lengths per second) is used predominantly, it is not as fast as jet-propelled movement (3.29 body lengths per second), but bipedal locomotion (2.25 body lengths per second) comes close (Huffard, 2006).

Octopuses can use their arms for various tasks such as "grooming" and "exploring", which do not have a set sequence of arm positions (Mather, 1998). Several other tasks have been named and qualitatively described (Borrelli et al., 2006; Huffard, 2006; Mather, 1998; Mather & Alupay, 2016; Packard & Sanders, 1971). One of the more often noted tasks is digging. Burying behavior in sand or mud has been observed in *Octopus burryi* (Hanlon & Hixon, 1980), *Octopus cyanea* (Roper & Hochberg, 1988), *Eledone moschata* (Boletzky, 1996; Mangold, 1983), and *E. cirrhosa* (Rodrigues et al., 2010). Observations in the field also show that *Macrotritopus defilippi* (Hanlon et al., 2010), *Thaumoctopus mimicus*, and *Wunderpus photogenicus* (Hanlon et al., 2007) also rapidly disappear into their sand environment, but the burying mechanism is unknown. There are very few descriptions of what exact actions occur in the arm during digging. The arms likely twist and the suckers sweep away larger grains of sediment (Boletzky, 1996). In *O. cyanea*, sediment is pushed to the side by ventral arms to form a cavity where the octopus can settle (Roper & Hochberg, 1988). It was noted that no water jets from the funnel were involved during the process. During sand-covering in *E. cirrhosa*, the arms push up small grained sediment around the body by twisting of the proximal and distal parts of the arm (Guerra et al., 2006). After making a hole, the octopus sinks in and the arms form a circle to throw out sediment from underneath its body, so the mantle and head sink deeper. In octopuses, digging is not only used for concealment, but also to build a burrow for shelter (Guerra et al., 2006). The sequence of behaviors for the arms or other potential effectors is not known, so unlike in the Sepiolids (Boletzky & Boletzky, 1970), sequences of action that might be species-typical are not known.

7.2.5 CONTRIBUTION OF BUOYANCY

An important aspect to consider for all modes of locomotion in water is buoyancy. Having lost an external shell which might hold gas, the coleoid cephalopods have developed four general methods to maintain their vertical position in the water column (see summary Fig. 15 in Packard, 1972). The first uses variable gas-filled spaces in an internalized chambered shell, as seen in the enclosed *Nautilus*-like shell of *Spirula* (Denton et al., 1967) and the flattened cuttlebones of *Sepia* (Birchall & Thomas, 1983; Denton & Gilpin-Brown, 1961). The volume of gas in these spaces is regulated by pumping salts from the fluid filling the spaces of the cuttlebone to maintain an osmotic pressure that withstands hydrostatic pressure from water flowing into the shell (Boyle & Rodhouse, 2005a). The superfamily of pelagic octopuses, Argonautoidea, is an interesting example of how gas-mediated buoyancy has evolved in recent coleoids (Bello, 2012). The females of *Argonauta* have an external brittle shell whose primary role is to hold and brood eggs. The shell also controls buoyancy by retaining air "gulped" at the sea surface before the octopus descends deeper (Finn & Norman, 2010). *Ocythoe tubercuclata* females have gas-filled swim bladders (Packard & Wurtz, 1994) and octopod relatives *Tremoctopus* and *Haliphron* have homologous structures whose position in the mantle suggest they originate from the digestive system (Bello, 2012).

The three remaining methods use materials that are less dense than seawater to maintain neutral buoyancy. The most common method among squids is to use ammonium to replace sodium in lowering the density of internal fluids (Boucher-Rodoni & Mangold, 1995; Clarke et al., 1979; Clarke, 1988; Voight et al., 1995). Ammonia is a product of nitrogen metabolism present in all cephalopods. This fluid is stored in vacuoles within the muscle, but in one family, the Cranchiidae, it is stored in a specialized coelom (Clarke et al., 1979; Clarke, 1988; Denton & Shaw, 1969). A second method is to store lipids or oils in a digestive gland such as the liver, as seen in the Gonatid squid (Clarke, 1988; Seibel et al., 2004). The third method, exhibited in pelagic octopods like *Japatella*, is to have layers of low-density gelatinous tissue surrounding the musculature, which decrease the number of dense ions like sulfate and chlorine in the body (Clarke et al., 1979; Denton & Shaw, 1961; Denton et al., 1967).

7.3 LOOKING FORWARD

It is evident from this overview that the muscular hydrostat plays an important role in the types of locomotion used by coleoid cephalopods. The basic
principles of how hydrostatic skeletons move and detailed descriptions of the
structures involved were pioneered by Kier (1992, 2012) and Kier and Smith
(1985). They found that the same leverage principles of agonist–antagonist
muscles used in hard skeletal bodies also applies to the soft bodies of cephalopods and other muscular hydrostat-based animals. In fact, despite the great
diversity of movement behaviors used by animals with different structures
and environmental pressures, there are underlying similarities on multiple
biological levels. One of the areas under investigation in hydrostat systems
is understanding the fundamental mathematical basis for their movement.
Some progress has been made in mathematically modeling body pattern
formation on cephalopod skin (Mather et al., 2014), models that are being
used by the same authors to also explain common dynamical principles in
hydrostat systems besides cephalopods, like the nematode, *Caenorhabditis
elegans*, and the human tongue.. In addition, recent developments in visualization techniques using ultrasound (King et al., 2005; Margheri et al., 2011)
and magnetic resonance imaging (Xavier et al., 2014) provide a method for
studying muscle dynamics *in situ* for these varieties of soft bodied organisms.

Despite the abundant information about the cephalopod musculature,
control of these structures to produce different movement patterns is still
not as well known. Detailed information about the neural structures and
some of the pathways (namely the giant fiber system [Young, 1938, 1939,
1965]) has been described, but considering the diversity of locomotor
modes, we still know very little about how movement is regulated. Coordination of mantle–funnel jets with different fin undulation and flapping
patterns in squid and cirrate octopods still need better understanding. Even
more complex is the coordination of arm movement used for crawling and
other modes of locomotion in benthic octopods. Grasso (2014) outlined the
center–periphery problem in coordinating the arm and sucker units. One
approach to studying this is through the idea of embodiment (Hochner,
2012, 2013) where the central nervous system and peripheral programs of
the arms and suckers interact with each other and are modulated by external
information from the sensory and mechanical systems, and the environments. This interaction and feedback, though complex, is thought to be
what underlies the combinations of many behaviors, including movement.
Motor primitives (Flash & Hochner, 2005) are the simple units of behavior
that in combination can lead to more complex locomotor programs. The

previously described "reaching" pattern is one such example of a unit. However, recent studies show that octopus arm movements may be more flexible than originally thought with respect to stereotyped movement patterns, as different individuals use different motor programs in the same situation (Richter et al., 2015).

The detailed descriptions of movement patterns from different cephalopod groups can be informative to other fields, including bio-inspired soft robotics. Several models for soft robotics have been based on the octopus arm because of its dexterity, flexibility, and variable stiffness, all controlled by its muscle arrangements (Cianchetti et al., 2011; Laschi et al., 2009). Furthermore, jet-propelled robots based on the cephalopod mantle (Renda et al., 2014) and web (Sfakiotakis et al., 2014) as well as undulations of the fin (Liu et al., 2012; Willy & Low, 2005) have all been mimicked for soft robotics. These models are often specialized to be efficient in only a few aspects and in isolation. By taking into account the diversity of movements by a combination of effectors, we may better obtain models for soft robotics. For example, the recent incorporation of compliant webbing in swimming movements of a multi-armed robot improved its velocity, thrust, and efficiency than having arms alone (Sfakiotakis et al., 2014).

7.4   CONCLUSIONS

Cephalopods are predominantly studied for their exceptional jet propulsion abilities and the principles behind their movement are thought to be based largely on the mantle–funnel complex. This review of coleoid cephalopod locomotion highlights the role of muscular hydrostats and the actions of multiple effector units—mantle–funnel complex, fin, arms, suckers, and the web in giving us a diversity of locomotor types. These structures, alone and in combination, produce several variations on swimming, crawling, and burying behavior. Using this overview as a basis, we can examine in more detail how the muscle structures adapt to these many modes of locomotion and better understand the general principles that underlie locomotion with muscular hydrostats.

ACKNOWLEDGMENTS

This research was supported in part by grant number 1246750 from the National Science Foundation.

KEYWORDS

- **locomotion**
- **cephalopods**
- **mantle-funnel**
- **muscular hydrostat**
- **fin**
- **jet propulsion**
- **multiple arms**

REFERENCES

Aitken, J. P.; O'Dor, R. K. Respirometry and Swimming Dynamics of the Giant Australian Cuttlefish, *Sepia apama* (Mollusca, Cephalopoda). *Mar. Freshw. Behav. Physiol.* **2004**, *37*, 217–234.

Albertin, C. B.; Bonnaud, L.; Brown, C. T.; Crookes-Goodson, W. J.; da Fonseca, R. R.; Di Cristo, C.; Dilkes, B. P.; Edsinger-Gonzales, E.; Freeman, R. M.; Hanlon, R. T.; et al. Cephalopod Genomics: A Plan of Strategies and Organization. *Stand. Genomic Sci.* **2012**, *7*, 175–188.

Aldred, R. G.; Nixon, M.; Young, J. Z. *Cirrothauma murrayi* Chun, a Finned Octopod. *Philos. Trans. R. Soc. B: Biol. Sci.* **1983**, *301*, 1–54.

Alexander, R. M. *Principles of Animal Locomotion*. Princeton University Press: Princeton, NJ, 2003.

Allen, J. J.; Bell, G. R. R.; Kuzirian, A. M.; Hanlon, R. T. Cuttlefish Skin Papilla Morphology Suggests a Muscular Hydrostatic Function for Rapid Changeability. *J. Morphol.* **2013**, *274*, 645–656.

Alves, C.; Boal, J. G.; Dickel, L. Short-distance Navigation in Cephalopods: A Review and Synthesis. *Cogn. Process.* **2008**, *9*, 239–247.

Anderson, E. J.; DeMont, M. E. The Mechanics of Locomotion in the Squid *Loligo pealei*: Locomotory Function and Unsteady Hydrodynamics of the Jet and Intramantle Pressure. *J. Exp. Biol.* **2000**, *203*, 2851–2863.

Anderson, E.; DeMont, M. E. The Locomotory Function of the Fins in the Squid *Loligo pealei*. *Mar. Freshw. Behav. Physiol.* **2005**, *38*, 169–189.

Anderson, E. J.; Grosenbaugh, M. A. Jet Flow in Steadily Swimming Adult Squid. *J. Exp. Biol.* **2005**, *208*, 1125–1146.

Anderson, E. J.; Quinn, W.; DeMont, M. Hydrodynamics of Locomotion in the Squid *Loligo pealei*. *J. Fluid Mech.* **2001**, *436*, 249–266.

Anderson, R. C.; Mather, J. A.; Steele, C. W. The Burying Behavior of the Sepiolid Squid *Euprymna scolopes* Berry, 1913 (Cephalopoda: Sepiolidae). *Annu. Rep. West. Soc. Malacol.* **1999**, *33*, 1–7.

Anderson, R. C.; Mather, J. A.; Steele, C. W. Burying and Associated Behaviors of *Rossia pacifica* (Cephalopoda: Sepiolidae). *Vie Milieu* **2004**, *54*, 13–19.

Bairati, A. The Collagens of the Mollusca. In *Biology of Invertebrate and Lower Vertebrate Collagens: NATO ASI Series*; Bairati, A., Garrone, R., Eds.; Plenum Press: New York, 1985; Vol. 93, pp 277–297.

Bairati, A.; Comazzi, M.; Gioria, M. A Comparative Microscopic and Ultrastructural Study of Perichondrial Tissue in Cartilage of *Octopus vulgaris* (Cephalopoda, Mollusca). *Tissue Cell* **1995**, *27*, 515–523.

Baldwin, J. Correlations between Enzyme Profiles in Cephalopod Muscle and Swimming Behavior. *Pac. Sci.* **1982**, *36*, 349–356.

Baldwin, J. Energy Metabolism of *Nautilus* Swimming Muscles. In *Nautilus*; Saunders, W., Landman, N., Eds.; Springer Netherlands, 2010; Vol 6, pp 325–329.

Bartol, I. K.; Mann, R.; Patterson, M. R. Aerobic Respiratory Costs of Swimming in the Negatively Buoyant Brief Squid *Lolliguncula brevis*. *J. Exp. Biol.* **2001**, *204*, 3639–3653.

Bartol, I. K.; Patterson, M. R.; Mann, R. Swimming Mechanics and Behavior of the Shallow-water Brief Squid *Lolliguncula brevis*. *J. Exp. Biol.* **2001**, *204*, 3655–3682.

Bartol, I. K.; Krueger, P. S.; Thompson, J. T.; Stewart, W. J. Swimming Dynamics and Propulsive Efficiency of Squids throughout Ontogeny. *Integr. Comp. Biol.* **2008**, *48*, 720–733.

Bartol, I. K.; Krueger, P. S.; Stewart, W. J.; Thompson, J. T. Hydrodynamics of Pulsed Jetting in Juvenile and Adult Brief Squid *Lolliguncula brevis*: Evidence of Multiple Jet "Modes" and their Implications for Propulsive Efficiency. *J. Exp. Biol.* **2009**, *212*, 1889–1903.

Bello, G. Exaptations in Argonautoidea (Cephalopoda: Coleoidea: Octopoda). *N. Jb. Geol. Paläont. Abh.* **2012**, *266*, 85–92.

Bidder, A.; Boycott, B. B. Pelagic Mollusca: Swimmers and Drifters. *Nature* **1956**, *177*, 1023–1025.

Birchall, J. D.; Thomas, N. L. On the Architecture and Function of Cuttlefish Bone. *J. Mater. Sci.* **1983**, *18*, 2081–2086.

Boletzky, S. von. Evolutionary Aspects of Development, Life Style, and Reproductive Mode in Incirrate Octopods (Mollusca, Cephalopoda). *Rev. Suisse Zool.* **1992**, *99*, 755–770.

Boletzky, S. von. Cephalopods Burying in Soft Substrata: Agents of Bioturbation? *Mar. Ecol.* **1996**, *17*, 77–86.

Boletzky, S. von; Boletzky, M. V. Das Eingraben in Sand Bei Sepiola Und *Sepietta* (Mollusca, Cephalopoda). *Rev. Suisse Zool.* **1970**, *77*, 536–548.

Bone, Q.; Howarth, J. V. The Role of L-Glutamate in Neuromuscular Transmission in Some Molluscs. *J. Mar. Biol. Assoc. U.K.* **1980**, *60*, 619–626.

Bone, Q.; Pulsford, A.; Chubb, A. D. Squid Mantle Muscle. *J. Mar. Biol. Assoc. U.K.* **1981**, *61*, 327–342.

Bone, Q.; Brown, E.; Travers, G. On the Respiratory Flow in the Cuttlefish *Sepia officinalis*. *J. Exp. Biol.* **1994**, *194*, 153–165.

Borrelli, L.; Gherardi, F.; Fiorito, G. *A Catalogue of Body Patterning in Cephalopoda, Stazione Zoologica A*. Dohrn Firenze University Press: Napoli, Italy, 2006.

Boucher-Rodoni, R.; Mangold, K. Ammonia Production in Cephalopods, Physiological and Evolutionary Aspects. *Mar. Freshw. Behav. Physiol.* **1995**, *25*, 53–60.

Boycott, B. B. The Functional Organization of the Brain of Cuttlefish *Sepia officinalis*. *Proc. R. Soc. B: Biol. Sci.* **1961**, *153*, 503–534.

Boyle, P. R. Receptor Units Responding to Movement in the *Octopus* Mantle. *J. Exp. Biol.* **1976**, *65*, 1–9.

Boyle, P. R. Neural Control of Cephalopod Behavior. In *The Mollusca: Neurobiology and Behavior Part 2*; Wilbur, K. M., Willows, A. O. D., Eds.; Academic Press: London, 1986; Vol 9, pp 1–99.

Boyle, P.; Rodhouse, P. Form and Function. In *Cephalopods: Ecology and Fisheries*; Blackwell Science: Oxford, 2005a; pp 7–35.

Boyle, P.; Rodhouse, P. Oceanic and Deep-sea Species. In *Cephalopods: Ecology and Fisheries*; Blackwell Science: Oxford, 2005b; pp 176–204.

Budelmann, B. U. The Statocysts of Squid. In *Squid as Experimental Animals*; Gilbert, D. L., Adelman, Jr., W. J., Arnold, J. M., Eds.; Springer US: New York, 1990; pp 421–439.

Budelmann, B. U. The Cephalopod Nervous System: What Evolution Has Made of the Molluscan Design. In *The Nervous Systems of Invertebrates: An Evolutionary and Comparative Approach*; Breidbach, O., Kutsch, W., Eds.; Birkhauser: Basel, 1995a; pp 115–138.

Budelmann, B. U. Cephalopod Sense Organs, Nerves and the Brain: Adaptations for High Performance and Life Style. *Mar. Freshw. Behav. Physiol.* **1995b**, *25*, 13–33.

Budelmann, B. U.; Young, J. Z. Central Pathways of the Nerves of the Arms and Mantle of *Octopus*. *Philos. Trans. R. Soc. B: Biol. Sci.* **1985**, *310*, 109–122.

Budelmann, B. U.; Schipp, R.; Boletzky, S. von. Cephalopoda. In *Microscopic Anatomy of Invertebrates*; Harrison, F. W.; Kohn, A., Eds.; Wiley-Liss: New York, 1997; Vol 6A, Mollusca II, pp 119–414.

Burford, B. P.; Robison, B. H.; Sherlock, R. E. Behaviour and Mimicry in the Juvenile and Subadult Life Stages of the Mesopelagic Squid *Chiroteuthis calyx*. *J. Mar. Biol. Assoc. UK.* **2014**, 1–15.

Bush, S. L.; Robison, B. H.; Caldwell, R. L. Behaving in the Dark: Locomotor, Chromatic, Postural, and Bioluminescent Behaviors of the Deep-sea Squid *Octopoteuthis deletron* Young 1972. *Biol. Bull.* **2009**, *216*, 7–22.

Byrne, R. A.; Kuba, M. J.; Meisel, D. V; Griebel, U.; Mather, J. A. Does *Octopus vulgaris* Have Preferred Arms? *J. Comp. Psychol.* **2006**, *120*, 198–204.

Calisti, M.; Giorelli, M.; Levy, G.; Mazzolai, B.; Hochner, B.; Laschi, C.; Dario, P. An Octopus-bioinspired Solution to Movement and Manipulation for Soft Robots. *Bioinspir. Biomimet.* **2011**, *6*(3), 036002.

Calisti, M.; Arienti, A.; Renda, F.; Levy, G.; Hochner, B.; Mazzolai, B.; Dario, P.; Laschi, C. Design and Development of a Soft Robot with Crawling and Grasping Capabilities. In *Proceedings of the IEEE International Conference of Robotics and Automation*, Minnesota, May 14–13, 2012; pp 4950–4955.

Calisti, M.; Corucci, F.; Arienti, A.; Laschi, C. Bipedal Walking of an Octopus-inspired Robot. In *Biommimetic and Biohybrid Systems*; Duff, A., Lepora, N. F., Mura, A., Prescott, T. J., Verschure, P. F. M. J., Eds.; Springer International Publishing: Switzerland, 2014; pp 35–46.

Chamberlain, Jr., J. A. The Role of Body Extension in Cephalopod Locomotion. *Palaeontology* **1980**, *23*, 445–461.

Chamberlain, Jr., J. A. Hydromechanical Design of Fossil Cephalopods. *Syst. Assoc. Spec. Vol. Ser.* **1981**, 289–336.

Chamberlain, Jr., J. A. Jet Propulsion of *Nautilus*: A Surviving Example of Early Paleozoic Cephalopod Locomotor Design. *Can. J. Zool.* **1988**, *68*, 806–814.

Chamberlain, Jr., J. A. Cephalopod Locomotor Design and Evolution: The Constraints of Jet Propulsion. In *Biomechanics in Evolution*; Rayner, J.; Wootton, R., Eds.; Cambridge University Press: Cambridge, UK, 1991; pp 57–98.

Chamberlain, Jr., J. A.; Saunders, W. B.; Landman, N. H. Locomotion of *Nautilus*. In *Nautilus*; Saunders, W., Landman, N., Eds.; Springer: Netherlands, 2010; Vol 6, pp 489–525.

Cianchetti, M.; Arienti, A.; Follador, M.; Mazzolai, B.; Dario, P.; Laschi, C. Design Concept and Validation of a Robotic Arm Inspired by the Octopus. *Mater. Sci. Eng., C* **2011**, *31*, 1230–1239.

Clarke, M. Evolution of Buoyancy and Locomotion in Recent Cephalopods. In *The Mollusca: Paleontology and Neontology of Cephalopods*; Wilbur, K., Clarke, M., Trueman, E., Eds.; Academic Press: New York, 1988; Vol 12, pp 203–213.

Clarke, M. R.; Denton, E. J.; Gilpin-Brown, J. B. On the Use of Ammonium for Buoyancy in Squids. *J. Mar. Biol. Assoc. U. K.* **1979**, *59*, 259–276.

Clarke, M.; Trueman, E. Introduction. In *The Mollusca: Paleontology and Neontology of Cephalopods*; Wilbur, K.; Clarke, M.; Trueman, E., Eds.; Academic Press: New York, 1988; Vol 12, pp 1–11.

Cole, A. G.; Hall, B. K. Cartilage Differentiation in Cephalopod Molluscs. *Zoology* **2009**, *112*, 2–15.

Collins, M. A.; Villaneuva, R. Taxonomy, Ecology and Behaviour of the Cirrate Octopods. *Oceanogr. Mar. Biol. Annu. Rev.* **2006**, *44*, 277–322.

Crimaldi, J.; Koehl, M.; Koseff, J. Effects of the Resolution and Kinematics of Olfactory Appendages on the Interception of Chemical Signals in a Turbulent Odor Plume. *Environ. Fluid Mech.* **2002**, *2*, 35–63.

DeMont, M. E.; Ford, M. D.; Mitchell, S. C. *Locomotion in Invertebrates*; eLS, 2005, doi: 10.1038/npg.els.0003641.

Denton, E. J.; Shaw, T. I. The Buoyancy of Gelatinous Marine Animals. *J. Physiol.* **1961**, *161*, 14–15.

Denton, E. J.; Gilpin-Brown, J. B. The Buoyancy of the Cuttlefish, *Sepia officinalis* (L.). *J. Mar. Biol. Assoc. U. K.* **1961**, *41*, 319–342.

Denton, E. J.; Shaw, T. I. A Buoyancy Mechanism Found in Cranchiid Squid. *Proc. R. Soc. London. Ser. B: Biol. Sci.* **1969**, *174*, 271–279.

Denton, E. J.; Gilpin-Brown, J. B.; Howarth, J. V. On the Buoyancy of *Spirula spirula*. *J. Mar. Biol. Assoc. UK* **1967**, 181–191.

Dickinson, M. H.; Farley, C. T.; Full, R. J.; Koehl, M. A; Kram, R.; Lehman, S. How Animals Move: An Integrative View. *Science* **2000**, *288*, 100–106.

Doguzhaeva, L. A.; Mutvei, H.; Stehli, F. G.; Jones, D. S. Attachment of the Body to the Shell in Ammonoids. In *Ammonoid Paleobiology*; Landman, N., Ed.; Plenum Press: New York, 1996; Vol 13, pp 43–63.

Ebel, K. Hydrostatics of Fossil Ectocochleate Cephalopods and Its Significance for the Reconstruction of Their Lifestyle. *Paläeont. Z.* **1999**, *73*, 277–288.

Engeser, T. S. Fossil "Octopods"—A Critical Review. In *The Mollusca, Volume 12: Paleontology and Neontology of Cephalopods*; Wilbur, K. M.; Clarke, M. R.; Trueman, E. R., Eds.; Academic Press: New York, 1988; pp 81–87.

Finke, E.; Pörtner, H. O.; Lee, P. G.; Webber, D. M. Squid (*Lolliguncula brevis*) Life in Shallow Waters: Oxygen Limitation of Metabolism and Swimming Performance. *J. Exp. Biol.* **1996**, *199*, 911–921.

Finn, J. K.; Norman, M. D. The Argonaut Shell: Gas-mediated Buoyancy Control in a Pelagic Octopus. *Proc. Biol. Sci.* **2010**, *277*, 2967–2971.

Finn, J. K.; Tregenza, T.; Norman, M. D. Defensive Tool Use in a Coconut-carrying Octopus. *Curr. Biol.* **2009**, *19*, R1069–R1070.

Flash, T.; Hochner, B. Motor Primitives in Vertebrates and Invertebrates. *Curr. Opin. Neurobiol.* **2005**, *15*, 660–666.

Flash, T.; Kier, W.; Hochner, B.; Tsakiris, D.; Laschi, C. Controlling Movement in the Octopus—From Biological to Robotic Arms. In *Society for the Neural Control of Movement 22nd Annual Meeting*; Venice, Italy, 2012; p 17.

Gilly, W. F. Horizontal and Vertical Migrations of *Dosidicus gigas* in the Gulf of California Revealed by Electronic Tagging. In *The Role of Squids in Open Ocean Ecosystems: Report of a GLOBEC-CLIOTOP/PFRP Workshop*, Hawaii, November 16–17, 2006; Olson, R. J., Young, J. W., Eds.; 2006; Vol 24, pp 3–6.

Gilly, W. F.; Zeidberg, L. D.; Booth, J. A. T.; Stewart, J. S.; Marshall, G.; Abernathy, K.; Bell, L. E. Locomotion and Behavior of Humboldt Squid, *Dosidicus gigas*, in Relation to Natural Hypoxia in the Gulf of California, Mexico. *J. Exp. Biol.* **2012**, *215*, 3175–3190.

Gosline, J. M.; DeMont, M. E. Jet-propelled Swimming in Squids. *Sci. Am.* **1985**, *256*, 96–103.

Gosline, J. M.; Steeves, J. D.; Anthony, D.; DeMont, M. E. Patterns of Circular and Radial Mantle Muscle Activity in Respiration and Jetting of the Squid *Loligo opalescens*. *J. Exp. Biol.* **1983**, *104*, 97–109.

Grasso, F. W. Octopus Sucker-arm Coordination in Grasping and Manipulation. *Am. Malacol. Bull.* **2008**, *24*, 13–23.

Grasso, F. W. The Octopus with Two Brains: How Are Distributed and Central Representations Integrated in the Octopus Central Nervous System? In *Cephalopod Cognition*; Darmaillacq, A.-S., Dickel, L., Mather, J., Eds.; Cambridge University Press: Cambridge, 2014; pp 94–124.

Grasso, F. W.; Basil, J. A. The Evolution of Flexible Behavioral Repertoires in Cephalopod Molluscs. *Brain. Behav. Evol.* **2009**, *74*, 231–245.

Graziadei, P. Receptors in the Suckers of *Octopus*. *Nature* **1962**, *195*, 57–59.

Graziadei, P. Muscle Receptors in Cephalopods. *Proc. R. Soc. B: Biol. Sci.* **1965**, *161*, 392–402.

Graziadei, P. The Nervous System of the Arms. In *The Anatomy of the Nervous System of Octopus vulgaris*; Young, J. Z., Ed.; Clarendon Press: Oxford, 1971; pp 45–62.

Graziadei, P. P. C.; Gagne, H. T. Sensory Innervation in the Rim of the *Octopus* Sucker. *J. Morphol.* **1976**, *150*, 639–679.

Guerin, J. Contribution a` l'étude des systémes cutané, musculaire et nerveux de l'appareil tentaculaire des céphalopodes. *Archs. Zool. Exp. Gen.* **1908**, *38*, 1–178.

Guerra, Á.; Rocha, F.; Gonzalez, A. E.; Gonzalez, J. L. First Observation of Sand-covering by the Lesser Octopus *Eledone cirrhosa*. *Iberus* **2006**, *24*, 27–31.

Gutfreund, Y.; Flash, T.; Yarom, Y.; Fiorito, G.; Segev, I.; Hochner, B. Organization of Octopus Arm Movements: A Model System for Studying the Control of Flexible Arms. *J. Neurosci.* **1996**, *16*, 7297–7307.

Gutnick, T.; Byrne, R. A.; Hochner, B.; Kuba, M. *Octopus vulgaris* Uses Visual Information to Determine the Location of its Arm. *Curr. Biol.* **2011**, *21*, 460–462.

Hanlon, R. T.; Hixon, R. Body Patterning and Field Observations of *Octopus burryi* Voss, 1950. *Bull. Mar. Sci.* **1980**, *30*(4), 749–755.

Hanlon, R. T.; Wolterding, M. R. Behavior, Body Patterning, Growth and Life History of *Octopus briareus* Cultured in the Laboratory. *Am. Malacol. Bull.* **1989**, *7*, 21–45.

Hanlon, R. T.; Messenger, J. B. *Cephalopod Behaviour*; Cambridge University Press: Cambridge, UK, 1996.

Hanlon, R. T.; Forsythe, J. W.; Joneschild, D. E. Crypsis, Conspicuousness, Mimicry, and Polyphenism as Antipredator Defences of Foraging Octopuses on Indo-Pacific Coral Reefs, with a Method of Quantifying Crypsis from Video Tapes. *Biol. J. Linn. Soc.* **1999**, *66*, 1–22.

Hanlon, R. T.; Conroy, L. -A.; Forsythe, J. W. Mimicry and Foraging Behaviour of Two Tropical Sand-flat Octopus Species off North Sulawesi, Indonesia. *Biol. J. Linn. Soc.* **2007**, *93*, 23–38.

Hanlon, R. T.; Watson, A. C.; Barbosa, A. A "Mimic Octopus" in the Atlantic: Flatfish Mimicry and Camouflage by *Macrotritopus defilippi*. *Biol. Bull.* **2010**, *218*, 15–24.

Harrop, J.; Vecchione, M.; Felley, J. D. *In Situ* Observations on Behaviour of the Ommastrephid Squid Genus *Illex* (Cephalopoda: Ommastrephidae) in the Northwestern Atlantic. *J. Nat. Hist.* **2014**, *48*(41–42), 2501–2516.

Haszprunar, G. The First Molluscs—Small Animals. *Boll. Zool.* **1992**, *59*, 1–16.

Haszprunar, G.; Wanninger, A. Molluscs. *Curr. Biol.* **2012**, *22*, R510–R514.

Hoar, J. A.; Sim, E.; Webber, D. M.; O'Dor, R. K. The Role of Fins in the Competition between Squid and Fish. In *Mechanics and Physiology of Animal Swimming*; Maddock, L., Bone, Q., Raynor, J. M. V., Eds.; Cambridge University Press: Cambridge, 1994; pp 27–43.

Hochberg, F.; Norman, M.; Finn, J. *Wunderpus photogenicus* N. Gen. and Sp., a New Octopus from the Shallow Waters of the Indo-Malayan Archipelago (Cephalopoda: Octopodidae). *Molluscan Res.* **2006**, *26*, 128–140.

Hochner, B. An Embodied View of Octopus Neurobiology. *Curr. Biol.* **2012**, *22*, R887–R892.

Hochner, B. How Nervous Systems Evolve in Relation to their Embodiment: What We Can Learn from Octopuses and Other Molluscs. *Brain. Behav. Evol.* **2013**, *82*, 19–30.

Hou, J.; Bonser, R. H. C.; Jeronimidis, G. Design of a Biomimetic Skin for an Octopus-inspired Robot—Part I: Characterising Octopus Skin. *J. Bionic Eng.* **2011**, *8*, 288–296.

Hou, J.; Bonser, R. H. C.; Jeronimidis, G. Development of Sensorized Arm Skin for an Octopus Inspired Robot—Part I: Soft Skin Artifacts. In *Biomimetic and Biohybrid Systems: First International Conference Living Machines*; Prescott, T. J., Lepora, N. F., Mura, A., Verschure, P. F. M. J., Eds.; Springer: Berlin, 2012; pp 3840–3845.

Hoving, H. -J. T.; Perez, J. A.; Bolstad, K. S. R.; Braid, H. E.; Evans, A. B.; Fuchs, D.; Judkins, H.; Kelly, J. T.; Marian, J. E. A. R.; Nakajima, R.; et al. The Study of Deep-sea Cephalopods. In *Advances in Marine Biology*; Vidal, E. A. G., Ed; Elsevier: Oxford, 2014; Vol 67, pp 235–359.

Huffard, C. L. Locomotion by *Abdopus aculeatus* (Cephalopoda: Octopodidae): Walking the Line between Primary and Secondary Defenses. *J. Exp. Biol.* **2006**, *209*, 3697–3707.

Huffard, C. L.; Godfrey-Smith, P. Field Observations of Mating in *Octopus tetricus* Gould, 1852 and *Amphioctopus marginatus* (Taki, 1964) (Cephalopoda: Octopodidae). *Molluscan Res.* **2010**, *30*, 81–86.

Huffard, C. L.; Boneka, F.; Full, R. J. Underwater Bipedal Locomotion by Octopuses in Disguise. *Science* **2005**, *307*, 1927.

Hunt, J. C.; Zeidberg, L. D.; Hamner, W. M.; Robison, B. H. The Behaviour of *Loligo opalescens* (Mollusca: Cephalopoda) as Observed by a Remotely Operated Vehicle (ROV). *J. Mar. Biol. Assoc. U. K.* **2000**, *80*, 873–883.

Jacobs, D. K. Shape, Drag, and Power in Ammonoid Swimming. *Paleobiology* **1992**, *18*, 203–220.

Jacobs, D. K.; Landman, N. H. *Nautilus*—a Poor Model for the Function and Behavior of Ammonoids? *Lethaia* **1993**, *26*, 101–111.

Jacobs, D. K.; Chamberlain, Jr., J. A.; Stehli, F. G.; Jones, D. S. Buoyancy and Hydrodynamics in Ammonoids. *Ammonoid Paleobiol.* **1996**, *13*, 169–224.

Johnsen, S.; Kier, W. M. Intramuscular Crossed Connective Tissue Fibres: Skeletal Support in the Lateral Fins of Squid and Cuttlefish (Mollusca: Cephalopoda). *J. Zool.* **1993**, *231*, 311–338.

Johnsen, S.; Balser, E. J.; Fisher, E. C.; Widder, E. A. Bioluminescence in the Deep-sea Cirrate Octopod *Stauroteuthis syrtensis* Verrill (Mollusca: Cephalopoda). *Biol. Bull.* **1999**, *197*, 26–39.

Jordan, M.; Chamberlain, J. A.; Chamberlain, R. B. Response of *Nautilus* to Variation in Ambient Pressure. *J. Exp. Biol.* **1988,** *137*, 175–190.

Jozet-Alves, C.; Darmaillacq, A.-S.; Boal, J. G. Navigation in Cephalopods. In *Cephalopod Cognition*; Darmaillacq, A.-S.; Dickel, L.; Mather, J., Eds.; Cambridge University Press: Cambridge, 2014; pp 150–176.

Kang, R.; Kazakidi, A.; Guglielmino, E.; Branson, D. T.; Tsakiris, D. P.; Ekaterinaris, J. A.; Caldwell, D. G. Dynamic Model of a Hyper-Redundant, Octopus-like Manipulator for Underwater Applications. In *Proceedings of the IEEE International Conference on Intelligent Robots and Systems*, San Francisco, CA September 25–30, 2011; pp 4054–4059.

Kang, R.; Branson, D. T.; Guglielmino, E.; Caldwell, D. G. Dynamic Modeling and Control of an Octopus Inspired Multiple Continuum Arm Robot. *Comput. Math. Appl.* **2012,** *64*, 1004–1016.

Kier, W. M. The Musculature of Squid Arms and Tentacles : Ultrastructural Evidence for Functional Differences. *J. Morphol.* **1985,** *185*, 223–239.

Kier, W. M. The Arrangement and Function of Molluscan Muscle. In *The Mollusca: Form and Function*; Wilbur, K. M., Trueman, E. R., Clarke, M. R., Eds.; Academic Press: New York, 1988; Vol 11, pp 211–252.

Kier, W. M. The Fin Musculature of Cuttlefish and Squid (Mollusca, Cephalopoda): Morphology and Mechanics. *J. Zool.* **1989,** *217*, 23–38.

Kier, W. M. Squid Cross-striated Muscle: The Evolution of a Specialized Muscle Fiber Type. *Bull. Mar. Sci.* **1991,** *49*, 389–403.

Kier, W. M. Hydrostatic Skeletons and Muscular Hydrostats. In *Biomechanics: Structures and Systems: A Practical Approach*; Biewener, A. A., Ed.; Oxford University Press: Oxford, 1992; Vol 92, pp 205–231.

Kier, W. M. Muscle Development in Squid: Ultrastructural Differentiation of a Specialized Muscle Fiber Type. *J. Morphol.* **1996,** *229*, 271–288.

Kier, W. M. The Diversity of Hydrostatic Skeletons. *J. Exp. Biol.* **2012,** *215*, 1247–1257.

Kier, W. M.; Smith, K. K. Tongues, Tentacles and Trunks: The Biomechanics of Movement in Muscular-hydrostats. *Zool. J. Linn. Soc.* **1985,** *83*, 307–324.

Kier, W. M.; Smith, A. The Morphology and Mechanics of Octopus Suckers. *Biol. Bull.* **1990,** *178*, 126–136.

Kier, W. M.; Leeuwen, J. A. Kinematic Analysis of Tentacle Extension in the Squid *Loligo pealei*. *J. Exp. Biol.* **1997,** *200*, 41–53.

Kier, W. M.; Curtin, N. A. Fast Muscle in Squid (*Loligo pealei*): Contractile Properties of a Specialized Muscle Fibre Type. *J. Exp. Biol.* **2002,** *205*, 1907–1916.

Kier, W. M.; Smith, A. M. The Structure and Adhesive Mechanism of Octopus Suckers. *Integr. Comp. Biol.* **2002,** *42*, 1146–1153.

Kier, W. M.; Thompson, J. T. Muscle Arrangement, Function and Specialization in Recent Coleoids. *Berliner Palaobiol. Abh.* **2003,** 141–162.

Kier, W. M.; Stella, M. P. The Arrangement and Function of Octopus Arm Musculature and Connective Tissue. *J. Morphol.* **2007,** *268*, 831–843.

Kier, W. M.; Smith, K. K.; Miyan, J. A. Electromyography of the Fin Musculature of the Cuttlefish *Sepia officinalis*. *J. Exp. Biol.* **1989,** *143*, 17–31.

King, A. J.; Henderson, S. M.; Schmidt, M. H.; Cole, A. G.; Adamo, S. A. Using Ultrasound to Understand Vascular and Mantle Contributions to Venous Return in the Cephalopod *Sepia officinalis* L. *J. Exp. Biol.* **2005,** *208*, 2071–2082.

Klug, C.; Korn, D. The Origin of Ammonoid Locomotion. *Acta Palaeontol. Pol.* **2004,** *49*, 235–242.

Kröger, B. Antipredatory Traits of the Ammonoid Shell-indications from Jurassic Ammonoids with Sublethal Injuries. *Paläont. Z.* **2002,** *76,* 223–234.

Kröger, B.; Vinther, J.; Fuchs, D. Cephalopod Origin and Evolution: A Congruent Picture Emerging from Fossils, Development and Molecules: Extant Cephalopods are Younger than Previously Realised and Were under Major Selection to Become Agile, Shell-less Predators. *Bioessays* **2011,** *33,* 602–613.

Laan, A.; Gutnick, T.; Kuba, M. J.; Laurent, G. Behavioral Analysis of Cuttlefish Traveling Waves and Its Implications for Neural Control. *Curr. Biol.* **2014,** *24,* 1737–1742.

Laschi, C.; Mazzolai, B.; Mattoli, V.; Cianchetti, M.; Dario, P. Design of a Biomimetic Robotic Octopus Arm. *Bioinspir. Biomim.* **2009,** *4*(1), 015006.

Levy, G.; Flash, T.; Hochner, B. Arm Coordination in Octopus Crawling Involves Unique Motor Control Strategies. *Curr. Biol.* **2015,** *25,* 1–6.

Li, T.; Nakajima, K.; Calisti, M.; Laschi, C.; Pfeifer, R. Octopus-inspired Sensorimotor Control of a Multi-arm Soft Robot. In *Proceedings of the IEEE International Conference on Mechatronics and Automation,* Chengdu, China, August 5–8, 2012; pp 948–955.

Liu, F.; Lee, K. M.; Yang, C. J. Hydrodynamics of an Undulating Fin for a Wave-like Locomotion System Design. *IEEE/ASME Trans. Mechatronics* **2012,** *17,* 554–562.

Maciá, S.; Robinson, M. P.; Craze, P.; Dalton, R.; Thomas, J. D. New Observations on Airborne Jet Propulsion (Flight) in Squid, with a Review of Previous Reports. *J. Moll. Stud.* **2004,** *70,* 297–299.

Maddock, L.; Young, J. Z. Quantitative Differences among the Brains of Cephalopods. *J. Zool.* **1987,** *212,* 739–767.

Mangold, K. *Eledone moschata.* In *Cephalopods Life Cycle. Species Accounts*; Boyle, P. R., Ed.; Academic Press: London, 1983; Vol I, pp 387–400.

Margheri, L.; Mazzolai, B.; Cianchetti, M.; Dario, P.; Laschi, C. Tools and Methods for Experimental In-Vivo Measurement and Biomechanical Characterization of an *Octopus vulgaris* Arm. In *Annual International Conference of the IEEE Engineering in Medicine and Biology Society*; September 2–6, 2009; pp 7196–7199.

Margheri, L.; Mazzolai, B.; Ponte, G.; Fiorito, G.; Dario, P.; Laschi, C. Methods and Tools for the Anatomical Study and Experimental *in vivo* Measurement of the *Octopus vulgaris* Arm for Biomimetic Design. In *Proceedings of the IEEE RAS & EMBS International Conference on Biomedical Robotics and Biomechatronics*; Tokyo, Japan, September 26–29, 2010; pp 467–472.

Margheri, L.; Ponte, G.; Mazzolai, B.; Laschi, C.; Fiorito, G. Non-invasive Study of *Octopus vulgaris* Arm Morphology Using Ultrasound. *J. Exp. Biol.* **2011,** *214,* 3727–3731.

Mather, J. A. Sand Digging in *Sepia officinalis*: Assessment of a Cephalopod Mollusc's "Fixed" Behavior Pattern. *J. Comp. Psychol.* **1986,** *100,* 315–320.

Mather, J. A. How Do Octopuses Use Their Arms? *J. Comp. Psychol.* **1998,** *112,* 306–316.

Mather, J. A.; Mather, D. L. Skin Colours and Patterns of Juvenile *Octopus vulgaris* (Mollusca: Cephalopoda) in Bermuda. *Vie Milieu* **1994,** *44,* 267–272.

Mather, J.; Alupay, J. S. An Ethogram for Benthic Octopods (Cephalopoda:Octopoda). *J. Comp. Psychol.* **2016,** *130*(2), 109–127. http://dx.doi.org/10.1037/com0000025.

Mather, J. A..; Griebel, U.; Byrne, R. A. Squid Dances: An Ethogram of Postures and Actions of *Sepioteuthis sepioidea* Squid with a Muscular Hydrostatic System. *Mar. Freshw. Behav. Physiol.* **2010,** *43,* 45–61.

Mather, J. A.; Alupay, J. S.; Iskarous, K. Unravelling the Kaleidoscope of Patterns on the Octopus Skin. In *Animal Behavior Society*; Princeton, NJ, 2014.

Mazzolai, B.; Laschi, C.; Cianchetti, M.; Patanè, F.; Bassi-Luciani, L.; Izzo, I.; Dario, P. Biorobotic Investigation on the Muscle Structure of an Octopus Tentacle. In *Annual International Conference of the IEEE Engineering in Medicine and Biology Society*; Pisa, Italy, 2007; pp 1471–1474.

McMahan, W.; Jones, B. Robotic Manipulators Inspired by Cephalopod Limbs. In *Proceedings of the Canadian Engineering Education Association*, 2011.

Messenger, J. B. The Effects on Locomotion of Lesions to the Visuo-motor System in *Octopus*. *Proc. R. Soc. B: Biol. Sci.* **1967a**, *167*, 252–281.

Messenger, J. B. The Peduncle Lobe: A Visuo-motor Centre in *Octopus*. *Proc. R. Soc. Lond. B: Biol. Sci.* **1967b**, *167*, 225–251.

Messenger, J. B. The Optic Tract Lobes. In *The Anatomy of the Nervous System of Octopus vulgaris*; Young, J. Z., Ed.; Oxford University Press: London, 1971; pp 481–506.

Messenger, J. B. The Nervous System of *Loligo*. IV. The Peduncle and Olfactory Lobes. *Philos. Trans. R. Soc. London Ser. B: Biol. Sci.* **1979**, *285*, 275–309.

Messenger, J. B. Neurotransmitters of Cephalopods. *Invertebr. Neurosci.* **1996**, *2*, 95–114.

Navet, S.; Bassaglia, Y.; Baratte, S.; Andouche, A.; Bonnaud, L. Shell Reduction and Locomotory Development in Cephalopods: The Recruitment of Engrailed and NK4 Genes in *Sepia officinalis*. *Ferrantia* **2010**, *59*, 156–164.

Nesher, N.; Feinstein, N.; Anglister, L.; Finkel, E.; Hochner, B. Characterization of the Cholinergic Motor Innervation in the Neuromuscular System of the Octopus Arm. *J. Mol. Neurosci.* **2012**, *48*, S85.

Nesher, N.; Levy, G.; Grasso, F. W.; Hochner, B. Self-recognition Mechanism between Skin and Suckers Prevents Octopus Arms from Interfering with Each Other. *Curr. Biol.* **2014**, *24(11)*, 1271–1275.

Neumeister, H.; Budelmann, B. U. Structure and Function of the *Nautilus* Statocyst. *Philos. Trans. R. Soc. Lond., B: Biol. Sci.* **1997**, *352*, 1565–1588.

Nixon, M.; Young, J. *The Brains and Lives of Cephalopods*; Oxford University Press: Oxford, 2003.

Norman, M. D.; Finn, J. Revision of the *Octopus horridus* Species-group, Including Erection of a New Subgenus and Description of Two Member Species from the Great Barrier Reef, Australia. *Invertebr. Taxon.* **2001**, *15*, 13–35.

O'Dor, R. K. Respiratory Metabolism and Swimming Performance of the Squid, *Loligo opalescens*. *Can. J. Fish. Aquat. Sci.* **1982**, *39*, 580–587.

O'Dor, R. K. The Forces Acting on Swimming Squid. *J. Exp. Biol.* **1988**, *137*, 421–442.

O'Dor, R. Telemetered Cephalopod Energetics: Swimming, Soaring, and Blimping. *Integr. Comp. Biol.* **2002**, *1070*, 1065–1070.

O'Dor, R. K.; Webber, D. M. The Constraints on Cephalopods: Why Squid Aren't Fish. *Can. J. Zool.* **1986**, *64*, 1591–1605.

O'Dor, R. K.; Webber, D. Invertebrate Athletes: Trade-offs between Transport Efficiency and Power Density in Cephalopod Evolution. *J. Exp. Biol.* **1991**, *160*, 93–112.

O'Dor, R. K.; Hoar, J. A. Does Geometry Limit Squid Growth? *ICES J. Mar. Sci.* **2000**, *57*, 8–14.

O'Dor, R. K.; Wells, J.; Wells, M. J. Speed, Jet Pressure and Oxygen Consumption Relationships in Free-swimming *Nautilus*. *J. Exp. Biol.* **1990**, *154*, 383–396.

O'Dor, R. K.; Hoar, J. A.; Webber, D. M.; Carey, F. G.; Tanaka, S.; Martins, H. R.; Porteiro, F. M. Squid (*Loligo forbesi*) Performance and Metabolic Rates in Nature. *Mar.. Behav. Physiol.* **1994**, *25*, 163–177.

O'Dor, R. K.; Adamo, S.; Aitken, J. P.; Andrade, Y.; Finn, J.; Hanlon, R. T.; Jackson, G. D. Currents as Environmental Constraints on the Behavior, Energetics and Distribution of Squid and Cuttlefish. *Bull. Mar. Sci.* **2002**, *71*, 601–617.

O'Dor, R.; Stewart, J.; Gilly, W. F.; Payne, J.; Borges, T. C.; Thys, T. Squid Rocket Science: How Squid Launch into Air. *Deep. Res. Part II Top. Stud. Oceanogr.* **2013**, *95*, 113–118.

Packard, A. Cephalopods and Fish: The Limits of Convergence. *Biol. Rev.* **1972**, *47*, 241–307.

Packard, A.; Sanders, G. Body Patterns of *Octopus vulgaris* and Maturation of the Response to Disturbance. *Anim. Behav.* **1971**, *19*, 780–790.

Packard, A.; Trueman, E. R. Muscular Activity of the Mantle of *Sepia* and *Loligo* (Cephalopoda) During Respiratory Movements and Jetting, and its Physiological Interpretation. *J. Exp. Biol.* **1974**, *61*, 411–419.

Packard, A.; Wurtz, M. An Octopus, *Ocythoe*, with a Swimbladder and Triple Jets. *Philos. Trans. R. Soc. B: Biol. Sci.* **1994**, *344*, 261–275.

Packard, A.; Bone, Q.; Hignette, M. Breathing and Swimming Movements in a Captive *Nautilus*. *J. Mar. Biol. Assoc. U. K.* **1980**, *60*, 313–327.

Payne, N. L.; Gillanders, B. M.; Seymour, R. S.; Webber, D. M.; Snelling, E. P.; Semmens, J. M. Accelerometry Estimates Field Metabolic Rate in Giant Australian Cuttlefish *Sepia apama* During Breeding. *J. Anim. Ecol.* **2011**, *80*, 422–430.

Poirier, R.; Chichery, R.; Dickel, L. Effects of Rearing Conditions on Sand Digging Efficiency in Juvenile Cuttlefish. *Behav. Process.* **2004**, *67*, 273–279.

Ponte, G.; Fiorito, G. Immunohistochemical Analysis of Neuronal Networks in the Nervous Nystem of *Octopus vulgaris*. In *Immunocytochemistry and Related Techniques, Neuromethods*; Merighi, A.; Lossi, L., Eds.; Springer: New York, 2015; Vol 101, pp 63–79.

Pörtner, H. O.; Zielinski, S. Environmental Constraints and the Physiology of Performance in Squids. *South Afr. J. Mar. Sci.* **1998**, *20*, 207–221.

Pörtner, H. O.; Webber, D. M.; O'Dor, R. K.; Boutilier, R. G. Metabolism and Energetics in Squid (*Illex illecebrosus, Loligo pealei*) during Muscular Fatigue and Recovery. *Am. J. Physiol.* **1993**, *265*, R157–R165.

Pörtner, H. O.; Finke, E.; Lee, P. G. Metabolic and Energy Correlates of Intracellular pH in Progressive Fatigue of Squid (*Lolliguncula brevis*) Mantle Muscle. *Am. J. Physiol.* **1996**, *271*, R1403–R1414.

Preuss, T.; Budelmann, B. U. Proprioceptive Hair Cells on the Neck of the Squid *Lolliguncula brevis*: A Sense Organ in Cephalopods for the Control of Head-to-body Position. *Philos. Trans. R. Soc. London Ser. B: Biol. Sci.* **1995**, *349*, 153–178.

Redmond, J.; Bourne, G.; Johansen, K. Oxygen Uptake by *Nautilus pompilius*. *J. Exp. Zool.* **1978**, *205*, 45–50.

Renda, F.; Boyer, F.; Laschi, C. Dynamic Model of a Jet-propelled Soft Robot Inspired by the Octopus Mantle. In *Biomimetic and Biohybrid Systems: Third International Conference Living Machines*; Duff, A., Lepora, N. F., Mura, A., Prescott, T. J., Verschure, P. F. M. J., Eds.; Springer International Publishing: Switzerland, 2014; pp 261–272.

Richter, J. N.; Hochner, B.; Kuba, M. J. Octopus Arm Movements under Constrained Conditions. Adaptation, Modification and Plasticity of Motor Primitives. **2015**, *218*(7), 1069–1076.

Ritterbush, K. A.; Hoffmann, R.; Lukeneder, A.; De Baets, K. Pelagic Palaeoecology: The Importance of Recent Constraints on Ammonoid Palaeobiology and Life History. *J. Zool.* **2014**, *292*, 229–241.

Robin, J. -P.; Roberts, M.; Zeidberg, L.; Bloor, I.; Rodriguez, A.; Briceño, F.; Downey, N.; Mascaró, M.; Navarro, M.; Guerra, A.; et al. Transitions during Cephalopod Life History:

The Role of Habitat, Environment, Functional Morphology and Behaviour. In *Advances in Marine Biology*; Elsevier, 2014; Vol 67, pp 361–437.

Rodrigues, M.; Garci, M. E.; Troncoso, J. S.; Guerra, A. Burying Behaviour in the Bobtail Squid *Sepiola atlantica* (Cephalopoda: Sepiolidae). *Ital. J. Zool.* **2010**, *77*, 247–251.

Roper, C. F. E.; Brundage, W. Cirrate Octopods with Associated Deep-sea Organisms: New Biological Data Based on Deep Benthic Photographs (Cephalopoda). *Smithson. Contrib. Zool.* **1972**, *121*, 1–46.

Roper, C. F. E.; Hochberg, F. Behavior and Systematics of Cephalopods from Lizard Island, Australia, Based on Color and Body Patterns. *Malacologia* **1988**, *29*, 153–193.

Roper, C. F. E.; Lu, C. C. Comparative Morphology and Function of Dermal Structures in Oceanic Squids (Cephalopoda). *Smithson. Contrib. Zool.* **1990**, *493*, 1–40.

Rosa, R.; Seibel, B. A. Metabolic Physiology of the Humboldt Squid, *Dosidicus gigas*: Implications for Vertical Migration in a Pronounced Oxygen Minimum Zone. *Prog. Oceanogr.* **2010a**, *86*, 72–80.

Rosa, R.; Seibel, B. A. Voyage of the Argonauts in the Pelagic Realm: Physiological and Behavioural Ecology of the Rare Paper Nautilus, *Argonauta nouryi*. *ICES J. Mar. Sci.* **2010b**, *67*, 1494–1500.

Rosenberger, L. J. Pectoral Fin Locomotion in Batoid Fishes: Undulation versus Oscillation. *J. Exp. Biol.* **2001**, *204*, 379–394.

Rowell, C.H.F. Activity of Interneurones in the Arm of *Octopus vulgaris* in Response to Tactile Stimulation. *J. Exp. Biol.* **1966**, *44*, 589–605.

Russell, F. S.; Steven, G. A. The Swimming of Cuttlefish. *Nature* **1930**, *125*, 893.

Seibel, B. A.; Thuesen, E. V.; Childress, J. J.; Gorodezky, L. A. Decline in Pelagic Cephalopod Metabolism with Habitat Depth Reflects Differences in Locomotory Efficiency. *Biol. Bull.* **1997**, *192*, 262–278.

Seibel, B. A.; Thuesen, E.; Childress, J. Flight of the Vampire: Ontogenetic Gait-transition in *Vampyroteuthis infernalis* (Cephalopoda: Vampyromorpha). *J. Exp. Biol.* **1998**, *201*, 2413–2424.

Seibel, B. A.; Thuesen, E. V.; Childress, J. J. Light- Limitation on Predator–prey Interactions: Consequences for Metabolism and Locomation of Deep-sea Cephalopods. *Biol. Bull.* **2000**, *198*, 284–298.

Seibel, B. A.; Goffredi, S. K.; Thuesen, E. V.; Childress, J. J.; Robison, B. H. Ammonium Content and Buoyancy in Midwater Cephalopods. *J. Exp. Mar. Biol. Ecol.* **2004**, *313*, 375–387.

Sfakiotakis, M.; Kazakidi, A.; Pateromichelakis, N.; Tsakiris, D. P. Octopus-inspired Eight-arm Robotic Swimming by Sculling Movements. In *Proceedings of the IEEE International Conference on Robotics and Automation;* May, 6–10, 2013a; pp 5155–5161.

Sfakiotakis, M.; Kazakidi, A.; Tsakiris, D. P. Turning Maneuvers of an Octopus-inspired Multi-arm Robotic Swimmer. In *Proceedings of the 21st Mediterranean Conference on Control & Automation*; Crete, Greece, June 25–28, 2013b; pp 1343–1349.

Sfakiotakis, M.; Kazakidi, A.; Chatzidaki, A.; Evdaimon, T.; Tsakiris, D. P. Multi-arm Robotic Swimming with Octopus-inspired Compliant Web. In *Proceedings of the IEEE/RSJ International Conference on Intelligent Robots and Systems*, Chicago, IL, September 14–18, 2014, pp 302–308.

Shadwick, R. E. Mechanical Organization of the Mantle and Circulatory System of Cephalopods. *Mar. Behav. Physiol.* **1994**, *25*, 69–85.

Smith, A. Cephalopod Sucker Design and the Physical Limits to Negative Pressure. *J. Exp. Biol.* **1996**, *199*, 949–958.

Sreeja, V.; Bijukumar, A. Ethological Studies of the Veined Octopus *Amphioctopus marginatus* (Taki) (Cephalopoda: Octopodidae) in Captivity, Kerala , India. *J. Threat. Taxa* **2013**, *5*, 4492–4497.

Staaf, D. J.; Gilly, W. F.; Denny, M. W. Aperture Effects in Squid Jet Propulsion. *J. Exp. Biol.* **2014**, *217*, 1588–1600.

Stewart, W. J.; Bartol, I. K.; Krueger, P. S. Hydrodynamic Fin Function of Brief Squid, *Lolliguncula brevis*. *J. Exp. Biol.* **2010**, *213*, 2009–2024.

Sumbre, G.; Gutfreund, Y.; Fiorito, G.; Flash, T.; Hochner, B. Control of Octopus Arm Extension by a Peripheral Motor Program. *Science* **2001**, *293(5536)*, 1845–1848.

Sumbre, G.; Fiorito, G.; Flash, T.; Hochner, B. Octopuses Use a Human-like Strategy to Control Precise Point-to-point Arm Movements. *Curr. Biol.* **2006**, *16*, 767–772.

Tansey, E. Neurotransmitters in the Cephalopod Brain: A Review. *Comp Biochem. Physiol., C. Comp. Pharmcol.* **1979**, *64*(2), 173–182.

Thompson, J. T.; Kier, W. M. Ontogenetic Changes in Mantle Kinematics During Escape-jet Locomotion in the Oval Squid, *Sepioteuthis lessoniana* Lesson, 1830. *Biol. Bull.* **2001**, *201*, 154–166.

Thompson, J. T.; Kier, W. M. Ontogeny of Squid Mantle Function : Changes in the Mechanics of Escape-jet Locomotion in the Oval Squid, *Sepioteuthis lessoniana* Lesson, 1830. *Biol. Bull.* **2002**, *203*, 14–26.

Thompson, J. T.; Kier, W. M. Ontogeny of Mantle Musculature and Implications for Jet Locomotion in Oval Squid *Sepioteuthis lessoniana*. *J. Exp. Biol.* **2006**, *209*, 433–443.

Thompson, J. T.; Shelton, R. M.; Kier, W. M. The Length-force Behavior and Operating Length Range of Squid Muscle Vary as a Function of Position in the Mantle Wall. *J. Exp. Biol.* **2014**, *217*, 2181–2192.

Trueblood, L. A.; Seibel, B. A. Slow Swimming, Fast Strikes: Effects of Feeding Behavior on Scaling of Anaerobic Metabolism in Epipelagic Squid. *J. Exp. Biol.* **2014**, *217*, 2710–2716.

Trueman, E. R. Swimming by Jet Propulsion. In *Aspects of Animal Movement*; Elder, H. Y., Trueman, E. R., Eds.; Cambridge University Press: Cambridge, 1980; pp 93–105.

Trueman, E. R. Locomotion in Molluscs. In *The Mollusca: Physiology, Part 1*; Saleuddin, A. S. M., Wilbur, K. M., Eds.; Academic Press: New York, 1983; Vol 4, pp 155–198.

Trueman, E. R.; Packard, A. Motor Performances of Some Cephalopods. *J. Exp. Biol.* **1968**, *49*, 495–507.

Trueman, E. R.; Clarke, M. R. Introduction. In *The Mollusca: Form and Function*; Wilbur, K. M., Trueman, E. R., Clarke, M. R., Eds.; Academic Press: New York, 1988; Vol 11, pp 1–9.

Van Leeuwen, J. L.; Kier, W. M. Functional Design of Tentacles in Squid: Linking Sarcomere Ultrastructure to Gross Morphological Dynamics. *Philos. Trans. R. Soc. B: Biol. Sci.* **1997**, *352*, 551–571.

Vecchione, M. Systematics and the Lifestyle and Performance of Cephalopods. *Mar. Freshw. Behav. Physiol.* **1995**, *25*, 179–191.

Vecchione, M.; Roper, C. F. E. Cephalopods Observed from Submersibles in the Western North Atlantic. *Bull. Mar. Sci.* **1991**, *49*, 433–445.

Vecchione, M.; Young, R. E. Aspects of the Functional Morphology of Cirrate Octopods: Locomotion and Feeding. *Vie Milieu* **1997**, *47*, 101–110.

Villanueva, R. Observations on the Behaviour of the Cirrate Octopod *Opisthoteuthis grimaldii* (Cephalopoda). *J. Mar. Biol. Assoc. U. K.* **2000**, *80*, 555–556.

Villanueva, R.; Segonzac, M.; Guerra, A. Locomotion Modes of Deep-sea Cirrate Octopods (Cephalopoda) Based on Observations from Video Recordings on the Mid-Atlantic Ridge. *Mar. Biol.* **1997a**, *129*, 113–122.

Villanueva, R.; Nozais, C.; Boletzky, S. von. Swimming Behaviour and Food Searching in Planktonic *Octopus vulgaris* Cuvier from Hatching to Settlement. *J. Exp. Mar. Bio. Ecol.* **1997b**, *208*, 169–184.

Voight, J. R. Observations of Deep-sea Octopodid Behavior from Undersea Vehicles. *Am. Malacol. Bull.* **2008**, *24*, 43–50.

Voight, J. R.; Pörtner, H. O.; O'Dor, R. K. A Review of Ammonia-mediated Buoyancy in Squids (Cephalopoda: Teuthoidea). *Mar. Freshw. Behav. Physiol.* **1995**, *25*, 193–203.

Wainwright, S.; Biggs, W.; Currey, J.; Gosline, J. *Mechanical Design in Organisms*. Princeton University Press: Princeton, NJ, 1982.

Wang, Z.; Hang, G.; Wang, Y. L. J. Swimming Mechanism of Squid/Cuttlefish and Its Application to Biomimetic Underwater Robots. *Chin. J. Mech. Eng.* **2008**, *44*(6), 1–9.

Ward, D. V; Wainwright, S. A. Locomotory Aspects of Squid Mantle Structure. *J. Zool.* **1972**, *167*, 437–449.

Webber, D.; Aitken, J.; O'Dor, R. Costs of Locomotion and Vertical Dynamics of Cephalopods and Fish. *Physiol. Biochem. Zool.* **2000**, *73*, 651–662.

Wells, M. J. Circulation in Cephalopods. In *The Mollusca: Physiology, Part 2*; Wilbur, K. M., Saleuddin, A. S. M., Eds.; Academic Press: New York, 1983; Vol 5; pp 239–290.

Wells, M. J. The Mantle Muscle and Mantle Cavity of Cephalopods. In *The Mollusca: Form and Function*; Wilbur, K. M., Trueman, E. R., Clarke, M. R., Eds.; Academic Press: New York, 1988; Vol 11; pp 287–300.

Wells, M. J. Oxygen Extraction and Jet Propulsion in Cephalopods. *Can. J. Zool.* **1990**, *68*, 815–824.

Wells, M. J. The Cephalopod Heart—The Evolution of a High-performance Invertebrate Pump. *Experientia* **1992**, *48*, 800–808.

Wells, M. J. The Evolution of a Racing Snail. *Mar.. Behav. Physiol.* **1994**, *25*, 1–12.

Wells, M. J.; Wells, J. Ventilatory Currents in the Mantle of Cephalopods. *J. Exp. Biol.* **1982**, *99*, 315–330.

Wells, M. J.; Smith, P. J. S. The Ventilation Cycle in *Octopus*. *J. Exp. Biol.* **1985**, *116*, 375–383.

Wells, M. J.; O'Dor, R. K. Jet Propulsion and the Evolution of the Cephalopods. *Bull. Mar. Sci.* **1991**, *49*, 419–432.

Wells, M. J.; O'Dor, R. K.; Mangold, K.; Wells, J. Oxygen Consumption in Movement by *Octopus*. *Mar. Behav. Physiol.* **1983**, *9*, 289–303.

Wild, E.; Wollesen, T.; Haszprunar, G.; Heß, M. Comparative 3D Microanatomy and Histology of the Eyes and Central Nervous Systems in Coleoid Cephalopod Hatchlings. *Org. Divers. Evol.* **2014**, *15*, 37–64.

Williamson, R. Factors Affecting the Sensory Response Characteristics of the Cephalopod Statocyst and their Relevance in Predicting Swimming Performance. *Biol. Bull.* **1991**, *180*, 221–227.

Williamson, R.; Budelmann, B. U. An Angular Acceleration Receptor System of Dual Sensitivity in the Statocyst of *Octopus vulgaris*. *Experientia* **1985**, *41*, 1321–1323.

Willy, A.; Low, K. H. Initial Experimental Investigation of Undulating Fin. *2005 IEEE/RSJ Int. Conf. Intell. Robot. Syst. IROS* **2005**, *1*, 2059–2064.

Wilson, D. M. Nervous Control of Movement in Cephalopods. *J. Exp. Biol.* **1960**, *37*, 57–72.

Xavier, J. C.; Allcock, A. L.; Cherel, Y.; Lipinski, M. R.; Pierce, G. J.; Rodhouse, P. G. K.; Rosa, R.; Shea, E. K.; Strugnell, J. M.; Vidal, E. A. G.; et al. Future Challenges in Cephalopod Research. *J. Mar. Biol. Assoc. U. K.* **2014**, 1–17.

Yekutieli, Y.; Sagiv-Zohar, R.; Aharonov, R.; Engel, Y.; Hochner, B.; Flash, T. Dynamic Model of the Octopus Arm. I. Biomechanics of the Octopus Reaching Movement. *J. Neurophysiol.* **2005**, *94*, 1443–1458.

Yekutieli, Y.; Sagiv-Zohar, R.; Hochner, B.; Flash, T. Dynamic Model of the Octopus Arm. II. Control of Reaching Movements. *J. Neurophysiol.* **2005**, *94*, 1459–1468.

York, C. A.; Bartol, I. K. Anti-predator Behavior of Squid Throughout Ontogeny. *J. Exp. Mar. Biol. Ecol.* **2016**, *480*, 26–35.

Young, J. Z. The Functioning of the Giant Nerve Fibres of the Squid. *J. Exp. Biol.* **1938**, *15*, 170–185.

Young, J. Z. Fused Neurons and Synaptic Contacts in the Giant Nerve Fibres of Cephalopods. *Philos. Trans. R. Soc. B: Biol. Sci.* **1939**, *229*, 465–503.

Young, J. Z. The Diameters of the Fibres of the Peripheral Nerves of *Octopus*. *Proc. R. Soc. Lond. Ser. B: Biol. Sci.* **1965**, *162*, 47–79.

Young, J. Z. *The Anatomy of the Nervous System of Octopus vulgaris*; Oxford University Press: London, UK, 1971.

Young, J. Z. The Organization of a Cephalopod Ganglion. *Philos. Trans. R. Soc. London, Ser. B: Biol. Sci.* **1972**, *263*, 409–429.

Young, J. Z. The Central Nervous System of *Loligo*. I. The Optic Lobe. *Philos. Trans. R. Soc. Lond. Ser. B: Biol. Sci.* **1974**, *267*, 263–302.

Young, J. Z. The Nervous System of *Loligo*. II. Suboesophageal Centres. *Philos. Trans. R. Soc. Lond. Ser. B: Biol. Sci.* **1976**, *274*, 101–167.

Young, J. Z. The Nervous System of *Loligo*. III. Higher Motor Centres: The Basal Supraoesophageal Lobes. *Philos. Trans. R. Soc. London, Ser. B: Biol. Sci.* **1977**, *276*, 351–398.

Young, J. Z. The Nervous System of *Loligo*. V. The Vertical Lobe Complex. *Philos. Trans. R. Soc. London Ser. B: Biol. Sci.* **1979**, *285*, 311–354.

Young, J. Z. The Distributed Tactile Memory System of *Octopus*. *Proc. R. Soc. Lond. Ser. B: Biol. Sci.* **1983**, *218*, 135–176.

Young, J. Z. The Statocysts of Cranchiid Squids (Cephalopoda). *J. Zool.* **1984**, *203*, 1–21.

Young, J. Z. Evolution of the Cephalopod Brain. In *The Mollusca: Paleontology and Neontology of Cephalopods/Paleontology and Neontology of Cephalopods*; Wilbur, K. M., Clarke, M. R., Trueman, E. R., Eds.; Academic Press: New York, 1988; Vol 12, pp 215–228.

Young, J. Z. The Angular Acceleration Receptor System of Diverse Cephalopods. *Philos. Trans. R. Soc. Lond., Ser. B: Biol. Sci.* **1989**, *325*, 189–237.

Young, R. E.; Vecchione, M. Analysis of Morphology to Determine Primary Sister-taxon Relationships within Coleoid Cephalopods. *Am. Malacol. Bull.* **1996**, *12*, 91–112.

Zelman, I.; Titon, M.; Yekutieli, Y.; Hanassy, S.; Hochner, B.; Flash, T. Kinematic Decomposition and Classification of Octopus Arm Movements. *Front. Comput. Neurosci.* **2013**, *7*, 60.

Zheng, T.; Godage, I. S.; Branson, D. T.; Kang, R.; Guglielmino, E.; Caldwell, D. G. Octopus Inspired Walking Robot: Design, Control and Experimental Validation. In *Proceedings of the IEEE International Conference on Robotics and Automation (ICRA)*; Karlsruhe, Germany, May 6–10, 2013; pp 816–821.

Zullo, L.; Sumbre, G.; Agnisola, C.; Flash, T.; Hochner, B. Nonsomatotopic Organization of the Higher Motor Centers in Octopus. *Curr. Biol.* **2009**, *19*, 1632–1636.

CHAPTER 8

KEY MOLECULAR REGULATORS OF METABOLIC RATE DEPRESSION IN THE ESTIVATING SNAIL *OTALA LACTEA*

CHRISTOPHER J. RAMNANAN[1*], RYAN A. BELL[2], and
JOHN-DOUGLAS MATTHEW HUGHES[1]

[1]Faculty of Medicine, University of Ottawa, Ottawa, ON, Canada K1H 8M5. E-mail: cramnana@uottawa.ca

[2]Ottawa Hospital Research Institute, Ottawa, ON, Canada

[*]Corresponding author.

CONTENTS

ABSTRACT

Estivation is a state of aerobic dormancy used by the desert land snail, *Otala lactea*, to endure harsh environmental conditions. Paramount to survival in the estivating state is the sustained and profound depression of metabolic rate, which facilitates survival from limited endogenous energy stores for extended periods of time. Metabolic rate depression requires coordinated suppression of ATP-generating and ATP-consuming cellular functions by stable regulatory mechanisms. One such mechanism that has been well-studied in this estivating species is reversible protein phosphorylation. Studies in *O. lactea* have established that protein phosphorylation has far-reaching regulatory capacity in facilitating the biochemical transition between active and estivating conditions. This mechanism plays a role in modifying the activities of rate-limiting enzymes of carbohydrate metabolism, enhancing intracellular tolerance to oxidative stress, suppressing ATP-intensive processes such as global protein turnover and ion pumping, and activating specific arms of the AMP-activated protein kinase (AMPK) signaling and the insulin signaling cascades. This review chapter will document the evidence suggesting that differential protein phosphorylation, brought about by specific protein kinases and protein phosphatases, is essential to regulating biochemical adaptations in *O. lactea* that are critical to survival.

8.1 INTRODUCTION

When food or water is limiting, the land snail *Otala lactea* enters a state of aerobic dormancy termed estivation. A number of behavioral, physiological, and biochemical adaptations are associated with this survival strategy. One particularly remarkable aspect of estivation is that estivating animals can profoundly suppress their metabolic rate, and this metabolic rate depression can be sustained for extensive periods of time. This adaptation facilitates energy conservation that permits the snails to live off finite energy stores for prolonged periods of time, which in turn facilitates survival in unfavorable living conditions for many months. The metabolic rate depression characteristic of estivation is common to other survival strategies that have evolved to help cope with difficult environmental conditions and has been observed and characterized to varying degrees in hibernating, freeze-tolerant, and anoxia-tolerant species (Storey, 2002; Storey & Storey, 2004). While different aspects of estivation have been characterized in several species of snails, the molecular biochemistry underlying the depression of metabolic rate has been particularly well-defined in *O. lactea*, especially in the recent decade. This chapter will discuss recent research highlights in the field of estivation, with a focus on biochemical markers that have been implicated as key regulators during the transition to the estivating state.

8.2   LACK OF WATER AND ELEVATED CARBON DIOXIDE ARE TRIGGERS FOR ESTIVATION

While *O. lactea* snails are native to the seasonally arid Mediterranean region, these animals have been introduced around the world to countries with seasonally challenging climates, including Canada. In fact, the studies described in this chapter involved *O. lactea* animals imported from Morocco and purchased from a local retailer in Ottawa, Ontario. *O. lactea* are typically active for only a few months of the year, particularly in challenging environments. During this time, these snails must accumulate body-fuel reserves that are sufficient for survival throughout many months in dormancy. Like many estivating animals, dormancy is triggered in *O. lactea* by sensing lack of food and water in the environment. When this occurs, these snails typically seek out sheltered sites that limit their exposure both to the elements and to predators. At the onset of estivation, snails secrete a calcified mucous membrane, called an epiphragm, which effectively limits evaporative water loss (Barnhart, 1983). Estivators may also elevate the solute concentrations of their body fluids, thereby utilizing the colligative properties of dissolved solutes such as urea to aid water retention (Withers & Guppy, 1996). Consistent with this, metabolic enzymes in estivating animals are frequently much more resistant to the denaturing effects of urea, when compared to enzymes in non-estivators (Cowan & Storey, 2002). Many animals (typically but not exclusively desert-dwelling organisms) also convert nitrogenous wastes into uric acid, since excretion and storage of nitrogen in this insoluble form requires minimal water (Dejours, 1989). The kidney has a relatively limited capacity for uric acid storage in several normal and estivating gastropods (e.g., Athawale & Reddy, 2002), where uric acid accumulates in specialized extrarenal cells and tissues (Giraud-Billoud et al., 2008; Vega et al., 2007). Extrarenal uric acid can protect cells and tissues against the oxidative stress associated with arousal from estivation (Giraud-Billoud et al., 2011, 2013). However, a role for uric acid in *O. lactea* has not yet been studied.

Evaporative water loss is also minimized by employing discontinuous breathing, as the snail's pneumostome, the specialized orifice leading to the lung, is open for brief periods, with long, irregular intervals between openings, allowing for intermittent respiration. Estivating *O. lactea* can take as little as 2–3 breaths per hour (Barnhart & McMahon, 1987). Thus, estivation in *O. lactea*, while aerobic, is characterized by long periods of apnea that are irregularly interrupted by short bouts of oxygen and carbon dioxide gas exchange (Barnhart, 1986). While physical adaptations such as the epiphragm and behavioral modifications such as the intermittent pneumostome opening contribute to water conservation, these adaptations also decrease ventilation of the lung. As a result, between breaths, snails experience a progressive hypoxia, extracellular and intracellular acidosis, and hypercapnia (Barnhart & McMahon, 1988). Between breaths, the partial pressure of carbon dioxide gradually rises and the partial pressure of oxygen gradually falls in the snail's tissues. It was determined that the interval length between breaths is determined by oxygen need (Barnhart, 1986). On the other hand, elevation of the partial pressure of carbon dioxide (or the resulting cellular acidosis that this creates) has been implicated as the molecular trigger for the suppression of metabolic rate (Guppy et al., 1994). Consistent with this notion, oxygen consumption (an index for metabolic rate) was reduced by 50% in active *O. lactea* that were exposed to carbon dioxide, but upon removal of carbon dioxide oxygen consumption quickly returned to normal levels in these snails (Barnhart & McMahon, 1988). Hence, elevated carbon dioxide can be considered a signal that triggers the substantial metabolic rate depression that occurs in estivating snails.

8.3 METABOLIC RATE DEPRESSION IS ESSENTIAL FOR SURVIVAL IN ESTIVATING ANIMALS

The transition into a hypometabolic state is advantageous to estivating animals because it extends the duration by which finite energy reserves can sustain life. This transition requires coordinated suppression of the rates of both energy consuming and energy producing pathways to create a new, lower balanced state of ATP turnover. Estivating snails typically show metabolic rates that are <30% of the corresponding resting rate in active snails (Bishop & Brand, 2000; Rees & Hand, 1990). The transition into a hypometabolic state requires coordinated suppression of the rates of both ATP-consuming and ATP-producing cell functions as well as a reorganization of the priorities for energy use to sustain essential cellular processes and suppress other energy-intensive functions such as growth, development, and reproduction (Storey & Storey, 2004). Carbohydrate oxidation was observed to be the primary energy reserve in two species of estivating mountain snails,

Oreohelix strigosa and *Oreohelix subrudis* (Rees & Hand, 1993). When carbohydrate stores were depleted after the first few months of estivation in these snails, only then was protein oxidation initiated; overall, there was little contribution at any stage from lipid breakdown (Rees & Hand, 1993). Carbohydrate reserves can therefore fuel life processes during estivation for months, and in extreme conditions, even years (Herreid, 1977), largely because these estivating snails can profoundly depress their metabolic rates.

Several levels of control can be involved in regulating entry into and/ or arousal from dormancy including changes in gene expression, protein synthesis, posttranslational modification, and allosteric control of enzymes. Physiologic states of natural metabolic rate depression are not typically accompanied by wholesale changes in gene transcription (Storey & Storey, 2004). Modification of gene expression patterns is an elaborate, time-consuming, and energy-intensive process, and it can take many hours for a change in the gene transcriptional level to manifest in a measurable change in the corresponding protein level. For example, substantial physiologic increases in glucoregulatory hormone concentrations can bring about rapid and substantial alterations in glucoregulatory gene expression in vivo, such that mRNA levels of target genes can be increased many fold or decreased up to 90% within 30 min, but several hours are required for these changes to manifest in actual modification of corresponding protein levels (Ramnanan et al., 2010a, 2011). This type of regulation would not be ideal for animals that move into a hypometabolic state in response to adverse conditions, since these animals typically need to make this transition quickly, in response to environmental and physiological cues. Thus, control at the level of gene expression is typically limited to just a few selected genes that are either up-regulated or down-regulated in terms of mRNA expression. Furthermore, the manufacturing of proteins is energetically demanding, and ATP is at a premium for dormant animals. Hence, large scale protein synthesis (and also large scale protein degradation) is not compatible with the animal's need for energy conservation during dormancy (and our studies on the topic will be discussed later in this chapter). In addition, dormant animals can be aroused rapidly when the environment returns to more favorable conditions; for example, *O. lactea* arouse within minutes of sensing water (Hermes-Lima et al., 1998). Since metabolic capabilities need to be maintained in a state or readiness to facilitate a rapid return to normal life, major metabolic restructuring at the level of altering protein levels does not typically occur during dormancy. Instead, the changes in gene expression and protein content are relatively subtle, and regulation comes instead from reversible and energy-efficient controls on metabolism. One such mechanism that has been well established to regulate a myriad of cellular functions across many hypometabolic states in several species is reversible protein phosphorylation.

8.4 REVERSIBLE PHOSPHORYLATION: KEY MECHANISM OF METABOLIC REGULATION

Reversible protein phosphorylation is an important and far-reaching theme in posttranslational regulation. The covalent binding of a phosphate group to an enzyme (mediated by protein kinases) or removal of a phosphate group (catalyzed by protein phosphatases) can have immediate, dramatic effects on enzyme activity and kinetic/regulatory properties, typically modifying enzymes from active (or more active) to inactive (or less active) conformations. Protein phosphorylation in animal cells typically targets serine or threonine residues (both having a free hydroxyl group) or tyrosine residues (featuring a phenolic group). The enzymes that mediate phosphorylation events are therefore typically classified as being either serine/threonine kinases or tyrosine kinases (and similar distinctions exist for the corresponding phosphatases). Moreover, serine/threonine kinases can be further subclassified based on their dependency on biochemical coactivators or second messengers, while serine/threonine phosphatases can be classified based on their substrate affinities, ion dependency, and sensitivity to natural or pharmacological inhibitors (Cohen, 1989).

The presence of saturating amounts of Mg^{2+}-ATP (a substrate and source of inorganic phosphate for kinase catalytic activity) and specific second messengers or molecular activators in experimental conditions can promote the activity of specific protein kinases. These kinases include: the cyclic AMP (cAMP)-dependent protein kinase (protein kinase A; PKA); the cyclic GMP (cGMP)-dependent protein kinase (protein kinase G; PKG): the calcium- and phorbol myristate acetate-activated protein kinase C (protein kinase C; PKC); the AMP-activated protein kinase (AMPK); and the calcium/calmodulin-dependent protein kinase (CaMK). Similarly, experimental conditions can be designed to be permissive for specific protein phosphatases, as protein phosphatases are known to be sensitive to different inhibitors and require certain ions for their function. Sodium fluoride (NaF) can inhibit all serine/threonine phosphatases, while freshly prepared sodium orthovanadate (Na_3VO_4) can inhibit global protein tyrosine phosphatase activity. Protein phosphatase of type-1 activity (PP1) and of type-2A (PP2A) are both ion-dependent and sensitive to the marine toxin okadaic acid, but can be differentiated due to PP2A being is 50–100-fold more sensitive to this inhibitor than PP1 (Cohen,

1989). Type-2B phosphatase activity (PP2B) is calcium- and calmodulin-dependent, whereas type-2C phosphatase activity (PP2C) is manganese- or magnesium-dependent. Moreover, these ion-dependent phosphatases can be inhibited by including chelating agents (EDTA or EGTA) in experimental conditions.

Covalent phosphorylation of a protein is a very stable mechanism for protein modification, and removal of phosphate(s) in vivo can only be achieved by the corresponding serine/threonine or tyrosine protein phosphatases. Modification of an enzyme via phosphorylation can have substantial effects on the flux of a metabolic pathway, at a much lower energetic cost than by other means of changing enzyme activities (i.e., protein synthesis or degradation). Another desirable feature of phosphorylation-mediated control is that the modification is readily reversible, and this allows the animal to respond quickly to environmental cues, and enter (or exit) the dormant state (and adjust metabolic rates) as rapidly (Storey, 2002). Metabolic pathways are commonly regulated by (1) controlling the rate of substrate entry into the pathway, and/or (2) controlling the rate of enzymatic reaction(s) that influence flux through the pathway. In every metabolic pathway, there is at least one nonequilibrium, highly exergonic (and essentially irreversible, under cellular conditions) reaction catalyzed by a low activity enzyme, the rate of which influences the rate of the entire pathway. These key pathway-controlling enzymes are likely targets for regulatory mechanisms such as feedback allosteric inhibition by end products and reversible posttranslational modification. As *O. lactea* relies primarily on aerobic catabolism of carbohydrates for energy to fuel life during dormancy, it could be expected that enzymatic activities influencing the rates of glycolysis and the Krebs cycle would be subject to modification to facilitate transition to the estivating condition.

8.5   ENZYMES OF CARBOHYDRATE METABOLISM ARE REGULATED BY PHOSPHORYLATION IN *O. LACTEA*

Several enzymes that can influence the rate of carbohydrate catabolism have been studied in *O. lactea*, and these proteins have all been implicated as targets of estivation-dependent covalent modification. These enzymes include (1) glycogen phosphorylase (GP) which regulates the catabolism of glycogen to provide the substrate for glycolysis; (2) PFK-1 which is recognized as the key control enzyme of glycolysis; (3) pyruvate kinase (PK) which catalyzes the essentially irreversible terminal reaction of glycolysis;

and (4) pyruvate dehydrogenase (PDH) which regulates the enzymatic reaction that is considered to be the entry point for carbohydrates into the Krebs cycle. Activities of GP, PFK-1, PK, and PDH were all reduced in estivating *O. lactea*, and in all cases these were linked to changes in the phosphorylation state of the enzymes (Brooks & Storey, 1992, 1997; Whitwam & Storey, 1990, 1991). As a consequence of change in phosphorylation state, these catabolic enzymes all displayed at least one of the following characteristics: reduced maximum activities (V_{max}), reduced affinity for substrates (increased K_m), or increased sensitivity to feedback inhibition by end products and inhibition by metabolites in general (decreased I_{50}). These parameters of enzyme kinetics will be explicitly defined later in the chapter. The decrease in activity of these regulatory enzymes of ATP-generating catabolism is in line with *O. lactea*'s priority to conserve carbohydrate stores during the estivating state.

8.6   THE ACTIVATION OF G6PDH DURING ESTIVATION FACILITATES PROTECTION AGAINST OXIDATIVE STRESS

Glucose-6-phosphate dehydrogenase (G6PDH), another enzyme of carbohydrate metabolism, is considered to be a key control or rate-limiting enzyme of the pentose phosphate pathway (or shunt). This enzyme displayed increased activity in the hepatopancreas of estivating *O. lactea* (Ramnanan & Storey, 2006a). The question arises as to why increased activity of this enzyme, or increased carbohydrate flux down this pathway, would be beneficial to an animal whose priority is to conserve carbohydrate stores during the dormant phase. The pentose phosphate pathway (alternatively termed the hexose monophosphate shunt) has several important functions, including (1) the production of pentose sugars for synthesis of nucleotides and nucleic acids, (2) serving as the entry point for dietary nucleotides and 5-carbon sugars into catabolic pathways, (3) rearranging the carbon skeletons of dietary carbohydrates into glycolytic/gluconeogenic intermediates, and (4) the generation of reducing equivalents in the form of NADPH (Ozer et al., 2002).

It is this latter function of the pentose phosphate pathway that is of particular relevance to the estivating snail. NADPH supplies the reducing power for the production of reduced glutathione and thioredoxin, two of the key antioxidant reducing agents in cells. Biosynthesis is not a priority for dormant animals, but protection against oxidative damage remains key for organisms that must remain viable over many weeks/months in a hypometabolic state

(Hermes-Lima et al., 1998; Storey & Storey, 2004). Indeed, although oxygen consumption is significantly reduced in estivation, and the generation of oxyradicals in tissues is generally proportional to oxygen consumption, the activities of a variety of antioxidant enzymes are elevated during estivation in *O. lactea*, including superoxide dismutase and catalase (Hermes-Lima & Storey, 1995; Hermes-Lima et al., 1998). The increase in antioxidant enzyme function during dormancy would prepare the estivating animal to deal with the large increase in oxyradical formation associated with arousal from dormancy, when oxygen is rapidly reintroduced in large amounts. Moreover, increased antioxidant capacity in estivating snails would be an adaptation that serves the animal well in dealing with intermittent sharp increases in tissue oxygenation (and oxyradical formation) brought about by discontinuous breathing patterns during estivation. For antioxidant defenses to be elevated during estivation, it follows that pools of reducing power must be available, and hence the regulation of the pentose phosphate pathway becomes important.

While preliminary analysis indicated no measurable changes in G6PDH enzyme kinetics in foot muscle or mantle tissues during estivation, G6PDH in hepatopancreas extracts from estivating *O. lactea* featured increased maximal enzyme velocities (V_{max}) in saturating substrate conditions, relative to G6PDH assayed from active snail extracts (Ramnanan & Storey, 2006a). This tissue-specific alteration of G6PDH in estivation is consistent with the notion that the hepatopancreas in snails, much like hepatic tissue in other animals, plays critical roles in biosynthesis, protection against xenobiotics, and protection against oxidative stress. Further analysis determined that the Michaelis constant (K_m), which represents the substrate concentration at which the reaction rate is one-half the V_{max}, for glucose-6-phosphate (G6P) was decreased during estivation in hepatopancreas. A reduced K_m can be interpreted to suggest that the enzyme's affinity for that substrate has increased. Given that G6P concentrations are known to be reduced by nearly 70% in the hepatopancreas of estivating *O. lactea* (Churchill and Storey, 1989), reduced K_m for G6P could be an adaptation that brings G6PDH enzyme kinetics in line with substrate availability. In any case, it appears that an estivation-dependent response in *O. lactea* hepatopancreas results in a more active form of G6PDH. Moreover, given that PFK (a key control enzyme of glycolysis) is regulated in an inverse manner to G6PDH [such that PFK displayed reduced V_{max} and increased K_m for its substrate fructose-6-phosphate (F6P)], it appears as these enzymes are regulated in a coordinated fashion to reduce G6P carbon flux through glycolysis and enhance

G6P carbon flux through the pentose phosphate pathway during the estivating state.

We then incubated hepatopancreas extracts in different conditions that stimulated specific endogenous protein kinases or specific protein phosphatases, prior to assay for G6PDH activity (Ramnanan & Storey, 2006a). Incubation of active snail hepatopancreas extracts in conditions that stimulated endogenous PKG activity increased G6PDH activity to levels comparable to those seen in estivating *O. lactea* hepatopancreas. Conversely, incubation of estivating snail hepatopancreas extracts in conditions that favored PP1 activity reduced G6PDH activity to levels seen in active snails. In addition, chromatographic isolation and profile of G6PDH activity from tissue extracts revealed two major peaks of enzyme activity, where the first peak featured enzyme kinetics consistent with the lower activity form of the enzyme and the second peak featured enzyme kinetics characteristic of the higher activity form of the enzyme (Ramnanan & Storey, 2006a). Notably, active snails featured a proportionally larger first peak of G6PDH activity and estivating snails featured a proportionally larger second peak of G6PDH activity. Finally, incubation of active snail hepatopancreas extracts in conditions that promoted endogenous kinases (particularly PKG) converted the subsequent chromatographic G6PDH profile into a peak pattern (i.e., larger second peak) resembling that of estivated snails. Similarly, incubation of hepatopancreas extracts from estivating snails in conditions that promoted PP1 activity was able to convert the subsequent chromatographic profile into a pattern that was similar to that of active snails (i.e., larger first peak). Taken together, these experiments support the notion that the pool of G6PDH enzymes in the hepatopancreas of estivating *O. lactea* features a larger proportion of relatively highly phosphorylated, higher activity enzyme with greater substrate affinity (Ramnanan & Storey, 2006a). While these studies in estivating *O. lactea* were the first to demonstrate hypometabolism-dependent changes in G6PDH activity were related to phosphorylation state, subsequent studies confirmed that G6PDH kinetics were similarly regulated in the hypometabolic states conferring anoxia tolerance in the marine mollusc *Littorina litorrea* (Lama et al., 2013), and the freshwater crayfish *Orconectes virilis* (Lant & Storey, 2011). Thus, the evidence suggests that carbohydrate flux through the pentose phosphate pathway, unlike flux through glycolysis, may benefit dormant animals by increasing the capacity for these animals to cope with oxidative stress, and this flux can be regulated by modifying the phosphorylation state of G6PDH.

8.7   ATP-DEPENDENT ION PUMPS ARE SUPPRESSED IN DORMANT SNAILS BY REVERSIBLE PHOSPHORYLATION

The Na^+K^+-ATPase has a critical function in the maintenance of plasma membrane potential difference in all animal cells, pumping sodium and potassium ions against their concentration gradients to maintain high sodium levels outside cells and high potassium levels inside cells. The Na^+K^+-ATPase consumes a considerable amount of cellular energy. In resting endotherms, this ion pump is responsible for 5–40% of total ATP consumption (Clausen, 1986). The activity of Na^+K^+-ATPase can be modified via reversible phosphorylation by several protein kinases (Lopina, 2001). It is essential that transmembrane sodium and potassium gradients are maintained, even during periods of dormancy, to permit cellular living conditions despite strongly suppressed rates of ATP turnover. This requires coordinated suppression of the rates of Na^+ and K^+ movements through ion channels (termed channel arrest) and oppositely directed ATP-driven ion pumps to match the rates of ATP availability from catabolic pathways (Hochachka, 1986). We hypothesized that Na^+K^+-ATPase activity would be decreased during the estivating state in tissues of *O. lactea*, thereby resulting in substantial ATP savings for the animal. Further, we proposed that the mechanism involved would be phosphorylation.

Our study of *O. lactea* Na^+K^+-ATPase determined that activity is strongly suppressed in multiple tissues during estivation, and this suppression was independent of any substantial alteration in protein content (Ramnanan & Storey, 2006b). The lower Na^+K^+-ATPase activity observed in estivating tissue extracts was coincident with reduced affinities for substrates (sodium ions, ATP) and co-substrates (magnesium ions). In addition, we assessed V_{max} over a range of assay temperatures to calculate Arrhenius activation energy. The Arrhenius activation energy associated with the estivating Na^+K^+-ATPase was 1.5-fold greater than that of the ion pump from active snails, consistent with a less active form of the enzyme. Finally, in vitro incubations promoting the activity of several protein kinases (PKA, PKC, and PKG) were shown to reduce Na^+K^+-ATPase activity in extracts of active snails to the levels seen in extracts isolated from estivating animals; conversely, stimulation of specific protein phosphatases (PP1 and PP2A) raised the activities measured in tissue extracts from estivated animals back to the levels seen in active snails. Phosphorylation can either inhibit or stimulate Na^+K^+-ATPase activity, depending on cellular context (Lopina, 2001). In *O. lactea*, it was clear that phosphorylation inhibits the enzyme, thereby contributing substantially to the overall decrease in ATP consumption in snail tissues that defines estivation (Ramnanan & Storey, 2006b). The reduction of Na^+K^+-ATPase activity during a hypometabolic state was also associated with a change in phosphorylation state in the hibernating ground squirrel (MacDonald & Storey, 1999), indicating that this mechanism could be employed across animal species that utilize metabolic rate depression as a survival strategy.

Another energetically taxing ion pump in living cells is the sarcoendoplasmic reticulum calcium ATPase (SERCA). This enzyme serves to maintain and/or restore calcium gradients at a considerable ATP cost. Analysis of SERCA kinetics in *O. lactea* demonstrated that estivating tissues featured a lower activity form of the SERCA enzyme (Ramnanan & Storey, 2008). Interestingly, the low activity form that was present in estivation demonstrated increased kinetic (substrate affinity for Mg-ATP was maintained over varying temperatures) and conformational (increased resistance to denaturation in presence of increasing urea concentrations) stability. Again, we proposed that phosphorylation could have been responsible for the estivation-dependent alterations in SERCA (Ramnanan & Storey, 2008). In vitro incubations of tissue extracts from active *O. lactea* snails in conditions that promoted several endogenous kinases decreased SERCA activity to levels comparable to the low SERCA activity observed in estivating animals. Conversely, incubation of extracts from estivating animals in conditions that stimulated endogenous PP2A (in foot muscle) and endogenous PP2C (in hepatopancreas) resulted in elevated SERCA activity similar to levels seen in active animals. Taken together, our data suggested that SERCA could be downregulated in hypometabolic conditions, and this decrease was the result of changes in phosphorylation state. Estivation-dependent phosphorylation and alteration of SERCA activity has since been observed in both freeze-tolerant and freeze-avoiding insects (McMullen et al., 2010) and anoxia-tolerant turtle tissues (Ramnanan et al., 2010b), which suggests a high degree of conservation of this mechanism across species that naturally depress their metabolic rates in response to unfavorable environmental conditions.

8.8   PROTEIN TURNOVER IS REGULATED BY PHOSPHORYLATION IN ESTIVATING *O. LACTEA*

Much like ion pumping, protein synthesis is a very energy and resource intensive cellular process, hydrolyzing four ATP equivalents for every peptide bond synthesized. Not only are proteins continually being synthesized under normal metabolic conditions, but the protein population also is continually being turned over, as proteins are eventually degraded in a specifically targeted, carefully regulated manner. It follows that a physiologic state of metabolic rate depression would include, as a key contributing component, a massive reduction in the rate of overall protein synthesis. Further, to maintain homeostasis, the overall rate of protein degradation would likely be reduced in concert in an estivating animal (Storey & Storey, 2004).

In vitro assay of protein synthesis revealed marked (~80%) reductions in *O. lactea* tissues early (2 days, the earliest time point measured) into the dormant period, and this suppressed level was maintained at later (14 days) stages of estivation (Ramnanan et al., 2009). Thus, the substantial suppression of protein synthesis could be considered as happening at an early stage of dormancy, thereby bringing about energetic savings at the onset of estivation. This was consistent with the reductions in protein synthesis previously observed in the estivating desert frog *Neobatrachus centralis* (Fuery et al., 1998) and the estivating snail *Helix aspersa* (Pakay et al., 2002). There are several levels by which protein synthesis rates could be regulated in estivating snails. We first looked at expression levels of key molecular regulators of ribosomal biogenesis. The transcriptional activator c-Myc and the transcriptional repressor Mitotic Arrest Deficient-1 (MAD1) are determining factors in the synthesis of the upstream binding factor (UBF), a key regulator of ribosomal DNA (rDNA) expression (Poortinga et al., 2004). We observed decreases in protein levels of c-Myc, increases in protein levels of MAD1, and reduction in UBF protein levels, consistent with reduced ribosome formation (Ramnanan et al., 2009). However, alterations in MAD1 and UBF were only apparent at 14 days, meaning that alterations in ribosomal biogenesis could not be the determining factor mediating the substantial decrease in protein synthesis evident after 2 days of estivation. These changes may, however, play a role in bringing about ribosomal machinery changes suited for longer term dormancy.

Protein synthesis can also be regulated by the level of covalent phosphorylation of key regulators of protein translation. In conditions of nutrient excess, the mTOR (mammalian target of rapomycin) protein kinase becomes hyperphosphorylated, which increases its protein synthesis-promoting activity (Gingras et al., 2001). The mTOR kinase then phosphorylates two key substrates that are, in turn, key regulators of protein translation. The first of these substrates is a regulatory kinase of approximately 70 kDa in size that phosphorylates and modifies several residues on the ribosomal S6 subunit protein. Phosphorylation of this 70 kDa S6 kinase (or p70S6K) by mTOR typically occurs during nutrient-rich conditions that favor growth and proliferation, and phosphorylated p70S6K results in enhanced rates of 5' terminal oligopyrimidine tract (TOP) translation. The second of these mTOR targets is the binding protein of the eukaryotic initiation factor eIF4E (4E-BP).

In fasting or nutrient poor conditions, 4E-BP1 binds to eIF4E, preventing the association of eIF4E with eIF4G. In nutrient-rich conditions, 4E-BP1 becomes phosphorylated, releasing eIF4E from its inhibitory binding, which permits eIF4E and eIF4G interaction to promote cap-dependent translation. Although we characterized no alteration in mTOR kinase phosphorylation (Ramnanan et al., 2007), decreased levels of p70S6K and 4E-BP1 phosphorylation were observed in estivating snail tissues (Ramnanan et al., 2009). These decreases were consistent with and indicative of decreased protein translation in the dormant state.

That altered p70S6K and 4E-BP1 phosphorylation were observed, independent of measurable changes in mTOR phosphorylation (and therefore mTOR activity) in estivation, raised the possibility that other translation factors were differentially regulated in dormancy as well in this model system. We then performed a thorough analysis of the phosphorylation state of protein translation factors implicated in the control of protein synthesis. The eukaryotic initiation factor eIF2α plays an important role in driving translational initiation by facilitating the necessary GTP-GDP exchange activity of the eIF2B complex (Proud, 2006). This function of eIF2α is inhibited when this factor is phosphorylated. Similarly, the elongation factor eEF2 mediates ribosome translocation along the mRNA strand after addition of an amino acid, and phosphorylation of eEF2 by an upstream kinase inhibits binding of this factor to the ribosome, thereby suppressing translation. In estivating *O. lactea*, levels of total eIF2α and eEF2 both decreased relatively slowly (reduced protein levels were not observed after 2 days of estivation but were observed after 14 days of dormancy). On the other hand, levels of eIF2α and eEF2 phosphorylation were evident relatively early (after 2 days) in estivation (Ramnanan et al., 2009). Thus, there appears to be at least two mechanisms by which these loci are regulated to inhibit energetically taxing protein translation: phosphorylation events that happen early in dormancy and reduced protein levels that happen at a later point in the estivating state. It is probable that the phosphorylation events involving these two factors evident early in estivation played some role in the substantial decreases of protein translation measured in vitro. In any case, by 14 days of estivation the ratio of phosphorylated protein to total protein for both eIF2α and eEF2 had substantially increased in both foot muscle and hepatopancreas (Ramnanan et al., 2009), indicating that the capability for protein translation initiation and elongation were markedly reduced.

Possible estivation-dependent regulation of eIF4E or eIF4GI, two components of the eIF4F complex that is responsible for the rate-limiting process of cap dependent mRNA recruitment to the ribosome, was also characterized (Ramnanan et al., 2009). Moreover, expression of eIF4E and eIF4GI proteins at the genetic level are determined by the interplay of the transcriptional activator c-Myc and the transcription repressor MAD1; c-Myc and MAD1 were observed to be upregulated and downregulated, respectively, in estivation by 14 days (Ramnanan et al., 2009). Generally, protein levels of eIF4E or eIF4GI were not altered in either foot muscle or hepatopancreas at either 2 days or 14 days of estivation, indicating that these loci were not regulated at the genetic level. However, the amount of phosphorylated eIF4E was decreased in foot muscle (but not hepatopancreas) and the level of phosphorylated eIF4GI was decreased in hepatopancreas (but not foot muscle). These decreases were only apparent after 14 days of estivation. Counter to the case with eIF2α and eEF2, phosphorylation of eIF4E and eIF4GI enhances the translational activity of the eIF4F complex (Proud, 2006). It appears that these two proteins are regulated in a tissue-specific manner in *O. lactea*, with the net result in either tissue being a reduction of translation (Ramnanan et al., 2009). While these phosphorylation events are temporally discordant with and do not contribute to the substantial reduction in protein synthesis rates observed in vitro from 2 days estivating snails, it may be that the reduced phosphorylation of eIF4E (in foot muscle) and of eIF4GI (in hepatopancreas) may be factors in sustaining the profound reductions in protein synthesis that will endure for the duration for dormancy.

Because we observed reductions in protein synthesis rates in vitro and characterized alterations in biochemical markers that would seem to suggest that protein translation and ribosomal biogenesis are downregulated in estivation, it would follow that protein degradation would be globally decreased in concert. This would permit an increase in the lifetime of intracellular proteins during the dormant phase. A coordinated decrease in protein degradation would prevent or at least delay animal tissues from entering a state of negative protein balance, which would facilitate a return to normal life function upon arousal from estivation. Protein carbonyl levels were not elevated in estivating *O. lactea* (Ramnanan et al., 2009), indicating that the level of oxidatively damaged proteins was not increased during estivation (an aerobic, oxidative condition). Moreover, expression of selected heat shock proteins was increased during estivation (particularly in hepatopancreas), suggesting that the estivating animal has a greater capacity to protect unfolded proteins from being degraded during dormancy (Ramnanan et al., 2009). The last question to address was whether protein degradation rates per se were downregulated in estivation.

The majority of intracellular protein degradation is mediated by the multicatalytic proteinase (MCP) complex (Orlowski & Wilk, 2003).

Proteolytic activity of the 20S proteasome (a significant component of the MCP complex) in vitro was markedly reduced in both hepatopancreas and foot muscle of estivating *O. lactea*. Decreased activity was generally associated with increased K_m values for substrates, suggesting lower substrate affinity (Ramnanan et al., 2009). These kinetic changes in 20S proteolytic activity were observed in the absence of any measurable decrease in 20S expression. As several subunits of the 20S proteasome are known to be phosphorylated, it was possible that the decrease in 20S proteasome function was mediated by an increase in phosphorylation state. Incubation of hepatopancreas extracts from active snails with 8-bromo-cGMP (a more stable, potent activator of PKG, as compared with endogenous, labile cGMP) decreased 20S proteasome activity to levels that approached the low proteasome activity seen in dormancy (Ramnanan et al., 2009). On the other hand, incubation of hepatoapancreas extracts from either active or estivating snails in conditions that permitted ion-independent phosphatase (PP1/PP2A) activity enhanced 20S proteasome activity. Further, this stimulation of the 20S proteasome was fully abolished when PP2A was completely inhibited with nanomolar amounts of okadaic acid. Thus, PKG and PP2A appear to mediate the phosphorylation and dephosphorylation, respectively, of the 20S proteosome in *O. lactea* in an estivation-dependent manner (Ramnanan et al., 2009). Thus, our data characterize a coordinated suppression of both global cellular protein synthesis and protein degradation rates that accompany transition to the estivating state, which serves the dormant *O. lactea* organism well in terms of conserving energy (Fig. 8.1).

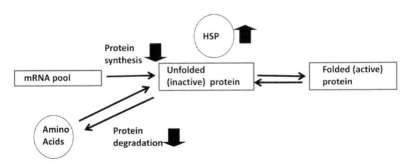

FIGURE 8.1   The global suppression of protein metabolism during estivation in *O. lactea*. Overall protein synthesis and protein degradation rates are both substantially suppressed, as indicated by various indices indicated in the text (differential phosphorylation states of protein translation initiation and elongation factors, reduced activities of protein synthesis, and multicatalytic protein as in vitro). In addition, prolonged lifetime of unfolded proteins may be facilitated by increased expression of chaperone heat shock proteins (HSPs), which serves the animal well upon arousal from the dormant state.

8.9   A ROLE FOR THE AKT SIGNALING PATHWAY IN ESTIVATION

The serine/threonine kinase Akt (also called protein kinase B; PKB) is well-known to be a key downstream regulator of growth, proliferation, and survival and has been perhaps best defined in terms of mediating the downstream response of the anabolic hormone insulin (Ramnanan et al., 2010a). Insulin binding to its receptor can trigger the phosphoinositide 3 kinase (PI3K)/3-phosphoinositide-dependent (PDK1) signaling cascade, which leads to the phosphorylation of Akt at several sites; assessment of Akt phosphorylation at the Ser473 residue is typically considered an index of Akt activation. Phosphorylated, active Akt mediates several different protein responses. In response to a physiologic rise in circulating insulin, for example, phosphory-lated Akt can phosphorylate and inactivate glycogen synthase (GS) kinase 3β (GSK3β), which prevents GSK3β from phosphorylating and inhibiting GS, permitting anabolic carbohydrate metabolism and glycogen deposition. Indeed, the time course and magnitude of Akt phosphorylation and activa-tion in vivo is closely related to the time course and magnitude of GSK3β phosphorylation (Ramnanan et al., 2010a, 2011–2013). As discussed previ-ously, Akt can also bring about mTOR phosphorylation which enhances protein synthesis (Proud, 2006). In addition, Akt can phosphorylate tran-scriptional regulators such as members of the forkhead box, class O (FOXO) family of transcription factors. Dephosphorylated FOXO proteins reside in the nucleus where they are free to drive the expression of genes involved in gluconeogenesis, cell cycle arrest, and pro-apoptotic regulators (Ramnanan et al., 2010a). Upon phosphorylation, FOXO proteins exit the nucleus and are sequestered in the cytoplasm where they are transcriptionally inert (Rivera et al., 2010). Similarly, Akt can also mediate phosphorylation of the Bcl-2-associated death (BAD) promoter proteins. Dephosphorylated BAD proteins form a complex with Bcl-2, blocking Bcl-2 from inhibiting the pro-apoptotic activity of the Bax protein. On the other hand, phosphorylation of BAD by Akt releases Bcl-2 from BAD-mediated inhibition, which permits Bcl-2 to suppress the pro-apoptotic activity of Bax (Zhang et al., 2011).

   We observed that Akt phosphorylation increased approximately 40% in both foot muscle and hepatopancreas, and these increases were correlated with twofold increases in Akt V_{max} and reduced affinity toward its synthetic peptide substrate (Ramnanan et al., 2007). These alterations in Akt enzyme kinetics were coincident with increased protein stability (as measured by increased resistance to denaturation in the presence of urea). Incubation of tissue extracts from active snails in conditions that stimulated endogenous protein kinases (likely PKA or PKG) led to increases in assayed Akt activity, while incubation of tissue extracts from dormant snails in conditions that promoted endogenous protein phosphatases (likely PP2A or PP2C) reduced assayed Akt activity. The question arises as to why Akt, a kinase with well-characterized anabolic, ATP-consuming downstream effects, would be activated in estivating *O. lactea*, an animal model where ATP is at a premium and biosynthetic pathways are generally suppressed as part of a wholesale move to a depressed metabolic state. We confirmed that mTOR phosphorylation was no different in estivating versus active snails, and as discussed previously, multiple lines of evidence supported the notion that both protein synthesis and protein degradation rates were reduced in concert in dormant snails. With regards to carbohydrate metabolism, Akt activation was discordant with changes in GSK3β, where decreased phosphorylation and increased kinetic activity were both observed in estivation, changes consistent with suppressed glycogen synthesis (Ramnanan et al., 2007). On the other hand, phosphorylated Akt was correlated with increases in measured phosphorylation of both FOXO and BAD proteins. Thus, it appears as though Akt activation in this estivating model is uncoupled from energetically taxing anabolic processes involving both protein and carbohydrate biosynthesis, but is correlated with mechanisms that enhance cell survival and suppress apoptosis, which would facilitate cell survival throughout the estivating state (Fig. 8.2).

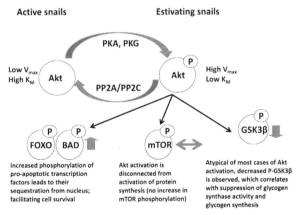

FIGURE 8.2   The stimulation of the master metabolic regulator AMPK during estivation in *O. lactea*. Increased activity and phosphorylation of AMPK in this model (likely mediated by either LKB1, a known upstream regulator of AMPK, or PKG, a kinase that has established regulatory capacity across the mollusk phylum) phosphorylates and regulates many metabolic enzymes, including acetyl-CoA carboxylase (ACC), a rate-determining enzyme of fatty acid synthesis. Increased ACC phosphorylation in estivation leads to decreased ACC activity and decreased formation of high activity ACC polymers, which is well-suited to the overriding priority of the animal to conserve energy stores during the hypometabolic period.

8.10   THE ROLE OF AMPK IN METABOLIC RATE DEPRESSION IN ESTIVATING *O. LACTEA*

Unlike Akt, a signaling system typically associated with conditions of nutrient excess or anabolic states, AMPK is characterized as a molecular regulator that is sensitive to elevated AMP levels (which demarcates catabolic or energy-depleted states) and responds to restore cellular energetic balance by suppressing ATP-consuming anabolic activities and promoting ATP-generating catabolic processes (Hardie & Carling, 1997). AMPK has been defined to suppress ATP-intensive processes including fat, cholesterol, protein, and glycogen biosynthesis in various animal and nonanimal cells and tissues. AMPK was therefore a likely candidate for estivation-dependent regulation in *O. lactea*.

Characterization of AMPK phosphorylation (an index of its activity) revealed substantial activation of the enzyme early (2 days) into the estivating condition, and AMPK phosphorylation was maintained at this elevated level at a later stage (14 days) in dormancy (Ramnanan et al., 2010c). Kinetic analysis of enzyme activity indicated that V_{max} was elevated as well in estivation. The observations that AMPK affinity for its peptide substrate increased and Arrhenius activation energy decreased in estivation supports the notion that AMPK functions at a higher level in dormancy in this model. While AMPK can be activated by either elevated AMP levels or phosphorylation, AMP levels do not measurably increase in estivating *O. lactea* tissues. As such, it is likely that phosphorylation of AMPK drives the increase in its activity seen in estivation. There are several protein kinases established as being upstream regulators of AMPK phosphorylation, and of these kinases, the phosphorylation of LKB1 was temporally associated with AMPK activation in this estivating model. Moreover, increased AMPK protein and activity were observed in LKB1 immunoprecipitation preparations from estivating extracts, indicative of an estivation-dependent increase in LKB1–AMPK interaction. In addition, our incubation experiments suggested that PKG and PP2A had the ability to phosphorylate and dephosphorylate, respectively, AMPK in a manner consistent with elevated AMPK activity in estivating snails and decreased AMPK activity in dormant snails (Ramnanan et al., 2010c).

After confirming that AMPK was activated in the estivating snail, likely by LKB1 in vivo, we next elucidated the consequences of elevated AMPK activity in estivating *O. lactea*. One of the best characterized substrates of AMPK activity is acetyl-CoA carboxylase (ACC), an enzyme that catalyzes the conversion of acetyl-CoA to malonyl-CoA. ACC exists in two forms:

ACCα is present in the cytoplasm and is the rate-limiting enzyme in fatty acid biosynthesis, while ACCβ co-localizes with mitochondria and serves to inhibit (via its product malonyl-CoA) the outer mitochondrial membrane enzyme carnitine palmitoyltransferase (CPT1), which regulates fatty acid transport into mitochondria for oxidation (Hardie & Pan, 2002). In the context of estivation, given that snails rely primarily on carbohydrates during the estivating period and repletion of fatty acid stores is not likely a cellular priority, it would be reasonable to predict that ACC would be deactivated in estivating *O. lactea*. Kinetic analysis of ACC activity revealed a less active form of the enzyme in estivation, one with decreased V_{max}, reduced substrate affinity for Mg^{2+}-ATP, and (in foot muscle only) reduced ability to be activated by citrate (Ramnanan et al., 2010c). Citrate is a powerful regulator of ACC activity in vivo and tends to accumulate in the fed (nutrient-rich) state, stimulating the aggregation of ACC monomers (where ACC has relatively low activity) into ACC polymers that promote increased ACC function. Tissue extracts (from active snail foot muscle) were incubated in different conditions before being subjected to gel-filtration chromatography (Ramnanan et al., 2010c). These experiments determined that ACC activity exists in two fractions in *O. lactea*, a large molecular weight polysome fraction and a small molecular weight monosome fraction. Incubation in conditions that stimulate AMPK shifts the ACC activity profile into one that featured the monosome fraction only, consistent with the principle that AMPK phosphorylates and deactivates ACC. On the other hand, incubation of extracts with a saturating concentration of citrate shifted the ACC activity profile into one that exclusively featured a polysome fraction (Ramnanan et al., 2010c). Finally, tissue extracts were incubated in conditions that included excess citrate and that were permissive to endogenous AMPK activity. In this setting, AMPK was able to prevent the ability of citrate to stimulate ACC activity and promote ACC polysome formation, providing insight into the mechanism by which AMPK modifies ACC activity in a physiologic state of metabolic rate depression (Fig. 8.3).

We then characterized the protein levels and phosphorylation state of AMPK targets in both foot muscle and hepatopancreas (Ramnanan et al., 2010c). In both of these tissues, phosphorylation of both ACCα and ACCβ were increased in estivation, suggesting that the enzymatic capability for fatty acid synthesis is suppressed and for fatty acid oxidation is increased during dormancy. However, given that net fatty acid oxidation does not occur in estivating snails until several months into dormancy (when carbohydrate reserves have been exhausted), it is unlikely that differential ACCβ regulation plays any early role in estivating *O. lactea* tissues. While gluconeogenic and lipogenic gene expression were both suppressed in hepatopancreas (as would be expected in a state where AMPK is activated), gene expression related to mitochondrial biogenesis in foot muscle was not increased (which is discordant with typical states associated with AMPK activation). It may be that mitochondrial biogenesis may not be a driving priority in estivation as it would be in other states (such as exercise stress) that are associated with increased AMPK function. In any case, increased phosphorylation of GS was observed in foot muscle, consistent with the decreased levels of GSK3β phosphorylation (and increased levels of GSK3β activity) previously described in estivating *O. lactea* (Ramnanan et al., 2007). It must be noted that AMPK is suggested to drive opposite regulation of GS (inactivation) and GP (activation) to maintain whole body homeostasis in other physiological (hyperglucagonemia and/or hyperglycemia) conditions (Rivera et al., 2010). Our studies in *O. lactea* suggest that activated AMPK can play some part in the coordinated suppression of both GS and GP (decreasing rates of both glycogen synthesis and breakdown) in the context of a physiologic state of metabolic rate depression,, and that this suppression of glycogen metabolism is in line with the overall priority of the animal to suppress metabolism.

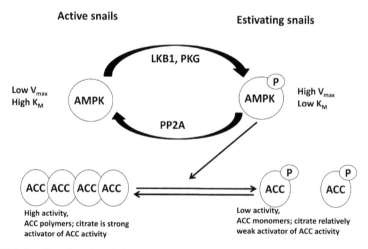

FIGURE 8.3   The phosphorylation and activation of Akt during estivation in *O. lactea*. In most physiologic systems, Akt activation leads to phosphorylation of targets that facilitate anabolism and growth. In the estivating snail, Akt is activated, and leads to phosphorylation and inactivation of pro-apoptotic transcription factors FOXO and BAD. However, activation of Akt does not result in increased levels of mTOR or GSK3β phosphorylation, and therefore does not facilitate protein synthesis or glycogen synthesis. This suggests that Akt can enhance pro-survival signaling in the absence of promoting energy-intensive anabolic cellular activities, in a state of hypometabolism.

8.11 DIFFERENTIAL REGULATION OF TYPE-1 PROTEIN PHOSPHATASE (PP1) ACTIVITY IN ESTIVATING SNAILS

To date, reversible phosphorylation has been clearly established as a dominant regulatory mechanism in estivating *O. lactea*. In this chapter, we have discussed in detail the role of phosphorylation in the estivation-dependent modification of proteins and enzymes involved in carbohydrate metabolism, ion pumping, protein synthesis,and degradation, transcriptional regulation, and signaling pathways (Akt, AMPK) with far-reaching downstream effects. If all of these processes are, at least in part, modified by differential phosphorylation state during the transition to the estivating condition, it stands to reason that protein dephosphorylation is differentially regulated as well. PP1 is a far-reaching phosphatase with known targets that bring about structural, metabolic, translational and transcriptional effects (Cohen, 1989). While PP1 did not have a major role in the estivating toad (Cowan et al., 2000), specific functions of this phosphatase have been elucidated in other forms of hypometabolism (MacDonald & Storey, 2002, 2007). Given its ubiquitous nature, we proposed that PP1 would be differentially regulated in estivating *O. lactea* (Ramnanan & Storey, 2009).

PP1 activity (V_{max}) was reduced in estivating tissues and several kinetic parameters were indicative of a less active enzyme in the estivating animal (Ramnanan & Storey, 2009). The PP1 catalytic subunit (PP1c) was subsequently purified and determined to be approximately 39 kDa in size, similar to sizes seen across other species, and with similar enzyme kinetics. However, the purified PP1c did not display altered enzyme kinetics between active and estivating snail tissue extracts, likely due to the fact that functional PP1c activity in vivo is largely determined by PP1c binding to regulatory subunits that target and localize PP1c activity (Cohen, 1989). This aspect of PP1c regulation is lost when assaying PP1 activity in crude extracts. As such, we prepared PP1 samples in a manner that would permit PP1c complexes with targeting proteins and subjected these preparations to gel-filtration chromatography (Ramnanan & Storey, 2009). This experiment yielded four major peaks of activity, including peaks associated with protein sizes of 257 ± 8 kDa (which featured high substrate affinity when assayed) and 76 ± 2 kDa (which featured low substrate affinity when assayed). Moreover, active snails had a greater proportion of PP1c activity associated with the 257 kDa/ high substrate affinity complex and a relatively small proportion of PP1c associated with the 76 kDa/low substrate affinity complex. Conversely, the distribution of the PP1c population shifted in estivating samples, with less phosphatase activity (and PP1c content) associated with the 257 kDa peak, and increased phosphatase activity (and PP1c content) associated with the 76 kDa peak. In addition, immunoblotting confirmed that the 76 kDa peak contained both PP1c and the nuclear inhibitor of PP-1, a known and relatively well-conserved regulatory targeting subunit of PP1c (Ramnanan & Storey, 2009). Finally, subcellular fractionation studies indicated that the PP1c activity in nuclear and glycogen-associated fractions were increased and decreased, respectively, in estivation (Ramnanan & Storey, 2009). The decrease in glycogen-associated PP1c fits with the overall suppression of glycogen metabolism described earlier. For example, in hepatic tissue, PKA and PP1 mediate opposing effects on the rate-determining enzymes of glycogenolysis and glycogen synthesis (Ramnanan et al., 2010a, 2011). PKA catalyzes the phosphorylation of GP and GS, favoring net glycogenolysis, and PP1 dephosphorylates GP and GS, resulting in net glycogen synthesis. That PP1c association with the glycogen fraction is decreased in estivation is consistent with the previous observation that PKA activity is suppressed during dormancy in this model, and is consistent with the observation that GS phosphorylation is relatively enhanced (and GS is relatively less active) in estivation (Ramnanan et al., 2010c). Decreased glycogen-associated PP1c in estivating *O. lactea* is consistent with a similar finding in hibernating ground squirrels (MacDonald & Storey, 2007).

8.12 SUMMARY AND FUTURE DIRECTIONS

While reversible protein phosphorylation has been established as a key intracellular theme biochemical adaptation in various hypometabolic states (Storey & Storey, 2004), the work done in the estivating land snail *O. lactea* over the previous decade has clearly established that this mechanism is critical in preparing for, and facilitating survival during, dormancy. In this chapter, we detailed the estivation-dependent changes in protein phosphorylation that have profound influences on carbohydrate metabolism, the production of reducing equivalents to support antioxidant capacity, the suppression of ATP-intensive processes such as ion pumping and protein translation, the coordinated reductions of protein synthesis and degradation rates, the differential targeting of protein phosphatase type-1, and selective activation of specific arms downstream of the Akt and AMPK signaling pathways. There are other candidate molecules that could have major regulatory capacity in dormant *O. lactea*. Given that cGMP levels are known to be elevated during the early hours of estivation (Brooks & Storey, 1996), and that cGMP and PKG has been implicated in the control of many of the estivating-dependent alterations characterized earlier in the chapter, ongoing experiments are defining PKG enzyme kinetics in estivating *O. lactea*. PKG seems to be central to cellular regulation across the phylum Mollusca (del Pilar Gomez & Nasi, 2005; Sung et al., 2004), including other species that utilize hypometabolic survival strategies (Larade & Storey, 2004). To date, preliminary studies have indicated that PKG may have altered enzyme kinetics and differential localization in estivating *O. lactea* (Ramnanan and Storey, unpublished data). Future studies may prove that PKG is, indeed, a master regulator in the move to a depressed metabolism in this remarkable species of desert snail.

KEYWORDS

- metabolic rate depression
- reversible protein phosphorylation
- metabolic pathways
- oxidative stress

REFERENCES

Athawale, M. S.; Reddy, S. R. Storage Excretion in the Indian Apple Snail *Pila globosa* (Swainson), during Aestivation. *Ind. J. Exp. Biol.* **2002,** *40*(11), 1304–1306.

Barnhart, M. C. Gas Permeability of the Epiphragm of a Terrestrial Snail, *Otala lactea. Physiol. Zool.* **1983,** *56,* 436–444.

Barnhart, M. C. Respiratory Gas Tensions and Gas Exchange in Active and Dormant Land Snails, *Otala lactea. Physiol. Zool.* **1986,** *59,* 733–745.

Barnhart, M. C.; McMahon, B. R. Discontinuous Carbon Dioxide Release and Metabolic Depression in Dormant Land Snails. *J. Exp. Biol.* **1987,** *128,* 123–138.

Barnhart, M. C.; McMahon, B. R. Depression of Aerobic Metabolism and Intracellular pH by Hypercapnia in Land Snails, *Otala lactea. J. Exp. Biol.* **1988,** *138,* 289–299.

Bishop, T.; Brand ,M. D. Processes Contributing to Metabolic Depression in Hepatopancreas Cells from the Snail *Helix Aspersa. J. Exp. Biol.* **2000,** *203,* 3603–3612.

Brooks, S. P. J.; Storey, K. B. Properties of Pyruvate Dehydrogenase from the Land Snail *Otala lactea*: Control of Enzyme Activity during Estivation. *Physiol. Zool.* **1992,** *65*(3), 620–633.

Brooks, S. P. J.; Storey, K. B. Protein Kinase Involvement in Land Snail Aestivation and Anoxia: Protein Kinase A kinetic Properties and changes in Second Messenger Compounds during Depressed Metabolism. *Mol. Cell. Biochem.* **1996,** *156,* 153–161.

Brooks, S. P. J.; Storey, K. B. Glycolytic controls in Estivation and Anoxia: A Comparison of Metabolic Arrest in Land and Marine Molluscs. *Comp. Biochem. Physiol.* **1997**, *118A*(4), 1103–1114.

Churchill, T. A.; Storey, K. B. Intermediary Energy Metabolism during Dormancy and Anoxia in the land snail *Otala lactea. Physiol. Zool.* **1989**, *62*, 1015–1030.

Clausen, T. Regulation of active Na+ K+-ATPase Transport in Skeletal Muscle. *Physiol. Rev.* **1986**, *66*, 542–580.

Cohen, P. The Structure and Regulation of Protein Phosphatase. *Ann. Rev. Biochem.* **1989**, *58*, 453–508.

Cowan, K. J.; Storey, K. B. Urea and KCl have Differential effects on Enzyme Activities in Liver and Muscle of Estivating Versus Nonestivating Species. *Biochem. Cell Biol.* **2002**, *80*(6), 745–755.

Cowan, K. J.; MacDonald, J. A.; Storey, J. M.; Storey,K. B. Metabolic Reorganization and Signal Transduction during Estivation in the Spadefoot Toad. *Exp. Biol. Online.* **2000**, *5*, 1.

Dejours, P. From Comparative Physiology of Respiration to Several Problems of Environmental Adaptations and to Evolution. *J. Physiol.* **1989**, *410*, 1–19.

del Pilar Gomez, M.; Nasi, E. Calcium-independent, cGMP-mediated Light Adaptation in Invertebrate Ciliary Photoreceptors. *J. Neurosci.* **2005**, *25*(8), 2042–2049.

Gingras, A.; Raught, B.; Sonnenburg, N. Regulation of Translation Initiation by FRAP/mTOR. *Genes Dev.* **2001**, *15*, 807–826.

Giraud-Billoud, M.; Abud, M. A.; Cueto, J. A.; Vega, I. A.; Castro-Vazquez, A. Uric Acid Deposits and Estivation in the Invasive Apple-snail, *Pomacea canaliculata. Comp. Biochem. Physiol. A* **2011**, *158*, 506–512.

Giraud-Billoud, M.; Koch, E.; Vega, I. A.; Gamarra-Luques, C.; Castro-Vazquez, A. Urate Cells and Tissues in the South American Apple-snail *Pomacea canaliculata. J. Mollusc. Stud.* **2008**, *74*, 259–266.

Giraud-Billoud, M.; Vega, I. A.; Rinaldi Tosi, M. E.; Abud, M. A.; Calderón, M. L.; Castro-Vazquez, A. Antioxidant and Molecular Chaperone defenses during Estivation and Arousal in the South American Apple-snail *Pomacea canaliculata. J. Exp. Biol.* **2013**, *216*, 614–622.

Guppy, M.; Fuery, C. J.; Flanigan, J. E. Biochemical Principles of Metabolic Depression. *Comp. Biochem. Physiol.* **1994**, *109*(B), 175–189.

Fuery, C. J.; Withers, P. C.; Hobbs, A. A.; Guppy, M. The Role of Protein Synthesis During Metabolic Depression in the Australian Desert Frog *Neobatrachus centralis. Comp. Biochem. Physiol. A* **1998**, *119*(2), 469–476.

Hardie, D. G.; Carling, D. The AMP-activated Protein Kinase: Fuel Gauge of the Mammalian Cell? *Eur. J. Biochem.* **1997**, *246*, 259–273.

Hardie, D. G.; Pan, D. A. Regulation of Fatty Acid Synthesis and Oxidation by the AMP-activated Protein Kinase. *Biochem. Soc. Trans.* **2002**, *30*, 1064–1070.

Hermes-Lima, M.; Storey, K. B. Antioxidant Defenses and Metabolic Depression in a Pulmonate Land Snail. *Am. J. Physiol.* **1995**, *268*, R1386–R1393.

Hermes-Lima, M.; Storey, J. M.; Storey K. B. Antioxidant Defenses and Metabolic Depression. The Hypothesis of Preparation for Oxidative Stress in Land Snails. *Comp. Biochem. Physiol.* **1998**, *120*(B), 437–448.

Herreid, C. F. Metabolism of Land Snails (*Otala lactea*) During Dormancy, Arousal and Activity. *Comp. Biochem. Physiol.* **1977**, *56*(A), 211–215.

Hochachka, P. W. Defence Strategies against Hypoxia and Hypothermia. *Science* **1986**, *231*, 234–241.

Lama, J. L.; Bell, R. A.; Storey, K. B. Glucose-6-phosphate Dehydrogenase Regulation in the Hepatopancreas of the Anoxia-tolerant Marine Mollusc, *Littorina littorea*. *Peer J.* **2013**, *1*, e21.

Lant, B.; Storey, K. B. Glucose-6-phosphate Dehydrogenase Regulation in Anoxia Tolerance of the Freshwater Crayfish *Orconectes virilis*. *Enzyme Res.* **2011**, *2011*, 524906.

Larade, K.; Storey, K. B. Anoxia-induced Transcriptional Upregulation of sarp-19: Cloning and Characterization of a Novel EF-hand Containing Gene Expressed in hepatopancreas of *Littorina littorea. Biochem. Cell Biol.* **2004**, *82*(2), 285–293.

Lopina, O. D. Interaction of Na,K-ATPase Catalytic subunit with Cellular Proteins and Other Endogenous Regulators. *Biochemistry (Mosc.)* **2001**, *66*, 1122–1131.

MacDonald, J. A.; Storey K. B. Regulation of Ground Squirrel Na+K+-ATPase by Reversible Phosphorylation During Hibernation. *Biochem. Biophys. Res. Commun.* **1999**, *254*, 424–429.

MacDonald, J. A.; Storey, K. B. Protein Phosphatase type-1 from Skeletal Muscle of the Freeze Tolerant Wood Frog. *Comp. Biochem. Physiol. B* **2002**, *131*, 27–36.

MacDonald, J. A.; Storey, K. B. The Effect of Hibernation on Protein Phosphatases from Ground Squirrel Organs. *Arch. Biochem. Biophys.* **2007**, *468*, 234–243.

McMullen, D. C.; Ramnanan, C. J.; Bielecki, A.; Storey, K. B. In Cold-hardy Insects, Seasonal, Temperature, and Reversible Phosphorylation Controls Regulate Sarco/Endoplasmic Reticulum Ca^{2+}-ATPase (SERCA). *Physiol. Biochem. Zool.* **2010**, *83*(4), 677–686.

Orlowski, M.; Wilk, S. Ubiquitin-independent Proteolytic Functions of the Proteasome. *Arch. Biochem. Biophys.* **2003**, *415*, 1–5.

Ozer, N.; Bilgi, C.; Ogus, H. I. Dog Liver Glucose-6-phosphate Dehydrogenase: Purification and Kinetic Properties. *Int. J. Biochem. Cell Biol.* **2002**, *34*, 253–262.

Pakay, J. L.; Withers, P. C.; Hobbs, A. A.; Guppy, M. In Vivo Downregulation of Protein Synthesis in the Snail *Helix aspersa* During Estivation. *Am. J. Physiol.* **2002**, *283*(1), R197–R204.

Poortinga, G.; Hannan, K. M.; Snelling, H.; Walkley, C. R.; Jenkins, A.; Sharkey, K.; Wall, M.; Brandenburger, Y.; Palatsides, M.; Pearson, R. B.; McArthur, G. A.; Hannan, R. D. MAD1 and c-MYC Regulate UBF and rDNA Transcription During Granulocyte Differentiation. *EMBO J.* **2004**, *23*(16), 3325–3335.

Proud, C. G. Regulation of Protein Synthesis by Insulin. *Biochem. Soc. Trans.* **2006**, *34*, 213–216.

Rees, B. B.; Hand, S. C. Heat Dissipation, Gas Exchange and Acid–Base Status in the Land Snail Oreohelix During Short-term Estivation. *J. Exp. Biol.* **1990**, *152*, 77–92.

Rees, B. B.; Hand S. C. Biochemical Correlates of Estivation Tolerance in the Mountain Snail *Oreohelix* (Pulmonata: Oreohelicidae). *Biol. Bull.* **1993**, *184*, 230–242.

Ramnanan, C. J.; Storey, K. B. Glucose-6-phosphate Dehydrogenase Regulation During Hypometabolism. *Biochem. Biophys. Res. Commun.* **2006a**, *339*(1), 7–16.

Ramnanan, C. J.; Storey, K. B. Suppression of Na+K+-ATPase Activity During Estivation in the Land Snail *Otala lactea. J. Exp. Biol.* **2006b**, *209*(Pt. 4), 677–688.

Ramnanan, C. J.; Groom, A. G.; Storey, K. B. Akt and its Downstream Targets Play Key Roles in Mediating Dormancy in Land Snails. *Comp. Biochem. Physiol., B* **2007**, *148*(3), 245–255.

Ramnanan, C. J.; Storey, K. B. The Regulation of Thapsigargin-sensitive Sarcoendoplasmic Reticulum Ca(2+)-ATPase Activity in Estivation. *J. Comp. Physiol., B* **2008**, *178*(1), 33–45.

Ramnanan, C. J.; Storey, K. B. Regulation of Type-1 Protein Phosphatase in a Model of Metabolic Arrest. *BMB Rep.* **2009**, *42*(12), 817–822.

Ramnanan, C. J.; Allen, M. E.; Groom, A. G.; Storey, K. B. Regulation of Global Protein Translation and Protein Degradation in Aerobic Dormancy. *Mol. Cell Biochem.* **2009**, *323*, 9–20.

Ramnanan, C. J.; Edgerton, D. S.; Rivera, N. Irimia-Dominguez, J.; Farmer, B.; Neal, D. W.; Lautz, M.; Donahue, E. P.; Meyer, C. M.; Roach, P. J.; Cherrington, A. D. Molecular Characterization of Insulin-mediated Suppression of Glucose Production In Vivo. *Diabetes* **2010a**, *59*, 1302–1311.

Ramnanan, C. J.; McMullen, D. C.; Bielecki, A.; Storey, K. B. Regulation of Sarcoendo-plasmic Reticulum Ca2+-ATPase (SERCA) in Turtle Muscle and Liver during Acute Exposure to Anoxia. *J. Exp. Biol.* **2010b**, *213*(1), 17–25.

Ramnanan, C. J.; McMullen, D. C.; Groom, A. G.; Storey, K. B. The Regulation of AMPK Signaling in a Natural State of Profound Metabolic Rate Depression. *Mol. Cell. Biochem.* **2010c**, *335*(1–2), 91–105.

Ramnanan C. J.; Edgerton, D. S.; Kraft, G.; Cherrington, A. D. Physiologic Action of Glucagon on Liver Glucose Metabolism. *Diabetes Obes. Metab.* **2011**, *13*(1), 118–125.

Ramnanan, C. J.; Cherrington, A. D.; Edgerton, D. S. Evidence Against a Role for Acute changes in CNS Insulin Action in the Rapid Regulation of Hepatic Glucose Production. *Cell Metab.* **2012**, *15*(5), 656–664.

Ramnanan, C. J.; Kraft, G.; Smith, M. S.; Farmer, B.; Neal, D.; Williams, P. E.; Lautz, M.; Farmer, T.; Donahue, E. P.; Cherrington, A. D.; Edgerton, D. S. Interaction Between the Central and Peripheral Effects of Insulin in Controlling Hepatic Glucose Metabolism in the Conscious Dog. *Diabetes* **2013**, *62*(1), 74–84.

Rivera, N.; Ramnanan, C. J.; An, Z.; Farmer, T.; Smith, M.; Farmer, B.; Irimia, J. M.; Snead, W.; Roach, P. J.; Cherrington, A. D. Insulin-induced Hypoglycemia Increases Hepatic Sensitivity to Glucagon in a Canine Model. *J. Clin. Invest.* **2010**, *120*(12), 4425–4435.

Storey, K. B. Life in the Slow Lane: Molecular Mechanisms of Estivation. *Comp. Biochem. Physiol., A* **2002**, *133*, 733–754.

Storey, K. B.; Storey, J. M. Metabolic Rate Depression in Animals: Transcriptional and Translational Controls. *Biol. Rev. Cambr. Phil. Soc.* **2004**, *79*, 207–233.

Sung, Y. J.; Walters, E. T.; Ambron, R. T. A Neuronal Isoform of Proteinkinase G Couples Mitogen-activated Protein Kinase Nuclear Import to Axotomy induced Long-term hyperexcitability in *Aplysia* Sensory Neurons. *J Neurosci.* **2004**, *24*(34), 7583–7595.

Vega, I. A.; Giraud-Billoud, M.; Koch, E.; Gamarra-Luques, C.; Castro-Vazquez, A. Uric Acid Accumulation within Intracellular Crystalloid Corpuscles of the Midgut Gland in *Pomacea canaliculata* (Caenogastropoda, Ampullariidae). *Veliger* **2007**, *48*, 276–283.

Withers, P. C.; Guppy, M. Do Australian Desert Frogs co-accumulate Counteracting Solutes with Urea during Aestivation? *J. Exp. Biol.* **1996**, *199*(8), 1809–1816.

Whitwam, R. E.; Storey, K. B. Pyruvate Kinase from the Land Snail *Otala lactea*: Regulation by Reversible Phosphorylation During Estivation and Anoxia. *J. Exp. Biol.* **1990**, *154*, 321–337.

Whitwam, R. E.; Storey, K. B. Regulation of Phosphofructokinase During Estivation and Anoxia in the Land Snail, *Otala lactea*. *Physiol. Zool.* **1991**, *64*(2), 595–610.

Zhang, Z.; Tang, N.; Hadden, T. J.; Rishi, A. K. Akt, FoxO and Regulation of Apoptosis. *Biochim. Biophys. Acta* **2011**, *1813*(11), 1978–1986.

CHAPTER 9

GASTROPOD ECOPHYSIOLOGICAL RESPONSE TO STRESS

MARIE-AGNÈS COUTELLEC[1] and THIERRY CAQUET[2]

[1]INRA, UMR 0985 Ecology and Ecosystem Health, Rennes, France.
E-mail: marie-agnes.coutellec@inra.fr

[2]INRA, UAR 1275 Research Division Ecology of Forests, Grasslands and Freshwater Systems, Champenoux, France.
E-mail: thierry.caquet-rennes@inra.fr

CONTENTS

ABSTRACT

Gastropods have colonized marine, freshwater, and terrestrial ecosystems, and the extraordinary diversification of their habitats makes them a group of high ecological importance. They may be exposed to various environmental stressors of natural or human origin, and thereby forced to acclimate or adapt to a highly changing and often stressful, if not toxic, environment. Using examples chosen from the most recent literature, this chapter reviews the various aspects of gastropod ecophysiological response to natural and anthropogenic stressors. Biological responses are presented in their diversity, from stress signaling to successive steps and components of the stress response including oxidative stress, detoxification systems, macromolecules alterations, and apoptosis. A specific section is devoted to the response of the immune system to stressors, including parasites. The chapter ends with some evolutionary perspectives associated with stress responses.

9.1   INTRODUCTION

Gastropods are by far the most speciose class of molluscs, widespread on nearly all possible environments on Earth (Hyman, 1967). As a unique feature in the phylum, gastropods have indeed colonized marine, freshwater, and terrestrial ecosystems, and the extraordinary diversification of their habitats makes them a group of high ecological importance. Gastropods are diverse in many other aspects, including reproduction (gonochorism, parthenogenesis, sequential or simultaneous hermaphroditism, self-fertilization ability), dispersal (from sedentary to pelagic life style), while retaining unique developmental and anatomical characteristics (e.g., torsion, radula). Last, gastropods are currently exposed to intense environmental pressure related to intensification of human activities, and thereby forced to acclimate or adapt to a highly changing and often stressful, if not toxic, environment.

Molluscs are particularly sensitive to chemical pollutants, which makes them well suited as bioindicators of environmental quality (Oehlman and Schulte-Oehlman, 2003). From the IUCN Red List (www.iucnredlist.org), 90% of molluscan species classified as near threatened, vulnerable, endangered, critically endangered, extinct in the wild, or extinct, are gastropods, with a total of 2541 species. Therefore, with regard to conservation, it is of utmost importance to improve the understanding of stress response in this group.

This chapter is devoted to gastropod ecophysiological response to environmental stress, including natural and anthropogenic stressors. Considering the broad scope of this topic, it was obviously impossible to provide a detailed analysis of all possible strategies developed by this group to deal with environmental variation and stress. As not treated elsewhere in this book, a specific section on immunological stress was also included. We have then opted for a general overview of the multiple facets of gastropod stress response, illustrated by examples chosen from the most recent literature. Following an introductory part devoted to stress definitions, biological responses are presented in their diversity, from stress signaling to successive steps and components of the stress response. The chapter ends with some evolutionary perspectives.

9.2   STRESS, STRESSORS, AND STRESS RESPONSE

9.2.1   MULTIPLE VIEWS ON STRESS IN BIOLOGY AND ECOPHYSIOLOGY

The term "stress" is probably among the most controversial biological terms. It originates in physics to describe pressure and deformation in a system. Following the work of Selye (1950) who recognized a similar suite of coordinated reactions to diverse noxious stimuli or "agents" in mammals as a "General Adaptation Syndrome", it has been widely applied into a biological context. Selye's model includes three stages. After an alarm initial stage, which includes a shock phase (changes in several organic systems) and a counter-shock phase (defensive response, including increased production of hormones), the organism attempts to resist or adapt to the stressor (resistance stage). The process eventually goes into an exhaustion stage, during which energy is depleted, presumably leading to death.

As stress can be applied to various levels of biological organization, the term is used in many different areas, especially in ecophysiology, ecology, toxicology, and ecotoxicology. In ancient works, "stress" has been used to refer both to the event or agent causing a response in an organism (e.g., change in temperature, attack by a predator) and to the response in itself. More recently, the triggering external stimuli have been named "stressors" while "stress" is the internal state brought about by a stressor (sometimes identified as a response syndrome), and "stress response" is a cascade of internal changes triggered by stress (Van Straalen, 2003).

Grime's definition of stress, "external constraints limiting the rates of resource acquisition, growth or reproduction of organisms" (Grime, 1989) has been widely used in ecology (see Borics et al., 2013), but it does not specify any component of temporal dynamics. "Biological stress" is used in the literature to describe any condition that forces living systems away from a physiological steady state (Kagias et al., 2012).

Kagias et al. (2012) defined "Physiological stress" as "the primary biological stress...defined as any external or internal condition that challenges the homeostasis of a cell or an organism." They also identified three dimensions in physiological stress: intrinsic developmental stress, environmental stress, and aging. "Intrinsic developmental stress" is associated with developmental events that may challenge the developing organism. For example, morphogenesis and changes in inner chemistry may yield stressful conditions that may trigger the activation of defense mechanisms. Indeed, transient upregulation of heat shock proteins (Hsps) followed by a subsequent downregulation has been described in tissues undergoing morphogenesis in several invertebrate species, including gastropods such as *Haliotis asinina* (Gunter and Degnan, 2007).

"Environmental stress" refers to the situation where environmental variations exceed certain levels and homeostasis is threatened. Aging is another source of stress that living organisms have to cope with. It may be viewed as the consequence of the stochastic accumulation of molecular damage over time (Hayflick, 2007; Holliday, 2006; Kirkwood and Melov, 2011; Kirkwood and Melov, 2011; Partridge, 2010; Rattan, 2006). The capacity of an individual to cope with aging and other stresses defines its longevity. Although gastropods are not frequently mentioned as model invertebrates for aging studies (Murthy and Ram, 2015), the pond snail *Lymnaea stagnalis* is a suitable model species for studying the process of neuronal aging and age-associated learning and memory impairment (Hermann et al., 2007; Watson et al., 2013).

Some authors make a difference between predictable and unpredictable stressors, especially in vertebrates (Wingfield, 1994; Wingfield et al., 1997). If environmental conditions change in a predictable way (e.g., seasonal reductions of temperature or food supply), individuals can prepare to such events that do not act in themselves as stressors. These individuals cannot be described as stressed but they may be more susceptible to the effects of real stress.

9.2.2   INTEGRATION OF STRESS AT VARIOUS LEVELS OF BIOLOGICAL ORGANIZATION

9.2.2.1   STRESS AND HOMEOSTASIS

Stress is frequently defined as the state of a biological system in which homeostasis (i.e., the mechanisms that maintain stability within the physiological systems and hold all the parameters of the organisms internal milieu within limits that allow an organism to survive) is threatened or not maintained (Moberg, 2000), and this definition applies to virtually all the biological systems. The shape of the relationship between external conditions and organisms internal conditions allows the distinction between "conformers," for which internal conditions are determined by environmental conditions, and "regulators" for which internal conditions are stable for a certain range of environmental conditions.

Stressors constantly challenge the homeostasis of biological systems (Charmandari et al., 2005). Various adaptations have evolved that provide organisms with the ability not only to survive, but also to reproduce under different, sometimes hostile conditions. Such adaptations are associated with specific molecular and cellular mechanisms, body structures and organism behaviors, tailored to a specific environment. Animals encounter ranges of environmental conditions in which performance is maximized as well as thresholds beyond which performance fails and tolerance becomes time limited (Hoffmann and Todgham, 2010). In an organism exposed to a stimulus (e.g., biotic or abiotic stressor), constitutive response processes are elicited to cope with the challenge. They operate within a dynamic range and are part of the perturbation response. If the response is efficient, homeostasis may be restored. Depending on the duration and strength of the perturbation, the dynamic capacity of the response may be exceeded, leading to the initiation of a stress response that involves both constitutive and inducible elements (Stefano et al., 2002).

9.2.2.2   ALLOSTASIS AS AN ALTERNATIVE TO HOMEOSTASIS

In order to take into account the existence of temporal rhythms (e.g., circadial or circannual) and of changes associated with the various life history stages of a species, the concept of "allostasis" was proposed for vertebrates and included in the Allostasis Model (Sterling and Eyer, 1988). Allostasis refers to the integrative adaptive processes maintaining "stability through change," that is, a stability that is not within the normal homeostatic range. Allostasis is theoretically able to account for evidence that physiological variables are not constant. It is based on the concept that the goal of regulation is not constancy, but rather fitness under natural selection, that implies preventing errors and minimizing costs. Both needs are best accomplished by using prior information to predict demand and then adjusting all parameters to meet it (Sterling, 2003). Although the definition has been refined by various authors (see e.g., McEwen and Wingfield, 2003; Sterling, 2012), in all cases, allostasis emphasizes the dynamic behavioral and physiological mechanisms that are used to anticipate or cope with environmental change to maintain organismal function.

Associated with allostatis is the concept of "allostatic load," that is, the wear and tear associated with the chronic overactivity or dysregulation of allostatic mechanisms (McEwen and Stellar, 1993). At this point, the activation of allostatic mechanisms causes serious negative physiological consequences (McEwen and Wingfield, 2010). Two theoretical types of allostatic load have been defined. Type I occurs when the energy demand exceeds the available energy supply, resulting in reduced performance. Type II occurs when the prolonged activation of the allostatic mechanisms causes pathology, even when sufficient energy is available. It is therefore necessary to focus on the costs of mounting a response to an environmental change, and on the distinction between responses that are beneficial, those that may impose a significant cost and those that may actually cause harm (Schulte, 2014).

9.2.2.3   TOWARD AN INTEGRATIVE FRAMEWORK

There is no consensus on the usefulness of replacing homeostasis by allostasis (Dallman, 2003; Day, 2005), and instead, Romero et al. (2009) proposed to integrate homeostasis, allostasis, and stress in a common framework, the Reactive Scope Model (Fig. 9.1).

According to the Reactive Scope Model, four levels or ranges may be defined for the hormones or other physiological mediators of the stress response: homeostatic failure, predictive homeostasis, reactive homeostasis, and homeostatic overload.

Homeostatic failure occurs when concentration/level of a mediator is too low to maintain homeostasis, leading rapidly to death. Predictive homeostasis refers to the situation where the concentration/level of the mediator necessary to cope with predictable environmental changes is met. Its range and temporal dynamics follow predictable life-history changes, circadian (area in dark gray in Fig. 9.1) and seasonal rhythms. The upper limit of the corresponding range usually lays a little bit above the peak values associated with the circadian cycle. Reactive Homeostasis occurs when unpredictable changes induce an emergency elevation of the mediator. This elevation corresponds to the classical "stress response." Together, the predictive and reactive homeostasis ranges form the normal reactive scope for that mediator. If the concentration/level of this mediator exceeds the upper limit of the reactive homeostasis range, the mediator itself may start to cause damage. Finally, homeostatic overload refers to the range above this threshold. "Chronic stress" refers to the periods here where the concentration/level of the mediator is in this range.

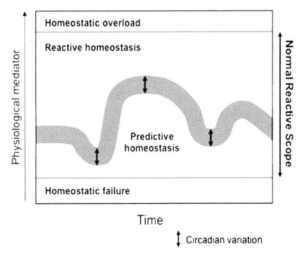

FIGURE 9.1 Schematic representation of the Reactive Scope Model (Adapted from "Using the Reactive Scope Model to Understand Why Stress Physiology Predicts Survival During Starvation in Galápagos Marine Iguanas" in General and Comparative Endocrinology, 2012 May 1;176(3):296-9, with permission from Elsevier.)

Reactive homeostasis is part of the normal physiology of the animal, but it consumes energy and resources that cannot be used in other systems, thereby increasing the allostatic load (i.e., the increase in workload required to maintain homeostasis). The result of this cost is an accumulation of wear and tear, or the depletion of energy or the directing of energy away from other tasks such as tissue maintenance.

9.2.2.4   STRESS RESPONSE AND BIOENERGETICS CONSEQUENCES

Various physiological reactions aim at minimizing the detrimental effects of the stressors. Usually, this involves reallocation of resources from nonessential functions to life-preserving processes (Buchanan, 2000). At the molecular level, studies of stress in animals have shown that various types of stressors induce a set of common responses, which include DNA and protein repair, apoptosis, lysis of molecules damaged by stress (e.g., proteolysis), and metabolic changes reflecting the transition from cellular growth to cellular repair (Sokolova and Lannig, 2008).

Energy reserves play a key role in the ability to elicit these responses. Indeed, stress can result in elevated basal metabolic demand due to the costs of upregulation of cellular protective mechanisms. Competition may therefore occur between these uses of reserves and other energy-demanding functions such as e.g., reproduction (Sokolova et al., 2012). The integration of all these dimensions has led to the concept of energy-limited tolerance to stress (Sokolova, 2013) that is presented below.

Stress responses have been recently reviewed by Kassahn et al. (2009), who also highlighted that oxygen imbalance (reduced partial pressure) and oxidative stress play a central role in the response to many stresses, through stress-inducible-signaling pathways (e.g., heat shock protein-Hsp expression modulated by the JNK pathway), redox-sensitive transcription factors (e.g., HFS1, which induces Hsp transcription, as a possible result of protein damaged by oxidative stress), and a group of genes known as "immediate early genes" (including transcription factors, cytokines, actin, fibronectin). The generic nature of these responses is interesting because it provides suitable conditions (larger datasets) to explore hypotheses about how stress responses relate across biological scales, in particular between molecular processes and higher phenotypic integration such as population divergence in life history traits.

Under aerobic conditions, substrates used for energy production include carbohydrates, lipids, and amino acids. The energetic costs involved in the defense strategies toward stressors may have consequences on the dynamics of these substrates in stressed organisms, suggesting that parameters linked with the status or use of these energy reserves may be used as biomarkers of stress (Huggett et al., 1992).

Polysaccharide level, especially glycogen, is one of the parameters that reflects the energetic and reserves status of an organism. Glycogen is used rapidly when organisms are under stress, and glycogen levels have been suggested as biomarker of general stress (Huggett et al., 1992; Vasseur and Cossu-Leguille, 2003). In gastropods, glycogen is considered to be the principal energetic reserve (Livingstone and de Zwaan, 1983). Usually, this reserve is found in specific storage cells, glycogen cells, which are widely distributed in the whole body of the snail (Geraerts, 1992; Hemminga et al., 1985).

Convergent results showed that glycogen reserves of gastropods may be severely affected by parasitism by digenetic trematodes (Becker, 1983; Pinheiro and Amato, 1994; Schwartz and Carter, 1982; Tunholi-Alves et al., 2014). Parasite infection may lead to an acceleration of gluconeogenesis through increased consumption of glucose from the hemolymph by the larval trematodes and an acceleration of the catalytic activity of the glycogenolysis pathway (Tielens et al., 1992).

Decreased glycogen levels have also been reported following exposure to hypoxia (e.g., in *Nassarius conoidalis*; Liu et al., 2014). Hypoxia induces specific impacts on the metabolic pathways in stressed aquatic invertebrates. Information about anaerobic metabolism of gastropods is fragmentary (Liu et al., 2014; Santini et al., 2001). A number of anaerobic end products have been reported, including lactate, octopine, alanine, succinate, acetate, propionate, butyrate, strombine, and alanopine (Livingstone and de Zwaan, 1983) but with differences between taxonomic groups (Larade and Storey, 2002), and even within the same genus (Liu et al., 2014). Environmental conditions (Prabhakara Rao and Prasada Rao, 1982; Wieser, 1980), hypoxia duration (de Zwaan and van Marrewijk, 1973; Kluytmans et al., 1975), and type (Gäde, 1975; Gäde et al., 1984) may also have an influence on this metabolism. Hypoxia or anoxia usually induces the increase of glycogen consumption (the "Pasteur Effect"), due to an activation of anaerobic glycolysis in order to compensate for the lower production or synthesis of ATP (Silva-Castiglioni et al., 2010). Under hypoxia, four anaerobic pathways have been identified in invertebrates: (1) glucose–lactate pathway (end product: lactate); (2) glucose–opine pathway (end product: opines); (3) glucose–succinate pathway (end products: succinate); and (4) aspartate–succinate pathway (end products: succinate and alanine; Hochachka and Somero, 1984; Livingstone, 1983). Lactate is an important end product in terrestrial and freshwater gastropod but not in marine species (see Wieser, 1980 and references therein). In marine species, lactate is replaced by alanine, succinate, acetate, and propionate (Liu et al., 2014). The accumulation of these various end-products may be used to demonstrate the exposure of gastropods to hypoxia in the field.

A decrease in glycogen content may also reveal the exposure to toxic compounds such as inorganic toxicants (e.g., arsenic and lead in *Biomphalaria glabrata*; Ansaldo et al., 2006) or organic pesticides (trichlorfon in *Lymnaea acuminata*; Mahendru and Agarwal, 1981; endosulfan, methyl parathion, quinalphos, and nuvanmay in *Bellamya dissimilis*; Padmaja and Rao, 1994; imidacloprid in *Helix aspersa* (*Cornu aspersum*); Radwan and Mohamed, 2013). Stimulation of the activity of polysaccharide-hydrolyzing enzymes has also been shown in *Lymnaea palustris* exposed to hexachloro-benzene (Baturo et al., 1995).

Since individuals only have a limited amount of carbohydrates, the next alternative source of energy to meet the increased energy demand associated with stress are proteins and, to a lesser extent, lipids. Radwan et al. (2008) showed a significant effect of two carbamate insecticides (methomyl and methiocarb) on the decrease of contents of total proteins and lipids in the tissues of *Eobania vermiculata*. Exposing *H. aspersa* (*C. aspersum*) snails to high concentrations of thiametoxam, Ait Hamlet et al. (2012) observed a significant negative effect of the two highest concentrations of the insec-ticide on the concentrations of total carbohydrates, total proteins, and total lipids. This effect was correlated with strong histological alteration of the hepatopancreas. In *Bellamya bengalensis*, Kumari (2013) showed that expo-sure to high detergent concentration caused a decrease in the protein content in the digestive gland, mantle, and foot of the exposed snails. In *L. stagnalis*, Bhide et al. (2006) reported that sublethal concentrations of baygon and nuvan induced depletion in the protein content. According to Padmaja and Rao (1994), the decrease in tissue proteins of snails exposed to pesticides could be due to various mechanisms such as the formation of lipoproteins which are utilized for the repair of damaged cells and tissue organelles or the direct utilization by cells to fulfill energy requirements.

However, decreasing lipid or protein content following acute exposure to a stressor is not always observed. For example, El-Gohary et al. (2011) and El-Shenawy et al. (2012) observed an increase in the lipid content of the digestive gland in *E. vermiculata* exposed to heavy metals or molluscicides, respectively. In *H. aspersa* (*C. aspersum*) exposed to imidacloprid, Radwan and Mohamed (2013) even observed an increase in the protein content of the digestive gland. This increase may be a consequence of an induction of the synthesis of proteic defense systems such as stress proteins.

Proposals have been made for modeling the bioenergetics of animals that may also integrate the costs associated with the responses to stressors and their consequences for fitness, such as the Scope for Growth (SfG) model (Winberg, 1960), the Dynamic Energy Budget (DEB) theory (Kooi-jman, 2010), the Ontogenetic Growth Model (OGM; Hou et al., 2008), the oxygen- and capacity-limited thermal tolerance (OCLTT) concept

(Pörtner, 2010, 2012), or the concept of energy-limited tolerance to stress (Sokolova, 2013). Some of these models have already been applied to assess the effects of individual or multiple stressors on bivalves (Widdows et al., 1987, 1995, 2002) and gastropods (Ducrot et al., 2007; Stickle et al., 1984; Wo et al., 1999).

The concept of energy-limited tolerance to stress has recently been proposed as a framework for assessing the effects of multiple environmental stressors that integrates some elements from DEB and OCLTT concepts (Sokolova, 2013). According to this framework, five main energy-demanding functions should be considered in an individual: basal maintenance, activity, reproduction/maturation, growth/development, and deposition of energy reserves. Relative allocation of energy to these different processes varies among species and life stages. These processes rely on ATP provided via aerobic or anaerobic metabolism.

Four levels or range may be defined regarding the response of an individual to stress. In the optimum range, the aerobic scope is maximal and aerobic ATP supply covers the basal maintenance and the investment in the other functions. The excess of energy is deposited in storage compounds (e.g., glycogen or lipids). During moderate stress (or pejus range), the maintenance costs increase in order to meet the needs for additional energy for protection from stress and for repairing damage. In some cases assimilation of food and/or capacity for aerobic metabolism may be impaired, therefore, leading to a decline in aerobic scope. Energy storage does no occur or it is reduced. During extreme stress (or pessimum range), the increase in ATP demand for maintenance and/or the impairment of aerobic metabolism overrides ATP supply via aerobic metabolism. Metabolism switches to partial anaerobiosis that fuels the essential maintenance costs and supports the time-limited survival of the organism. In some species, survival in the pessimum range can be enhanced by a metabolic rate depression that reduces rates of energy turnover at the expense of shutting down the ATP-demanding functions that are not essential for immediate survival. The energy balance is temporarily disrupted during transitions between the optimum, pejus, and pessimum ranges but is eventually reinstated, as the organism becomes acclimated to these conditions. In the lethal range, the balance of supply and demand of ATP is permanently disrupted, resulting in negative aerobic scope and death of the organism. This model may also serve as a theoretical basis to identify possible bioenergetics markers for the assessment of the impacts of stressors (Sokolova, 2013).

9.2.3   THE CASE OF MULTIPLE STRESSORS

A lot of studies have been performed to evaluate the effects of individual stressors on various gastropod species, usually in laboratory experiments where one stressor (e.g., temperature, salinity, toxic substance) is manipulated whereas all the other conditions are kept constant. However, under natural environmental conditions organisms may be simultaneously exposed to changes in various variables, including (a)biotic stressors that may therefore have an impact on the physiology of the exposed individuals (Todgham and Stillman, 2013).

These changes may have an additive, an antagonistic or a (non)linear synergistic effect on various parameters of interest (Crain et al., 2008; Darling and Côté, 2008; Holmstrup et al., 2010). Under additive hypothesis small shifts in multiple stressors may have a small effect on performance, whereas under synergistic hypothesis, small shifts in multiple stressors may have a great impact on physiological performance, thus generating unpredictable responses in terms of species' distributions and abundances. Inferences deduced from studies performed on single stressors are potentially misleading regarding the outcome of multiple stressors, including toxic stressors, under natural environmental conditions (see e.g., McBryan et al., 2013; Whitehead, 2013).

In addition, a first stressor may either cause the organisms to be more susceptible to a second stressor (cross-susceptibility; Sinclair et al., 2013) or increase tolerance to a second stressor (cross-tolerance; Horowitz, 2007; Sinclair et al., 2013). At the cellular level, "cross-talk" occurs when multiple pathways, each stimulated by a separate stressor, converge on one physiological function. "Cross-tolerance" is the situation where each stressor may modulate the same pathway, but producing distinct physiological outcomes (Todgham and Stillman, 2013). Both cross-talk and cross-tolerance may be adaptive, but with differing consequences for responding to future changes, due to their mechanisms of action.

The case of exposure to multiple stressors is highly relevant from a fundamental point of view and for environmental management. Several experimental studies have been performed on the combined effects of at least two stressors in gastropods: alkaline stress and copper (Paulson et al., 1983), temperature and salinity (Deschaseaux et al., 2010, 2011), temperature and rainfall (Dong et al., 2014; Williams et al., 2011), temperature and pH (Byrne et al., 2010, 2011), parasitism by a trematode and temperature, salinity, or hypoxia (Lee and Cheng, 1971; McDaniel, 1969; Sousa and Gleason, 1989; Tallmark and Norrgren, 1976; Vernberg and Vernberg, 1963), acidified conditions, elevated temperature, and solar UV radiation (Davis et al., 2013), metals and predacious cues (Lefcort et al., 2013), combination of metals (Byzitter et al., 2012) or other toxicants (Nevo and Lavie, 1989), or starvation and temperature (Jeno and Brokordt, 2014).

Cross-susceptibility was frequently observed. For example, larval trematode infections have been shown experimentally to cause increased mortality at high water temperatures in various molluscan hosts, including the marine snails *Nassarius obsoletus* (Vernberg and Vernberg, 1963), *Nassarius reticulatus* (Tallmark and Norrgren, 1976), and *Littorina littorea* (McDaniel, 1969), and in the freshwater snail, *B. glabrata* (Lee and Cheng, 1971). Cross-susceptibility may also be two ways. In *L. palustris*, for example, a brief exposure to heavy metals (cadmium, lead, zinc) impaired the ability to avoid predacious cues and conversely after preexposure to crushed conspecifics, individuals not only failed to avoid metal treated water, but also they actually moved toward it (Lefcort et al., 2013).

When addressing the effects of multiple stressors, it is necessary to distinguish the cases where the induced stress remains moderate and compatible with the survival of organisms from those where stress is unsustainable and leads to death. As it was already mentioned, cellular defense and maintenance of homeostasis implies the diversion of energy from other function such as growth or reproduction. The rates of intake and assimilation of energy are limited for all organisms, as well as their metabolic capacity to convert ingested food to ATP (Guderley and Pörtner, 2010). Exposure to stressors such as toxic substances may induce a reduction in food consumption and/or assimilation (Edwards, 1980), therefore, leading to a decrease in the availability of energy for defense systems. When the energy reserves are not sufficient, the synthesis and activity of defense substances may be reduced. For example, starvation decreased the content of stored energy substrates of juveniles of *Concholepas concholepas*, an intertidal snail, as well as their ability to synthesize Hsp70 during emersion under thermal stress, especially at high temperatures (Jeno and Brokordt, 2014).

9.3 STRESS RESPONSE, A SEQUENTIAL PROCESS

Deleterious effects induced by environmental change and stress affect organisms at various levels of biological organization, including molecular and cellular processes, endocrine systems and physiology, immune functions, lifespan, and fitness. As noted earlier (see Section 9.1), homeostasis maintenance is critical for the ability of an organism to cope with stress. In this respect, the ability to assure oxidative metabolism is of primary importance (Kassahn et al., 2009). Some processes are evolutionarily highly conserved, such as the cellular stress response and associated proteins (Kültz, 2003).

Stress response is a sequential process which involves stress-signaling pathways, followed with the activation of antioxidant systems and cellular/ molecular mechanisms that are common to many stressors, and which include the repair of DNA and protein damage, cell cycle arrest or apoptosis, changes in cellular metabolism associated with cellular repair, as well as the release of stress hormones (Kassahn et al., 2009). Among gastropods, these processes have been studied in three main "stress" contexts, that is, natural environmental variation in intertidal snails, parasitism, and water chemical contaminants in freshwater snails. Indeed, once in the body, many xenobiotics may induce signal transduction events leading to various cellular, physiological, and pharmacological responses including homeostasis, proliferation, differentiation, apoptosis, or necrosis.

9.3.1 STRESS SIGNALING

Environmental information is processed by extra-and intracellular-signaling pathways and membrane transporters. The mitogen-activated protein kinases (MAPKs) are important-signaling molecules involved in relaying extracellular signals to intracellular targets (Storey and Storey, 2004). Members of the MAPK superfamily include the cJun-N-terminal kinases (JNKs) and p38 MAPK pathways, which are typically responsive to environmental stresses (Kyriakis and Avruch, 2001). Stress-activated protein kinases (SAPK)/Jun amino-terminal kinases (JNK) are members of the MAPK family and are activated by a variety of environmental stresses, inflammatory cytokines, growth factors, and G-protein-coupled receptors (GPCR) agonists. Stress signals are delivered to this cascade by small GTPases of the Rho family (Rac, Rho, cdc42). As with the other MAPKs, the membrane proximal kinase is a MAPKKK, typically MEKK1–4, or a member of the mixed lineage kinases that phosphorylates and activates MKK4 (SEK) or MKK7, the SAPK/JNK kinases. Alternatively, MKK4/7 can be activated by a member of the germinal center kinase family in a GTPase-independent manner. SAPK/JNK translocates to the nucleus where it can regulate the activity of multiple transcription factors (see more on this topic at: http://www.cellsignal.com/common/content/content.jsp?id=pathways-mapk-sapk#sthash.BrLF8Hrm.dpuf).

Once activated during stress, these pathways mediate gene expression. For example, JNKs phosphorylate c-Jun protein, in combination with c-Fos, forms the AP-1 early response transcription factor (activator protein 1). This transcription factor regulates the expression of various genes, and controls several cellular processes and fate, including apoptosis. Other JNK-mediated transcription factors included ATF-2 (which regulates c-Jun transactivation, especially in response to genotoxic agents; Van Dam et al., 1995), Elk-1 (drug addiction, memory, breast cancer, depression), Myc, Smad3 (see TGF beta pathway), tumor suppressor p53, NFAT4, DPC4 and MADD, a cell death domain protein (Cowan and Storey, 2003). Cross-talks occur between heat shock response and JNK (Kassahn et al., 2009); for example, Hsp72 accumulation downregulates JNK and increases thermotolerance (through tolerance to caspase independent apoptosis) in humans (Gabai et al., 2000). Comparatively, p38 kinase pathway is a complex pathway known to be activated in mammalian response to various extracellular stimuli, including UV light, heat, osmotic shock, inflammatory cytokines, and growth factors (Zarubin and Han, 2005).

In gastropods, MAPKs have been characterized in a few representatives, the marine species *Aplysia californica* (cDNA, REF) and *Littorina littorea* (cDNA and protein characterization; p38, JNK, ERK; Iakovleva et al., 2006); and freshwater species such as *L. stagnalis* (protein characterization, activity assay; ERK; Plows et al., 2004) and *B. glabrata* (cDNA; Yoshino et al., 2001). Larade and Storey (2006) studied the response of JNK and p38-signaling pathways, as well as ERK pathway (extracellular signal-regulated protein kinases) to short-term anoxia in the common periwinkle *L. littorea*, a species adapted to high environmental variation, with respect to temperature, water, salinity, and oxygen. Anoxia was found to induce p38 phosphorylation in the digestive gland cells, by a twofold increase. Downstream effects of p38 activation were supported by the observed increase in phosphorylated Hsp27 (involved in cytoskeleton) and CREB (cAMP Response-Element Binding protein, which, among other, regulates c-fos expression), which suggests the involvement of MAPKAKP-2. By contrast, no changes in JNK and ERK pathways were detected. In this species, a specific ERK family member with high molecular weight was identified (p115 MAPK), that was transiently activated by freezing and anoxia (MacDonald and Storey, 2006). The role of ERK in regulating phagocytosis and immune response was assessed in *L. stagnalis* hemocytes (Plows et al., 2004). Recently, the role of post-transcriptional regulators of protein expression such as microRNAs has also been advocated (see Box 9.1). Altogether, these few examples, beyond the merit of existing, also demonstrate the critical need to improve knowledge on stress signaling in gastropods.

BOX 9.1 miRNA and Gastropod Stress Response

As a particularly class of post-transcriptional regulators of protein expression, microRNAs are short noncoding RNAs (about 22 nucleotids) known to have pervasive roles in regulation of cellular processes, including biological development, cell differentiation, apoptosis, or cell cycle control. These regulators are transcribed as long RNAs which are first cleaved to pre-miRNAs by the nuclear processing enzyme *drosha*, and then processed into mature miRNAs by the cytoplasmic ribonuclease *Dicer*. Their involvement in mediating stress response was recently reviewed (Leung and Sharp, 2010 and references therein). For example, tumor suppressor p53, which can be induced upon DNA damage, induces the transcription of miRNAs of the 34 family, as well as their further processing through its association with a cofactor of *drosha*. These miRNAs in turn promote cell growth arrest and apoptosis. DNA damage also induces repression of another miRNA involved in the regulation of p53. More generally, upon stress, a miRNA can either be involved in homeostasis restoration or act as an enforcer of a new gene expression pathway.

The effects of anoxia and freezing on miRNAs have been recently explored in the common periwinkle *L. littorea* (Biggar et al., 2012). By focusing on highly conserved miRNAs with previously established roles in metabolic rate depression in other organisms, these authors were able to identify effects on miRNAs of particular relevance to stress response: miR-210 (hypoxia-inducible, likely to be regulated by transcription factor HIFalpha), miR-29b (involved in the PI3K/Akt pathway itself related to p53 regulation and apoptosis), miR-34a (transcription induced by p53), miR-125b (cold-inducible in fish, involved in p53 regulation and activating the antioxidant defense-related NF-κB pathway), etc. (see Biggar et al., 2012). To our knowledge, this is first attempt to decipher the role of miRNAs in gastropod response to environmental stress, and this pioneer study paves the way for future breakthroughs in this domain.

9.3.2 OXIDATIVE STRESS AND ANTIOXIDANT SYSTEMS

As a common consequence of many stressful conditions, oxidative stress is a major actor in eliciting responses to stressors. Signal transduction pathways (JNK, p38 MAPK) and transcription factors sensitive to redox imbalance include immediate early genes (c-fos, fosB, c-Jun, JunB, c-myc, egr-1, KC and JE cytokines, actin, fibronectin), HIF1-alpha, NF-κB, and p53 (Kassahn et al., 2009).

In aerobic organisms, the production of reactive oxygen species (ROS, a term which encompasses both initial species produced by oxygen reduction as well as secondary reactive products; Winterbourn, 2008) is a normal process which results from the inherent dangerousness of oxygen ("oxygen paradox"). The superoxide anion radical, hydrogen peroxide, and the extremely reactive hydroxyl radical are common products of life in an aerobic environment, and these agents appear to be responsible for oxygen toxicity (Davies, 1995). The oxidative burden is contributed by various cellular processes. ROS are predominantly generated within mitochondria, as a consequence of oxidative phosphorylation, but also in the plasma membrane (e.g., NAPDH oxidases), in the peroxisomes (lipid metabolism), as a product of many cytosolic enzymes such as cyclooxygenases (Balaban et al., 2005). Despite their toxic effects, ROS also act as essential physiological regulators of various intracellular-signaling pathways (D'Autréaux and Toledano, 2007; Finkel, 2011; Poulsen et al., 2000). Notably, ROS play a critical role in innate immune defense, including in gastropods (see specific section below), and hydrogen peroxide was also shown to be involved in metamorphosis, such as in the nudibranch *Phestilla sibogae* (velar loss; Pires and Hadfield, 1991). Therefore, the management of oxidative stress represents a balance between meeting the functional requirements for ROS (e.g., as signaling molecules) and preventing or repairing oxidative damage (Dowling and Simmons, 2009; Monaghan et al., 2009).

The maintenance of intracellular redox homeostasis is dependent on a complex system of antioxidant molecules. These antioxidants include low molecular weight molecules such as glutathione, as well as an array of protein antioxidants that each has specific subcellular localizations and chemical reactivities (Finkel, 2011). Oxidative stress occurs when levels of ROS exceed the capacity of antioxidant defenses. Oxidative damage can affect most macromolecules, namely DNA, proteins, and lipids.

In gastropods, oxidative stress has been largely studied as part of the immune response (see specific section below), as well as in ecotoxicology (see Box 9.2). For example, in the latter field, the response of *L. stagnalis* to various herbicides was investigated by Russo and her collaborators. In this species, hemocyte ROS production (H_2O_2) was decreased by atrazine (Russo and Lagadic, 2004), whereas the peroxidizing herbicide fomesafen had the opposite effect (Russo et al., 2007). More detailed work on this topic is presented in Box 9.2.

BOX 9. 2 Toxicants as Exogenous Sources of Oxidative Stress

Environmental contaminants are often sources of stress to organisms and may impair population demography and fate. Furthermore, because these molecules are likely to interact with natural factors of stress, their effects need to be jointly estimated. Population ecotoxicology typically deals with this issue and a battery of tests have been developed to assess the toxicity of chemicals that are based on stress response at molecular, cellular, and organismic levels.

Oxidative stress is known to elicit the toxicity of many pollutants (e.g., aromatic hydrocarbons, pharmaceuticals, metals and pesticides; Isaksson, 2010; Regoli et al., 2002; Valavanidis et al., 2006). As active pro-oxidants, some pollutants (such as most heavy metals, as well as bipyridyl herbicides) directly increase the level of ROS. In other cases, pollutants can increase oxidative stress by inhibiting gene expression of antioxidants and, thereby, change the pro-oxidant/antioxidant balance (Limón-Pacheco and Gonsebatt, 2009). Therefore, antioxidant defense systems and biotransformation pathways are commonly used as biomarkers in ecotoxicology, including gastropods (e.g., Bouétard et al., 2013; Gust et al., 2013a,b). Antioxidant mechanisms are induced to counteract ROS production and overcome the stressful condition. When antioxidant defenses are overwhelmed by ROS generation, different forms of toxicity arise, including lipid peroxidation of cellular membranes, protein degradation, enzyme inactivation and damage to DNA (Regoli et al., 2000).

Aerobic organisms have developed defenses against oxidative damaging, which include enzymatic systems and antioxidant (nonenzymatic) molecules. In gastropods, most of these systems have been described and studied in link with immune, stress, and toxicological responses. A specific section of this chapter is devoted to immunity, so that only environmental stress and toxicological responses will be considered here. Some findings in gastropods are presented, in the framework of cellular defenses as well as detoxification processes.

Within cells, superoxide anions are reduced by superoxide dismutase (SOD) and produce hydrogen peroxide and singlet oxygen. Hydrogen peroxide is in turn converted to water by catalase (CAT) or glutathione peroxidase (GPX). GPX, using glutathione as cofactor, is a protagonist of the glutathione cycle which also involves glutathione reductase and the second phase detoxication enzymes glutathione S-transferases (GSTs). Singlet oxygen is quenched by other antioxidants, notably vitamin E. Peroxiredoxin enzymes also play an important ROS scavenging role in the mitochondria (Balaban et al., 2005). Depending on tissues and species, the transcription factor retinoid X receptor (RXR) is recognized to play different roles in response to oxidative stress, such as being an activator for the phase II detoxication induction (Kang et al., 2005), an antiapoptotic factor, and an inhibitor of intracellular ROS generation by up-regulating CAT activity (Shan et al., 2008).

Many gastropods are naturally exposed and thus adapted to recurrent conditions generating oxidative stress. This is typically the case of inter-tidal snails, such as *L. littorea*, which is regularly exposed to high environmental variation (incl. heat and freezing, hypoxia, desiccation). Such organisms are thus expected to have elaborated an efficient antioxidant system. A nice example of this efficiency is given in the study of Pannunzio and Storey (1997). These authors showed that in *L. littorea*, anoxia significantly reduced enzymatic activities in the digestive gland (CAT, SOD, glutathione *S*-transferase, glutathione reductase, and glutathione peroxidase), in consistency with anoxia-induced deprivation of oxygen free radicals. The responsive ability of the digestive gland antioxidant defenses was evidenced by a rapid recovery of the four glutathione-linked activities when oxygen was reintroduced. On the other hand, the total content of glutathione (GHS, reduced, and GSSG, oxidized form) increased significantly, which may reflect the anticipated need for GSH during the recovery period in this facultative anaerobe species. Last, the ratio of the two forms (GHS/GSSG) remained stable during anaerobiosis, and increased only during the aerobic recovery period, which is consistent with a protective response to the oxidative damage associated with the reintroduction of oxygen, as well as with an involvement in detoxification of pro-oxidant products accumulated during anaerobiosis.

Terrestrial snails also have to cope with adverse environmental conditions, which led them to adapt through estivation. Arousal from estivation, by triggering abrupt changes from a hypometabolic state to an active one (reset of oxygen metabolism), sets up a condition of increased ROS formation, against which anticipatory mechanisms have evolved. For example, in the land snail *Otala lactea*, arousal triggered a strong induction of SOD, CAT, GST, and GPX activity, followed with a rapid return to normal activity for SOD and GPX (Hermes-Lima and Storey, 1995). Estivation induced the enzymatic preparation for oxidative stress in different snails species, with a protective key-role identified for the selenium dependent GPX (reviewed in Feirrera-Cravo et al., 2010). In the aquatic snail *Pomacea canaliculata*, urate, which has antioxidant properties, was found to increase during estivation and its oxidation product, allantoin, to increase after arousal (Giraud-Billoud et al., 2013).

Using data from a broad taxonomic range of anoxia-tolerant species, Hermes-Lima and Zenteno-Savin (2002) identified that an increase in the baseline activity of key antioxidant enzymes, as well as "secondary" enzymatic defenses and/or glutathione levels in preparation for a putative oxidative stressful situation arising from tissue reoxygenation seem to be the preferred evolutionary adaptation.

However, the huge diversity of responses produced by hypoxia-tolerant animals demonstrates the lack of a general mechanism, as tolerance may result from different strategies, from high constitutive expression of antioxidant defenses, to high inducibility of antioxidant defenses during anoxia (preparation for oxidative stress upon reoxygenation). These processes were recently reviewed and the role of transcription factors and miRNAs as antioxidant regulators discussed into details (Welker et al., 2013).

The transcription and/or activity of antioxidant enzymes have been largely studied for their biomarker value in ecotoxicology, notably in *L. stagnalis*, in response to pesticides (Bouétard et al., 2013) and to waste water effluents and to pharmaceutical mixtures (Gust et al., 2013a,b), in *Lymnaea natalensis* exposed to various metals (Siwela et al., 2010), in the marine gastropod *Onchidium struma* exposed to copper (SOD, CAT; Li et al., 2009), in the terrestrial snails *Achatina fulica* faced to cadmium and zinc (Chandran et al., 2005) and *Chilina gibbosa* exposed to the organophosphate insecticide azinphos-methyl (Bianco et al., 2013). *B. glabrata* exhibited genetic variation in SOD and CAT responses to the same insecticide, between albino and pigmented strains (Kristoff et al., 2008). However, in all cases, mechanisms underlying observational and correlational data could not be clearly specified, despite the multimarker approach implemented. Such datasets provide nevertheless basic information that can be useful to guide further investigations on the early response of snails to oxidative stress.

9.3.3   DETOXIFICATION OF XENOBIOTICS

The exposure of organisms to toxic substances leads to various catabolic reactions that aim to degrade and eliminate them. Successive phases can be described, which correspond to different metabolic pathways. Drug metabolism includes the activation of so-called phase I (biotransformation), phase II (conjugation) enzymes as well as phase III transporters (Xu et al., 2005).

9.3.3.1   PHASE I ENZYMES (BIOTRANSFORMATION ACTIVITY)

Phase I enzymes mainly belong to the superfamily of cytochrome P450 (CYP). Types of P450-mediated reactions include hydroxylation, epoxidation, oxidative deamination, S-, N-, and O-dealkylations, and dehalogenation. The end results of P450 reactions are most often more hydrophilic and presumably more excretable products (Snyder, 2000). P450 enzymes may reside in the mitochondria and/or in the membrane of the endoplasmic reticulum (ER) (microsomal P450s). Detoxification is mainly performed by microsomal P450s, although some mitochondrial P450s show activity toward exogenous compounds (Rewitz et al., 2006).

In gastropods, these enzymes have been described and studied in the context of natural and xenobiotic stress. Due to their role in early response to chemical stress (phase I), CYPs have been proposed and used as exposure biomarkers (defined as "biochemical, physiological, or histological indicators of either exposure to or effects of xenobiotic chemicals"; Huggett et al., 1992), including in gastropods. For example, in the marine species *Avicularia gibbosula*, the content of CYP450 in the ER from the digestive gland cells was found significantly increased following an oilspill pollution (Yawetz et al., 1992).

The measure of CYP-related biotransformation activity is traditionally assessed using artificial substrates such as ethoxyresorufin (ethoxyresorufin-*O*-dealkylase activity, EROD) and related molecules, that is, pentoxyresorufin (pentoxyresorufin-*O*-dealkylase activity, PROD) and ethoxycoumarin (ethoxycoumarin-*O*-deethylase activity, ECOD), or benzo(*a*)pyrene (BaP-hydroxylase activity, BaPH). The value of CYP-related biotransformation activities as exposure biomarker was investigated in freshwater gastropods (*Valvata piscinalis, Potamopyrgus antipodarum*; Gagnaire et al., 2009). In *H. aspersa* (*C. aspersum*), naphthalene was found to modify some of these activities (EROD, ECOD) in an organ-dependent manner, that is, increase in the kidney and decrease in the digestive gland (Ismert et al., 2002). A CYP oxidase system was described in *L. stagnalis* (Wilbrink et al., 1991a) and *L. palustris* exhibited increased microsomal CYP450-related activity (BaPH) when exposed to the herbicide atrazine (Baturo, and Lagadic, 1996).

Altogether, results from these assays suggest that CYPs could be relevant exposure markers in gastropods, although P450-mediated detoxification is generally limited in molluscs (Gooding and LeBlanc, 2001). However, due to the current lack of knowledge on the diversity of the CYP superfamily in this taxonomic group, methodologies often rely on vertebrate models and assumptions (e.g., use of antibodies to vertebrate CYP1A) and may often lead to departures from toxicological expectations, thus reducing the value of CYPs as biomarkers. Actually, the diversity of CYPs may be still higher in invertebrates than in vertebrates. In molluscs, a recent survey of available genetic resources has been performed in bivalves, providing a benchmark for future studies in ecotoxicology as well as for the understanding of CYP genes evolution (Zanette et al., 2010). Based on sequences from four species (*Mytilus calfornianus, Mytilus galloprovincialis, Crassostrea gigas*, and *Crassostrea virginica*) the authors identified 123 expressed sequence tags (ESTs) homologous to members of various vertebrate CYP families. The same approach could be developed in gastropods, as genetic resources are becoming substantial in a growing number of species, notably, *A. californica* (Moroz et al., 2006, see also http://www.genome.gov/Pages/Research/Sequencing/SeqProposals/AplysiaSeq.pdf), *L. stagnalis* (Bouétard et al., 2012; Feng et al., 2009; Sadamoto et al., 2012), *Lottia gigantea* (Veenstra, 2010), *B. glabrata* (Adema et al., 2010, available from https://www.vectorbase.org), *Radix balthica* (Feldmeyer et al., 2011), as well as *Crepidula fornicata, Crepidula plana, Physa acuta*, and *Physa gyrina* (see Romiguier et al., 2014). For example, NGS-based transcriptomic data obtained in *L. stagnalis* provided more than a hundred transcripts matching with various CYPs (M.A. Coutellec, unpublished).

It is also to be noted that besides activation related to biotransformation activity, some compounds may also affect the endogenous functions of CYPs. A remarkable example of such possible effect is found in marine gastropods, in which tributyltin (TBT, previously used as antifouling biocide agent) is known to cause imposex, that is, pseudohermaphroditism in females (at least 195 species; deFur et al., 1999; Oehlmann et al., 2007; Sternberg et al., 2010). Among the alternative hypotheses for TBT-induced imposex, the inhibition of CYP aromatase (CYP19) has received much attention. This steroidogenic enzyme converts testosterone to estrogen and has been shown to be inhibited by TBT, especially in vertebrates. However, in gastropods, conclusions are controversial and still subject to debate (Oehlmann et al., 2007; Rewitz et al., 2006; Sternberg et al., 2010; see Box 9.3).

BOX 9.3 Mechanism of Imposex Induction in Prosobranchs

Since the 1970, female masculinization has been documented in marine prosobranchs all over the world (Sternberg et al., 2010). This syndrome was named "imposex" because it involves the presence of one or more male reproductive organs surimposed onto the normal female apparatus (Smith,

1971). Tributyltin (TBT), a biocide compound used in antifouling paints, was rapidly identified as agent causing imposex and subsequent negative impact at the population level, sometimes down to population extirpation (Gibbs and Bryan, 1996; Minchin et al., 1996). Therefore, the total ban of TBT was voted in 1998, implying a worldwide prohibition on application (within 5 years) and presence (within 10 years) (see Sternberg et al., 2010).

Due to the general lack of knowledge on invertebrate endocrine systems, mechanisms underlying TBT-induced imposex are still not understood. Three main hypotheses have been proposed (all reviewed in details in Sternberg et al., 2010; see references therein):

1. The *steroid hypothesis*, under which TBT increases the level of testosterone by inhibiting steroid metabolizing enzymes, such as aromatase (CYP19, which converts testosterone to estrogen), sulfotransferase (SULTs) (which catalyses the sulfoconjugation and inactivation of steroid hormones), and acyl coenzyme A–steroid acyltransferase (ATAT, which catalyses the fatty acid esterification of testosterone).
2. The *neuroendocrine hypothesis*, under which TBT causes the aberrant secretion of neuropeptide APGWamide, involved in the regulation of male sexual differentiation, and putatively acting as penis morphogenic factor.
3. The *RXR agonist hypothesis*, which relies on the high affinity of TBT as ligand of RXR, and resulting disruption of retinoic acid-dependent-signaling pathways.

Sternberg et al. (2010) evaluated these hypotheses in the light of available results obtained from various prosobranchs. Within the *steroid hypothesis*, the ATAT inhibition would get strongest empirical support. However, the comparison of complete genomes representative of lophotrochozoans, arthropods, and vertebrates revealed that steroidogenic enzymes evolved independently in these three metazoan phyla, and that genes orthologous to the vertebrate steroidogenic enzymes are not present in lophotrochozoans (Markov et al., 2009). Consistently, recent reviews highlight the lack of undisputable evidence for a hormonal role of vertebrate sexual steroids in molluscs, which globally invalidates the steroid hypothesis (see Scott, 2013).

Recent work on prosobranchs provides strong support to the *RXR agonist hypothesis* (Lima et al., 2011; Pascoal et al., 2013; Stange et al., 2012). In particular, full transcriptomic data from *Nucella lapillus* individuals exposed to TBT suggest the RXR/PPAR (peroxisome profilerator-activated receptor) axis as the most plausible mechanism of TBT induction of imposex (Pascoal et al., 2013). Still, much has to be done to clearly demonstrate the molecular mode of action of TBT on prosobranchs.

9.3.3.2   PHASE II ENZYMES (CONJUGATION ACTIVITY)

In vertebrates, the phase II metabolizing or conjugating enzymes consist of many families of enzymes including SULTs and UDP-glucuronosyltransferases, DT-diaphorase or NAD(P)H: quinone oxidoreductase (NQO) or NAD(P)H: menadione reductase (NMO), epoxide hydrolases (EPH), GSTs and *N*-acetyltransferases (NAT) (Xu et al., 2005). In gastropods, several of these enzymes have been studied in an ecotoxicological context.

A focus will be given on GSTs, which are members of a superfamily of multifunctional proteins involved in cellular detoxification and protection against oxidative damage. GSTs catalyze the conjugation of a wide range of endogenous and exogenous electrophilic substrates to glutathione (Armstrong, 1997). The conjugates are too hydrophilic to diffuse freely from the cell, and must be pumped out actively by a transmembrane ATPase (phase III system). This results in the unidirectional excretion of the xenobiotic from the cell (Sheehan et al., 2001).

GSTs have been characterized in a few gastropod species (Wilbrink et al., 1991b), in link with parasitism (*Bulinus truncatus*; Abdalla et al., 2006), dietary toxins (*Cyphoma gibbosum*; Whalen et al., 2008), endocrine disruption (*Thais clavigera*; Rhee et al., 2008), or environmental contamination (*N. obsoletus* and *Cerithium floridanum*; Lee et al., 1988). Furthermore, a survey of Genbank nucleotide database (2015, March) indicates 291 gastropod GST sequences, obtained from seven species: *A. californica* (141 sequences), *L. gigantea* (122), the viviparid *Cypangopaludina cathayensis* (3), abalones *Haliotis discus* and *H. diversicolor* (18), *Th. clavigera* (4), and *C. gibbosum* (2). Although a very small fraction of the Gastropoda diversity is represented, this dataset actually constitutes most of the molluscan GST sequences available to date, since only 181 additional sequences are published, from cephalopods (57) and bivalves (124). It is to be noted that several GST cDNA sequences were also obtained in *L. stagnalis* (RNAseq; Bouétard et al., 2012).

GST activity has been studied in response to various environmental contaminants. It is not our purpose to give an exhaustive list of these studies here, yet a few findings can be highlighted. For example, in *Nucella lapillus*, copper inhibited GST activity with a lowest effect concentration of 0.044 mg/L, while cadmium had the opposite effect (although this was not significant). Several hypotheses were proposed to explain the inhibitory effect of copper, including ROS production and interaction with the enzyme, or depletion of glutathione. Comparatively, this enzymatic activity appeared generally much lower and unaffected by these metals at the tested concentrations in *Monodonta lineata*, despite a similar acute sensitivity in the two species (Cunha et al., 2007). Specific patterns may relate to differences in bioaccumulation capacities, which seem particularly high in species of the *Monodonta* genus.

In the context of biological control and potential impact on non-target species, the biocide used against mosquito, *Bacillus thuringiensis* var. *israelensis*, was found to elicit GST activity in the freshwater snail *Physa marmorata* (Mansouri et al., 2013). In *L. palustris*, GST activity was inhibited by the herbicide atrazine, whereas the effect of the chlorinated insecticide HCB changed from slight inhibition to significant increase over 24-h exposure (Baturo and Lagadic, 1996).

Although biomarkers are generally not studied alone, but in a set of potentially responsive markers, based on previous knowledge of the toxicant mode of action or properties, only limited information can be provided by these observational approaches. A deeper understanding of the underlying mechanisms would evidently require a proper toxicological methodology, including gene knock-down technology.

9.3.3.3   PHASE III TRANSPORTERS (EXCRETION)

Phase III transporters, including P-glycoprotein (P-gp), multidrug resistance-associated protein (MRP), and organic anion-transporting polypeptide 2 are expressed in many tissues such as the liver, intestine, kidney, and brain, where they provide a formidable barrier against drug penetration, and play crucial roles in drug absorption, distribution, and excretion. P-gp and MRP utilize the energy from the hydrolysis of ATP to substrate transport across the cell membrane and are called ATP–binding cassette (ABC) transporters (see references in Xu et al., 2005).

In this transporter family, several members of the ABCB (P-gp, MDR, or MXR) and ABCC (MRP) subfamilies function as highly promiscuous transporters, capable of trafficking a diverse array of moderately hydrophobic xenobiotics across cell membranes (Bodo et al., 2003).

Information on these transporters is very scarce in molluscs. The implication of multidrug resistance in the response to environmental contamination has been mainly investigated in bivalves (see Bard, 2000). A recent study provided new insight into the regulatory pathway of ABCB in *M. galloprovincialis*, confirming the role of phosphorylation activity of the cAMP-dependent protein kinase PKA in mediating P-gp activity, as in mammalian models (Franzellitti and Fabbri, 2013). Among gastropods, P-gp has been previously characterized in *Monondota turbinata* (Kurelec et al., 1995). In this class of molluscs, ABC transporters have been recently studied in the particular context of dietary toxins and diet selection. The tropical ovulid *C. gibbosum* and nudibranch *Tritonia hamnerorum* are known to feed exclusively on allelochemically defended gorgonian corals. Using immunohistochemistry and functional approaches, Whalen et al. (2010) showed that in *T. hamnerorum* (the diet of which is more specialized than that of *C. gibbosum*), P-gp expressed specifically in the midgut epithelia and in the epidermis, two tissues in close contact with gorgonian toxins. This location is thus consistent with a role of P-gp in protection against ingested prey toxins. Comparatively, no immunoreactivity was observed for this transporter in *C. gibbosum*, despite its demonstrated occurrence as two ABCB isoforms. Besides, constitutive expression was found in the latter species for one MRP (ABCC) isoform. The differences may relate to the degree of specialism/generalism of the studied species in terms of prey diet, as MRP and P-gp are known to differ in substrate selectivity (see human resistance to anticancer agents; Kruh and Belinsky, 2003). In *L. stagnalis*, a transcriptomic analysis based on RNAseq highlighted several ABC transporter genes as upregulated in individuals exposed to the pro-oxidant herbicide diquat (Bouétard et al., 2012; M.-A. Coutellec, unpublished data).

9.3.3.4   THE CASE OF METALLOTHIONEINS

Metallothioneins (MTs) are a class of nonenzymatic cysteine-rich proteins of low molecular weight which lack aromatic amino acid residues. They are widespread in most living phyla and have unique metal-binding properties, which combine high thermodynamic stability and kinetic lability (Sigel et al., 2009). These proteins have various biological functions, with a pivotal role in the homeostatic control of essential metals (Cu, Zn) as well as the detoxification of nonessential metals (Cd, Hg, Pb, etc.) or excess of essential metals. Other roles have also been ascribed to MTs, including free-radical scavenging or oxidative stress protection, anti-inflammatory, antiapoptotic, proliferative, angiogenic, neuroprotectant (Blindauer and Leszczyszyn, 2010; Sigel et al., 2009). Consistent with their multiple activities, MTs are often present as multiple isoforms in a given species and are also highly diverse across phyla. In view of their high diversity, MTs have been classified into three classes: class I (mammalian homolog, also present in molluscs and crustaceans), class II (nonhomolog to mammalian sequences, a highly heterogeneous group), and class III (phytochelatins, plant polypeptides derived from glutathione polymerization). A new classification has been proposed, in which classes I and II MTs are classified into 15 families, molluscan MTs belonging to family 2 (see Vergani, 2009). However, it is clear that MTs forms a superfamily of proteins within which phylogenetic relationships are still unresolved.

Due to their properties, MTs are now part of a core suite of biomarkers recognized at European level in monitoring programs (Amiard et al., 2006). Mollusks have high metal accumulation capacities, and therefore MTs have been widely used as biomarkers in this phylum, especially in marine bivalves, but also in intertidal gastropods (Bebianno et al., 1992, 2003). However, their biomarker value is still questioned, since MTs can also be induced by other contaminants, including organic and inorganic compounds.

In molluscs, MTs contains about 75 amino acids, except in gastropods, which MTs are shorter, due to an atypical N-terminal region (Vergani, 2009). MTs have been particularly studied in helicid snails by R. Dallinger's group. In these snails, two isoforms occur, a Cd-binding peptide involved in cadmium detoxification, and a Cu-binding peptide, involved in homeostatis metal regulation. In the Roman snail *Helix pomatia*, Cd-MT transcription is highly inducible by cadmium in the digestive gland and the gut tissue, whereas Cu-MT is constitutively expressed in rhogocytes (pore cells, present at high density in the mantle tissue of the lung), which are specialized in the synthesis of hemocyanin (a Cu-containing respiratory protein; Chabicovsky et al., 2003). Studies on metal accumulation associated to snail exposure from contaminated soils showed that copper accumulation is much more limited than accumulation of nonessential metals (see Dallinger et al., 2005, and references therein). In fact, copper is stored in rhogocytes in two ways, as bound to Cu-MT and as granular precipitations. The Cu-MT is a stable copper pool, whereas the granular form is highly inducible by Cu exposure, and thus most probably involved in metal excess detoxification. The noninducibility of Cu-MT by Cu exposure seems to be specific to the study species and contrasts with other findings, including in gastropods (*L. littorea*; Mason and Borja, 2002). In *H. pomatia*, Dallinger et al. (2005) showed that the two pools (Cu-MT and granules) do not coexist in the same cells, suggesting that rhogocytes have two distinct physiological states. Due to their phagocytic properties (see Section 9.5.1.), it is possible that these cells are responsible for the sequestration and excretion of copper excess in snails.

9.3.4   ALTERATION OF MACROMOLECULES: PROTEIN MISFOLDING, GENOTOXICITY, LIPID PEROXIDATION

Oxidative stress is known to alter most macromolecules, that is, nucleic acids, proteins, and lipids, with highly damaging consequences for the cell. Lipid peroxidation (LPO) leads to the formation of lipid peroxides and a disturbance of biomembranes (Gutteridge and Halliwell, 1990).

9.3.4.1   PROTEIN MISFOLDING

ROS also induce protein misfolding. Accumulation of misfolded proteins in stressed cells activates heat shock factors and results in the expression of Hsps (Kim et al., 2007). Hsps have the ability to restore the damaged proteins to their functional three-dimensional structure, to prevent aggregation of unfolded proteins, or to facilitate their degradation in case of irreversible damage (Grune et al., 2011; Qiu et al., 2006).

Hsp70, which may act with its co-chaperone Hsp40 (Kim et al., 2007) has been commonly assessed as a biomarker inducible by pollutants in aquatic snails (e.g., platinum in *Marisa cornuarietis*; Osterauer et al., 2010; pesticides in *L. stagnalis*; Bouétard et al., 2013; pharmaceuticals and wastewater effluents in the same species; Gust et al., 2013a,b), and terrestrial gastropods (heavy metals in *Deroceras reticulatum*; Köhler et al., 1996). *L. stagnalis* is used as model in neurophysiology. In this species, hypoxia suppresses sensory and motor behaviors, and Hsp70, induced as an early response to hypoxia, prevents degradation of synaptic proteins syntaxin and synaptotagmin I (Fei et al., 2007). In land snails, these enzymes have been extensively studied either as exposure biomarkers to dessication, overheat and solar radiations, as well as for their expected implication in physiological change related to estivation. *Xeropicta derbentina* is a land snail from Eastern Mediterranean recently introduced in Southern France, where it has been very successful, especially for the last decades. The dynamics of thermal Hsp70 induction was experimentally investigated over 8 h of heat-exposure (45°C) and compared to that of another Mediterranean land snail, *Theba pisana* (Scheil et al., 2011). In *X. derbentina*, besides a circadian evolution of Hsp expression noticed in control snails (peak at noon-time), the experiment demonstrated two temperature-induced expression peaks, one during the heat-exposure phase (after 2 h) and one during the post-heating phase (after 8 h of recovery at 24.7°C). The decrease in expression following the first peak was interpreted as an element of the noncompensation phase due to stress–protein impairment, which was supported by histopathological data from the digestive gland. Beyond results, this experiment demonstrates that Hsp induction and activity is a rapid and dynamic process, and that studies based on single time point assays can be highly misleading. Furthermore, prior conditions encountered by organisms may also have consequences on results, and this may be particularly problematic when individuals are brought from the field to the lab. This may be illustrated by the results obtained for *Th. pisana*, the second species used in the study of Scheil et al. (2011, 2012). Despite being clearly more sensitive to heat stress, test individuals of this species subjected to similar conditions exhibited a very weak induction of Hsp70, which might reflect a preliminary depletion of this enzyme in the field, as snails were collected in summer. Altogether, these results suggest that extreme care is required to design experiments dealing with the issue of molecular responses to short-term stress.

A field experiment designed to identify proximal determinants of Hsp70 levels in *X. derbentina* showed that both morphological and behavioral factors (i.e., shell size and color morph, and diurnal location of snails above ground as well as shell aperture orientation relative to the sun) were effective and interacted. The fully white morph was predominant in larger snails, which also climbed higher above the ground and had lower Hsp70 levels, smaller snails tended to be darker (higher frequency of darker morphs), stayed closer to the ground level, and had higher Hsp70 levels (Di Lellis et al., 2012). Possible explanations for the observed discrepancy may relate to higher thermal thresholds for Hsp induction in more tolerant individuals, a size-dependence of the ability to tolerate desiccation (smaller snails being more sensitive), higher constitutive expression of Hsp associated to growth and maturity acquisition in smaller (also younger) snails as compared to adult ones (Di Lellis et al., 2012). Consistent results were obtained from a more narrow range of size, that is, avoiding very young and very old snails (Dieterich et al., 2015). On the other hand, no difference in thermal capacity was observed between dark and pale shell coloration morphs of the land snail *Theba pisana* (Scheil et al., 2012).

Similar differences in Hsp levels were also demonstrated between two land snails, the desert species, *Sphincterochila zonata*, and a Mediterranean, desiccation-sensitive species of the same genus, *S. cariosa*, in link with their ability to tolerate desiccation (Arad et al., 2010). Western-blot analysis of Hsps expression (Hps70, Hps90, and sHps) showed both species and organ-specific patterns of variation during estivating and active snails, as well as during arousal after estivation. In active snails, lower expression of Hsp70 occurred in the foot, kidney, and digestive gland of the desert-adapted species *S. zonata*, compared to the level observed in the Mediterranean (desiccation sensitive) species (*S. cariosa*). Likewise, small Hsps (Hsp25 and Hsp30) expressed to a higher level in the kidney and digestive gland of active individuals of the latter species. By contrast, higher expression of Hsp90 was detected in the kidney of *S. zonata*. Transition from estivation to activity (arousal) is a stressful period, during which oxidative stress may be associated with the increase in metabolic rate and in oxygen consumption. Therefore, it is possible that Hsp induction might reflect a specific response to oxidative stress during this process. Interestingly, the authors of this study observed a stronger increase in Hsp72 expression in *S. cariosa* kidney and digestive gland during arousal, as compared to *S. zonata*. This might suggest a higher inducibility of Hsp70 in the sensitive species. However, a reversed relationship was detected in the foot. It is to be noted that, during estivation, endogenous expression of Hsp72 and Hsp90 in the kidney was higher in *S. zonata* than in *S. cariosa*. This result suggests a role of these enzymes in protecting organisms against the deleterious effects of high concentration of solutes such as urea, as a defense mechanism against water loss during dormancy. The higher osmolality of pallial fluid observed in *S. zonata* supports this hypothesis. As a chapter is fully devoted to regulatory pathways involved in estivation (Ramnanan, Bell, and Hughes, this volume), it is not the place here to further elaborate on this phenomenon.

9.3.4.2 LIPID PEROXIDATION

Oxidative stress may trigger LPO, which leads in turn to the formation of lipid peroxides and a disturbance of biomembranes. LPO is a complex phenomenon which encompasses various mechanisms, including free-radical-mediated oxidation. This mechanism proceeds by a chain mechanism, starting with one initiating free radical which can oxidize many molecules. The chain propagation is carried by lipid peroxyl radicals independent of the type of chain-initiating free radicals. LPO induces biomembrane disturbance, such as alteration of structure, integrity, fluidity, permeability, and functionality, and modifies low-density lipoproteins to proinflammatory forms, and generates potentially toxic, mutagenic, and carcinogenic products (Niki, 2009). Although many LPO products exert cytotoxicity, sublethal concentrations of LPO products induce cellular adaptive responses and enhance tolerance against subsequent oxidative stress through upregulation of antioxidant compounds and enzymes (Niki, 2009). Lipid hydroperoxides, the primary products of LPO, are substrates of GPX.

ROS-mediated damages to lipids can be quantified at different stage in the peroxidation process (Gutteridge and Halliwell, 1990). Conjugated dienes are an initial product of free radical attack on lipids, lipid hydroperoxides represent an intermediate product, and malondialdehyde is a terminal product of lipid breakdown that can be measured using the thiobarbituric acid reactive substances (TBARs) assay. These measures were applied by Pannunzio and Storey (1997) to study the effect of anoxia on *L. littorea*, along with a set of antioxidant enzymatic and nonenzymatic assays. Under anoxia, the only responsive marker in the digestive gland was the lipid hydroperoxide content, which was strongly reduced, and consistent with oxygen deprivation. By contrast, the stable and low level of conjugated dienes in this tissue could be explained by the likely activation of an efficient mechanism against their accumulation. TBARs were also stable in the digestive gland, but at a high level, which would indicate that aldehydes other than those produced from lipid degradation contribute to the TBARs content, or that the terminal products are not readily cleared from the tissue, particularly during anoxia. Comparatively, the foot muscle tissue exhibited a different pattern, with a significant increase of the two first stages of LPO, followed with a return to normal levels of hydrogen hydroperoxides only, during the recovery period. This was interpreted by the authors as reflecting the capacity of the foot tissue to reverse lipid hydroperoxides but not primary conjugates. These results were further interpreted in the light of the antioxidant response measured in parallel. Indeed, upon oxygen return, the three LPO indicators remained stable (did not increase) in the digestive gland, which reflects the efficiency of an antioxidant system naturally adapted to anaerobic to aerobic transitions (see Section 9.2.2). The abrupt and transient increase in TBARs, as evidence for arousal-induced oxidative stress and efficient response ability in estivating snails was also observed in the digestive gland of *O. lactea* (Hermes-Lima and Storey, 1995).

9.3.4.3 DNA DAMAGE

If DNA damage can be caused by many endogenous and exogenous agents, a significant portion of the damage is caused by ROS (Brazilai and Yamamoto, 2004). The occurrence of DNA lesions triggers that activation of an intricate web of signaling pathways known as the DNA damage response.

Some environmental contaminants are also genotoxic, for example, polycyclic aromatics hydrocarbons (PAHs), heavy metals, endocrine disrupters such as TBT (Hagger et al., 2006; Sarkar et al., 2014). DNA damage induced by xenobiotics encompasses simple and double-strand break and the formation of micronuclei, DNA–adducts, DNA–protein cross-links, as well as chromosomal aberration. Therefore, assessment methods dedicated to such defects have been developed in the context of ecotoxicology (biomarkers of DNA integrity and DNA-strand breaks).

The Comet Assay (or Single Cell Gel Electrophoresis [SCGE]) is a simple method for measuring DNA-strand breaks (Box 9.4). Based on results obtained in cephalopods (Raimundo et al., 2010) and bivalves (*Ruditapes philippinarum*; Hartl et al., 2004), it seems that the digestive gland is not a proper organ for the use of Comet Assay. It was shown to yield levels of single-strand breakage likely too high (from autolytic processes) for a valid application of the assay without proper cell sorting and viability check (de Lapuente et al., 2015). Therefore, hemocytes are the most common target for genotoxicity assessment in vivo and in vitro in gastropods using the Comet Assay. In gastropods, this method has been successfully applied as biomarker of genotoxicity, mainly using marine species exposed to contaminants, for example, *Bullacta exarata* (An et al., 2012), *L. littorea* (Noventa et al., 2011), *Nerita chamaeleon* (Sarkar et al., 2015), *M. granulata* (Sarkar et al., 2014), *N. lapillus* (together with a micronucleus test; Hagger et al., 2006), *Patella vulgata* (Lewis et al., 2010), *Planaxis sulcatus* (Bhagat and Ingole, 2015), but also in terrestrial species such as *C. aspersum* and *Helix vermiculata* (Angeletti et al., 2013; Ianistcki et al., 2009).

As a different class of DNA alteration, epigenetics, which refers to modifications in gene expression that are influenced by DNA methylation and/ or chromatin structure, RNA editing, and RNA interference without any changes in DNA sequences (Bird, 2002), should deserve particular attention with respect to stress. Molecular epigenetic studies in molluscs are rare, and Fneich et al. (2013) were the first to describe DNA methylation in a gastropod snail, *B. glabrata*. This domain is still only in its infancy, and we are aware of only one publication dealing with stress-mediated epigenetic modification, that is, the effect of immune challenge on *P. canaliculata* neurons (see Ottaviani et al., 2013).

BOX 9.4 The Comet Assay as a Standard Method for Determining Genotoxicity

The Comet Assay, or Single Cell Gel Electrophoresis (SCGE), has become a standard method for determining in vivo/in vitro genotoxicity (see review in Collins, 2014; de Lapuente et al., 2015). The assay is based on quantification of the denatured DNA fragments migrating out of the cell nucleus during electrophoresis. The image obtained with this technique looks like a "comet" with a "head" consisting of intact DNA, and a "tail" which contains damaged or broken pieces of DNA. The amount of DNA liberated from the head of the comet during electrophoresis depends on the level of effect of the stressor under evaluation. Although the first demonstration of "comets" (though they did not use the word) was by Östling and Johanson (1984), the assay became popular following its improvement by Singh et al. (1988) and Olive et al. (1990).

Several versions of the assay are currently in use (de Lapuente et al., 2015). Basically, after a suspension of cells has been obtained, the basic steps include preparation of microscopic slides layered with cells embedded in an agarose gel, lysis of cells to liberate the DNA, DNA unwinding, electrophoresis, neutralization of the alkali, DNA staining and scoring. Various image analysis systems are available for assessing the resultant images. Basically, when dealing with relatively low damage levels the distance of DNA migration from the body of the nuclear core is used to measure the extent of DNA damage. This technique is not very useful in situations where DNA damage is relatively high, as with increasing extent of DNA damage the tail increases in fluorescent staining intensity but not in length. Collins (2004) proposed a scoring method that has been shown to give quantitative resolution which is sufficient for many purposes. Probably the most popular method for comet evaluation is referred to as "tail moment" calculated as measure of tail length × measure of DNA in the tail; Olive et al., 1990). It incorporates relative measurements of both the smallest detectable size of migrating DNA (reflected by the length of the comet tail) and the number of broken pieces of DNA (represented by the staining intensity of DNA in the tail).

The Comet Assay can detect DNA single-strand breaks as initial damage and those developed from alkali-labile sites under alkaline condition (pH > 12.6), and that formed during repair of base adducts or alkylated bases, which are not initial DNA damage (Collins, 2004). The sensitivity and specificity of the assay are greatly enhanced if the nucleoids are incubated with bacterial repair endonucleases that recognize specific kinds of damage in the DNA and convert lesions to DNA breaks, increasing the amount of DNA in the comet tail. DNA repair can be monitored by incubating cells after treatment with damaging agent and measuring the damage remaining at intervals. Alternatively, the repair activity in a cell extract can be measured by incubating it with nucleoids containing specific damage (Collins, 2014).

9.4 APOPTOSIS

Apoptosis is the primary cell death program by which cells are physiologically eliminated without inducing inflammation. In Metazoans, apoptosis is involved in important stress-unrelated functions, such as ontogenetic processes related to embryonic development and metamorphosis (organ ontogenesis, cellular maintenance, and repair associated to developmental plasticity and error).

Contrary to necrosis, apoptosis involves the nuclear condensation and organized fragmentation (200 bp fragments), cleavage of chromosomal DNA into internucleosomal fragments and packaging of the deceased cell into apoptotic bodies without plasma membrane breakdown (Edinger and Thompson, 2004). Apoptosis is an evolutionary highly conserved process (see Meier et al., 2000). A recent review of the available knowledge on these pathways in Mollusca was performed by Kiss (2010). Although the major part of this knowledge comes from bivalve models, the set of information gathered should be useful for further investigations in gastropods. Two different pathways can mediate apoptosis, the extrinsic (Fas and other tumor necrosis factor receptor [TNFR] superfamily members and ligands) and intrinsic (mitochondria-associated) pathways, and among which cross-talks are possible.

With an illustrative purpose, Figure 9.2 presents the apoptosis pathway as described by KEGG orthology for human (Kanehisa and Goto, 2000; http://www.kegg.jp/pathway/map04210), with *L. stagnalis* transcripts expressed in the digestive gland highlighted in red (M.A. Coutellec, unpublished). These data were obtained from RNAseq analysis, which led to a total of 202 contigs matching with KEGG terms involved in both extrinsic (e.g., TNF, TRAIL, FADD, caspase8) Bcl2 and intrinsic pathway (IAP). Differential expression induced by the pro-oxidant herbicide diquat was observed for several apoptosis-associated transcripts, namely, FASL, TRAILR, CASP7, transcripts annotated as IAP repeat-protein 2/3 and 7/8, E3 ubiquitin–protein ligase XIAP, and CASP9.

Apoptosis is a genetically controlled mechanism of cell death involved in the regulation of tissue homeostasis. The two major pathways of apoptosis are the extrinsic (Fas and other TNFR superfamily members and ligands) and the intrinsic (mitochondria-associated) pathways, both of which are found in the cytoplasm. The extrinsic pathway is triggered by death receptor engagement, which initiates a signaling cascade mediated by caspase-8 activation. Caspase-8 both feeds directly into caspase-3 activation and stimulates the release of cytochrome *c* by the mitochondria. Caspase-3 activation leads to the degradation of cellular proteins necessary to maintain cell survival and integrity.

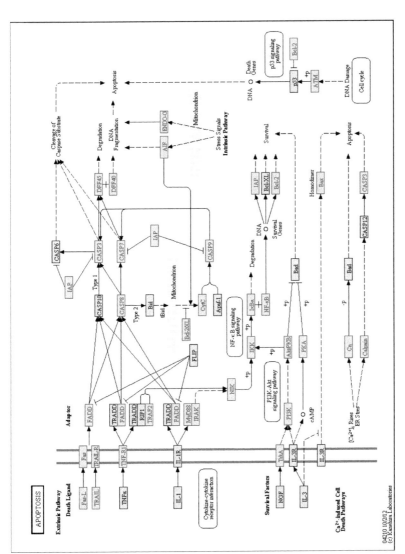

FIGURE 9.2 Apoptosis pathway. (From Romero LM, "Using the reactive scope model to understand why stress physiology predicts survival during starvation in Galápagos marine iguanas", Gen Comp Endocrinol. 2012 May 1;176(3):296-9. doi: 10.1016/j.ygcen.2011.11.004. Epub 2011 Nov 12. http://www.kegg.jp/kegg/kegg1.html; http://www.kegg.jp/pathway/map04210; Used with permission. (KEGG annotation resources, http://www.kegg.jp/pathway/map04210) with transcripts identified in *Lymnaea stagnalis* [in red]).

The intrinsic pathway occurs when various apoptotic stimuli trigger the release of cytochrome c from the mitochondria (independently of caspase-8 activation). Cytochrome c interacts with Apaf-1 and caspase-9 to promote the activation of caspase-3. Recent studies point to ER as a third subcellular compartment implicated in apoptotic execution. Alterations in Ca^{2+} homeostasis and accumulation of misfolded proteins in the ER cause ER stress. Prolonged ER stress can result in the activation of BAD and/or caspase-12, and execute apoptosis.

In gastropods, cellular responses have been particularly well studied in hemolymph. Phagocytosis parameters and lysosomal fragility as well as apoptotic processes were investigated in *L. stagnalis* hemocytes in response to pesticides (Russo and Madec, 2007; Russo et al., 2008). Interestingly, early apoptotic events could be related to mitochondrial membrane alterations and exposure of phosphatidylserine to the outer face of the plasma membrane, suggesting an implication of the intrinsic pathway in this pesticide-induced apoptosis. Again, such results provide a good basis for further investigations on gastropod cellular responses to stress, and vice versa. For example, *C. gigas* hemocytes were transfected with gastropod sequences of the Hsp70 promoter and a Ras-related gene (Rho) to study noradrenaline-mediated apoptosis in mollusk hemocytes (Lacoste et al., 2001).

Tissues nonspecifically implicated in immune or stress response are also subject to cell death processes. Membrane-associated hallmarks of both apoptosis and necrosis were detected in *L. stagnalis* neurons treated with the anesthesic lidocaine (Onizuka et al., 2012). Similarly, hydrogen peroxide induced both processes in *Aplysia kurodai* sensory neurons, as reflected by apoptotic (nuclear shrinkage and chromatin condensation) as well as necrotic (organelle swelling) features (Lim et al., 2002).

Interestingly, antipredatory or deterrent proteins have been identified in gastropods as being cytotoxic and inducing cell death. Among them, the glycoprotein Cyplasin was isolated from the California sea hare *A. californica* and studied for its potential biotechnological application to cancer treatment (Petzelt et al., 2002). Other molecules involved in the defensive system of this sea hare have been characterized, which may indirectly trigger apoptosis-related cascades of reactions. The L-amino acid oxidase Escapin, was isolated from the ink of *A. californica*. Ink results from the combined secretion of ink gland and opaline gland. Escapin reacts extremely quickly with L-lysin, which is present at high concentration in the secretion, to produce a diverse mixture of molecules, some of which are strongly bioactive. Intermediate and end reaction products result from the initial Escapin-induced deamination of L-lysin (leading to alpha-keto acids) and interaction with H_2O_2 (EIP-K, EEP-K) with bactericidal effects (see details in Ko et al., 2008). Various L-amino acid oxidases (LAAOs) express specifically in different organs (purple ink, egg masses, and albumen gland). In another sea hare, *Dolabella auricularia*, smaller polypeptides with LAAO activity have been found in the skin, body wall, and coelomic fluid (Iijima et al., 2003). A phylogenetic analysis of orthologs and paralogs of *A. californica* Cyplasin showed a closer relationship of the latter with Aplysianin A (isolated from *A. kurodai*; Jimbo et al., 2003), whereas the sequence of Escapin was closer to *Aplysia punctata* ink toxin APIT1 (Butzke et al., 2004). In *A. californica*, Escapin occurs in the ink, where it functions as defensive molecule against attack, while the occurrence of its paralog Aplysianin (*A. californica* homolog) in egg masses is likely to have a protective function against egg consumption (Derby, 2007). A LAAO was also described in the mucus of the terrestrial snail *A. fulica*. Cell death caused by this enzyme was shown to be due to both H_2O_2-related cytotoxicity (necrosis or apoptosis) and to caspase-mediated apoptosis (Kanzawa et al., 2004).

Thus, both interspecific and intraspecific organ-dependent variation occurs in gastropod LAAOs, which discovery opens the way to further exciting studies on antipredatory functions of this defensive pathway.

9.5 STRESS AND THE IMMUNE SYSTEM

Available information on the response of gastropod immune system to stress mostly concern model species (e.g., *L. stagnalis*; van der Knaap et al., 1993) or species of commercial (e.g., abalones; Hooper et al., 2007) or medical interest (e.g., *B. glabrata*; Coustau et al., 2015). Since gastropods may be the intermediate hosts of various parasites, a lot of immunological studies also focus on the interaction and the larval stages of digeneans (digenetic trematodes) such as schistosomes. Besides, the immunobiology of the vast majority of gastropod species have never been studied (Loker, 2010) and gastropod immunobiology still remains largely unknown.

9.5.1 BRIEF OVERVIEW OF GASTROPOD IMMUNE SYSTEM

Gastropods lack the acquired immune system of vertebrates. Their immune systems rely on three basic mechanisms of immune defense: physical barriers, cellular defenses, and humoral mechanisms (Ellis et al., 2011). Cellular and humoral defenses, which are provided by the hemolymph, are similar to the innate immune system of vertebrates (Rowley and Powell, 2007). The mucus produced by the epithelium that covers gastropod body forms the first barrier against pathogens and foreign elements. Bioactive substances that contribute to the efficiency of this barrier have been isolated from mucus such as achacin in the giant African land snail *A. fulica* (Ehara et al., 2002) or dolabellanin B from the sea hare *D. auricularia* (Iijima et al., 2003).

Rhogocytes (also known as pore cells), which are involved in synthesis or processing of respiratory proteins and MTs (Dallinger et al., 2005; see Section 9.3.3.4), have also been implicated in ingestion of small foreign particles (Albrecht et al., 2001). Cells with phagocytic activity may be found in the connective tissue (*L. stagnalis*; Sminia et al., 1979) or in organs such as the digestive gland where they may form a fixed phagocyte system (*H. pomatia*; Reade, 1968) or be distributed throughout the entire gland (*Planorbarius corneus*; Ottaviani, 1990). Antigen-trapping cells have been described in blood sinus and kidney in *H. pomatia* (Renwrantz et al., 1981). Hemocyte islets are also found in the kidney of *P. canaliculata* and they show phagocytic activity and spheroid formation after bacterial or yeast injections (Cueto et al., 2013). In *B. glabrata*, fixed phagocytic cells are located in various tissues (Matricon-Gondran and Letocart, 1999). The relative contribution of these fixed cells to defense is not well understood and most data concern hemolymph-related defense mechanisms.

9.5.1   CELLULAR COMPONENT

Cellular defense in gastropods is coordinated by circulating hemolymph cells that functionally resemble mammalian macrophages (van der Knaap et al., 1993). These cells are usually designated as hemocytes (or haemocytes) but other names may be found in the literature, including amebocytes (Prowse and Tait, 1969; Sminia and Barendsen, 1980), leukocytes (Müller, 1956 in Sminia, 1981), granulocytes (Barracco et al., 1993), macrophages (Yamaguchi et al., 1999), lymphocytes (Kress, 1968), fibrocytes (Foley and Cheng, 1974), and hyaline cells (George and Ferguson, 1950).

No standard nomenclature of gastropod hemocytes has been defined so far. They are frequently classified into subpopulations based upon functional, morphological, and staining characteristics analyzed by light and electron microscopy. The use of flow cytometry or immunostaining has increased the opportunities of distinguishing between various additional cell types

(Dikkeboom et al., 1988a; Franceschi et al., 1991; Johnston and Yoshino, 2001).

Two types of immunocytes have generally been described in gastropods, granulocytes, and hyalinocytes (sometimes called agranulocytes), respectively, although the corresponding nomenclature and even the number of cell populations may vary according to species and between authors for a given species (Table 9.1). Granulocytes contain cytoplasmic granules, have a low nuclear-to-cytoplasmic ratio and are effective in phagocytising foreign materials; in contrast, hyalinocytes are smaller cells, have a high nuclear-to-cytoplasmic ratio, few cytoplasmic granules and a poor capacity to phagocytose foreign materials (Martin et al., 2007). Hemocyte subpopulations that differ both chemically and functionally are probably regulated in their activities or behaviors through specific receptors and the signals conveyed by their interaction with appropriate ligands (Humphries and Yoshino, 2003).

TABLE 9.1 Examples of Nomenclature of Gastropod Hemolymph Circulating Cells Recorded in the Literature.

Clade/Species	Hemocyte Cell Types (Synonyms or Subpopulations)	References
Vetigastropoda		
Haliotis discus discus	Blast-like cells and hyalinocytes	Donaghy et al. (2010)
Haliotis tuberculata	Large and small (blast-like cells) hyalinocytes	Travers et al. (2008a)
Megathura crenulata	Hemocytes	Martin et al. (2007)
Turbo cornutus	Blast-like cells, type I and II hyalinocytes, and granulocytes	Donaghy et al. (2010)
Architenioglossa		
Ampullaria cuprina	Type I, II, and III (spreading cells) hemocytes	Wojtaszek et al. (1998)
Bellamya bengalensis	Agranulocytes (blast-like cells, round hyalinocytes, and spindle hyalinocytes), semigranulocytes (semigranular asterocytes and round semigranulocytes), and granulocytes (round granulocytes, spindle granulocytes, and granular asterocytes)	Ray et al. (2013)
Pila globosa	Agranulocytes and granulocytes	Mahilini and Rajendran (2008)

TABLE 9.1 *(Continued)*

Clade/Species	Hemocyte Cell Types (Synonyms or Subpopulations)	References
Pila globosa	Agranulocytes (blast-like cells, round hyalinocytes and spindle hyalinocytes), semigranulocytes (semigranular asterocytes and round semigranulocytes), and granulocytes (round granulocytes, spindle granulocytes, and granular asterocytes)	Ray et al. (2013)
Pomacea canaliculata	Group I (small) and group II (large) hemocytes	Accorsi et al. (2013, 2014)
Pomacea canaliculata	Nongranular, granular with few granules, and granular will electron dense granules hemocytes	Cueto et al. (2007), Shozawa and Suto (1990)
Viviparus ater	Spreading hemocytes	Franchini and Ottaviani (1990), Ottaviani (1989)
Littorhinomorpha		
Littorina littorea	Hyalinocytes	Neves et al. (2015)
Littorina littorea	Hyalinocytes (Juvenile round cells, intermediate cells, and large mature hemocytes)	Gorbushin and Iakovleva (2006)
Oncomelania hupensis	Round cells with filiformfilopodia, acidophilic round cells, basophilic round cells without filiformfilopodia, and spindle cells	Zhang et al. (2007a)
Oncomelania nosophora	Type I (macrophage-like) and type II (lymphocytes-like) cells	Sasaki et al. (2003)
Neogastropoda		
Babylonia areolata	Granulocytes and hyalinocytes (Type I and II)	Di et al. (2011, 2013)
Heterobranchia		
Acteon tornatilis	Granulocytes and hyalinocytes	Yonow and Renwrantz (1986)
Doto coronata, D. pinnatifida, D. fragilis	Granulocytes and hyalinocytes	Kress (1968)
Hygrophila		
Biomphalaria glabrata	Granulocytes and hyalinocytes	Cheng (1975), Cheng and Auld (1977)
Biomphalaria glabrata	Amebocytes	Sminia and Barendsen (1980)

TABLE 9.1 *(Continued)*

Clade/Species	Hemocyte Cell Types (Synonyms or Subpopulations)	References
Biomphalaria glabrata	Small, medium and large hemocytes	Martins-Souza et al. (2009)
Biomphalaria glabrata	Granulocytes, hyalinocytes, and round cells	Noda and Loker (1989a)
Biomphalaria tenagophila	Granulocytes and hyalinocytes	Barracco et al. (1993)
Biomphalaria tenagophila	Small, medium, and large hemocytes	Martins-Souza et al. (2009)
Bulinus guernei	Granulocytes	Krupa et al. (1977)
Bulinus truncatus	Amebocytes	Sminia and Barendsen (1980)
Indoplanorbis exustus	Agranulocytes and granulocytes	Mahilini and Rajendran (2008)
Lymnaea stagnalis	Spreading and round amebocytes	Stang-Voss (1970)
Lymnaea stagnalis	Amebocytes	Sminia (1972), Sminia and Barendsen (1980)
Lymnaea stagnalis	Phagocytes (hemocytes)	Van der Knaap et al. (1993)
Lymnaea stagnalis	Round cells and granulocytes	Russo and Lagadic (2004)
Lymnaea truncatula	Round cells and spreading cells	Monteil and Matricon-Gondran (1993)
Planorbarius corneus	Spreading (SH) and round hemocytes (RH)	Ottaviani (1983), Ottaviani and Franchini (1988)
Sigmurethra		
Helix aspersa maxima	Type I (=granulocytes) and type II (=hyalinocytes) hemocytes	Adamowicz and Bolaczek (2003)
Helix pomatia	Granulocytes	Renwrantz (1979)
Helix pomatia	Type I, II, III (spreading cells), and IV hemocytes	Wojtaszek et al. (1998)
Incilaria bilineata	Type I (macrophage-like), II (lymphocyte-like), and III (fibroblast-like) hemolymph cells	Furuta et al. (1990)
Incilaria fruhstorferi	Type I (macrophage-like), II (lymphocyte-like), and III (fibroblast-like) hemolymph cells	Furuta et al. (1986)

TABLE 9.1 *(Continued)*

Clade/Species	Hemocyte Cell Types (Synonyms or Subpopulations)	References
Trachea vittata	Agranulocytes and granulocytes	Mahilini and Rajendran (2008)
Aplysiomorpha		
Aplysia californica	Hemocytes	Martin et al. (2007)

The hemopoiesis of molluscs has not yet been fully understood and two major theories have been proposed. Cheng (1981) and Auffret (1988) suggested that hyalinocytes and granulocytes might differentiate from two distinct cell precursors. However, juvenile cells containing granules (also named granuloblasts) have been extremely rarely observed, prompting Hine (1999) to suggest that one cell type might give rise to hyalinocytes that would further mature to become granulocytes. Hematopoietic organs (amebocytes producing organ or APO) have been described in pulmonate gastropods (Jeong et al., 1983; Lie et al., 1975; Rondelaud et al., 1982) and in *M. cornuarietis*, an ampullariid snail (Yousif et al., 1980). However, it has also been shown in littorinimorph and pulmonate gastropods that hemocyte proliferation occurs in peripheral vascular locations or in the circulation (Gorbushin and Iakovleva, 2006; Sminia, 1974; Sminia et al., 1983; Souza and Andrade, 2006).

Gastropod hemocytes can recognize and subsequently eliminate, or sequester, invading pathogens through various processes, including phagocytosis, encapsulation, and the production of lysosomal enzymes and bacteriostatic substances (see e.g., Nunez et al., 1994; van der Knaap et al., 1993; Yoshino and Vasta, 1996). They can also produce cytotoxic molecules such as reactive oxygen and nitrogen intermediates that play an important role in the destruction of microorganisms and parasites (Adema et al., 1994; Conte and Ottaviani, 1995; Dikkeboom et al., 1988b; Hahn et al., 2000, 2001; Zelck et al., 2005).

During these processes, increased amounts of oxygen are consumed, and the cells undergo an oxidative burst and produce a variety of cytotoxic ROS, including superoxide, hydrogen peroxide, hydroxyl radicals and possibly singlet molecular oxygen (Adema et al., 1994; Dikkeboom et al., 1988b). Several enzymes participate in the generation of ROS, including NADPH-oxidase complex generating superoxide (Adema et al., 1993), SOD converting it to hydrogen peroxide and peroxidases that catalyze the transformation of hydrogen peroxide.

9.5.1.2 HUMORAL COMPONENT

Humoral factors play a fundamental role in the immune responses in molluscs. In addition to phagocytosis, hemocytes are able to secrete soluble antimicrobial peptides (AMPs) and other cytotoxic substances into the hemolymph. Together with other nonspecific humoral defense molecules, including lectins, bactericidins, nitric oxide (NO), lysozymes, and serine proteases, these form the humoral component of invertebrate immunity.

9.5.1.2.1 Bioactive Peptides

AMPs represent the most universal immune effectors and several compounds have been identified in molluscs since the mid-1990s, mostly in bivalves (see review in Li et al., 2011). However, as compared to many other invertebrates, the study of gastropod AMPs remains in its infancy (Loker, 2010). Among them, defensin has been identified in the marine gastropods *Haliotis discus hannai* (Hong et al., 2008) and *Haliotis discus discus* (De Zoysa et al., 2010). Abhisin, a 40 amino acids AMP was identified in *H. discus discus* (De Zoysa et al., 2009) and another peptide, littorein, has been observed in the plasma of the common periwinkle *L. littorea* (Defer et al., 2009).

Results from laboratory experiments on the antiviral activity of hemolymph and lipophilic extract of the digestive gland of the abalone *Haliotis laevigata* also suggest that abalone have at least two antiviral compounds with different modes of action (Dang et al., 2011). In addition, evidence for the existence of AMPs was found in *B. glabrata* (Mitta et al., 2005).

9.5.1.2.2 Lectins

Lectins are carbohydrate-binding proteins that bind to specific carbohydrate structures endogenous to the host or presented by microbial invaders. They play diverse roles in nonself-recognition and clearance of invaders as pattern recognition receptors (PRRs; Wang et al., 2011). Many lectins have been isolated from mollusc eggs, being involved in the immune protection of the eggs from bacterial invasions (Prokop and Köhler, 1967). For example, *H. pomatia* agglutinin was first isolated from perivitelline fluid of eggs for which it provides efficient antibacterial protection (Sanchez et al., 2006). Several lectins, especially the C-type lectins, have been well documented (Wang et al., 2011) and they are found to mediate various innate immune responses such as pathogen recognition (Janeway and Medzhitov, 2002), agglutination (Song et al., 2011), opsonization (Yang et al., 2011), and phagocytosis (Canesi et al., 2002). A galectin present on the surface of circa 60% of *B. glabrata* hemocytes has been characterized, and in recombinant form binds to the tegument of *Schistosoma mansoni* sporocysts in a carbohydrate-inhibitable manner, suggesting it is a hemocyte-bound pattern recognition molecule (Yoshino et al., 2008). The presence of "counter receptors" on hemocytes, such as integrin-like molecules has also been suggested (Davids et al., 1998). These receptors could be bound by soluble forms of galectin, such that the galectin could also serve in cross-bridging hemocytes to a parasite surface (Yoshino et al., 2008).

Fibrinogen-related proteins (FREPs) are a family of lectins that contain one or two Ig domains and a fibrinogen domain, and they can bind soluble and surface antigens of parasites that infect the snail (Adema et al., 1997). They are considered to play a key role in the compatibility polymorphism that determines parasite/snail compatibility on an individual basis (see review in Coustau et al., 2015).

9.5.1.2.3 Lysozymes

Lysozyme is an enzyme existing in diverse organisms. It catalyzes the hydrolysis of β-1, 4-glycosidic linkage between *N*-acetylmuramic acid and *N*-acetylglucosamine of peptidoglycan, a major component of bacterial cell wall and causes bacterial cell lysis. Six distinct lysozyme types have been identified and three are found in animals, commonly designated as the c-type (chicken-type), the g-type (goose-type), and the i-type (invertebrate-type; Callewaert and Michiels, 2010). Although the data regarding gastropods are far less abundant than for bivalves, there is evidence that the three types may be found in these organisms. For example, c-type lysozyme has been isolated in the abalone *H. discus hannai* (Ding et al., 2011), g-type in the freshwater snails *O. hupensis* (Zhang et al., 2012) and *P. acuta* (Guo and He, 2014), and i-type in the marine conch *Lunella coronata* (Ito et al., 1999).

9.5.1.2.4   LBP/BPI Proteins

LBP lipopolysaccaride-binding protein and BPI (bactericidal/permeability-increasing protein) are components of the immune system that have been principally studied in mammals for their involvement in defense against Gram-negative bacterial pathogens (Krasity et al., 2011). Following specific binding to LPS, they increase the permeability of the bacterial membranes, and contributes to the elimination of bacteria (Elsbach et al., 1994). LBP/BPI genes have been detected in *B. glabrata* hemocytes and egg masses (Hathaway et al., 2010; Mitta et al., 2005). Baron et al. (2013) showed that a LBP/BPI (BgLBP/BPI1) of maternal origin is the major protein in *B. glabrata* eggs. It displays a strong biocidal activity against oomycetes in addition to its antibacterial activity, therefore, protecting *B. glabrata* offspring from lethal bacteria and fungi infections.

9.5.1.2.5   Phenoloxidase

Phenoloxidase (PO) catalyses the hydroxylation of L-tyrosine to L-DOPA (monophenoloxidase activity, MPO), as well as the oxidation of the diphenols to their respective quinones (diphenoloxidase activity, DPO). These quinones are capable of binding free amino groups in proteins to form protein crosslinkages, eventually leading to the formation of insoluble, chemically resistant protein polymers leading to the entrapment of foreign material in a capsule. This process has been referred to as sclerotization (sclerotin formation) or melanization (melanin formation; Waite, 1990).

PO is present in mollusc plasma in an inactive state (prophenoloxidase, proPO) that can be activated to PO by the endogenous proPO-activation system or some exogenous elicitors such as laminarin, SDS, LPS, etc. (Asokan et al., 1997). Numerous studies have demonstrated the importance of PO in the immunological defenses of bivalve molluscs (see e.g., Aladaileh et al., 2007; Butt and Raftos, 2008; Hellio et al., 2007). In gastropods, PO plays a role in sclerotization of shell (Nellaiappan and Kalyani, 1989). It can be released from circulating hemocytes into hemolymph when the animals are stimulated by physical injury or infestation (Asokan et al., 1997; Bahgat et al., 2002). It has also been found in the reproductive tract of *B. glabrata* and in its egg masses following transfer from the maternal organism (Bai et al., 1996, 1997; Hathaway et al., 2010). However, its role in immune defense of gastropods remains controversial (Scheil et al., 2013). Bahgat et al. (2002) did not show difference in PO activity in hemocytes of *B. glabrata* strains susceptible or resistant to infection with miracidia of *S. mansoni*. PO-like activity was identified in the hemolymph of *L. stagnalis* (Leicht et al., 2013) but its response to immune elicitors was not consistent (Seppälä and Leicht, 2013). Furthermore, using histochemical methods, Vorontsova et al. (2015) recently showed that dopamine oxidation in this species involved peroxidase rather than PO activity.

9.5.1.2.6   Nitric Oxide

Nitric oxide (NO) plays an important role as a signal molecule throughout the animal kingdom, especially as an intercellular messenger in the central nervous system (Bruckdorfer, 2005). In gastropods, NO is involved in the control of feeding and locomotion in *Clione limacina* (Moroz et al., 2000), in the regulation of feeding in *A. californica* (Lovell et al., 2000), in food-attraction conditioning in *H. pomatia* (Teyke, 1996), in chemosensory acti-vation of feeding in *L. stagnalis* (Elphick et al., 1995; Moroz et al., 1993), in the oscillation of olfactory neurons in the procerebral lobe in *Limax maximus* (Gelperin, 1994), and in the swimming rate in *Melibe leonine* (Newcomb and Watson, 2002).

Although it is not toxic itself, this short-lived radical, generated by nitric oxide synthase (NOS), also plays an important role in the elimination of pathogens as part of the innate immune response (Rodríguez-Ramos et al., 2010). Together with superoxide anions it forms peroxynitrile anion which is a highly toxic compound with antibacterial and antiviral activity (Beckman et al., 1996; Fang, 1997). The generation of NO (or its stable end products) in response to immunological challenge has been described in hemocytes of *V. ater* (Conte and Ottaviani, 1995; Ottaviani et al., 1993), *L. stagnalis* (Wright et al., 2006), and *B. glabrata* (Zahoor et al., 2009).

9.5.1.2.7   Other Factors

Other factors involved in immune response have been described in gastro-pods, such as aplysianin/achacin-like protein in the sea hare *A. kurodai* (Kisugi et al., 1989) and in *B. glabrata* egg masses (Hathaway et al., 2010), protease inhibitors, Gram-negative bacteria-binding protein (GNBP), and scavenger receptor cysteine-rich protein, C1q domain-containing protein, and protease inhibitor in the perivitelline fluid of the eggs of *P. canaliculata* (Sun et al., 2012).

Non-targeted transcriptomic studies have recently yielded a lot of original information. Mitta et al. (2005) have identified several hundred novel transcripts including 31 immune-relevant transcripts corresponding to various functional groups using random sequencing of a *B. glabrata* hemocyte cDNA library. For the first time transcripts displaying similarities with the mammalian cytokine MIF (macrophage migration inhibitory factor) and a PGRP (peptidoglycan recognition protein) were identified in a gastropod. Additional nontargeted studies helped identifying numerous candidates genes belonging to various functional groups including those coding for pattern recognition proteins, cell adhesion molecules, immune regulators, cellular defense effectors, proteases and protease inhibitors, or oxidative stress and stress-related proteins as well as candidates involved in regulatory networks and signaling pathways (review in Coustau et al., 2015; see also Adema et al., 2010; Deleury et al., 2012; Hanelt et al., 2008; Mitta et al., 2005).

9.5.1.3   HEMOCYTE-SIGNALING PATHWAYS

Deciphering the intracellular signal transduction pathways likely to be activated by exposure of hemocytes to exotic stimuli is a key to understanding hemocyte effector functions (Loker, 2010). Experimental studies with protein kinase C (PKC) activator showed that PKC is involved in H_2O_2 generation in *B. glabrata* (Bender et al., 2005), of superoxide anions in *L. littorea* (Gorbushin and Iakovleva, 2007), and of NO in *L. stagnalis* (Wright et al., 2006).

PKC activity is dependent on its phosphorylation status and studies have shown that phosphorylation of PKC's hydrophobic motif regulates the enzymes stability, phosphatase sensitivity, subcellular localization and catalytic function (Bornancin and Parker, 1997; Edwards and Newton, 1997). The fact that inhibitors of MAPKs also prevent hemocyte spreading or H_2O_2 production suggest that PKC activation is likely to result in activation via phosphorylation of MAPK-like ERK or p38 (Skála et al., 2014; Wright et al., 2006; Zelck et al., 2007). Natural stimuli such as laminarin also activate PKC and H_2O_2 production in *L. stagnalis* hemocytes (Lacchini et al., 2006). A role for phosphatidylinositol 3-kinase in controlling phagocytic activity has been shown in *L. stagnalis* hemocytes (Plows et al., 2006) and G-protein-coupled membrane receptors have also been reported from *L. littorea* (Gorbushin et al., 2009). Targeting by pathogens of components of gastropod-signaling pathways such as p38 has been documented in abalones challenged by *Vibrio harveyi* (Travers et al., 2009).

Toll-like receptors (TLRs) are well-characterized PRRs of innate immunity, known to induce immune responses by interacting with evolutionarily conserved pathogen-associated molecular patterns (PAMPs). Homologs of TLRs are present in the *Lottia* genome, a Rel-like NF-κB transcription factor is known from abalones (Jiang and Wu, 2007), and AbTLR a TLR homolog from disk abalone (*H. discus discus*) was identified and characterized at molecular level (Elvitigala et al., 2013). Therefore, additional Toll-pathway homologs are likely to be present in gastropods, although their functional relevance remains to be assessed.

Acting upstream of Toll-signaling pathways is pattern recognition molecules like PGRPs and GNBP (or B-1-3 glucan recognition/binding protein or LGBP). Both short and long form PGRP-encoding genes are present in *B. glabrata* and at least three different GNBPs are also known from this species (Zhang et al., 2007b).

9.5.2 RESPONSE OF THE IMMUNE SYSTEM TO STRESS

9.5.2.1 THE CASE OF DIGENEAN TREMATODES

Along with the usual background of viral or bacterial challenge, gastropods face other pathogens that are unequivocally gastropod specialists, the best known being the digenetic trematodes, also known as digeneans or "flukes" (Loker, 2010). Studies on the interactions between these parasites and their gastropod host are probably the source of the major part of the available knowledge on gastropod immunity. This is undoubtedly the case for the investigations on the interactions between the freshwater snail *B. glabrata* and its trematode parasite *S. mansoni* (Coustau et al., 2015).

Detailed studies on the response of *B. glabrata* to different stressors, including infection by *S. mansoni* or other trematodes, have shown that exposed individuals may modulate their immunological response according to the stressor (Adema et al., 2010; Hanington et al., 2010a). In case of infection by a trematode, the snail-immune response may fail to clear the infection due to a combination of trematode-mediated avoidance and inhibition of snail defense mechanisms (Coustau and Yoshino, 1994; Douglas et al., 1993; Lie and Heyneman, 1977; Loker et al., 1986; Loker and Hertel, 1987; Noda and Loker, 1989a,b; Roger et al., 2008a). The identification of immune genes that could play a role in *B. glabrata* immune processes has been the focus of many studies (Bouchut et al., 2006a,b, 2007; Deleury et al., 2012; Guillou et al., 2007a; Ittiprasert et al., 2010; Lockyer et al., 2007, 2008; Nowak et al., 2004; Raghavan et al., 2003; Vergote et al., 2005).

The dialog between the host snail and its parasite has been intensively studied. The comparative analysis of *B. glabrata* strains that are susceptible (M-line or NMRI strains) or resistant (13-16-R1 and BS-90 strains) to specific *S. mansoni* stocks have clearly demonstrated a strong genetic basis for the susceptibility of *B. glabrata* to *S. mansoni* (Lewis et al., 2001; Richards et al., 1992). Significant progress toward the identification of resistance genes have been made (Blouin et al., 2013; Bonner et al., 2012; Ittiprasert et al., 2013; Knight et al., 1999) but without complete achievement (Coustau et al., 2015).

Differences were shown in the short-term expression of genes in hemocytes from parasite-exposed and control groups of both schistosome-resistant and schistosome-susceptible strains. Genes involved in immune/stress response, signal transduction, and matrix/adhesion were differentially expressed between the two strains (Lockyer et al., 2013). Results suggest that resistant snails recognize parasites and mount an appropriate defense response. A lack of capacity to recognize and react to the parasite or an active suppression of hemocytes response by the parasite early in infection are the two main hypothesis that may explain the absence of defense in susceptible snails.

Studies on the mechanisms involved in the compatibility polymorphism characteristics in certain *B. glabrata/S. mansoni* populations allowed the identification of two repertoires of polymorphic and/or diversified molecules that were shown to interact: the parasite antigens SmPoMucs (*S. mansoni* polymorphic mucins) and *B. glabrata* FREP immune receptors (Roger et al., 2008a–c; Mitta et al., 2012; Moné et al., 2010). SmPoMucs are only expressed by larval schistosome stages that interact with the snail intermediate host. They are highly glycosylated and polymorphic. Each individual parasite possesses a specific and unique pattern of SmPoMucs. In *B. glabrata*, FREPs are also highly diversified (Adema et al., 1997; Hanington et al., 2010b; Loker et al., 2004), and they play a key role in the fate of the interaction between the snail and its trematode parasites. For example, FREP3 plays a central role in resistance to digenetic trematodes. It is up-regulated in *B. glabrata* infected with *S. mansoni* or *Echinostoma paraensei*, and functions as an opsonin favoring phagocytosis by hemocytes (Hanington et al., 2010b). Subsequent studies showed that FREP3 is important for successful defense against schistosome infections in *B. glabrata* and that its suppression by trematode parasites facilitate their establishment within the snail (Hanington et al., 2012). The cytokine-like molecule, BgMIF (*B. glabrata* macrophage MIF) may also play a role in the anti-parasite response of *B. glabrata* (Baeza Garcia et al., 2010).

The availability of resistant and susceptible strains of *B. glabrata* has also stimulated innovative studies that provided support to several lines of evidence regarding the host–parasite interaction. Hemocytes from resistant snails possess a different allelic form of the Cu/Zn SOD and produce higher levels of H_2O_2, a substance lethal to *S. mansoni* sporocysts (Bayne, 2009; Bender et al., 2005, 2007; Goodall et al., 2004, 2006; Hahn et al., 2001). It remains unclear if there is a link between the two phenomena. Other putative immune effectors have been identified, including LBP, BPI, and AMPs (Guillou et al., 2007b; Mitta et al., 2005), but their functions remain to be determined. In addition, the first cytolytic β pore-forming toxin from a mollusc, biomphalysin, has recently been identified and characterized in *B. glabrata* (Galinier et al., 2013; Moné et al., 2010). Biomphalysin is only expressed in hemocytes and plays a sentinel role in preventing pathogen invasion. It has a hemolytic activity and binds parasite membrane, being highly cytotoxic.

Sporocysts of *S. mansoni* bear larval transformation proteins that may be involved in scavenging of ROS, therefore, offering a protection against hemocyte attack (Guillou et al., 2007b; Wu et al., 2009). A reciprocal coevolution has been demonstrated between ROS and ROS scavengers produced by sympatric populations of *B. glabrata* and *S. mansoni* (Moné et al., 2011). Other compounds such as glycoconjugates have been detected at the surface of sporocysts, which may serve as host-mimicking molecules (Peterson et al., 2009) and divert attack by humoral factors (Roger et al., 2008b,c).

As recently reviewed by Coustau et al. (2015), advances in the field of trematode/snail interaction clearly showed the multigenic and variable nature of the mechanisms underlying the success or failure of parasite development. Interactions include recognition mechanisms (e.g., lectin/glycan interactions) and effector and anti-effector systems (e.g., biomphalysin, LBP/BPI, and ROS/ROS scavengers). The relative importance of these different factors varies greatly among population of hosts and parasites. The consequences of additional phenomena such as interaction with environmental factors (e.g., temperature; Ittiprasert and Knight, 2012), immune priming associated with previous encounters with parasites (Portela et al., 2013), or epigenetic mechanisms (Perrin et al., 2013) remain to be studied in detail (Coustau et al., 2015). The current availability of the genome of *S. mansoni* (Berriman et al., 2009; Protasio et al., 2012) and the perspectives concerning the genome of *B. glabrata* (undergoing assembling and annotation; see https://www.vectorbase.org) offer great perspectives for forthcoming studies.

9.5.2.2   OTHER STRESSORS

Although there is more variation in neuroendocrine systems across inverte-brate groups than there is within the vertebrates, similar molecules mediate some aspects of the acute stress response across phyla. In gastropods, the acute stress response originates in the endocrine system, with corticotropin-releasing hormone stimulating the release of adrenocorticotrophic hormone (ACTH), leading to the release of biogenic amines, noradrenaline, and dopa-mine into the hemolymph, which then mediate secondary effects (Malham et al., 2003). Although there is evidence that ACTH, a hormone that induces the release of stress hormones in vertebrates (Charmandari et al., 2005), may be present in molluscs (Ottaviani and Franceschi, 1996), its role in the response to acute stress in gastropods remains to be demonstrated (Adamo, 2012).

In the European abalone *H. tuberculata*, acute mechanical stress (mechan-ical shaking) increased the concentration of noradrenaline and dopamine (Malham et al., 2003). Concomitantly, a transient decrease was shown for the number of circulating hemocytes, their migratory activity, their phagocytic activity, and their respiratory burst response. This decrease was followed by an increase in some immune functions (e.g., phagocytic ability) a few hours later before returning to baseline, suggesting that acute stress can also have a delayed immunoenhancing effect in abalone. Evidence from studies in bivalves suggest that stress-related circulating biotic amines may inhibit hemocyte function following interaction with hemocyte β-adrenergic-like receptors that are coupled to a cAMP/protein kinase A pathway, similar to the receptors found in vertebrates (Lacoste et al., 2001). The existence of an identical pathway in gastropod has not yet been demonstrated.

One of the key questions when addressing the effects of stress on gastro-pods is the possible consequences on the immunocompetence of the stressed organisms. Immunocompetence, the general capacity of an individual to exhibit an immune response (Schmid-Hempel, 2003), is commonly assessed through the measurement of phagocytic activity (Hooper et al., 2007). Phago-cytosis is probably the measure that has received the greatest amount of investigation when assessing the impact of changing environmental condi-tions on the mollusc immune response. It may be measured either through the proportion of hemocytes that are phagocytically active in a population or using the phagocytic index, that is, the number of bacteria or particles engulfed by each immune cell.

Phagocytic activity may exhibit natural seasonal variation caused by changes in organism physiology especially sexual maturation and spawning, sometimes leading to a natural increased susceptibility to infection at some key moments of the life cycle. For example, a reduction in phagocytosis and phenoloxidase activity during gametogenesis has been reported in *H. tuberculata*, associated with an enhanced susceptibility to *V. hayeri* (Travers et al., 2008b). Therefore, measuring a change in phagocytic activity does not give information on the origin of this change neither on other possible cellular immune dysfunction. The measurement of additional parameters such as the abundance, morphology, or viability of hemocytes may provide information on the mechanisms behind phagocytic activity changes. In addition, although an increase in phagocytic activity might feasibly suggest an increase in the activity of the hemocytes themselves without any influence of hemocyte number as shown by Seppälä and Jokela (2010a) in *L. stagnalis* exposed to elevated temperature, this increase may also be an indirect effect of increased hemocyte numbers caused by an alteration in hemocyte proliferation or mobilization of cells from peripheral tissues (Hooper et al., 2011). Therefore, in measuring total hemocyte counts, it is possible to demonstrate a change in the number of hemocytes in context with changes in other cellular immunological measures (Perez and Fontanetti, 2011).

A number of authors have shown that gastropod immune system is sensitive to "natural" environmental changes. Significant reduction in hemocyte abundance, phagocytic activity, or respiratory burst have been recorded in gastropod species following exposure to changes in parameters such as temperature (Cheng et al., 2004a), salinity (Cheng et al., 2004b; Martello et al., 2000), air exposure (Cardinaud et al., 2014), and hypoxia (Cheng et al., 2004c), as well as changes in concentrations of ammonia (Cheng et al., 2004d) and nitrite (Cheng et al., 2004e).

Similar observations were made following exposure to anthropogenically induced stressors, including butyltins (Zhou et al., 2010), polyaromatic hydrocarbons (Gopalakrishnan et al., 2009), and pesticides (Russo and Lagadic, 2004; Russo et al., 2007). Experiments were also performed recently on primary cultured hemocytes with various chemicals, including metals and pharmaceuticals (Gaume et al., 2012; Ladhar-Chaabouni et al., 2014; Latire et al., 2012; Minguez et al., 2014; Mottin et al., 2010), but their results are not necessarily representative of the phenomena that may occur in whole animals.

In some cases, an increase in hemocyte number, ROS production and/ or phagocytic activity following exposure to mechanic stress (Cardinaud et al., 2014; Hooper et al., 2011), municipal effluent (Gust et al., 2013b), metals (Itziou et al., 2011), or pesticides (Russo and Lagadic, 2000) was also observed.

Immunodulation by acute stress may lead to an increased in the suscep-tibility of gastropods to pathogens. For example, handling caused an altera-tion of all the immune parameter levels and a metabolic depression in *H. tuberculata*, associated with an enhanced susceptibility to *V. harveyi* infec-tion (Cardinaud et al., 2014). A series of studies by Cheng et al. (2004a–e) reported an increase in susceptibility to *Vibrio parahaemolyticus* in the Taiwan abalone *H. diversicolor supertexta* exposed to elevated temperature, fluctuating salinity, high concentrations of organic compounds, and reduced oxygen levels.

The case of chronic exposure to toxic stress raises additional concerns. Chronic (several week-long) exposure to sublethal concentrations of toxic compounds was shown to modulate the immunocompetence in abalones (*H. diversicolor*, *H. diversicolor supertexta*) exposed to benzo[*a*]pyrene or TBT (Gopalakrishnan et al., 2011; Zhou et al., 2010). Short-term recovery following cessation of exposure was observed for benzo[*a*]pyrene but not for TBT (Gopalakrishnan et al., 2011).

Chronic stress is immunosuppressive in invertebrates with a decline in the expression of immune-related genes and in immunological function, and a loss of disease resistance (Ellis et al., 2011). However, the involved mech-anisms are not well understood, especially regarding the links with stress hormones. In vertebrates, the common explanation is that elevated levels of stress hormones are initially adaptive but become pathological if they remain elevated for too long, due to the exhaustion of molecular resources, build-up of toxic compounds, and dysregulated pathways (Dhabhar, 2002; Hawlena et al., 2011; Romero et al., 2009). However, no evidence is avail-able today regarding these processes in gastropods.

9.5.3 CURRENT PERSPECTIVES IN MOLLUSC IMMUNE SYSTEM

As stated earlier the immune system in molluscs is traditionally considered to lack specific immunity and to rely only on an innate immune system. However, increasing evidences have recently revealed the existence of specific or "primed" immunity in bivalves (Cong et al., 2008; Yue et al., 2013), and maternal transfer of immunity (i.e., immunity transferred via eggs from mother to offspring) is one of the highlighted instances to favor the presence of a kind of "specific" immunity in invertebrates (Yue et al., 2013; Wang et al., 2015). Many maternally derived immune factors have been identified in mollusc eggs or embryos (Wang et al., 2015). Antibacte-rial and lysozyme activities as well as agglutination against pathogens have been demonstrated in terrestrial and marine gastropod eggs (Fiolka and Witkowski, 2004; Kamiya et al., 1986). Maternal transfer of lectins from mother to eggs/offspring has been reported in some gastropod species such as *H. pomatia* (agglutinin; Sanchez et al., 2006), *B. glabrata* (C-type lectin; Hathaway et al., 2010), *Pomacea scalaris* (scalarin; Ituarte et al., 2012), *Pila ovata* (anti-B-like agglutinin; Uhlenbruck et al., 1973), and *A. kurodai* (D-galactose-binding lectine; Kawsar et al., 2009). Primarily maternally derived lysozyme was found in snail *H. aspersa maxima* and *Achatina achatina* eggs (Fiolka and Witkowski, 2004).

Various factors have been shown to have an influence on the maternal transfer of immunity to offspring, being responsible for a "trans-generational immune priming" (TGIP). In molluscs, TGIP has been evidenced so far mostly in bivalves following exposure to pathogens (Gueguen et al., 2003; Oubella et al., 1994), changeable-environment conditions (Lacoste et al., 2002), and environmental pollutants (Pipe and Coles, 1995). Some biological factors such as age (Xu et al., 2011) and reproduction (Duchemin et al., 2007) may also have an influence on this phenomenon. Among gastropods, the effects of heat (in *L. stagnalis*; Leicht et al., 2013), nutrition (in *L. stagnalis*; Seppälä and Jokela, 2010b), and exposure to benzo[*a*]pyrene (in *H. diversicolor*; Gopalakrishnan et al., 2009) on TGIP have also been reported. Although the specific mechanisms underlying TGIP clearly deserve more investigation, it may be considered as a beneficial survival strategy of invertebrates.

In addition to the maternal transfer of immune factors from mother to eggs, it seems that maternal experience of pathogen or other immune stimulation also has a profound influence on the phenotype of offspring, probably through epigenetic inheritance or genomic imprinting (Yue et al., 2013). The first evidence of epigenetic modifications triggered by an immune challenge in a gastropod species was recently reported in *P. canaliculata* by Ottaviani et al. (2013).

9.6   STRESS AND LIFE-HISTORY TRADE-OFFS

Life history trade-offs are likely to have determinants of various origins: ecological (resource allocation), physiological (endocrine systems, oxidative stress), and genetic (genetic correlations due to linkage disequilibrium, pleiotropy, correlational selection).

9.6.1 DELAYED EFFECTS OF STRESS

"Latent" or "carry over" effects reflects the effects that originate in embryonic and larval experiences, but are expressed only in juveniles or adults. They are of particular interest in the ecological literature (see e.g., Altwegg and Reyer, 2003; Goater, 1994; Marshall et al., 2003; Ng and Keough, 2003; Pahkala et al., 2001; Phillips, 2002, 2004). Latent effects do not include the effects of adult conditioning on offspring quality, that is, "maternal effects" (Pechenik, 2006).

Latent effects have been observed for several groups of marine invertebrates (review in Pechenik, 2006), including gastropods. Most studies concern the consequences of starvation, pollution, or delayed metamorphosis (Table 9.2).

TABLE 9.2   Results of Experimental Studies on the Latent Effect of Stress on Gastropods.

Clade/Species	Treatment in larval or embryonic stage	Latent effect	Sublethal consequences	References
Vetigastropoda				
Haliotis asinina	Exposure to larval settlement cue (coralline algae)	Yes	Gene expression in metamorphosing postlarvae	Williams and Degnan (2009)
Coenogastropoda				
Crepidula fornicata	Starvation	Yes	Slower juvenile growth rate	Pechenik et al. (1996a,b, 2002)
Crepidula fornicata	Cadmium	No		Pechenik et al. (2001)
Crepidula onyx	Starvation	Yes	Reduced size, total organic content, and energy reserves of newly metamorphosed juveniles	Chiu et al. (2007, 2008)
			Reduced growth and filtration rate of juveniles	
Crepidula onyx	Hypoxia and low food availability	Yes	Reduced growth rate, dry weight, and filtration rate of juveniles	Li and Chiu (2013)

TABLE 9.2   *(Continued)*

Clade/Species	Treatment in larval or embryonic stage	Latent effect	Sublethal consequences	References
Crepidula onyx	Hypoxia and high food availability	No		Li and Chiu (2013)
Crepidula fornicata, Crepidula onyx, Crepipatella fecunda	Temporary reduction of salinity	No		Diederich et al. (2011)
Crepipatella dilatata	Reduction of salinity for brooding females	Yes	Reduced growth and reduced rates of oxygen consumption and feeding	Chaparro et al. (2014)
Nassarius festivus	Hypoxia	Yes	Reduced size of juveniles	Chan et al. (2008)
Heterobranchia				
Phestilla sibogae	Delayed metamorphosis with fed larvae	No		Miller and Hadfield (1990)
Phestilla sibogae	Delayed metamorphosis with starved larvae	Yes	Lower mean juvenile weight Decreased juvenile survival Decreased weight at reproductive maturity Longer mean time to reproductive maturity	Miller (1993)
Siphonaria australis	Elevated UVB, salinity and temperature	Yes	Reduced survival, larval growth, and length of velar cilia	Fischer and Phillips (2014)
Euopistobranchia				
Haminaea vesicular	Hypoxia	Yes	Reduced size of juveniles	Strathmann and Strathmann (1995)
Melanochlamys diomedea	Hypoxia	Yes	Reduced size of juveniles	Strathmann and Strathmann (1995)

The mechanisms for delayed or latent effects are not known in details. Depletion of energy stores in stressed larvae leading to a reduced availability of energy at metamorphosis is probably one of the main causes for latent effect (Pechenik, 2006). However, some results indicate that this is not always the case. Several nonmutually exclusive hypothesis have been proposed: (1) larval stress would reduce the size of the juvenile feeding apparatus (e.g., abnormal gill size in juvenile *C. fornicata* issued from starved larvae; Pechenik et al., 2002), which in turns decreases the ability of the individual to feed after its metamorphosis thus causing a reduction in growth rate (Chiu et al., 2007, 2008; Marshall et al., 2003; Pechenik et al., 2002; Wendt, 1996); (2) interference with transcriptional or translational processes (Pechenik, 2006; Pechenik et al., 1998; Williams and Degnan, 2009); (3) direct damages to DNA and/or enzymes (Heintz et al., 2000); and (4) epigenetic effects (Jablonka and Raz, 2009). So far, the results obtained for gastropods only support the two first hypotheses.

Experimental studies with *C. onyx* showed that the latent effects of larval starvation varied depending on the timing of starvation (Chiu et al., 2008). The juveniles developed from larvae that had experienced starvation in the first two days of larval life had reduced growth and lower filtration rates than those developed from control larvae. Starvation experienced later in larval life caused a reduction in shell length, lipid content, and RNA:DNA ratio of larvae at metamorphosis. Furthermore, the corresponding juveniles performed poorly in terms of growth in shell length and total organic carbon content. Latent effects of some stressors may also vary according to the nutritional status of larvae. For example, Li and Chiu (2013) showed that hypoxia only exert a latent effect on juveniles under low larval food condition. When the food concentration during the larval stage was doubled there was no discernible effect on juveniles.

Williams and Degnan (2009) tested for latent effects of inductive cues (three different species of coralline algae) on gene expression in metamorphosing postlarvae of the tropical abalone, *H. asinina*. They showed that the expression profiles of 11 of 17 metamorphosis-related genes differed according to which species of algae the larvae settled upon. Several genes continue to be differentially expressed for at least 40 h after removal of the algae, clearly demonstrating a carryover effect of inductive cues on gene expression.

To conclude, the reasons why some stressors cause latent effects and others do not remain largely unknown. In addition, results are also often conflicting within a species since offsprings from different parents may exhibit different susceptibility, suggesting a genetic basis for this difference. Vulnerability depends on a variety of factors (e.g., intensity and duration of exposure, defense and repair mechanisms, physiological tolerance), including the previous history of individuals, and even their parents. As stated by Fischer and Phillips (2014), though early life stages are particularly vulnerable to environmental stress and can be bottlenecks for populations, focusing too narrowly on immediate responses likely underestimate cumulative effects on populations and communities.

9.6.2 MEDIATORY ROLE OF OXIDATIVE STRESS

Among physiological mechanisms underlying life-history trade-offs, oxidative damage is thought to play a central role and to represent a universal constraint on life history evolution (Dowling and Simmons, 2009; Monaghan et al., 2009). The amount of cellular and molecular damage caused by oxidative stress increases with age, and this would in turn induce ageing and reduce lifespan (free radical theory of ageing, Harman, 1956; oxidative stress theory of ageing; Beckman and Ames; 1998). The theory is mostly supported by correlational data, which moreover come from vertebrates. For example, using a dataset from 10 mammalian and 2 avian species, Lambert et al. (2007) found a negative relationship between mitochondria ROS production and maximum lifespan. Similarly, lower ROS production and more efficient antioxidant defenses were observed in long-lived than in short-lived garter snakes (Robert and Bronikowski, 2010). Moreover, an age-related decreased activity of the proteasome (the enzymatic complex which is responsible for the degradation of abnormal and oxidized proteins) is observed in various mammalian tissues (Löw, 2011). In gastropods, though data are very scant, consistent results were obtained in land snails, at both intra- and inter-specific levels. *C. aspersum* subsp. *aspersa*, which is shorter-lived than its counterpart *C. aspersum maxima* (giant form) also exhibits higher ROS production upon bacterial challenge (Russo and Madec, 2011). Also, considering the positive relationship between lifespan and body size, the same authors found a similar trend between species (higher ROS production in smaller species, *O. lactea, Th. pisana, Cepaea nemoralis*, compared to the larger species *C. aspersum* and *H. pomatia*; Russo and Madec, 2013).

    The causative role of oxidative stress in the ageing process, that is, that increased ROS production and/or decline in antioxidant defense are responsible for the accumulation of cellular damage, is far from being demonstrated. It is even strongly invalidated by recent findings, such as the increased levels of pro-oxidants triggered by increased antioxidant activity, or lifespan shortening by antioxidant dietary supply (see Gems and Doonan, 2009 and references therein). Also unsupportive of the oxidative stress theory of ageing is the lack of relationship between ROS production or antioxidant defenses and exceptional longevity, such as in the naked mole rats; see Austad, 2010). With respect to reproduction, opposing conclusions are also drawn from observations, for example, increased oxidative damage in plasma proteins during lactation in the red squirrel (Fletcher et al., 2012) on the one hand, and reduced oxidative damage in reproducing female house mouse (Garratt et al., 2012) on the other hand. Therefore, to be solved, this issue clearly calls for new datasets, and we feel that gastropod models represent a promising and relevant study system in this respect.

9.7 CONCLUDING REMARKS AND EVOLUTIONARY PERSPECTIVES

Through this chapter, we have tried to give to the reader an overview of the diversity of stress responses in gastropods, by considering natural (biotic and abiotic) and anthropogenic stressors. Most examples cited revolved around oxidative stress, as oxygen imbalance is pivotal in eliciting and regulating stress responses at the organismal, cellular, and molecular level (Kassahn et al., 2009). Furthermore, oxidative stress plays a central role in life history, through its effects on life history trade-offs (Monaghan et al., 2009). However, in many instances, stress response pathways could not be fully described, due either to the merely correlational nature of available data, or more basically to the lack of mandatory genetic resources, and the need to resort on the knowledge based on phylogenetically remote models such as vertebrates. Thanks to recent technological advances in sequencing (NGS, next generation sequencing), the latter issue is hopefully becoming a less important obstacle to researchers, especially those working with nonmodel organisms, such as most gastropods. As mentioned earlier in this chapter, gastropod full transcriptomes are continuously appearing in the literature, which contain a wealth of information for those interested in stress molecular responses of their preferred species. However, for a full analysis of the pathways involved in these responses, an annotated genome would be evidently more appropriate. Currently, two species have their genome sequenced and annotated, *L. gigantea* (selected for its small genome size) and *A. californica*, but it is to be noted that a *B. glabrata* genome project is underway and already provides a large set of resources relevant to malacologists. Also, the genome of *L. stagnalis* is currently being sequenced, under the scientific coordination of a multidisciplinary consortium. As this species has been for long a model of choice in invertebrate zoology and biology, and also serves as model to neurophysiologists, this new resource is keenly anticipated. Furthermore, at a deeper evolutionary scale, comparative studies based on full genomes provide a unique tool for hypothesis validation in gastropods (see e.g., evolution of steroid hormone-signaling pathway in Metazoans; Markov et al., 2009).

Throughout this chapter, variation in stress response has been raised, at both intra- and interspecific levels. While the latter level of variation can be attributed to phylogenetic determinants and/or to ecological niche differences, the former may result from shorter term (micro) evolutionary forces, such as natural selection or population isolation and associated random genetic drift, and of their interaction with environment on population genetic diversity and ability to adapt to new conditions.

Population response to stress has both a plastic and a genetic component. The latter forms the basis for adaptive evolution. Until recently, evolutionary processes related to adaptation have been mostly investigated through integrated polygenic characters (life-history traits, morphology). However, a growing number of studies focuses on adaptive evolution at the level of gene expression, and this is noticeably due to the huge breakthrough in genetical and population genomics that has been permitted by NGS (e.g., Jeukens et al., 2010; Stapley et al., 2010). High genetic variation in gene expression has been found in various taxa (Fay and Wittkopp, 2008; Gibson, 2008; Schadt et al., 2003; Whitehead and Crawford, 2006). Therefore, the role of regulatory changes in evolutionary adaptation evolution is a question that can now be investigated at the genome scale through population genomics. Gene expression is assumed to evolve mostly under stabilizing selection, although this hypothesis may be difficult to disentangle from neutrality (Gilad et al., 2006; Whitehead and Crawford, 2006).

Under the present context of global change and intensification of environmental pressures, the evolutionary component of population response to stress has also become an emerging issue in environmental sciences and in ecotoxicology (Bickham, 2011; Klerks et al., 2011; Lankau et al., 2011). Although evidences for evolutionary impact of pollutants are accumulating (see Coutellec and Barata, 2013 and references therein), very few studies have focused on gastropods so far, and basically on *L. stagnalis* and pesticides (Bouétard et al., 2014; Coutellec and Lagadic, 2006; Coutellec et al., 2011, 2013). Interestingly, this species exhibits high population genetic variation in copper tolerance (acute toxicity; Côte et al., 2015), as well as in transcriptomic expression, both constitutive and induced by the pro-oxidant herbicide diquat (Bouétard et al., 2013). It is expected that similar studies on gastropods from marine, terrestrial, and freshwater habitats will help disentangling stress response pathways and their evolution in this ecologically important group.

KEYWORDS

- **gastropods**
- **environmental stress**
- **homeostasis**
- **stress signaling**
- **immune system**
- **life history**

REFERENCES

Abdalla, A. M.; El-Mogy, M.; Farid, N. M.; El-Sharabasy, M. Two Glutathione *S*-Transferase Isoenzymes Purified from *Bulinus truncatus* (Gastropoda: Planorbidae). *Comp. Biochem. Physiol. B: Biochem. Mol. Biol.* **2006**, *143*, 76–84.

Accorsi, A.; Ottaviani, E.; Malagoli, D. Effects of Repeated Hemolymph Withdrawals on the Hemocyte Populations and Hematopoiesis in *Pomacea canaliculata*. *Fish Shellfish Immunol.* **2014**, *38*, 56–64.

Accorsi, A.; Bucci, L.; de Eguileor, M.; Ottaviani, E.; Malagoli, D. Comparative Analysis of Circulating Hemocytes of the Freshwater Snail *Pomacea canaliculata*. *Fish Shellfish Immunol.* **2013**, *34*, 1260–1268.

Adamo, S. E. The Effects of the Stress Response on Immune Function in Invertebrates: An Evolutionary Perspective on an Ancient Connection. *Horm. Behav.* **2012**, *62*, 324–330.

Adamowicz, A.; Bolaczek, M. Blood Cells Morphology of the Snail *Helix aspersa maxima* (Helicidae). *Zool. Pol.* **2003**, *48*, 93–101.

Adema, C. M.; Hertel, L. A.; Miller, R. D.; Loker, E. S. A Family of Fibrinogen-related Proteins that Precipitates Parasite-derived Molecules is Produced by an Invertebrate after Infection. *Proc. Natl. Acad. Sci. U.S.A.* **1997**, *94*, 8691–8696.

Adema, C. M.; van Deutekom-Mulder, E. C.; van der Knaap, W. P.; Sminia, T. NADPH-oxidase Activity: The Probable Source of Reactive Oxygen Intermediate Generation in Hemocytes of the Gastropod *Lymnaea stagnalis*. *J. Leukoc. Biol.* **1993**, *54*, 379–383.

Adema, C. M.; van Deutekom-Mulder, E. C.; van der Knaap, W. P. W.; Sminia, T. Schistosomicidal Activities of *Lymnaea stagnalis* Haemocytes: The Role of Oxygen Radicals. *Parasitology* **1994**, *109*, 479–485.

Adema, C. M.; Hanington, P. C.; Lun, C. M.; Rosenberg, G. H.; Aragon, A. D.; Stout, B. A.; Lennard Richard, M. L.; Gross, P. S.; Loker, E. S. Differential Transcriptomic Responses of *Biomphalaria glabrata* (Gastropoda, Mollusca) to Bacteria and Metazoan Parasites, *Schistosoma mansoni* and *Echinostoma paraensei* (Digenea, Platyhelminthes). *Mol. Immunol.* **2010**, *47*, 849–860.

Ait Hamlet, S.; Bensoltane, S.; Djekoun, M.; Yassi, F.; Berrebbah, H. Histological Changes and Biochemical Parameters in the Hepatopancreas of Terrestrial Gastropod *Helix aspersa* as Biomarkers of Neonicotinoid Insecticide Exposure. *Afr. J. Biotechnol.* **2012**, *11*, 16277–16283.

Aladaileh, S.; Nair, S. V.; Raftos, D. A. Induction of Phenoloxidase and Other Immunological Activities in Sydney Rock Oysters Challenged with Microbial Pathogen-associate Molecular Patterns. *Fish Shellfish Immunol.* **2007**, *23*, 1196–1208.

Albrecht, U.; Keller, H.; Gebauer, W.; Markl, J. Rhogocytes (Pore Cells) as the Site of Hemocyanin Biosynthesis in the Marine Gastropod *Haliotis tuberculata*. *Cell Tissue Res.* **2001**, *304*, 455–462.

Altwegg, R.; Reyer, H.-U. Patterns of Natural Selection on Size at Metamorphosis in Water Frogs. *Evolution* **2003**, *57*, 872–882.

Amiard, J.-C.; Amiard-Triquet, C.; Barka, S.; Pellerin, J.; Rainbow, P. S. Metallothioneins in Aquatic Invertebrates: Their Role in Metal Detoxification and Their Use as Biomarkers. *Aquat. Toxicol.* **2006**, *76*, 160–202.

An, L.; Zheng, B.; Wang, L.; Zhang, Y.; Chen, H.; Zhao, X.; Zhang, L.; Lei K. Biomarker Responses and Genotoxicity in the Mud Snail (*Bullacta exarata*) as Indicators of Coastal Contamination. *Mar. Pollut. Bull.* **2012**, *6*, 303–309.

Angeletti, D.; Sebbio, C.; Carer, C.; Cimmaruta, R.; Nascetti, G.; Pepe, G.; Mosesso, P. Terrestrial gastropods (*Helix* spp.) as Sentinels of Primary DNA Damage for Biomonitoring Purposes: A Validation Study. *Environ. Mol. Mutagen.* **2013**, *54*, 204–212.

Ansaldo, M.; Nahabedian, D. E.; Holmes-Brown, E.; Agote, M.; Ansay, C. V.; Verrengia Guerrero, N. R.; Wider, E. A. Potential Use of Glycogen Level as Biomarker of Chemical Stress in *Biomphalaria glabrata*. *Toxicology* **2006**, *224*, 119–127.

Arad, Z.; Mizrahi, T.; Goldenberg, S.; Heller, J. Natural Annual Cycle of Heat Shock Protein Expression in Land Snails: Desert Versus Mediterranean Species of *Sphincterochila*. *J. Exp. Biol.* **2010**, *213*, 3487–3495.

Armstrong, R. N. Structure, Catalytic Mechanism, and Evolution of the Glutathione Transferase. *Chem. Res. Toxicol.* **1997**, *10*, 2–18.

Asokan, R.; Arumugam, M.; Mullainadhan, P. Activation of Prophenoloxidase in the Plasma and Haemocytes of the Marine Mussel *Perna viridis* Linnaeus. *Dev. Comp. Immunol.* **1997**, *21*, 1–12.

Auffret, M. Bivalve Hemocyte Morphology. *Am. Fish. Soc. Spec. Publ.* **1988**, *18*, 169–177.

Austad, S. N. Cats, "Rats," and Bats: The Comparative Biology of Aging in the 21st Century. *Integr. Comp. Biol.* **2010**, *50*, 783–792.

Baeza Garcia, A.; Pierce, R. J.; Gourbal, B.; Werkmeister, E.; Colinet, D.; Reichhart, J. M.; Dissous, C.; Coustau, C. Involvement of the Cytokine MIF in the Snail Host Immune Response to the Parasite *Schistosoma mansoni*. *PLoS Pathog.* **2010**, *69*, e1001115.

Bahgat, M.; Doenhoff, M.; Kirschfink, M.; Ruppel, A. Serine Protease and Phenoloxidase Activities in Hemocytes of *Biomphalaria glabrata* Snails with Varying Susceptibility to Infection with the Parasite *Schistosoma mansoni*. *Parasitol. Res.* **2002**, *88*, 489–494.

Bai, G.; Li, J.; Christensen, B. M.; Yoshino, T. P. Phenoloxidase Activity in the Reproductive System and Egg Masses of the Gastropod, *Biomphalaria glabrata*. *Comp. Biochem. Physiol. B: Biochem. Mol. Biol.* **1996**, *114*, 353–359.

Bai, G.; Brown, J. F.; Watson, C.; Yoshino, T. P. Isolation and Characterization of Phenoloxidase from Egg Masses of the Gastropod Mollusc, *Biomphalaria glabrata*. *Comp. Biochem. Physiol. B: Biochem. Mol. Biol.* **1997**, *118*, 463–469.

Balaban, R. S.; Nemoto, S.; Finkel, T. Mitochondria, Oxidants, and Aging. *Cell* **2005**, *120*, 483–495.

Bard, S. M. Multixenobiotic Resistance as a Cellular Defense Mechanism in Aquatic Organisms. *Aquat. Toxicol.* **2000**, *48*, 357–389.

Baron, O. L.; van West, P.; Industri, B.; Ponchet, M.; Dubreuil, G.; Gourbal, B.; Reichhart, J.-M.; Coustau, C. Parental Transfer of the Antimicrobial Protein LBP/BPI Protects *Biomphalaria glabrata* Eggs Against Oomycete Infections. *PLoS Pathog.* **2013**, *9*, e1003792.

Barracco, M. A.; Steil, A. A.; Gargioni, R. Morphological Characterization of the Hemocytes of the Pulmonate Snail *Biomphalaria tenagophila*. *Mem. Inst. Oswaldo Cruz* **1993**, *88*, 73–83.

Baturo, W.; Lagadic, L. Benzo[*a*]pyrene Hydroxylase and Glutathione *S*-transferase Activities as Biomarkers in *Lymnaea palustris* (mollusca, gastropoda) Exposed to Atrazine and Hexachlorobenzene in Freshwater Mesocosms. *Environ. Toxicol. Chem.* **1996**, *15*, 771–781.

Baturo, W.; Lagadic, L.; Caquet, T. Growth, Fecundity and Glycogen Utilization in *Lymnaea palustris* Exposed to Atrazine and Hexachlorobenzene in Freshwater Mesocosms. *Environ. Toxicol. Chem.* **1995**, *14*, 503–511.

Bayne, C. J. Successful Parasitism of Vector Snail *Biomphalaria glabrata* by the Human Blood Fluke (Trematode) *Schistosoma mansoni*: A 2009 Assessment. *Mol. Biochem. Parasitol.* **2009**, *165*, 8–18.

Bebianno, M. J.; Cravo, A.; Miguel, C.; Morais, S. Metallothionein Concentrations in a Population of *Patella aspersa*: Variation with Size. *Sci. Total Environ.* **2003**, *301*, 151–161.

Bebianno, M. J.; Langston, W. J.; Simkiss, K. Metallothionein Induction in *Littorina littorea* (Mollusca: Prosobranchia) on Exposure to Cadmium. *J. Mar. Biol. Assoc. U.K.* **1992**, *72*, 329–342.

Becker, W. Purine Metabolism in *Biomphalaria glabrata* under Starvation and Infection with *Schistosoma mansoni*. *Comp. Biochem. Physiol. B Biochem. Mol. Biol.* **1983**, *76*, 215–219.

Beckman, J. S.; Koppenol, W. H. Nitric Oxide, Superoxide and Peroxynitrite: the Good, the Bad and the Ugly. *Am. J. Physiol. Cell Physiol.* **1996**, *271*, 1424–1437.

Beckman, K. B.; Ames, B. N. Mitochondrial Aging: Open Questions. *Ann. N. Y. Acad. Sci.* **1998**, *854*, 118–127.

Bender, R. C.; Broderick, E. J.; Goodall, C. P.; Bayne, C. J. Respiratory Burst of *Biomphalaria glabrata* Hemocytes: *Schistosoma mansoni*-resistant Snails Produce More Extracellular H_2O_2 Than Susceptible Snails. *J. Parasitol.* **2005**, *91*, 275–279.

Bender, R. C.; Goodall, C. P.; Blouin, M. S.; Bayne, C. J. Variation in Expression of *Biomphalaria glabrata* SOD1: A Potential Controlling Factor in Susceptibility/Resistance to *Schistosoma mansoni*. *Dev. Comp. Immunol.* **2007**, *31*, 874–878.

Berriman, M.; Haas, B. J.; LoVerde, P. T.; Wilson, R. A.; Dillon, G. P.; Cerqueira, G. C.; Mashiyama, S. T.; Al-Lazikani, B.; Andrade, L. F.; Ashton, P. D.; Aslett, M. A.; Bartholomeu, D. C.; Blandin, G.; Caffrey, C. R.; Coghlan, A.; Coulson, R.; Day, T. A.; Delcher, A.; DeMarco, R.; Djikeng, A.; Eyre, T.; Gamble, J. A.; Ghedin, E.; Gu, Y.; Hertz-Fowler, C.; Hirai, H.; Hirai, Y.; Houston, R.; Ivens, A.; Johnston, D. A.; Lacerda, D.; Macedo, C. D.; McVeigh,

P.; Ning, Z.; Oliveira, G.; Overington, J. P.; Parkhill, J.; Pertea, M.; Pierce, R. J.; Protasio, A. V.; Quail, M. A.; Rajandream, M. A.; Rogers, J.; Sajid, M.; Salzberg, S. L.; Stanke, M.; Tivey, A. R.; White, O.; Williams, D. L.; Wortman, J.; Wu, W.; Zamanian, M.; Zerlotini, A.; Fraser-Liggett, C. M.; Barrell, B. G.; El-Sayed, N. M. The Genome of the Blood Fluke *Schistosoma mansoni*. *Nature* 2009, *460*, 352–358.

Bhagat, J.; Ingole, B. S. Genotoxic Potency of Mercury Chloride in Gill Cells of Marine Gastropod *Planaxis sulcatus* Using Comet Assay. *Environ. Sci. Pollut. Res.* 2015. 22, 10758–10768.

Bhide, M.; Gupta, P.; Khan, A.; Dubey, U.; Thakur, P.; Nema P.; Jain S. Morphological and Biochemical Studies on the Different Developmental Stages of a Freshwater Snail, *Lymnaea stagnalis* (Lymnaeidae) after Treatment with Some Pesticides. *J. Environ. Biol.* 2006, *27*, 359–366.

Bianco, K.; Yusseppone, M. S.; Otero, S.; Luquet, C.; Ríos de Molina, M. del C.; Kristoff, G. Cholinesterases and Neurotoxicity as Highly Sensitive Biomarkers for an Organophosphate Insecticide in a Freshwater Gastropod (*Chilina gibbosa*) with Low Sensitivity Carboxylesterases. *Aquat. Toxicol.* 2013, *144–145*, 26–35.

Bickham, J. The Four Cornerstones of Evolutionary Toxicology. *Ecotoxicology* 2011, *20*, 497–502.

Biggar, K. K.; Kornfeld, S. F.; Maistrovski, Y.; Storey, K. B. MicroRNA Regulation in Extreme Environments: Differential Expression of MicroRNAs in the Intertidal Snail *Littorina littorea* During Extended Periods of Freezing and Anoxia. *Genom. Proteom. Bioinform.* 2012, *10*, 302–309.

Bird, A. DNA Methylation Patterns and Epigenetic Memory. *Genes Dev.* 2002, *16*, 6–21.

Blindauer, C. A.; Leszczyszyn, O. I. Metallothioneins: Unparalleled Diversity in Structures and Functions for Metal Ion Homeostasis and More. *Nat. Prod. Rep.* 2010, *27*, 720–741.

Blouin, M. S.; Bonner, K. M.; Cooper, B.; Amarasinghe, V.; O'Donnell, R. P.; Bayne, C. J. Three Genes Involved in the Oxidative Burst are Closely Linked in the Genome of the Snail, *Biomphalaria glabrata*. *Int. J. Parasitol.* 2013, *43*, 51–55.

Bodo, A.; Bakos, E.; Szeri, F.; Varadi, A.; Sarkadi, B. The Role of Multidrug Transporters in Drug Availability, Metabolism and Toxicity. *Toxicol. Lett.* 2003, *140–141*, 133–143.

Bonner, K. M.; Bayne, C. J.; Larson, M. K.; Blouin, M. S. Effects of Cu/Zn Superoxide Dismutase (sod1) Genotype and Genetic Background on Growth, Reproduction and Defense in *Biomphalaria glabrata*. *PLoS Negl. Trop. Dis.* 2012, *6*, e1701.

Borics, G.; Várbíró, G.; Padisák, J. Disturbance and Stress: Different Meanings in Ecological Dynamics? *Hydrobiologia* 2013, *711*, 1–7.

Bornancin, F.; Parker, P. J. Phosphorylation of Protein Kinase C Alpha on Serine 657 Controls the Accumulation of Active Enzyme and Contributes to its Phosphatase-resistant State. *J. Biol. Chem.* 1997, *272*, 3544–3549.

Bouchut, A.; Coustau, C.; Gourbal, B.; Mitta, G. Compatibility in the *Biomphalaria glabrata/ Echinostoma caproni* Model: New Candidate Genes Evidenced by a Suppressive Subtractive Hybridization Approach. *Parasitology* 2007, *134*, 575–588.

Bouchut, A.; Sautiere, P. E.; Coustau, C.; Mitta, G. Compatibility in the *Biomphalaria glabrata/Echinostoma caproni* Model: Potential Involvement of Proteins from Hemocytes Revealed by a Proteomic Approach. *Acta Trop.* 2006a, *98*, 234–246.

Bouchut, A.; Roger, E.; Coustau, C.; Gourbal, B.; Mitta, G. Compatibility in the *Biomphalaria glabrata/Echinostoma caproni* Model: Potential Involvement of Adhesion Genes. *Int. J. Parasitol.* 2006b, *36*, 175–184.

Bouétard, A.; Besnard, A.-L.; Vassaux, D.; Lagadic, L.; Coutellec, M.-A. Impact of the Redox-cycling Herbicide Diquat on Transcript Expression and Antioxidant Enzymatic Activities of the Freshwater Snail *Lymnaea stagnalis*. *Aquat. Toxicol.* **2013**, *126*, 256–265.

Bouétard, A.; Côte, J.; Besnard, A.-L.; Collinet, M.; Coutellec, M.-A. Environmental versus Anthropogenic Effects on Population Adaptive Divergence in the Freshwater Snail *Lymnaea stagnalis*. *PLoS ONE* **2014**, *9*, e106670.

Bouétard, A.; Noirot, C.; Besnard, A.-L.; Bouchez, O.; Choisne, D.; Robe, E.; Klopp, C.; Lagadic, L.; Coutellec, M.-A. Pyrosequencing-based Transcriptomic Resources in the Pond Snail *Lymnaea stagnalis*, with a Focus on Genes Involved in Molecular Response to Diquat-induced Stress. *Ecotoxicology* **2012**, *21*, 2222–2234.

Brazilai, A.; Yamamoto, K. DNA Damage Responses to Oxidative Stress. *Science* **2004**, *3*, 1109–1115.

Bruckdorfer, R. The Basics about Nitric Oxide. *Mol. Aspects Med.* **2005**, *26*, 3–31.

Buchanan, K. L. Stress and the Evolution of Condition-dependent Signals. *TREE* **2000**, *15*, 156–160.

Butt, D.; Raftos, D. Phenoloxidase-associated Cellular Defence in the Sydney Rock Oyster, *Saccostrea glomerata*, Provides Resistance against QX Disease Infections. *Dev. Comp. Immunol.* **2008**, *32*, 299–306.

Butzke, D.; Machuy, N.; Thiede, B.; Hurwitz, R.; Goedert, S.; Rudel, T. Hydrogen Peroxide Produced by *Aplysia* Ink Toxin Kills Tumor Cells Independent of Apoptosis via Peroxiredoxin I Sensitive Pathways. *Cell Death Differ.* **2004**, *11*, 608–617.

Byrne, M.; Ho, M. A.; Wong, E.; Soars, N.; Selvakumaraswamy, P.; Sheppard Brennand, H.; Dworjanyn, S. A.; Davis, A. R. Unshelled Abalone and Corrupted Urchins, Development of Marine Calcifiers in a Changing Ocean. *Proc. R. Soc. Lond. B* **2011**, *278*, 2376–2383.

Byrne, M.; Soars, N. A.; Ho, M. A.; Wong, E.; McElroy, D.; Selvakumaraswamy, P.; Dworjanyn, S. A.; Davis, A. R. Fertilization in a Suite of Coastal Marine Invertebrates from SE Australia is Robust to Near-future Ocean Warming and Acidification. *Mar. Biol.* **2010**, *157*, 2061–2069.

Byzitter, J.; Lukowiak, K.; Karnik, V.; Dalesman, S. Acute Combined Exposure to Heavy Metals (Zn, Cd) Blocks Memory Formation in a Freshwater Snail. *Ecotoxicology* **2012**, *21*, 860–868.

Callewaert, L.; Michiels, C. W. Lysozymes in the Animal Kingdom. *J. Biosci.* **2010**, *35*, 127–160.

Canesi, L.; Gallo, G.; Gavioli, M.; Pruzzo, C. Bacteria–Hemocyte Interactions and Phagocytosis in Marine Bivalves. *Microsc. Res. Technol.* **2002**, *57*, 469–476.

Cardinaud, M.; Offret, C.; Huchette, S.; Moraga, D.; Paillard, C. The Impacts of Handling and Air Exposure on Immune Parameters, Gene Expression, and Susceptibility to Vibriosis of European Abalone *Haliotis tuberculata*. *Fish Shellfish Immunol.* **2014**, *36*, 1–8.

Chabicovsky, M.; Niederstätter, H.; Thaler, R.; Hödl, E.; Parson, W.; Rossmanith, W.; Dallinger, R. Localization and Quantification of Cd- and Cu-specific Metallothionein Isoform mRNA in Cells and Organs of the Terrestrial Gastropod *Helix pomatia*. *Toxicol. Appl. Pharmacol.* **2003**, *190*, 25–36.

Chan, H. Y.; Xu, W. Z.; Shin, P. K. S.; Cheung S. G. Prolonged Exposure to Low Dissolved Oxygen Affects Early Development and Swimming Behaviour in the Gastropod *Nassarius festivus* (Nassariidae). *Mar. Biol.* **2008**, *153*, 735–743.

Chandran R.; Sivakumar, A. A.; Mohandass, S.; Aruchami, M. Effect of Cadmium and Zinc on Antioxidant Enzyme Activity in the Gastropod, *Achatina fulica*. *Comp. Biochem. Physiol. C: Toxicol. Pharmacol.* **2005**, *140*, 422–426.

Chaparro O. R.; Segura, C. J.; Osores, S. J. A.; Pechenik, J. A.; Pardo L. M.; Cubillos V. M. Consequences of Maternal Isolation from Salinity Stress for Brooded Embryos and Future Juveniles in the Estuarine Direct-developing Gastropod *Crepipatella dilatata*. *Mar. Biol.* **2014**, *161*, 619–629.

Charmandari, E.; Tsigos, C.; Chrousos, G. Endocrinology of the Stress Response. *Annu. Rev. Physiol.* **2005**, *67*, 259–284.

Cheng, T. C. Functional Morphology and Biochemistry of Molluscan Phagocytes. *Ann. N. Y. Acad. Sci.* **1975**, *266*, 343–379.

Cheng, T. C. Bivalves. In *Invertebrate Blood Cells Volume 1*; Ratcliffe, N. A., Rowley, A. F., Eds.; Academic Press: London, 1981; pp 233–300.

Cheng, T. C.; Auld, K. R. Hemocytes of the Pulmonate Gastropod *Biomphalaria glabrata*. *J. Invert. Pathol.* **1977**, *30*, 119–122.

Cheng, W.; Hsiao, I. S.; Hsu, C. H.; Chen, J. C. Change in Water Temperature on the Immune Response of Taiwan Abalone *Haliotis diversicolor supertexta* and Its Susceptibility to *Vibrio parahaemolyticus*. *Fish Shellfish Immunol.* **2004a**, *17*, 235–243.

Cheng, W.; Juang, F. M.; Chen, J. C. The Immune Response of Taiwan Abalone *Haliotis diversicolor supertexta* and Its Susceptibility to *Vibrio parahaemolyticus* at Different Salinity Levels. *Fish Shellfish Immunol.* **2004b**, *16*, 295–306.

Cheng, W.; Li, C. H.; Chen, J. C. Effect of Dissolved Oxygen on the Immune Response of *Haliotis diversicolor supretexta* and Its Susceptibility to *Vibrio parahaemolyticus*. *Aquaculture* **2004c**, *232*, 103–115.

Cheng, W.; Hsiao, I. S.; Chen, J. C. Effect of Ammonia on the Immune Response of Taiwan Abalone *Haliotis diversicolor supertexta* and Its Susceptibility to *Vibrio parahaemolyticus*. *Fish Shellfish Immunol.* **2004d**, *17*, 193–202.

Cheng, W.; Hsiao, I. S.; Chen, J. C. Effect of Nitrite on Immune Response of Taiwan Abalone *Haliotis diversicolor supertexta* and Its Susceptibility to *Vibrio parahaemolyticus*. *Dis. Aquat. Org.* **2004e**, *60*, 157–164.

Chiu, J. M. Y.; Wang, H.; Thiyagarajan, V.; Qian, P. Y. Differential Timing of Larval Starvation Effects on Filtration Rate and Growth in Juvenile *Crepidula onyx*. *Mar. Biol.* **2008**, *154*, 91–98.

Chiu, J. M. Y.; Ng, T. Y. T.; Wang, W. X.; Thiyagarajan, V.; Qian, P. Y. Latent Effects of Larval Food Limitation on Filtration Rate, Carbon Assimilation and Growth in Juvenile Gastropod *Crepidula onyx*. *Mar. Ecol. Prog. Ser.* **2007**, *343*, 173–182.

Collins, A. R. The Comet Assay for DNA Damage and Repair: Principles, Applications, and Limitations. *Mol. Biotechnol.* **2004**, *26*, 249–261.

Collins, A. R. Measuring Oxidative Damage to DNA and Its Repair with the Comet Assay. *Biochim. Biophys. Acta* **2014**, *1840*, 794–800.

Cong, M.; Song, L.; Wang, L.; Zhao, J.; Qiu, L.; Li, L.; Zhang, H. The Enhanced Immune Protection of Zhikong Scallop *Chlamys farreri* on the Secondary Encounter with *Listonella anguillarum*. *Comp. Biochem. Physiol. B: Biochem. Mol. Biol.* **2008**, *151*, 191–196.

Conte, A.; Ottaviani, E. Nitric Oxide Synthase Activity in Molluscan Hemocytes. *FEBS Lett.* **1995**, *365*, 120–124.

Côte, J.; Bouétard, A.; Pronost, Y.; Besnard, A.-L.; Coke, M.; Piquet, F.; Caquet, T.; Coutellec, M.-A. Genetic Variation of *Lymnaea stagnalis* Tolerance to Copper: A Test of Selection

Hypotheses and Its Relevance for Ecological Risk Assessment. *Environ. Pollut.* **2015**, *205*, 209–217.

Coustau, C.; Yoshino, T. P. *Schistosoma mansoni*: Modulation of Hemocyte Surface Polypeptides Detected in Individual Snails, *Biomphalaria glabrata*, Following Larval Exposure. *Exp. Parasitol.* **1994**, *79*, 1–10.

Coustau, C.; Gourbal, B.; Duval, D.; Yoshino, T. P.; Adema, C. M.; Mitta, G. Advances in Gastropod Immunity from the Study of the Interaction Between the snail *Biomphalaria glabrata* and Its Parasites: A Review of Research Progress Over the Last Decade. *Fish Shellfish Immunol.*, **2015**, *46*, 5–16.

Coutellec, M.-A; Barata, C. Special Issue on Long-term Ecotoxicological Effects: An Introduction. *Ecotoxicology* **2013**, *22*, 763–766.

Coutellec, M.-A.; Lagadic, L. Effects of Self-Fertilization, Environmental Stress and Exposure to Xenobiotics on Fitness-related Traits of the Freshwater Snail *Lymnaea stagnalis*. *Ecotoxicology* **2006**, *15*, 199–213.

Coutellec, M.-A.; Besnard, A.-L.; Caquet, T. Population Genetics of *Lymnaea stagnalis* Experimentally Exposed to Cocktails of Pesticides. *Ecotoxicology* **2013**, *22*, 879–888.

Coutellec, M.-A.; Collinet, M.; Caquet, T. Parental Exposure to Pesticides and Progeny Reaction Norm to a Biotic Stress Gradient in the Freshwater Snail *Lymnaea stagnalis*. *Ecotoxicology* **2011**, *20*, 524–534.

Cowan, K. J.; Storey, K. B. Mitogen-activated Protein Kinases: New Signaling Pathways Functioning in Cellular Responses to Environmental Stress. *J. Exp. Biol.* **2003**, *206*, 1107–1115.

Crain, C. M.; Kroeker, K.; Halpern, B. S. Interactive and Cumulative Effects of Multiple Human Stressors on Marine Systems. *Ecol. Lett.* **2008**, *11*, 1304–1315.

Cueto, J. A.; Fogal, T.; Castro-Vazquez, A. Ultrastructural Characterization of Circulating Hemocytes of *Pomacea canaliculata*. *Biocell* **2007**, *32*, 103.

Cueto, J. A.; Vega, I. A.; Castro-Vazquez, A. Multicellular Spheroid Formation and Evolutionary Conserved Behaviors of Apple Snail Hemocytes in Culture. *Fish Shellfish Immunol.* **2013**, *34*, 443–453.

Cunha, I.; Mangas-Ramirez, E.; Guilhermino, L. Effects of Copper and Cadmium on Cholinesterase and Glutathione *S*-transferase Activities of Two Marine Gastropods (*Monodonta lineata* and *Nucella lapillus*). *Comp. Biochem. Physiol. C: Toxicol. Pharmacol.* **2007**, *145*, 648–657.

D'Autréaux, B.; Toledano, M. B. ROS as signalling molecules: mechanisms that generate specificity in ROS homeostasis. *Nat. Rev. Mol. Cell Biol.* **2007**, *8*, 813–824.

Dallinger, R.; Chabicovsky, M.; Hödl, E.; Prem, C.; Hunziker, P.; Manzl, C. Copper in *Helix pomatia* (Gastropoda) is Regulated by One Single Cell Type: Differently Responsive Metal Pools in Rhogocytes. *Am. J. Physiol. Regul. Integr. Comp. Physiol.* **2005**, *289*, R1185–1195.

Dallman, M. F. Stress by Any Other Name…? *Horm. Behav.* **2003**, *43*, 18–30.

Dang, V. T.; Benkendorff, K.; Speck, P. In Vitro Antiviral Activity against Herpes Simplex Virus in the Abalone *Haliotis laevigata*. *J. Gen. Virol.* **2011**, *92*, 627–637.

Darling, E. S.; Côté, I. M. Quantifying the Evidence for Ecological Synergies. *Ecol. Lett.* **2008**, *11*, 1278–1286.

Davids, B. J.; Yoshino, T. P. Integrin-like RGD-dependent Binding Mechanism Involved in the Spreading Response of Circulating Molluscan Phagocytes. *Dev. Comp. Immunol.* **1998**, *22*, 39–53.

Davies, K. J. Oxidative Stress: the Paradox of Aerobic Life. *Biochem. Soc. Symp.* **1995**, *61*, 1–31.

Davis, A. R.; Coleman, D.; Broad, A.; Byrne, M.; Dworjanyn, S. A.; Przeslawski, R. Complex Responses of Intertidal Molluscan Embryos to a Warming and Acidifying Ocean in the Presence of UV Radiation. *PLoS ONE* **2013**, *8*, e55939.

Day, T. A. Defining Stress as a Prelude to Mapping Its Neurocircuitry: No Help from Allostasis. *Prog. Neuropsychopharmacol. Biol. Psychiatry* **2005**, *29*, 1195–1200.

de Lapuente J.; Lourenço, J.; Mendo, S. A.; Borràs, M.; Martins M. G.; Costa, P. M.; Pacheco, M. The Comet Assay and Its Applications in the Field of Ecotoxicology: A Mature Tool that Continues to Expand Its Perspectives. *Frontiers Genet.* **2015**, *6*, 180.

De Zoysa, M.; Nikapitiya, C.; Whang, I.; Lee, J. S.; Lee J. Abhisin: A Potential Antimicrobial Peptide Derived from Histone H2A of Disk Abalone (*Haliotis discus discus*). *Fish Shellfish Immunol.* **2009**, *27*, 639–646.

De Zoysa, M.; Whang, I.; Lee, Y.; Lee, S.; Lee, J. S.; Lee, J. Defensin from Disk Abalone *Haliotis discus discus*: Molecular Cloning, Sequence Characterization and Immune Response Against Bacterial Infection. *Fish Shellfish Immunol.* **2010**, *28*, 261–266.

de Zwaan, A.; van Marrewijk, W. J. A. Anaerobic Glucose Degradation in the Sea Mussel *Mytilus edulis* L. *Comp. Biochem. Physiol.* **1973**, *44*, 429–439.

Defer, D.; Bourgougnon, N.; Fleury, Y. Detection and Partial Characterisation of an Antimicrobial Peptide (Littorein) from the Marine Gastropod *Littorina littorea*. *Int. J. Antimicrob. Agents* **2009**, *34*, 188–190.

deFur, P. L; Crane, M.; Ingersoll, C. G.; Tattersfield, L. J. *Endocrine Disruption in Invertebrates: Endocrinology, Testing and Assessment.* Society of Environmental Toxicology and Chemistry (SETAC): Pensacola, FL, 1999.

Deleury, E.; Dubreuil, G.; Elangovan, N.; Wajnberg, E.; Reichhart, J. M.; Gourbal, B.; Duval, D.; Baron, O. L.; Gouzy, J.; Coustau C. Specific Versus Non-specific Immune Responses in an Invertebrate Species Evidenced by a Comparative De Novo Sequencing Study. *PLoS One* **2012**, *7*, e32512.

Derby, C. D. Escape by Inking and Secreting: Marine Molluscs Avoid Predators Through a Rich Array of Chemicals and Mechanisms. *Biol. Bull.* **2007**, *213*, 274–289.

Deschaseaux, E. S. M.; Taylor, A. M.; Maher, W. A. Measure of Stress Response Induced by Temperature and Salinity Changes on Hatched Larvae of Three Marine Gastropod Species. *J. Exp. Mar. Biol. Ecol.* **2011**, *397*, 121–128.

Deschaseaux, E. S. M.; Taylor, A. M.; Maher, W. A.; Davis, A. R. Cellular Responses of Encapsulated Gastropod Embryos to Multiple Stressors Associated with Climate Change. *J. Exp. Mar. Biol. Ecol.* **2010**, *383*, 130–136.

Dhabhar, F. Stress-induced Augmentation of Immune function—The Role of Stress Hormones, Leukocyte Trafficking, and Cytokines. *Brain Behav. Immun.* **2002**, *16*, 785–798.

Di, G.; Zhang, Z.; Ke, C. Phagocytosis and Respiratory Burst Activity of Haemocytes from the Ivory Snail, *Babylonia areolata*. *Fish Shellfish Immunol.* **2013**, *35*, 366–374.

Di, G. L.; Zhang, Z. X.; Ke, C. H.; Guo, J. R.; Xue, M.; Ni, J. B.; Wang, D. X. Morphological Characterization of the Haemocytes of the Ivory Snail, *Babylonia areolata* (Neogastropoda Buccinidae). *J. Mar. Biol. Assoc. U.K.* **2011**, *91*, 1489–1497.

Di Lellis, A.; Seifan, M.; Troschinski, S.; Mazzia, C.; Capowiez, Y.; Triebskorn, R.; Köhler, H.-R. Solar Radiation Stress in Climbing Snails: Behavioural and Intrinsic Features Define the Hsp70 Level in Natural Populations of *Xeropicta derbentina* (Pulmonata). *Cell Stress Chaperones* **2012**, *17*, 717–727.

Diederich, C. M.; Jarrett, J. N.; Chaparro, O. R.; Segura, C. J.; Arellano, S. M.; Pechenik J. A. Low Salinity Stress Experienced by Larvae Does Not Affect Post-metamorphic Growth or Survival in Three Calyptraeid Gastropods. *J. Exp. Mar. Biol. Ecol.* **2011**, *397*, 94–105.

Dieterich, A.; Troschinski, S.; Schwarz, S.; Di Lellis, M. A.; Henneberg, A.; Fischbach, U.; Ludwig, M.; Gärtner, U.; Triebskorn, R.; Köhler, H. R. Hsp70 and Lipid Peroxide Levels Following Heat Stress in *Xeropicta derbentina* (Krynicki 1836) (Gastropoda, Pulmonata) with Regard to Different Colour Morphs. *Cell Stress Chaperones* **2015**, *20*, 159–168.

Dikkeboom, R.; Tijnagel J. M.; van der Knaap, W. P. Monoclonal Antibody Recognized Hemocyte Subpopulations in Juvenile and Adult *Lymnaea stagnalis*: Functional Characteristics and Lectin Binding. *Dev. Comp. Immunol.* **1988a**, *12*, 17–32.

Dikkeboom, R.; Bayne, C. J.; van der Knaap, W. P. W.; Tijnagel, J. M. Possible Role of Reactive Forms of Oxygen in In Vitro Killing of *Schistosoma mansoni* Sporocysts by Hemocytes of *Lymnaea stagnalis*. *Parasitol. Res.* **1988b**, *75*, 148–154.

Ding, J.; Li, J.; Bao, Y.; Li, L.; Wu, F.; Zhang, G. Molecular Characterization of a Mollusk Chicken-type Lysozyme Gene from *Haliotis discus hannai* Ino, and the Antimicrobial Activity of its Recombinant Protein. *Fish Shellfish Immunol.* **2011**, *30*, 163–172.

Donaghy, L.; Hong, H. K.; Lambert, C.; Park, H.-S.; Shim, W. J.; Choi, K.-S. First Characterisation of the Populations and Immune-related Activities of Hemocytes from Two Edible Gastropod Species, the Disk Abalone, *Haliotis discus discus* and the Spiny Top Shell, *Turbo cornutus*. *Fish Shellfish Immunol.* **2010**, *28*, 87–97.

Dong, Y.-W.; Han, G.-D.; Huang, X.-W. Stress Modulation of Cellular Metabolic Sensors: Interaction of Stress from Temperature and Rainfall on the Intertidal Limpet *Cellana toreuma*. *Mol. Ecol.* **2014**, *23*, 4541–4554.

Douglas, J. S.; Hunt, M. D.; Sullivan, J. T. Effects of *Schistosoma mansoni* Infection on Phagocytosis and Killing of *Proteus vulgaris* in *Biomphalaria glabrata* Hemocytes. *J. Parasitol.* **1993**, *79*, 280–283.

Dowling, D. K.; Simmons, L. W. Reactive Oxygen Species as Universal Constraints in Life-history Evolution. *Proc. Biol. Sci.* **2009**, *276*, 1737–1745.

Duchemin, M. B.; Fournier, M.; Auffret, M. Seasonal Variations of Immune Parameters in Diploid and Triploid Pacific Oysters, *Crassostrea gigas* (Thunberg). *Aquaculture* **2007**, *264*, 73–81.

Ducrot, V.; Péry, A. R. R.; Mons, R.; Queau, H.; Charles, S.; Garric, J. Dynamic Energy Budget as a Basis to Model Population-level Effects of Zinc-spiked Sediments in the Gastropod, *Valvata piscinalis*. *Environ. Toxicol. Chem.* **2007**, *26*, 1774–1783.

Edinger, A. L.; Thompson, C. B. Death by Design: Apoptosis, Necrosis and Autophagy. *Curr. Opin. Cell Biol.* **2004**, *16*, 663–669.

Edwards, A. S.; Newton, A. C. Phosphorylation at Conserved Carboxyl-terminal Hydrophobic Motif Regulates the Catalytic and Regulatory Domains of Protein Kinase C. *J. Biol. Chem.* **1997**, *272*, 18382–18390.

Edwards, S. F. Crude Oil Effects on Mortality, Growth and Feeding of Young Oyster Drills, *Urosalpinx cinerea* (Say). *Veliger* **1980**, *23*, 125–130.

Ehara, T.; Kitajima, S.; Kanzawa, N.; Tamiya, T.; Tsuchiya, T. Antimicrobial Action of Achacin is Mediated by L-Amino Acid Oxidase Activity. *FEBS Letters* **2002**, *531*, 509–512.

El-Gohary, A.; Laila, R. A.; Genena, M. A. M. Biochemical Effect of Three Molluscicide Baits Against the Two Land Snails, *Monacha cantiana* and *Eobania vermiculata* (Gastropoda: Helicidae). *Int. J. Agric. Res.* **2011**, *6*, 682–690.

El-Shenawy, N. S.; Mohammadden, A.; Hessenan Al-Fahmie, Z. Using the Enzymatic and Non-enzymatic Antioxidant Defense System of the Land Snail *Eobania vermiculata* as

Biomarkers of Terrestrial Heavy Metal Pollution. *Ecotoxicol. Environ. Saf.* **2012**, *84*, 347–354.

Ellis, R. P.; Parry, H.; Spicer, J. I.; Hutchinson, T. H.; Pipe, R. K.; Widdicombe, S. Immuno-logical Function in Marine Invertebrates: Responses to Environmental Perturbation. *Fish Shellfish Immunol.* **2011**, *30*, 1209–1222.

Elphick, M. R.; Kemenes, G.; Staras, K.; O'Shea, M. Behavioral Role for Nitric Oxide in Chemosensory Activation of Feeding in a Mollusc. *J. Neurosci.* **1995**, *15*, 7653–7664.

Elsbach, P.; Weiss, J.; Levy, O. Integration of Antimicrobial Host Defenses: Role of the Bactericidal/Permeability-increasing Protein. *Trends Microbiol.* **1994**, *2*, 324–328.

Elvitigala, D. A. S.; Premachandra, H. K. A.; Whang, I.; Nam, B.-H.; Lee, J. Molecular Insights of the First Gastropod TLR Counterpart from Disk Abalone (*Haliotis discus discus*), Revealing Its Transcriptional Modulation Under Pathogenic Stress. *Fish Shellfish Immunol.* **2013**, *35*, 334–342.

Fang, F. C. Mechanism of Nitric Oxide-related Antimicrobial Activity. *J. Clin. Invest.* **1997**, *99*, 2818–2825.

Fay, J. C.; Wittkopp, P. J. Evaluating the Role of Natural Selection in the Evolution of Gene Regulation. *Heredity* **2008**, *100*, 191–199.

Fei, G.; Guo, C.; Sun, H. S.; Feng, Z. P. Chronic Hypoxia Stress-induced Differential Modulation of Heat-shock Protein 70 and Presynaptic Proteins. *J. Neurochem.* **2007**, *100*, 50–61.

Feirrera-Cravo, M.; Welker, A. F.; Hermes-Lima, M. The Connection Between Oxidative Stress and Estivation in Gastropods and Anurans. *Prog. Mol. Subcell. Biol.* **2010**, *49*, 47–61.

Feldmeyer, B.; Wheat, C. W.; Krezdorn, N.; Rotter, B.; Pfenninger, M. Short Read Illumina Data for the De Novo Assembly of a Non-model Snail Species Transcriptome (*Radix balthica*, Basommatophora, Pulmonata), and a Comparison of Assembler Performance. *BMC Genomics* **2011**, *12*, 317.

Feng, Z. P.; Zhang, Z.; van Kesteren, R. E.; Straub, V. A.; van Nierop, P.; Jin, K.; Nejatbakhsh, N.; Goldberg, J. I.; Spencer, G. E.; Yeoman, M. S.; Wildering, W.; Coorssen, J. R.; Croll, R. P.; Buck, L. T.; Syed, N. I.; Smit, A. B. Transcriptome Analysis of the Central Nervous System of the Mollusc *Lymnaea stagnalis*. *BMC Genomics* **2009**, *10*, 451.

Finkel, T. Signal Transduction by Reactive Oxygen Species. *J. Cell Biol.* **2011**, *194*, 7–15.

Fiolka, M. J.; Witkowski, A. Lysozyme-like Activity in Eggs and in Some Tissues of Land Snails *Helix aspersa maxima* and *Achatina achatina*. *Folia Biol.* **2004**, *52*, 3–4.

Fischer J.; Phillips N. E. Carry-over Effects of Multiple Stressors on Benthic Embryos are Mediated by Larval Exposure to Elevated UVB and Temperature. *Global Change Biol.* **2014**, *20*, 2108–2116.

Fletcher, Q. E.; Speakman, J. R.; Boutin, S.; McAdam, A. G.; Woods, S. B., Humphries, M. H. Seasonal Stage Differences Overwhelm Environmental and Individual Factors as Determinants of Energy Expenditure in Free-ranging Red Squirrels. *Funct. Ecol.* **2012**, *26*, 677–687.

Fneich, S.; Dheilly, N.; Adema, C.; Rognon, A.; Reichel, M.; Bulla, J.; Grunau, C.; Cosseau, C. 5-Methyl-cytosine and 5-Hydroxy-Methyl-Cytosine in the Genome of *Biomphalaria glabrata*, a Snail Intermediate Host of *Schistosoma mansoni*. *Parasites Vectors* **2013**, *6*, 167.

Foley, D. A.; Cheng, T. C. Morphology, Hematologic Parameters and Behavior of Hemo-lymph Cells of the Quahaug Clam, *Mercenaria mercenaria*. *Biol. Bull.* **1974**, *146*, 343–356.

Franceschi, C.; Cossarizza, A.; Monti, D.; Ottaviani, E. Cytotoxicity and Immunocyte Markers in Cells from the Freshwater *Planorbarius corneus* (L.) (*Gastropoda Pulmonata*) Implication on the Evolution of Natural Killer Cells. *Eur. J. Immunol.* **1991**, *21*, 489–493.

Franchini, A.; Ottaviani, E. Fine Structure and Acid Phosphatase Localization of Hemocytes in the Freshwater Snail *Viviparus ater* (Gastropoda, Prosobranchia). *J. Invert. Pathol.* **1990**, *55*, 28–34.

Franzellitti, S.; Fabbri, E. Cyclic-AMP Mediated Regulation of ABCB mRNA Expression in Mussel Haemocytes. *PLoS ONE* **2013**, *8*, e61634.

Furuta, E.; Yamaguchi, K.; Shimozawa, A. The Ultrastructure of Hemolymph Cells of the Land Slug, *Incilaria fruhstorferi* Collinge (Gastropoda: Pulmonata). *Anat. Anz.* **1986**, *162*, 215–224.

Furuta, E.; Yamaguchi, K.; Shimozawa, A. Hemolymph Cells and the Platelet-like Structures of the Land Slug, *Incilaria bilineata* (Gastropoda: Pulmonata). *Anat. Anz.* **1990**, *170*, 99–109.

Gabai, V. L.; Yaglom, J. A.; Volloch, V.; Meriin, A. B.; Force, T.; Koutroumanis, M.; Massie, B.; Mosser, D. D.; Sherman, M. Y. Hsp72-mediated Suppression of c-Jun N-terminal Kinase is Implicated in Development of Tolerance to Caspase-independent Cell Death. *Mol. Cell Biol.* **2000**, *20*, 6826–6836.

Gäde, G. Anaerobic Metabolism of the Common Cockle, *Cardium edule*. *Arch. Int. Physiol. Biochem.* **1975**, *83*, 879–886.

Gäde, G.; Carlsson, K. H.; Meinardus, G. Energy Metabolism in the Foot of the Marine Gastropod *Nassa mutabilis* During Environmental and Functional Anaerobiosis. *Mar. Biol.* **1984**, *80*, 49–56.

Gagnaire, B.; Gagné, F.; André, C.; Blaise, C.; Abbaci, K.; Budzinski, H.; Dévier, M.-H.; Garric J. Development of Biomarkers of Stress Related to Endocrine Disruption in Gastropods: Alkali-labile Phosphates, Protein-bound Lipids and Vitellogenin-like Proteins. *Aquat. Toxicol.* **2009**, *92*, 155–167.

Galinier, R.; Portela, J.; Mone, Y.; Allienne, J. F.; Henri, H.; Delbecq, S.; Mitta, G.; Gourbal, B.; Duval, D. Biomphalysin, A New Beta Pore-forming Toxin Involved in *Biomphalaria glabrata* Immune Defense against *Schistosoma mansoni*. *PLoS Pathog.* **2013**, *9*, e1003216.

Garratt, M.; McArdle, F.; Stockley, P.; Vasilaki, A.; Beynon, R.; Jackson, M. J.; Hurst, J. M. Tissue-dependent Changes in Oxidative Damage with Male Reproductive Effort in House Mice. *Funct. Ecol.* **2012**, *26*, 423–433.

Gaume, B.; Bourgougnon, N.; Auzoux-Bordenave, S.; Roig, B.; Le Bot, B.; Bedoux, G. In Vitro Effects of Triclosan and Methyl-triclosan on the Marine Gastropod *Haliotis tuberculata*. *Comp. Biochem. Physiol. C: Toxicol. Pharmacol.* **2012**, *156*, 87–94.

Gelperin, A. Nitric Oxide Mediates Network Oscillations of Olfactory Interneurons in a Terrestrial Mollusc. *Nature* **1994**, *369*, 61–63.

Gems, D.; Doonan, R. Antioxidant Defense and Aging in *C. elegans*: Is the Oxidative Damage Theory of Aging Wrong? *Cell Cycle* **2009**, *8*, 1681–1687.

George, W. C.; Ferguson, J. H. The Blood of Gastropod Molluscs. *J. Morphol.* **1950**, *86*, 315–324.

Geraerts, W. P. Neurohormonal Control of Growth and Carbohydrate Metabolism by the Light Green Cells in *Lymnaea stagnalis*. *Gen. Comp. Endocrinol.* **1992**, *83*, 433–444.

Gibbs, P. E.; Bryan, G. W. TBT-induced Imposex in Neogastropod Snails: Masculinization to Mass Extinction. In *Tributyltin: Case Study of an Environmental Contaminant*; de Mora, S. J., Ed.; Cambridge University Press, Cambridge, 1996; pp 212–236.

Gibson, G. The Environmental Contribution to Gene Expression Profiles. *Nat. Rev. Genet.* **2008**, *9*, 575–581.

Gilad, Y.; Oshlack A.; Smyth, G. K.; Speed T. P.; White, K. P. Expression Profiling in Primates Reveals a Rapid Evolution of Human Transcription Factors. *Nature* **2006**, *440*, 242–245.

Giraud-Billoud, M.; Vega, I. A.; Tosi, M. E.; Abud, M. A.; Calderón, M. L.; Castro-Vazquez, A. Antioxidant and Molecular Chaperone Defences During Estivation and Arousal in the South American Apple Snail *Pomacea canaliculata*. *J. Exp. Biol.* **2013**, *216*, 614–622.

Goater, C. Growth and Survival of Postmetamorphic Toads: Interactions among Larval History, Density, and Parasitism. *Ecology* **1994**, *75*, 2264–2274.

Goodall, C. P.; Bender, R. C.; Broderick, E. J.; Bayne, C. J. Constitutive Differences in Cu/Zn Superoxide Dismutase mRNA Levels and Activity in Hemocytes of *Biomphalaria glabrata* (Mollusca) that are Either Susceptible or Resistant to *Schistosoma mansoni* (Trematoda). *Mol. Biochem. Parasitol.* **2004**, *137*, 321–328.

Goodall, C. P.; Bender, R. C.; Brooks, J. K.; Bayne, C. J. *Biomphalaria glabrata* Cytosolic Copper/Zinc Superoxide Dismutase (SOD1) Gene: Association of SOD1 Alleles with Resistance/Susceptibility to *Schistosoma mansoni*. *Mol. Biochem. Parasitol.* **2006**, *147*, 207–210.

Gooding, M. P.; LeBlanc, G. A. Biotransformation and Disposition of Testosterone in the Eastern Mud Snail *Ilyanassa obsoleta*. *Gen. Comp. Endocrinol.* **2001**, *122*, 172–180.

Gopalakrishnan, S.; Thilagam, H.; Huang, W. B.; Wang, K. J. Immunomodulation in the Marine Gastropod *Haliotis diversicolor* Exposed to Benzo(a)pyrene. *Chemosphere* **2009**, *75*, 389–397.

Gopalakrishnan, S.; Huang, W. B.; Wang, Q. W.; Wu, M. L.; Liu, J.; Wang, K. J. Effects of Tributyltin and Benzo[a]pyrene on the Immune-associated Activities of Hemocytes and Recovery Responses in the Gastropod Abalone, *Haliotis diversicolor*. *Comp. Biochem. Physiol. C: Toxicol. Pharmacol.* **2011**, *154*, 120–128.

Gorbushin, A. M.; Iakovleva, N. V. Haemogram of *Littorina littorea*. *J. Mar. Biol. Assoc. U.K.* **2006**, *86*, 1175–1181.

Gorbushin, A. M.; Iakovleva, N. V. Functional Characterization of *Littorina littorea* (Gastropoda: Prosobranchia) Blood Cells. *J. Mar. Biol. Assoc. U.K.* **2007**, *87*, 741–746.

Gorbushin, A. M.; Klimovich, A. V.; Iakovleva, N. V. *Himasthla elongata*: Effect of Infection on Expression of the LUSTR-like Receptor mRNA in Common Periwinkle Haemocytes. *Exp. Parasitol.* **2009**, *123*, 24–30.

Grime, J. P. The Stress Debate: Symptom of Impending Synthesis? *Biol. J. Linn. Soc.* **1989**, *37*, 3–17.

Grune, T.; Catalgol, B.; Licht, A.; Ermak, G.; Pickering, A. M.; Ngo J. K.; Davies, K. J. HSP70 Mediates Dissociation and Reassociation of the 26S Proteasome During Adaptation to Oxidative Stress. *Free Radic. Biol. Med.* **2011**, *51*, 1355–1364.

Guderley, H.; Pörtner, H. O. Metabolic Power Budgeting and Adaptive Strategies in Zoology: Examples from Scallops and Fish. *Can. J. Zool.* **2010**, *88*, 753–763.

Gueguen, Y.; Cadoret, J. P.; Flament, D.; Barreau Roumiguière, C.; Girardot, A. L.; Garnier, J.; Hoareau, A.; Bachère, E.; Escoubas, J. M. Immune Gene Discovery by Expressed Sequence Tags Generated from Hemocytes of the Bacteria-challenged Oyster, *Crassostrea gigas*. *Gene* **2003**, *303*, 139–145.

Guillou, F.; Mitta, G.; Galinier, R.; Coustau, C. Identification and Expression of Gene Transcripts Generated During an Anti-parasitic Response in *Biomphalaria glabrata*. *Dev. Comp. Immunol.* **2007a**, *31*, 657–671.

Guillou, F.; Roger, E.; Mone, Y.; Rognon, A.; Grunau, C.; Theron, A.; Mitta, G.; Coustau, C.; Gourbal, B. E. Excretory–Secretory Proteome of Larval *Schistosoma mansoni* and *Echinostoma caproni*, Two Parasites of *Biomphalaria glabrata*. *Mol. Biochem. Parasitol.* **2007b**, *155*, 45–56.

Gunter, H. M.; Degnan, B. M. Developmental Expression of Hsp90, Hsp70 and HSF During Morphogenesis in the Vetigastropod *Haliotis asinina*. *Dev. Genes Evol.* **2007**, *217*, 603–612.

Guo, Y.; He, H. Identification and Characterization of a Goose-type Lysozyme from Sewage Snail *Physa acuta*. *Fish Shellfish Immunol.* **2014**, *39*, 321–325.

Gust, M.; Fortier, M.; Garric, J.; Fournier, M.; Gagné F. Immunotoxicity of Surface Waters Contaminated by Municipal Effluents to the Snail *Lymnaea stagnalis*. *Aquat. Toxicol.* **2013a**, *126*, 393–403.

Gust, M.; Fortier, M.; Garric, J.; Fournier, M.; Gagné, F. Effects of Short-term Exposure to Environmentally Relevant Concentrations of Different Pharmaceutical Mixtures on the Immune Response of the Pond Snail *Lymnaea stagnalis*. *Sci. Total Environ.* **2013b**, *445–446*, 210–218.

Gutteridge, J. M.; Halliwell, B. The Measurement and Mechanism of Lipid Peroxidation in Biological Systems. *Trends Biochem. Sci.* **1990**, *15*, 129–135.

Hagger, J. A.; Depledge, M. H.; Oehlmann, J.; Jobling, S.; Galloway, T. S. Is There a Causal Association Between Genotoxicity and the Imposex Effect? *Environ. Health Perspect.* **2006**, *114*, 20–26.

Hahn, U. K.; Bender, R. C.; Bayne, C. J. Production of Reactive Oxygen Species by Hemocytes of *Biomphalaria glabrata*: Carbohydrate-specific Stimulation. *Dev. Comp. Immunol.* **2000**, *24*, 531–541.

Hahn, U. K.; Bender, R. C.; Bayne, C. J. Killing of *Schistosoma mansoni* Sporocysts by Hemocytes from Resistant *Biomphalaria glabrata*: Role of Reactive Oxygen Species. *J. Parasitol.* **2001**, *87*, 292–299.

Hanelt, B.; Lun, C. M.; Adema, C. M. Comparative ORESTES-sampling of Transcriptomes of Immune-challenged *Biomphalaria glabrata* Snails. *J. Invertebr. Pathol.* **2008**, *99*, 192–203.

Hanington, P. C.; Forys, M. A.; Loker, E. S. A Somatically Diversified Defense Factor, FREP3, Is a Determinant of Snail Resistance to Schistosome Infection. *PLoS Negl. Trop. Dis.* **2012**, *6*, e1591.

Hanington, P. C.; Lun, C. M.; Adema, C. M.; Loker, E. S. Time Series Analysis of the Transcriptional Responses of *Biomphalaria glabrata* Throughout the Course of Intramolluscan Development of *Schistosoma mansoni* and *Echinostoma paraensei*. *Int. J. Parasitol.* **2010a**, *40*, 819–831.

Hanington, P. C.; Forys, M. A.; Dragoo, J. W.; Zhang, S. M.; Adema, C. M.; Loker, E. S. Role for a Somatically Diversified Lectin in Resistance of an Invertebrate to Parasite Infection. *Proc. Natl. Acad. Sci. U.S.A.* **2010b**, *107*, 21087–21092.

Harman, D. Aging: A Theory Based on Free Radical and Radiation Chemistry. *J. Gerontol.* **1956**, *11*, 298–300.

Hartl, M. G.; Coughlan, B. M.; Sheehan, D.; Mothersill, C.; van Pelt, F. N.; O'Reilly, S. J.; Heffron, J. J., O'Halloran, J.; O'Brien, N. M. Implications of Seasonal Priming and Reproductive Activity on the Interpretation of Comet Assay Data Derived from the Clam, *Tapes semidecussatus* Reeves 1864, Exposed to Contaminated Sediments. *Mar. Environ. Res.* **2004**, *57*, 295–310.

Hathaway, J. J.; Adema, C. M.; Stout, B. A.; Mobarak, C. D.; Loker, E. S. Identification of Protein Components of Egg Masses Indicates Parental Investment in Immunoprotection of

Offspring by *Biomphalaria glabrata* (Gastropoda, Mollusca). *Dev. Comp. Immunol.* **2010,** *34,* 425–435.

Hawlena, D.; Kress, H.; Dufresne, E. R.; Schmitz, O. J. Grasshoppers Alter Jumping Biomechanics to Enhance Escape Performance Under Chronic Risk of Spider Predation. *Funct. Ecol.* **2011,** *25,* 279–288.

Hayflick, L. Entropy Explains Aging, Genetic Determinism Explains Longevity, and Undefined Terminology Explains Misunderstanding Both. *PLoS Genet.* **2007,** *3,* e220.

Heintz, R. A.; Rice, S. D.; Wertheimer, A. C.; Bradshaw, R. F.; Thrower, F. P.; Joyce, J. E.; Short, J. W. Delayed Effects on Growth and Marine Survival of Pink Salmon *Oncorhynchus gorbuscha* after Exposure to Crude Oil During Embryonic Development. *Mar. Ecol. Prog. Ser.* **2000,** *208,* 205–216.

Hellio, C.; Bado-Nilles, A.; Gagnaire, B.; Renault, T.; Thomas-Guyon, H. Demonstration of a True Phenoloxidase Activity and Activation of a ProPO Cascade in Pacific Oyster, *Crassostrea gigas* (Thunberg) In Vitro. *Fish Shellfish Immunol.* **2007,** *22,* 433–440.

Hemminga M. A.; Maaskant J. J.; Jager J. C.; Joose J.. Glycogen Metabolism in Isolated Glycogen Cells of the Freshwater Snail *Lymnaea stagnalis*. *Comp. Biochem. Physiol.* **1985,** *82A,* 239–246.

Hermann, P. M.; Lee, A.; Hulliger, S.; Minvielle, M.; Ma, B.; Wildering, W. C. Impairment of Long-term Associative Memory in Aging Snails (*Lymnaea stagnalis*). *Behav. Neurosci.* **2007,** *121,* 1400–1414.

Hermes-Lima, M.; Storey, K. B. Antioxidant Defenses and Metabolic Depression in a Pulmonate Land Snail. *Am. J. Physiol.* **1995,** *268,* R1386–R1393.

Hermes-Lima, M.; Zenteno-Savín, T. Animal Response to Drastic Changes in Oxygen Availability and Physiological Oxidative Stress. *Comp. Biochem. Physiol. C Toxicol. Pharmacol.* **2002,** *133,* 537–556.

Hine, P. M. The Inter-relationships of Bivalve Haemocytes. *Fish Shellfish Immunol.* **1999,** *9,* 367–385.

Hochachka, P. W.; Somero, G. N. Limiting Oxygen Availability. In *Biochemical Adaptation*; Hochachka, P. W., Somero, G. N., Eds.; Princeton University Press: Princeton, NJ, 1984; pp 145–181.

Hoffmann, G. E.; Todgham, A. E. Living in the Now: Physiological Mechanisms to Tolerate a Rapidly Changing Environment. *Annu. Rev. Physiol.* **2010,** *72,* 127–145.

Holliday, R. Aging is No Longer an Unsolved Problem in Biology. *Ann. N.Y. Acad. Sci.* **2006,** *1067,* 1–9.

Holmstrup, M.; Bindesbøl, A.-M.; Oostingh, G. J.; Duschl, A.; Scheil, V.; Köhler, H.-R.; Loureiro, S.; Soares; A. M. V. M.; Ferreira, A. L. G.; Kienle, C.; Gerhardt, A.; Laskowski, R.; Kramarz, P. E.; Bayley, M.; Svendsen, C.; Spurgeon, D. J. Interactions Between Effects of Environmental Chemicals and Natural Stressors: A Review. *Sci. Total Environ.* **2010,** *408,* 3746–3762.

Hong, X.; Sun, X.; Zheng, M.; Qu, L.; Zan, J.; Zhang J. Characterization of Defensin Gene from Abalone *Haliotis discus hannai* and Its Deduced Protein. *Chin. J. Oceanol. Limnol.* **2008,** *26,* 375–379.

Hooper, C.; Day, R.; Slocombe, R.; Handlinger, J.; Benkendorff, K. Stress and Immune Responses in Abalone: Limitations in Current Knowledge and Investigative Methods Based on Other Models. *Fish Shellfish Immunol.* **2007,** *22,* 363–379.

Hooper, C.; Day, R.; Slocombe, R.; Benkendorff, K.; Handlinger, J. Effect of Movement Stress on Immune Function in Farmed Australian Abalone (Hybrid *Haliotis laevigata* and *Haliotis rubra*) *Aquaculture* **2011**, *315*, 348–354.

Horowitz, M. Heat Acclimation and Cross-tolerance against Novel Stressors: Genomic-physiological Linkage. *Prog. Brain Res.* **2007**, *162*, 373–392.

Hou, C.; Zuo, W.; Moses, M. E.; Woodruff, W. H.; Brown, J. H.; West, G. B. Energy Uptake and Allocation During Ontogeny. *Science* **2008**, *322*, 736–739.

Huggett, R. J.; Kimerle, R. A.; Mehrle, P. M., Jr; Bergman, H. L. *Biomarkers: Biochemical, Physiological, and Histological Markers of Anthropogenic Stress*. Lewis: Boca Raton, FL, 1992.

Humphries, J. E.; Yoshino, T. P. Cellular Receptors and Signal Transduction in Molluscan Hemocytes: Connections with the Innate Immune System of Vertebrates. *Integr. Comp. Biol.* **2003**, *43*, 305–312.

Hyman, L. H. *The Invertebrates, Vol. VI, Mollusca I*. McGraw-Hill: New York, 1967.

Iakovleva, N. V; Gorbushin, A. M.; Zelck, U. E. Partial Characterization of Mitogen-activated Protein Kinases (MAPK) from Haemocytes of the Common Periwinkle, *Littorina littorea* (Gastropoda: Prosobranchia). *Fish Shellfish Immunol.* **2006**, *20*, 665–668.

Ianistcki, M.; Dallarosa, J.; Sauer, C.; Teixeira, C. E.; Da Silva, J. Genotoxic Effect of Polyclic Aromatic Hydrocarbons in the Metropolitean Area of Porto Alegre, Brazil, Evaluated by *H. aspersa* (Müller, 1774). *Environ. Pollut.* **2009**, *157*, 2037–2042.

Iijima, R.; Kisugi, J.; Yamazaki, M. A Novel Antimicrobial Peptide from the Sea Hare *Dolabella auricularia*. *Dev. Comp. Immunol.* **2003**, *27*, 305–311.

Isaksson C. Pollution and Its Impact on Wild Animals: A Meta-analysis on Oxidative Stress. *Ecohealth* **2010**, *7*, 342–350.

Ismert, M.; Oster, T.; Bagrel, D. Effects of Atmospheric Exposure to Naphthalene on Xenobiotic-metabolising Enzymes in the Snail *Helix aspersa*. *Chemosphere* **2002**, *46*, 273–280.

Ito, Y.; Yoshikawa, A.; Hotani, T.; Fukuda, S.; Sugimura, K.; Imoto, T. Amino Acid Sequences of Lysozymes Newly Purified from Invertebrates Imply Wide Distribution of a Novel Class in the Lysozyme Family. *Eur. J. Biochem.* **1999**, *259*, 456–461.

Ittiprasert, W.; Knight, M. Reversing the Resistance Phenotype of the *Biomphalaria glabrata* Snail Host *Schistosoma mansoni* Infection by Temperature Modulation. *PLoS Pathog.* **2012**, *8*, e1002677.

Ittiprasert, W.; Miller, A.; Myers, J.; Nene, V.; El-Sayed, N. M.; Knight, M. Identification of Immediate Response Genes Dominantly Expressed in Juvenile Resistant and Susceptible *Biomphalaria glabrata* Snails Upon Exposure to *Schistosoma mansoni*. *Mol. Biochem. Parasitol.* **2010**, *169*, 27–39.

Ittiprasert, W.; Miller, A.; Su, X. Z.; Mu, J.; Bhusudsawang, G.; Ukoskit, K.; Knight, M. Identification and Characterisation of Functional Expressed Sequence Tags-derived Simple Sequence Repeat (eSSR) Markers for Genetic Linkage Mapping of *Schistosoma mansoni* Juvenile Resistance and Susceptibility Loci in *Biomphalaria glabrata*. *Int. J. Parasitol.* **2013**, *43*, 669–677.

Ituarte, S.; Dreon, M. S.; Ceolin, M.; Heras, H. Agglutinating Activity and Structural Characterization of Scalarin, the Major Egg Protein of the Snail *Pomacea scalaris* (d'Orbigny, 1832). *PLoS ONE* **2012**, *7*, e50115.

Itziou, A.; Kaloyianni, M.; Dimitriadis, V. K. In Vivo and In Vitro Effects of Metals in Reactive Oxygen Species Production, Protein Carbonylation, and DNA Damage in Land Snails *Eobania vermiculata*. *Arch. Environ. Contam. Toxicol.* **2011**, *60*, 697–707.

Jablonka, E.; Raz, G. Transgenerational Epigenetic Inheritance: Prevalence, Mechanisms, and Implications for the Study of Heredity and Evolution. *Q. Rev. Biol.* **2009,** *84,* 131–176.

Janeway, Jr., C. A.; Medzhitov, R. Innate Immune Recognition. *Sci. Signal.* **2002,** *20,* 197–216.

Jeno, K.; Brokordt K. Nutritional Status Affects the Capacity of the Snail *Concholepas concholepas* to Synthesize Hsp70 When Exposed to Stressors Associated with Tidal Regimes in the Intertidal Zone. *Mar. Biol.* **2014,** *161,* 1039–1049.

Jeong, K. H.; Lie, K. J.; Heyneman, D. The Ultrastructure of the Amoebocyte-producing Organ in *Biomphalaria glabrata. Dev. Comp. Immunol.* **1983,** *7,* 217–228.

Jeukens, J.; Renaut, S.; St-Cyr, J.; Nolte, A. W.; Bernatchez L. The Transcriptomics of Sympatric Dwarf and Normal Lake Whitefish (*Coregonus clupeaformis* spp., Salmonidae) Divergence as Revealed by Next-generation Sequencing. *Mol. Ecol.* **2010,** *19,* 5389–5403.

Jiang, Y.; Wu, X. Characterization of a Rel/NF-kappa B Homologue in a Gastropod Abalone, *Haliotis diversicolor supertexta. Dev. Comp. Immunol.* **2007,** *31,* 121–131.

Jimbo, M.; Nakanishi, F.; Sakai, R.; Muramoto, K.; Kamiya, H. Characterization of L-Amino Acid Oxidase and Antimicrobial Activity of Aplysianin A, a Sea Hare-derived Antitumor– Antimicrobial Protein. *Fish. Sci.* **2003,** *69,* 1240–1246.

Johnston, L. A.; Yoshino, T. P. Larval *Schistosoma mansoni* Excretory–secretory Glyco-proteins (ESPs) Bind to Hemocytes of *Biomphalaria glabrata* (Gastropoda) via Surface Carbohydrate Binding Receptors. *J. Parasitol.* **2001,** *87,* 786–793.

Kagias, K.; Nehammer, C.; Pocock R. Neuronal Responses to Physiological Stress. *Front. Genet.* **2012,** *3,* 222.

Kamiya, H.; Muramoto, K.; Yamazaki, M. Aplysianin-A, an Antibacterial and Antineoplastic Glycoprotein in the Albumen Gland of a Sea Hare, *Aplysia kurodai. Experientia* **1986,** *42,* 1065–1067.

Kanehisa, M.; Goto, S. KEGG: Kyoto Encyclopedia of Genes and Genomes. *Nucleic Acids Res.* **2000,** *28,* 27–30.

Kang, K. W.; Lee, S. J.; Kim, S. G. Molecular Mechanism of nrf2 Activation by Oxidative Stress. *Antioxid. Redox Signal.* **2005,** *7,* 1664–1673.

Kanzawa, N.; Shintani. S.; Ohta K.; Kitajima, S.; Ehara, T.; Kobayashi, H.; Kizaki, H.; Tsuchiya T. Achacin Induces Cell Death in HeLa Cells through Two Different Mechanisms. *Arch. Biochem. Biophys.* **2004,** *422,* 103–109.

Kassahn, K.; Crozier, R. H.; Pörtner, H. O.; Caley, M. J. Animal Performance and Stress: Responses and Tolerance Limits at Different Levels of Biological Organisation. *Biol. Rev.* **2009,** *84,* 277–292.

Kawsar, S. M.; Matsumoto, R.; Fujii, Y.; Yasumitsu, H.; Dogasaki, C.; Hosono, M.; Nitta, K.; Hamako, J.; Matsui, T.; Kojima, N. Purification and Biochemical Characterization of a D-Galactose Binding Lectin from Japanese Sea Hare (*Aplysia kurodai*) Eggs. *Biochemistry (Moscow)* **2009,** *74,* 709–716.

Kim, H. J.; Hwang, N. R.; Lee, K. J. Heat Shock Responses for Understanding Diseases of Protein Denaturation. *Mol. Cells.* **2007,** *23,* 123–131.

Kirkwood, T. B. Understanding Ageing from an Evolutionary Perspective. *J. Intern. Med.* **2008,** *263,* 117–127.

Kirkwood, T. B.; Melov, S. On the Programmed/Non-programmed Nature of Ageing within the Life History. *Curr. Biol.* **2011,** *21,* R701–707.

Kiss, T. Apoptosis and its Functional Significance in Molluscs. *Apoptosis* **2010,** *15,* 313–321.

Kisugi, J.; Ohye, H.; Kamiya, H.; Yamazaki, M. Biopolymers from Marine Invertebrates. X. Mode of Action of an Antibacterial Glycoprotein, Aplysianin E, from Eggs of a Sea Hare, *Aplysia kurodai*. *Chem. Pharm. Bull.* **1989**, *37*, 3050–3053.

Klerks, P. L.; Xie, L.; Levinton, J. S. Quantitative Genetics Approaches to Study Evolutionary Processes in Ecotoxicology: A Perspective from Research on the Evolution of Resistance. *Ecotoxicology* **2011**, *20*, 513–523.

Kluytmans, J. H. F. M.; Veenhof, P. R.; de Zwaan, A. Anaerobic Production of Volatile Fatty Acids in the Sea Mussel *Mytilus edulis* (L.). *J. Comp. Physiol.* **1975**, *104*, 71–78.

Knight, M.; Miller, A. N.; Patterson, C. N.; Rowe, C. G.; Michaels, G.; Carr, D.; Richards, C. S.; Lewis, F. A. The Identification of Markers Segregating with Resistance to *Schistosoma mansoni* Infection in the Snail *Biomphalaria glabrata*. *Proc. Natl. Acad. Sci. U.S.A.* **1999**, *96*, 1510–1515.

Ko, K. C.; Wang, B.; Tai, P. C.; Derby, C. D. Identification of Potent Bactericidal Compounds Produced by Escapin, an L-Amino Acid Oxidase in the Ink of the Sea Hare *Aplysia californica*. *Antimicrob. Agents Chemother.* **2008**, *52*, 4455–4462.

Köhler, H. R.; Rahman, B.; Gräff, S.; Berkus, M.; Triebskorn, R. Expression of the Stress-70 Protein Family (HSP70) Due to Heavy Metal Contamination in the Slug, *Derocera reticulatum*: A Approach to Monitor Sublethal Stress Conditions. *Chemosphere* **1996**, *33*, 1327–1340.

Kooijman, S. A. L. M. *Dynamic Energy Budget Theory for Metabolic Organisation*, 3rd ed.; Cambridge University Press: Cambridge, 2010.

Krasity, B. C.; Troll, J. V.; Weiss, J. P.; McFall-Ngai, M. J. LBP/BPI Proteins and their Relatives: Conservation over Evolution and Roles in Mutualism. *Biochem. Soc. Trans.* **2011**, *39*, 1039–1044.

Kress, A. Untersuchungen zur Histologie, Autotomie und Regenation dreir Dotoarten *Doto coronata*, *Doto pinnatifida*, *Doto fragilis* (Gastropoda, Opisthobranchiata). *Rev. Suisse Zool.* **1968**, *75*, 235–303.

Kristoff, G.; Verrengia Guerrero, N. R.; Cochón, A. C. Effects of Azinphos-methyl Exposure on Enzymatic and Non-enzymatic Antioxidant Defenses in *Biomphalaria glabrata* and *Lumbriculus variegatus*. *Chemosphere* **2008**, *72*, 1333–1339.

Kruh, G. D.; Belinsky, M. G. The MRP Family of Drug Efflux Pumps. *Oncogene* **2003**, *22*, 7537–7552.

Krupa, P. L.; Lewis, L. M.; Vecchio, P. D. *Schistosoma haematobium* in *Bulinus guerneri*: Electron Microscopy of Hemocyte-sporocyst Interactions. *J. Inv. Pathol.* **1977**, *30*, 35–45.

Kültz, D. Evolution of the Cellular Stress Proteome: From Monophyletic Origin to Ubiquitous Function. *J. Exp. Biol.* **2003**, *206*, 3119–3124.

Kumari, P. R. Detergent induced protein alterations in freshwater gastropod *Bellamya bengalensis* (Lamarck). *Indian J. Sci. Res.* **2013**, *4*, 57–60.

Kurelec, B.; Lucic, D.; Pivcevic, B.; Krca, S. Induction and Reversion of Multixenobiotic Resistance in the Marine Snail *Monodonta turbinata*. *Mar. Biol.* **1995**, *123*, 305–312.

Kyriakis, J. M.; Avruch, J. Mammalian Mitogen-activated Protein Kinase Signal Transduction Pathways Activated by Stress and Inflammation. *Physiol. Rev.* **2001**, *81*, 807–869.

Lacchini, A. H.; Davies, A. J.; Mackintosh, D.; Walker, A. J. Beta-1, 3-Glucan Modulates PKC Signalling in *Lymnaea stagnalis* Defence Cells: A Role for PKC in H_2O_2 Production and Downstream ERK Activation. *J. Exp. Biol.* **2006**, *209*, 4829–4840.

Lacoste, A.; Jalabert, F.; Malham, S. K.; Cueff, A.; Poulet, S. A. Stress and Stress Induced Neuroendocrine Changes Increase the Susceptibility of Juvenile Oysters (*Crassostrea gigas*) to *Vibrio splendidus*. *Appl. Environ. Microbiol.* **2001**, *67*, 2304–2309.

Lacoste, A.; Malham, S. K.; Gélébart, F.; Cueff, A.; Poulet, S. A. Stress-induced Immune Changes in the Oyster *Crassostrea gigas*. *Dev. Comp. Immunol.* **2002**, *26*, 1–9.

Ladhar-Chaabouni, R.; Machreki-Ajmi, M.; Serpentini, A.; Lebel, J. M.; Hamza-Chaffai, A. Does a Short-term Exposure to Cadmium Chloride Affects Haemocyte Parameters of the Marine Gastropod *Haliotis tuberculata*? *Environ. Sci. Pollut. Res. Int.* **2014**. DOI: 10.1007/s11356-014-3387-5.

Lambert, A. J.; Boysen, H. M.; Buckingham, J. A.; Yang, T.; Podlutsky, A.; Austad, S. N.; Kunz, T. H.; Buffenstein, R.; Brand, M. D. Low Rates of Hydrogen Peroxide Production by Isolated Heart Mitochondria Associate with Long Maximum Lifespan in Vertebrate Homeotherms. *Aging Cell* **2007**, *6*, 607–618.

Lankau, R.; Jørgensen, P. S.; Harris, D. J.; Sih, A. Incorporating Evolutionary Principles into Environmental Management and Policy. *Evol. Appl.* **2011**, *4*, 315–325.

Larade, K.; Storey, K. B. A Profile of the Metabolic Responses to Anoxia in Marine Invertebrates at Cell and Molecular Responses to Stress. In *Sensing, Signalling and Cell Adaptation*; Storey, K. B., Storey, J. M., Eds.; Elsevier Press: Amsterdam, **2002**; pp 27–46.

Larade, K.; Storey, K. B. Analysis of Signal Transduction Pathways During Anoxia Exposure in a Marine Snail: A Role for p38 MAP Kinase and Downstream Signaling Cascades. *Comp. Biochem. Physiol., B: Biochem. Mol. Biol.* **2006**, *143*, 85–91.

Latire, T.; Le Pabic, C.; Mottin, E.; Mottier, A.; Costil, K.; Koueta, N.; Lebel, J. M.; Serpentini, A. Responses of Primary Cultured Haemocytes from the Marine Gastropod *Haliotis tuberculata* Under 10-Day Exposure to Cadmium Chloride. *Aquat. Toxicol.* **2012**, *109*, 213–221.

Lee, F. O.; Cheng, T. C. *Schistosoma mansoni* Infection in *Biomphalaria glabrata*: Alterations in Heart Rate and Thermal Tolerance in the Host. *J. Invertebr. Pathol.* **1971**, *18*, 412–418.

Lee, R. F.; Keeran, W. S.; Pickwell, G. V. Marine Invertebrate Glutathione *S*-transferase: Purification, Characterization and Induction. *Mar. Environ. Res.* **1988**, *24*, 97–100.

Lefcort, H.; Wehner, E. A.; Cocco, P. L. Pre-exposure to Heavy Metal Pollution and the Odor of Predation Decrease the Ability of Snails to Avoid Stressors. *Arch. Environ. Contam. Toxicol.* **2013**, *64*, 273–280.

Leicht, K.; Jokela, J.; Seppälä, O. An Experimental Heat Wave Changes Immune Defense and Life History Traits in a Freshwater Snail. *Ecol. Evol.* **2013**, *3*, 4861–4871.

Leung, A. K. L.; Sharp, P. A. MicroRNA Functions in Stress Responses. *Mol. Cell* **2010**, *40*, 205–215.

Lewis, C.; Guitart, C.; Pook, C.; Scarlett, A.; Readman, J. W.; Galloway, T. S. Integrated Assessment of Oil Pollution Using Biological Monitoring and Chemical Fingerprinting. *Environ. Toxicol. Chem.* **2010**, *29*, 1358–1366.

Lewis, F. A.; Patterson, C. N.; Knight, M.; Richards, C. S. The Relationship between *Schistosoma mansoni* and *Biomphalaria glabrata*: Genetic and Molecular Approaches. *Parasitology* **2001**, *123 (Suppl.)*, S169–179.

Li, A.; Chiu, J. M. Y. Latent Effects of Hypoxia on the Gastropod *Crepidula onyx*. *Mar. Ecol. Progr. Ser.* **2013**, *480*, 145–154.

Li, H.; Parisi, M.-G.; Parrinello, N.; Cammarata, M.; Roch, P. Molluscan Antimicrobial Peptides, A Review from Activity-based Evidences to Computer-assisted Sequences. *ISJ* **2011**, *8*, 85–97.

Li, X. B.; Hou, X. L.; Mao, Q.; Zhao; Y. L.; Cheng, Y. X.; Wang, Q. Toxic Effects of Copper on Antioxidative and Metabolic Enzymes of the Marine Gastropod, *Onchidium struma*. *Arch. Environ. Contam. Toxicol.* **2009**, *56*, 776–784.

Lie, K. J.; Heyneman, D. Studies on Resistance in Snails: Interference by Nonirradiated Echinostome Larvae with Natural Resistance to *Schistosoma mansoni* in *Biomphalaria glabrata*. *J. Invertebr. Pathol.* **1977**, *29*, 118–125.

Lie, K. J.; Heyneman, D.; Yau, P. The Origin of Ameobocytes in *Biomphalaria glabrata*. *J. Parasitol.* **1975**, *63*, 574–576.

Lim, C.-S.; Lee, J.-C.; Kim, S. D.; Chang, D.-J.; Kaang, B.-K. Hydrogen Peroxide-induced Cell Death in Cultured *Aplysia* Sensory Neurons. *Brain Res.* **2002**, *941*, 137–145.

Lima, D.; Reis-Henriques, M. A.; Silva, R.; Santos, A. I.; Castro, L. F.; Santos, M. M. Tributyltin-induced Imposex in Marine Gastropods Involves Tissue-specific Modulation of the Retinoid X Receptor. *Aquat. Toxicol.* **2011**, *101*, 221–227.

Limón-Pacheco, J.; Gonsebatt, M. E. The Role of Antioxidants and Antioxidant-related Enzymes in Protective Responses to Environmentally Induced Oxidative Stress. *Mutat. Res.* **2009**, *674*, 137–147.

Liu, C. C.; Shin, P. K. S.; Cheung, S. G. Comparisons of the Metabolic Responses of Two Subtidal Nassariid Gastropods to Hypoxia and Re-oxygenation. *Mar. Pollut. Bull.* **2014**, *82*, 109–116.

Livingstone, D. R. Invertebrate and Vertebrate Pathways of Anaerobic Metabolism: Evolutionary Considerations. *J. Geol. Soc. Lond.* **1983**, *140*, 27–37.

Livingstone, D. R.; de Zwaan, A. Carbohydrate Metabolism of Gastropods. In *The Mollusca, Metabolic Biochemistry and Molecular Biomechanics, Vol. 1*; Hochachka, P.W., Ed.; Academic Press: New York, 1983; pp 177–242.

Lockyer, A. E.; Spinks, J. N.; Walker, A. J.; Kane, R. A.; Noble, L. R.; Rollinson, D.; Dias-Neto E.; Jones, C. S. *Biomphalaria glabrata* Transcriptome: Identification of Cell-signalling, Transcriptional Control and Immune-related Genes from Open Reading Frame Expressed Sequence Tags (ORESTES). *Dev. Comp. Immunol.* **2007**, *31*, 763–782.

Lockyer, A. E.; Spinks, J.; Kane, R. A.; Hoffmann, K. F.; Fitzpatrick, J. M.; Rollinson, D.; Noble, L. R.; Jones, C. S. *Biomphalaria glabrata* Transcriptome: cDNA Microarray Profiling Identifies Resistant- and Susceptible-specific Gene Expression in Haemocytes from Snail Strains Exposed to *Schistosoma mansoni*. *BMC Genomics* **2008**, *9*, 634.

Lockyer, A. E.; Emery, A. M.; Kane, R. A.; Walker, A. J.; Mayer, C. D.; Mitta, G.; Coustau, C.; Hanelt, B.; Rollinson, D.; Noble, L. R.; Jones, C. S. Early Differential Gene Expression in Haemocytes from Resistant and Susceptible *Biomphalaria glabrata* Strains in Response to *Schistosoma mansoni*. *PLoS ONE* **2013**, *7*, e51102.

Loker, E. S. Gastropod Immunobiology. In *Madame Curie Bioscience Database* [Internet]. Landes Bioscience: Austin, TX **2010**. Available from http://www.ncbi.nlm.nih.gov/books/NBK45994/.

Loker, E. S.; Hertel, L. A. Alterations in *Biomphalaria glabrata* Plasma Induced by Infection with the Digenetic Trematode *Echinostoma paraensei*. *J. Parasitol.* **1987**, *73*, 503–513.

Loker, E. S.; Bayne, C. J.; Yui, M. A. *Echinostoma paraensei*: Hemocytes of *Biomphalaria glabrata* as Targets of Echinostome Mediated Interference with Host Snail Resistance to *Schistosoma mansoni*. *Exp. Parasitol.* **1986**, *62*, 149–154.

Loker, E. S.; Adema, C. M.; Zhang, S. M.; Kepler, T. B. Invertebrate Immune Systems–Not Homogeneous, Not Simple, Not Well Understood. *Immunol. Rev.* **2004**, *198*, 10–24.

Lovell, P. J.; Kabotyanski, E. A.; Sadreyev, R. I.; Boudko, D. Y.; Byrne, J. H.; Moroz, L. L. Nitric Oxide Activates Buccal Motor Programs in *Aplysia californica*. *Soc. Neurosci. Abstr.* **2000**, *26*, 918.

Löw, P. The Role of Ubiquitin–Proteasome System in Ageing. *Gen. Comp. Endocrinol.* **2011**, *172*, 39–43.

Lushchak, V. I. Environmentally Induced Oxidative Stress in Aquatic Animals. *Aquat. Toxicol.* **2011**, *101*, 13–30.

MacDonald, J. A.; Storey, K. B. Identification of a 115 kDa MAP-kinase Activated by Freezing and Anoxic Stresses in the Marine Periwinkle, *Littorina littorea*. *Arch. Biochem. Biophys.* **2006**, *450*, 208–214.

Mahendru, V. K.; Agarwal, R. A. Changes in Carbohydrate Metabolism in Various Organs of the Snail, *Lymnaea acuminata* Following Exposure to Trichlorfon. *Acta Pharmacol.* **1981**, *48*, 377–381.

Mahilini, H. M.; Rajendran, A. Categorization of Hemocytes of Three Gastropod Species *Trachea vittata* (Müller), *Pila globosa* (Swainson) and *Indoplanorbis exustus* (Dehays). *J. Invert. Pathol.* **2008**, *97*, 20–26.

Malham, S. K.; Lacoste, A.; Gelebart, F.; Cueff, A.; Poulet, S. A. Evidence for a Direct Link Between Stress and Immunity in the Mollusc *Haliotis tuberculata*. *J. Exp. Zool. A Comp. Exp. Biol.* **2003**, *295A*, 136–144.

Mansouri, M.; Bendali-Saoudi, F.; Benhamed, D.; Soltani, N. Effect of *Bacillus thuringiensis* var *israelensis* Against *Culex pipiens* (Insecta: Culicidae). Effect of *Bti* on Two Non-target Species *Eylais hamata* (Acari: Hydrachnidia) and *Physa marmorata* (Gastropoda: Physidae) and Dosage of Their GST Biomarker. *Ann. Biol. Res.* **2013**, *4*, 85–92.

Markov, G. V.; Tavares, R.; Dauphin-Villemant, C.; Demeneix, B. A.; Baker, M. E.; Laudet, V. Independent Elaboration of Steroid Hormone Signaling Pathways in Metazoans. *Proc. Natl. Acad. Sci. U.S.A.* **2009**, *106*, 11913–11918.

Marshall, D. J.; Pechenik, J. A.; Keough, M. J. Larval Activity Levels and Delayed Metamorphosis Affect Post-larval Performance in the Colonial Ascidian *Diplosoma listerianum*. *Mar. Ecol. Prog. Ser.* **2003**, *246*, 153–162.

Martello, L. B.; Friedman, C. S.; Tjeerdema, R. S. Combined Effects of Pentachlorophenol and Salinity Stress on Phagocytic and Chemotactic Function in Two Species of Abalone. *Aquat. Toxicol.* **2000**, *49*, 213–225.

Martin, G. G.; Oakes, C. T.; Tousignant, H. R.; Crabtree, H.; Yamakawa, R. Structure and Function of Haemocytes in Two Marine Gastropods, *Megathura crenulata* and *Aplysia californica*. *J. Moll. Stud.* **2007**, *73*, 355–365.

Martins-Souza, R. L.; Pereira, C. A. J.; Coelho, P. M. Z.; Martins-Filho O. A.; Negrão-Corrêa D. Flow Cytometry Analysis of the Circulating Haemocytes from *Biomphalaria glabrata* and *Biomphalaria tenagophila* Following *Schistosoma mansoni* Infection. *Parasitology* **2009**, *136*, 67–76.

Mason, A. Z.; Borja, M. R. A Study of Cu Turnover in Proteins of the Visceral Complex of *Littorina littorea* by Stable Isotopic Analysis Using Coupled HPLC-ICP–MS. *Mar. Environ. Res.* **2002**, *54*, 351–355.

Matricon-Gondran, M.; Letocart, M. Internal Defenses of the Snail *Biomphalaria glabrata*. I. Characterization of Hemocytes and Fixed Phagocytes. *J. Invertebr. Pathol.* **1999**, *74*, 224–234.

McBryan, T. L.; Anttila, K.; Healy, T. M.; Schulte, P. M. Responses to Temperature and Hypoxia as Interacting Stressors in Fish: Implications for Adaptation to Environmental Change. *Integr. Comp. Biol.* **2013**, *53*, 648–659.

McDaniel, S. J. *Littorina littorea*: Lowered Heat Tolerance Due to *Cryptocotyle lingua*. *Exp. Parasitol.* **1969**, *25*, 13–15.

McEwen, B. S.; Stellar, E. Stress and the Individual. Mechanisms Leading to Disease. *Arch. Intern. Med.* **1993**, *153*, 2093–2101.

McEwen, B. S.; Wingfield, J. C. The Concept of Allostasis in Biology and Biomedicine. *Horm. Behav.* **2003**, *43*, 2–15.

McEwen, B. S.; Wingfield, J. C. What's in a Name? Integrating Homeostasis, Allostasis and Stress. *Horm. Behav.* **2010**, *57*, 105.

Meier, P.; Finch, A.; Evan, G. Apoptosis in development. *Nature* **2000**, *407*, 796–801.

Miller, S. E. Larval Period and Its Influence on Post-larval Life History: Comparison of Lecitotrophy and Facultative Planktotrophy in the Aeolid Nudibranch *Phestilla sibogae*. *Mar. Biol.* **1993**, *117*, 635–645.

Miller S. E.; Hadfield, M. G. Development Arrest During Larval Life and Life-span Extension in a Marine Mollusc. *Science* **1990**, *248*, 356–358.

Minchin, D.; Stroben, E.; Oehlmann, J.; Bauer, B.; Duggan, C. B.; Keatinge, M. Biological Indicators Used to Map Organotin Contamination in Cork Harbour, Ireland. *Mar. Poll. Bull.* **1996**, *32*, 188–195.

Minguez, L.; Halm-Lemeille, M. P.; Costil, K.; Bureau, R.; Lebel, J. M.; Serpentini, A. Assessment of Cytotoxic and Immunomodulatory Properties of Four Antidepressants on Primary Cultures of Abalone Hemocytes (*Haliotis tuberculata*). *Aquat. Toxicol.* **2014**, *153*, 3–11.

Mitta, G.; Adema, C. M.; Gourbal, B.; Loker, E. S.; Théron A. Compatibility Polymorphism in Snail/Schistosome Interactions: From Field to Theory to Molecular Mechanisms. *Dev. Comp. Immunol.* **2012**, *37*, 1–8.

Mitta, G.; Galinier, R.; Tisseyre, P.; Allienne, J. F.; Girerd-Chambaz, Y.; Guillou, F.; Bouchut, A.; Coustau, C. Gene Discovery and Expression Analysis of Immune-relevant Genes from *Biomphalaria glabrata* Hemocytes. *Dev. Comp. Immunol.* **2005**, *29*, 393–407.

Moberg, G. P. Biological Response to Stress: Implications for Animal Welfare. In *The Biology of Animal Stress*; Moberg, G. P., Mench, J. A., Eds.; CAB International: Wallingford, UK, 2000; pp 1–21.

Monaghan, P.; Metcalfe, N. B.; Torres, R. Oxidative Stress as a Mediator of Life History Trade-offs: Mechanisms, Measurements and Interpretation. *Ecol. Lett.* **2009**, *12*, 75–92.

Moné, Y.; Gourbal, B.; Duval, D.; Du Pasquier, L.; Kieffer-Jaquinod, S.; Mitta, G. A Large Repertoire of Parasite Epitopes Matched by a Large Repertoire of Host Immune Receptors in an Invertebrate Host/Parasite Model. *PLoS Negl. Trop. Dis.* **2010**, *4*, e813.

Moné, Y.; Ribou, A. C.; Cosseau, C.; Duval, D.; Theron, A.; Mitta, G.; Gourbal, B. An Example of Molecular Co-Evolution: Reactive Oxygen Species (ROS) and ROS Scavenger Levels in *Schistosoma mansoni*/*Biomphalaria glabrata* Interactions. *Int. J. Parasitol.* **2011**, *41*, 721–730.

Monteil, J. F.; Matricon-Gondran, M. Structural and Cytochemical Study of the Hemocytes in Normal and Trematode-infected *Lymnaea truncatula*. *Parasitol. Res.* **1993**, *79*, 675–682.

Moroz, L. L.; Park, J.-H.; Winlow, W. Nitric Oxide Activates Buccal Motor Patterns in *Lymnaea stagnalis*. *NeuroReport* **1993**, *4*, 643–646.

Moroz, L. L.; Norekian, T. P.; Pirtle, T. J.; Robertson, K. J.; Satterlie, R. A. Distribution of NADPH-Diaphorase Reactivity and Effects of Nitric Oxide on Feeding and Locomotory Circuitry in the Pteropod Mollusc, *Clione limacina*. *J. Comp. Neurol.* **2000**, *427*, 274–284.

Moroz, L. L.; Edwards, J. R.; Puthanveettil, S. V.; Kohn, A. B.; Ha, T.; Heyland, A.; Knudsen, B.; Sahni, A.; Yu, F.; Liu, L.; Jezzini, S.; Lovell, P.; Iannucculli, W.; Chen, M.; Nguyen,

T.; Sheng, H.; Shaw, R.; Kalachikov, S.; Panchin, Y. V; Farmerie, W.; Russo, J. J.; Ju, J.; Kandel, E.R. Neuronal Transcriptome of *Aplysia*: Neuronal Compartments and Circuitry. *Cell* **2006**, *127*, 1453–1467.

Mottin, E.; Caplat, C.; Mahaut, M. L.; Costil, K.; Barillier, D.; Lebel, J. M.; Serpentini, A. Effect of *In Vitro* Exposure to Zinc on Immunological Parameters of Haemocytes from the Marine Gastropod *Haliotis tuberculata*. *Fish Shellfish Immunol.* **2010**, *29*, 846–853.

Murthy, M.; Ram, J. L. Invertebrates as Model Organisms for Research on Aging Biology. *Invert. Reprod. Dev.* **2015**, *59*, 1–4.

Nellaiappan, K.; Kalyani, R. Mantle Phenoloxidase Activity and Its Role in Sclerotization in a Snail *Achatina fulica*. *Arch. Int. Physiol., Biochem. Biophys.* **1989**, *97*, 45–51.

Neves, R. A. F.; Figueiredo, G. M.; Valentina, J.-L.; da Silva Scarduad, P. M.; Hégarete, H. Immunological and Physiological Responses of the Periwinkle *Littorina littorea* During and after Exposure to the Toxic Dinoflagellate *Alexandrium minutum*. *Aquat. Toxicol.* **2015**, *160*, 96–105.

Nevo, E.; Lavie, B. Differential Viability of Allelic Isozymes in the Marine Gastropod *Cerithium scabridum* Exposed to the Environmental Stress of Non-ionic Detergent and Crude Oil-surfactant Mixtures. *Genetica* **1989**, *78*, 205–213.

Newcomb, J. M.; Watson III, W. H. Modulation of Swimming in the Gastropod *Melibe leonina* by Nitric Oxide. *J. Exp. Biol.* **2002**, *205*, 397–403.

Ng, T. Y. T.; Keough, M. J. Delayed Effects of Larval Exposure to Cu in the Bryozoan *Watersipora subtorquata*. *Mar. Ecol. Prog. Ser.* **2003**, *257*, 77–85.

Niki, E. Lipid Peroxidation: Physiological Levels and Dual Biological Effects. *Free Radic. Biol. Med.* **2009**, *47*, 469–484.

Noda S.; Loker, E. S. Effects of Infection with *Echinostoma paraensei* on the Circulating Haemocyte Population of the Host Snail *Biomphalaria glabrata*. *Parasitology* **1989a**, *98*, 35–41.

Noda, S.; Loker, E. S. Phagocytic Activity of Hemocytes of M-line *Biomphalaria glabrata* Snails: Effect of Exposure to the Trematode *Echinostoma paraensei*. *J. Parasitol.* **1989b**, *75*, 261–269.

Noventa, S.; Pavoni, B.; Galloway, T. S. Periwinkle (*Littorina littorea*) as a Sentinel Species: A Field Study Integrating Chemical and Biological Analyses. *Environ. Sci. Technol.* **2011**, *45*, 2634–2640.

Nowak, T. S.; Woodards, A. C.; Jung, Y.; Adema, C. M.; Loker, E. S. Identification of Transcripts Generated During the Response of Resistant *Biomphalaria glabrata* to *Schistosoma mansoni* Infection Using Suppression Subtractive Hybridization. *J. Parasitol.* **2004**, *90*, 1034–1040.

Nunez, P. E.; Adema, C.M.; de Jong-Brink, M. Modulation of the Bacterial Clearance Activity of Haemocytes from the Freshwater Mollusc, *Lymnaea stagnalis*, by the Avian Schistosome, *Trichobilharzia ocellata*. *Parasitology* **1994**, *109*, 299–310.

Oehlmann, J.; Di Benedetto, P.; Tillmann, M.; Duft, M.; Oetken, M.; Schulte-Oehlmann U. Endocrine Disruption in Prosobranch Molluscs: Evidence and Ecological Relevance. *Ecotoxicology* **2007**, *16*, 29–43.

Oehlman, J.; Schulte-Oehlman, S. Molluscs as Bioindicators. In *Bioindicators and Biomonitors*; Markert, B. A., Breure, A. M., Zechmeister, H. G., Eds.; Elsevier Science: New York, Amsterdam, 2003; pp 577–635.

Olive, P. L.; Banath, J. P.; Durand, R. E. Heterogeneity in Radiation-induced DNA Damage and Repair in Tumor and Normal Cells Measured Using the "Comet" Assay. *Radiat. Res.* **1990**, *122*, 86–94.

Onizuka, S.; Tamura, R.; Yonaha, T.; Oda N.; Kawasaki, Y.; Shirasaka, T.; Shiraishi, S.; Tsuneyoshi, I. Clinical Dose of Lidocaine Destroys the Cell Membrane and Induces Both Necrosis and Apoptosis in an Identified *Lymnaea* Neuron. *J. Anesth.* **2012**, *26*, 54–61.

Osterauer, R.; Köhler, H. R.; Triebskorn, R. Histopathological Alterations and Induction of hsp70 in Ramshorn Snail (*Marisa cornuarietis*) and Zebrafish (*Danio rerio*) Embryos after Exposure to PtCl(2). *Aquat. Toxicol.* **2010**, *99*, 100–107.

Östling, O.; Johanson, K. J. Microelectrophoretic Study of Radiation-induced DNA Damages in Individual Mammalian Cells. *Biochem. Biophys. Res. Commun.* **1984**, *123*, 291–298.

Ottaviani, E. The Blood Cells of the Freshwater Snail *Planorbis corneus* (Gastropoda, Pulmonata). *Dev. Comp. Immunol.* **1983**, *7*, 209–216.

Ottaviani, E. Haemocytes of the Freshwater Snail *Viviparus ater* (Gastropoda, Prosobranchia). *J. Moll. Stud.* **1989**, *55*, 379–382.

Ottaviani, E. Immunocytochemical Study on Bacterial Elimination from the Freshwater Snail *Planorbarius corneus* (L.) (Gastropoda, Pulmonata). *Zool. Jb. Anat.* **1990**, *120*, 57–62.

Ottaviani, E.; Franceschi, C. The Neuroimmunology of Stress from Invertebrates to Man. *Prog. Neurobiol.* **1996**, *48*, 421–440.

Ottaviani, E.; Franchini, A. Ultrastructural Study of Haemocytes of the freshwater snail *Planorbarius corneus* (L.) (Gastropoda, Pulmonata). *Acta Zool.* (*Stockh.*), **1988**, *69*, 157–162.

Ottaviani, E.; Paemen, L. R.; Cadet, P.; Stefano, G. B. Evidence for Nitric Oxide Production and Utilization as a Bacteriocidal Agent by Invertebrate Immunocytes. *Eur. J. Pharmacol.—Environ. Toxicol. Pharmacol.* **1993**, *248*, 319–324.

Ottaviani, E.; Alice Accorsi, A.; Rigillo G.; Malagoli D.; Blom, J. M. C.; Tascedda, F. Epigenetic Modification in Neurons of the Mollusc *Pomacea canaliculata* after Immune Challenge. *Brain Res.* **2013**, *1537*, 18–26.

Oubella, R.; Paillard, C.; Maes, P.; Auffret, M. Changes in Hemolymph Parameters in the Manila Clam *Ruditapes philippinarum* (Mollusca, Bivalvia) Following Bacterial Challenge. *J. Invertebr. Pathol.* **1994**, *64*, 33–38.

Padmaja, J. R.; Rao, M. B. Effect of an Organochlorine and Three Organophosphate Pesticides on Glucose, Glycogen, Lipid and Protein Contents in Tissues of the Freshwater Snail, *Bellammya dissimillis* (Müller). *Bull. Environ. Contam. Toxicol.* **1994**, *53*, 142–148.

Pahkala, M.; Laurila, A.; Merila, J. Carry-over Effects of Ultraviolet-B Radiation on Larval Fitness in *Rana temporaria. Proc. R. Soc. Lond. B* **2001**, *268*, 1699–1706.

Pannunzio, T. M.; Storey, K. B. Antioxidant Defenses and Lipid Peroxidation During Anoxia Stress and Aerobic Recovery in the Marine Gastropod *Littorina littorea. J. Exp. Mar. Biol. Ecol.* **1998**, *221*, 277–292.

Partridge, L. The New Biology of Ageing. *Philos. Trans. R. Soc. Lond., B: Biol. Sci.* **2010**, *365*, 147–154.

Pascoal, S.; Carvalho, G.; Vasieva, O.; Hughes, R.; Cossins, A.; Fang, Y.; Ashelford, K.; Olohan, L.; Barroso, C.; Mendo S.; Creer, S. Transcriptomics and In Vivo Tests Reveal Novel Mechanisms Underlying Endocrine Disruption in an Ecological Sentinel, *Nucella lapillus. Mol. Ecol.* **2013**, *22*, 1589–1608.

Paulson, P. C.; Pratt, J. R.; Cairns, Jr., J. Relationship of Alkaline Stress and Acute Copper Toxicity in the Snail *Goniobasis livescens* (Menke). *Bull. Environ. Contam. Toxicol.* **1983**, *31*, 719–726.

Pechenik, J. A. Larval Experience and Latent Effects—Metamorphosis is not a New Beginning. *Integr. Comp. Biol.* **2006**, *46*, 323–333.

Pechenik, J. A.; Estrella, M. S.; Hammer, K. Food Limitation Stimulates Metamorphosis of Competent Larvae and Alters Postmetamorphic Growth Rate in the Marine Prosobranch Gastropod *Crepidula fornicata*. *Mar. Biol.* **1996a**, *127*, 267–275.

Pechenik, J. A.; Gleason, T.; Daniels, D.; Champlin, D. Influence of Larval Exposure to Salinity and Cadmium Stress on Juvenile Performance of Two Marine Invertebrates (*Capitella* sp. I and *Crepidula fornicata*). *J. Exp. Mar. Biol. Ecol.* **2001**, *264*, 101–114.

Pechenik, J. A.; Hilbish, T. J.; Eyster, L. S.; Marshall, D. Relationship Between Larval and Juvenile Growth Rates in Two Marine Gastropods *Crepidula plana* and *C. fornicata*. *Mar. Biol.* **1996b**, *125*, 119–127.

Pechenik, J. A.; Jarrett, J. N.; Rooney, J. Relationships Between Larval Nutritional Experience, Larval Growth Rates, Juvenile Growth Rates, and Juvenile Feeding Rates in the Prosobranch Gastropod *Crepidula fornicata*. *J. Exp. Mar. Biol. Ecol.* **2002**, *280*, 63–78.

Pechenik, J. A.; Wendt, D. E.; Jarrett, J. N. Metamorphosis is Not a New Beginning. *Bioscience* **1998**, *48*, 901–910.

Perez, D. G.; Fontanetti, C. S. Hemocitical Responses to Environmental Stress in Invertebrates: A Review. *Environ. Monitor. Assess.* **2011**, *177*, 437–447.

Perrin, C.; Lepesant, J. M.; Roger, E.; Duval, D.; Fneich S.; Thuillier, V.; Alliene, J. F.; Mitta, G.; Grunau, C.; Cosseau, C. *Schistosoma mansoni* Mucin Gene (SmPoMuc) Expression: Epigenetic Control to Shape Adaptation to a New Host. *PLoS Pathog.* **2013**, *9*, e1003571.

Peterson, N. A.; Hokke, C. H.; Deelder, A. M.; Yoshino, T. P. Glycotype Analysis in Miracidia and Primary Sporocysts of *Schistosoma mansoni*: Differential Expression During the Miracidium-to-sporocyst Transformation. *Int. J. Parasitol.* **2009**, *39*, 1331–1344.

Petzelt, C.; Joswig, G.; Stammer, H.; Werner, D. Cytotoxic Cyplasin of the Sea Hare, *Aplysia punctata*, cDNA Cloning, and Expression of Bioactive Recombinants in Insect Cells. *Neoplasia* **2002**, *4*, 49–59.

Phillips, N. E. Effects of Nutrition-mediated Larval Condition on Juvenile Performance in a Marine Mussel. *Ecology* **2002**, *83*, 2562–2574.

Phillips, N. E. Variable Timing of Larval Food Has Consequences for Early Juvenile Performance in a Marine Mussel. *Ecology* **2004**, *85*, 2341–2346.

Pinheiro, J.; Amato, S. B. *Eurytrema coelomaticum* (Digenea, Dicrocoeliidae): the Effect of Infection on Carbohydrate Contents of Its Intermediate Snail Host, *Bradybaena similaris* (Gastropoda, Xantonychidae). *Mem. Inst. Oswaldo Cruz* **1994**, *89*, 407–410.

Pipe, R. K.; Coles, J. A. Environmental Contaminants Influencing Immune Function in Marine Bivalve Molluscs. *Fish Shellfish Immunol.* **1995**, *5*, 581–595.

Pires, A.; Hadfield, M. G. Oxidative Breakdown Products of Catecholamines and Hydrogen Peroxide Induce Partial Metamorphosis in the Nudibranch *Phestilla sibogae* Bergh (Gastropoda: Opisthobranchia). *Biol. Bull.* **1991**, *180*, 310–317.

Plows, L. D.; Cook, R. T.; Davies, A. J.; Walker, A. J. Activation of Extracellular-signal Regulated Kinase is Required for Phagocytosis by *Lymnaea stagnalis* Haemocytes. *Biochim. Biophys. Acta* **2004**, *1692*, 25–33.

Plows, L. D.; Cook, R. T.; Davies, A. J.; Walker, A. J. Phagocytosis by *Lymnaea stagnalis* Haemocytes: A Potential Role for Phosphatidylinositol 3-Kinase But Not Protein Kinase A. *J. Invertebr. Pathol.* **2006**, *91*, 74–77.

Portela, J.; Duval, D.; Rognon, A.; Galinier, R.; Boissier, J.; Coustau, C.; Mitta, G.; Théron, A.; Gourbal, B. Evidence for Specific Genotype-dependent Immune Priming in the Lophotrochozoan *Biomphalaria glabrata* Snail. *J. Innate Immun.* **2013**, *5*, 261–276.

Pörtner, H. O. Oxygen- and Capacity-limitation of Thermal Tolerance: A Matrix for Integrating Climate-related Stressor Effects in Marine Ecosystems. *J. Exp. Biol.* **2010**, *213*, 881–893.

Pörtner, H. O. Integrating Climate-related Stressor Effects on Marine Organisms: Unifying Principles Linking Molecule to Ecosystem-level Changes. *Mar. Ecol. Progr. Ser.* **2012**, *470*, 273–290.

Poulsen, H. E.; Jensen, B. R.; Weimann, A.; Jensen, S. A.; Sorensen, M.; Loft, S. Antioxidants, DNA Damage and Gene Expression. *Free Rad. Res.* 2000, *33 (Suppl.)*, S33–S39.

Prabhakara Rao, Y.; Prasada Rao, D. G. V. End Products of Anaerobic Metabolism in *Cerithidea (Cerithideospilla) cingulata* (Gmellin 1970) and *Cerithium coralium* Kiener 1841. *Can. J. Zool.* **1982**, *61*, 1304–1310.

Prokop, O.; Köhler, W. Agglutination Reactions of Micro-organisms with *Helix pomatia* Protein Gland Extract. (Anti-Ahel-agglutination). *Z. Immunitätsforsch. Allerg. klin. Immunol.* 1967, *133*, 176–179.

Protasio, A. V.; Tsai, I. J.; Babbage, A.; Nichol, S.; Hunt, M.; Aslett, M. A.; De Silva, N.; Velarde, G. S.; Anderson, T. J.; Clark, R. C.; Davidson, C.; Dillon, G. P.; Holroyd, N. E.; LoVerde, P. T.; Lloyd, C.; McQuillan, J.; Oliveira, G.; Otto, T. D.; Parker-Manuel, S. J.; Quail, M. A.; Wilson, R. A.; Zerlotini, A.; Dunne, D. W.; Berriman, M. A Systematically Improved High Quality Genome and Transcriptome of the Human Blood Fluke *Schistosoma mansoni. PLoS Neglec. Trop. Dis.* **2012**, *6*, e1455.

Prowse, R. H.; Tait, N. N. In Vitro Phagocytosis by Amebocytes from the Haemolymph of *Helix aspersa* (Müller). *Immunology* 1969, *17*, 437–443.

Qiu, X.-B.; Shao, Y.-M.; Miao, S.; Wang, L. The Diversity of the DnaJ/Hsp40 Family, the Crucial Partners for Hsp70 Chaperones. *Cell. Mol. Life Sci.* **2006**, *63*, 2560–2570.

Radwan, M. A.; Mohamed, M. S. Imidacloprid Induced Alterations in Enzyme Activities and Energy Reserves of the Land Snail, *Helix aspersa. Ecotoxicol. Environ. Saf.* **2013**, *95*, 91–97.

Radwan, M. A.; Essawy, A. E.; Abdelmeguied, N. E.; Hamed, S. S.; Ahmed, A. E. Biochemical and Histochemical on the Digestive Gland of *Eobania vermiculata* Snails Treated with Carbamate Pesticides. *Pestic. Biochem. Physiol.* **2008**, *90*, 154–167.

Raghavan, N.; Miller, A. N.; Gardner, M.; FitzGerald, P. C.; Kerlavage, A. R.; Johnston, D. A.; Lewis, F. A.; Knight, M. Comparative Gene Analysis of *Biomphalaria glabrata* Hemocytes Pre- and Post-exposure to Miracidia of *Schistosoma mansoni. Mol. Biochem. Parasitol.* **2003**, *126*, 181–191.

Raimundo, J.; Costa, P. M.; Vale, C.; Costa, M. H.; Moura, I. DNA Damage and Metal Accumulation in Four Tissues of Feral *Octopus vulgaris* from Two Coastal Areas in Portugal. *Ecotoxicol. Environ. Saf.* **2010**, *73*, 1543–1547.

Rattan, S. I. Theories of Biological Aging: Genes, Proteins, and Free Radicals. *Free Radic. Res.* 2006, *40*, 1230–1238.

Ray, M.; Bhunia, N. S.; Bhunia, A. S.; Ray, S. A Comparative Analyses of Morphological Variations, Phagocytosis and Generation of Cytotoxic Agents in Flow Cytometrically Isolated Hemocytes of Indian Molluscs. *Fish Shellfish Immunol.* **2013**, *34*, 244–253.

Reade, P. C. Phagocytosis in Invertebrates. *Aust. J. Exp. Biol. Med. Sci.* 1968, *46*, 219–229.

Regoli, F.; Nigro, M.; Bompadre, S.; Winston G. W. Total Oxidant Scavenging Capacity (TOSC) of Microsomal and Cytosolic Fractions from Antarctic, Arctic and Mediterranean Scallops: Differentiation Between Three Potent Oxidants. *Aquat. Toxicol.* **2000,** *49,* 13–25.

Regoli, F.; Gorbi, S.; Frenzilli, G.; Nigro, M.; Corsi, I.; Focardi, S.; Winston, G. W. Oxidative Stress in Ecotoxicology: From the Analysis of Individual Antioxidants to a More Integrated Approach. *Mar. Environ. Res.* **2002,** *54,* 419–423.

Renwrantz, L. An investigation of Molecules and Cells in the Hemolymph of *Helix pomatia* with Special Reference to Immunobiologically Active Components. *Zool. Jb. Zool. Physiol.* **1979,** *83,* 283–333.

Renwrantz, L.; Schfinke, W.; Harm, H.; Erl, H.; Liebsch, H.; Gerken, J. Discriminative Ability and Function of the Immunobiological Recognition System of the Snail, *Helix pomatia. J. Comp. Physiol.* **1981,** *141,* 477–488.

Rewitz, K. F.; Styrishave, B.; Løbner-Olsen, A.; Andersen, O. Marine Invertebrate Cytochrome P450: Emerging Insights from Vertebrate and Insects Analogies. *Comp. Biochem. Physiol. C: Toxicol. Pharmacol.* **2006,** *143,* 363–381.

Rhee, J. S.; Raisuddin, S.; Hwang, D. S.; Horiguchi, T.; Cho, H. S.; Lee, J. S. A Mu-class Glutathione S-Transferase (GSTM) from the Rock Shell *Thais clavigera. Comp. Biochem. Physiol. C: Toxicol. Pharmacol.* **2008,** *148,* 195–203.

Richards, C. S.; Knight, M.; Lewis, F. A. Genetics of *Biomphalaria glabrata* and its Effect on the Outcome of *Schistosoma mansoni* Infection. *Parasitol. Today* **1992,** *8,* 171–174.

Robert, K. A.; Bronikowski, A. M. Evolution of Senescence in Nature: Physiological Evolution in Populations of Garter Snake with Divergent Life Histories. *Am. Nat.* **2010,** *175,* 147–159.

Rodríguez-Ramos, T.; Carpio, Y.; Bolívar, J.; Espinosa, G.; Hernández-López, J.; Gollas-Galván, T.; Ramos, L.; Pendón, C.; Estrada, M. P. An Inducible Nitric Oxide Synthase (NOS) is Expressed in Hemocytes of the Spiny Lobster *Panulirus argus*: Cloning, Characterization and Expression Analysis. *Fish Shellfish Immunol.* **2010,** *29,* 469–479.

Roger, E.; Gourbal, B.; Grunau, C.; Pierce, R. J.; Galinier, R.; Mitta, G. Expression Analysis of Highly Polymorphic Mucin Proteins (SmPoMuc) from the Parasite *Schistosoma mansoni. Mol. Biochem. Parasitol.* **2008a,** *157,* 217–227.

Roger, E.; Grunau, C.; Pierce, R. J.; Hirai, H.; Gourbal, B.; Galinier, R.; Emans, R.; Cesari, I. M.; Cosseau, C.; Mitta, G. Controlled Chaos of Polymorphic Mucins in a Metazoan Parasite (*Schistosoma mansoni*) Interacting with Its Invertebrate Host (*Biomphalaria glabrata*). *PLoS Negl. Trop. Dis.* **2008b,** *2,* e330.

Roger, E.; Mitta, G.; Mone, Y.; Bouchut, A.; Rognon, A.; Grunau, C.; Boissier, J.; Théron, A.; Gourbal, B. E. Molecular Determinants of Compatibility Polymorphism in the *Biomphalaria glabrata/Schistosoma mansoni* Model: New Candidates Identified by a Global Comparative Proteomics Approach. *Mol. Biochem. Parasitol.* **2008c,** *157,* 205–216.

Romero, L. M. Using the Reactive Scope Model to Understand Why Stress Physiology Predicts Survival During Starvation in Galápagos Marine Iguanas. *Gen. Comp. Endocrinol.* **2012,** *176,* 296–299.

Romero, L. M.; Dickens, M. J.; Cyr, N. E. The Reactive Scope Model—A New Model Integrating Homeostasis, Allostasis, and Stress. *Horm. Behav.* **2009,** *55,* 375–389.

Romiguier, J.; Gayral, P.; Ballenghien, M.; Bernard, A.; Cahais, V.; Chenuil, A.; Chiari, Y.; Dernat, R.; Duret, L.; Faivre, N.; Loire, E.; Lourenco, J. M.; Nabholz, B.; Roux, C.; Tsagkogeorga, G.; Weber, A. A.; Weinert, L. A.; Belkhir, K.; Bierne, N.; Glémin, S.; Galtier,

N. Comparative Population Genomics in Animals Uncovers the Determinants of Genetic Diversity. *Nature* **2014**, *515*, 261–263.

Rondelaud, D.; Barthe, D. Relationship of the Amebocyte-producing Organ with the Generalized Amoebocytic Reaction in *Lymnaea truncatula* Muller Infected by *Fasciola hepatica* L. *Int. J. Parasitol.* **1982**, *68*, 967–969.

Rowley, A. F.; Powell, A. Invertebrate Immune Systems-specific, Quasi-specific, or Nonspecific? *J. Immunol.* **2007**, *179*, 7209–7214.

Russo, J.; Lagadic, L. Effects of Parasitism and Pesticide Exposure on Characteristics and Functions of Hemocyte Populations in the Freshwater Snail *Lymnaea palustris* (Gastropoda, Pulmonata). *Cell Biol. Toxicol.* **2000**, *16*, 15–30.

Russo, J.; Lagadic, L. Effects of Environmental Concentrations of Atrazine on Hemocyte Density and Phagocytic Activity in the Pond Snail *Lymnaea stagnalis* (Gastropoda, Pulmonata). *Environ. Pollut.* **2004**, *127*, 303–311.

Russo J.; Madec, L. Haemocyte Apoptosis as a General Cellular Immune Response of the Snail, *Lymnaea stagnalis*, to a Toxicant. *Cell Tissue Res.* **2007**, *328*, 431–441.

Russo J.; Madec, L. Dual Strategy for Immune Defense in the Land Snail *Cornu aspersum* (Gastropoda, Pulmonata). *Physiol. Biochem. Zool.* **2011**, *84*, 212–221.

Russo J.; Madec, L. Linking Immune Patterns and Life History Shows Two Distinct Defense Strategies in Land Snails (Gastropoda, Pulmonata). *Physiol. Biochem. Zool.* **2013**, *86*, 193–204.

Russo, J.; Madec, L.; Brehelin, M. Effect of a Toxicant on Phagocytosis Pathways in the Freshwater Snail *Lymnaea stagnalis*. *Cell Tissue Res.* **2008**, *333*, 147–158.

Russo, J.; Lefeuvre-Orfila, L.; Lagadic, L. Hemocyte-specific Responses to the Peroxidizing Herbicide Fomesafen in the Pond Snail *Lymnaea stagnalis* (Gastropoda, Pulmonata). *Environ. Pollut.* **2007**, *146*, 420–427.

Sadamoto, H.; Takahashi, H.; Okada, T.; Kenmoku, H.; Toyota, M.; Asakawa, Y. *De novo* Sequencing and Transcriptome Analysis of the Central Nervous System of Mollusc *Lymnaea stagnalis* by Deep RNA Sequencing. *PLoS ONE* **2012**, *7*, e42546.

Sanchez, J. F.; Lescar, J.; Chazalet, V.; Audfray, A.; Gagnon, J.; Alvarez, R.; Breton, C.; Imberty, A.; Mitchell, E. P. Biochemical and Structural Analysis of *Helix pomatia* Agglutinin. A Hexameric Lectin with a Novel Fold. *J. Biol. Chem.* **2006**, *281*, 20171–20180.

Santini, G.; Bruschini, C.; Pazzagli, L.; Pieraccini, G.; Moneti, G.; Chelazzi, G. Metabolic Responses of the Limpet *Patella caerulea* (L.) to Anoxia and Dehydration. *Comp. Biochem. Physiol., A* **2001**, *130*, 1–8.

Sarkar, A.; Bhagat, J.; Ingole, B.; Markad, V.; Rao, D. P. Genotoxicity of Cadmium Chloride in Marine Gastropod *Nerita chamaeleon* using Comet Assay and Alkaline Unwinding Assay. *Environ. Toxicol.* **2015**, *30*, 177–187.

Sarkar, A.; Bhagat, J.; Sarkar, S. Evaluation of Impairment of DNA in Marine Gastropod, *Morula granulata* as a Biomarker of Marine Pollution. *Ecotoxicol. Environ. Saf.* **2014**, *106*, 253–261.

Sasaki, Y.; Furuta, E.; Kirinoki, M.; Seo, N.; Matsuda, N. Comparative Studies on the Internal Defense System of Schistosome-resistant and -susceptible Amphibious Snail *Oncomelania nosophora* 1. Comparative Morphological and Functional Studies on Hemocytes from both Snails. *Zool. Sci.* **2003**, *20*, 1215–1222.

Schadt, E. E.; Monks, S. A.; Drake, T. A.; Lusis, A. J.; Che, N.; Colinayo, V.; Ruff, T. G.; Milligan, S. B.; Lamb, J. R.; Cavet, G.; Linsley, P. S.; Mao, M.; Stoughton, R. B.; Friend, S. H. Genetics of Gene Expression Surveyed in Maize, Mouse and Man. *Nature* **2003**, *422*, 297–302.

Scheil, A. E.; Köhler, H. R.; Triebskorn, R. Heat Tolerance and Recovery in Mediterranean Land Snails after Pre-exposure in the Field. *J. Moll. Stud.* **2011**, *77*, 165–174.

Scheil, A. E.; Gärtner, U.; Köhler, H.-R. Colour Polymorphism and Thermal Capacities in *Theba pisana* (O.F. Müller 1774). *J. Therm. Biol.* **2012**, *37*, 462–467.

Scheil, A. E.; Hilsmann, S.; Triebskorn, R.; Köhler, H.-R. Shell Colour Polymorphism, Injuries and Immune Defense in Three Helicid Snail Species, *Cepaea hortensis*, *Theba pisana* and *Cornu aspersum maximum*. *Results Immunol.* **2013**, *3*, 73–78.

Schmid-Hempel P. Variation in Immune Defence as a Question of Evolutionary Ecology. *Proc. Roy. Soc. London Ser. B* **2003**, *270*, 375–466.

Schulte, P. M. What is Environmental Stress? Insights from Fish Living in a Variable Environment. *J. Exp. Biol.* **2014**, *217*, 23–34.

Schwartz, C. F.; Carter, C. E. Effect of *Schistosoma mansoni* on Glycogen Synthase and Phosphorylase from *Biomphalaria glabrata* (Mollusca). *J. Parasitol.* **1982**, *68*, 236–242.

Scott, A. P. Do Mollusks Use Vertebrate Sex Steroids as Reproductive Hormones? II. Critical Review of the Evidence that Steroids have Biological Effects. *Steroids* **2013**, *78*, 268–281.

Selye, H. Stress and the General Adaptation Syndrome. *Brit. Med. J.* **1950**, *1*, 1386–1392.

Seppälä, O.; Leicht, K. Activation of the Immune Defence of the Freshwater Snail *Lymnaea stagnalis* by Different Immune Elicitors. *J. Exp. Biol.* **2013**, *216*, 2902–2907.

Seppälä, O.; Jokela, J. Immune Defence under Extreme Ambient Temperature. *Biol. Lett.* **2010a**, *7*, 119–122.

Seppälä, O.; Jokela, J. Maintenance of Genetic Variation in Immune Defense of a Freshwater Snail: Role of Environmental Heterogeneity. *Evolution* **2010b**, *64*, 2397–2407.

Shan, P.; Pu, J.; Yuan, A.; Shen, L.; Shen, L.; Chai, D.; He, B. RXR Agonists Inhibit Oxidative Stress-Induced Apoptosis in H9c2 Rat Ventricular Cells. *Biochem. Biophys. Res. Commun.* **2008**, *375*, 628–633.

Sheehan, D.; Meade, G.; Foley, V. M.; Dowd, C. A. Structure, Function and Evolution of Glutathione Transferases: Implications for Classification of Non-mammalian Members of an Ancient Enzyme Superfamily. *Biochem. J.* **2001**, *360*, 1–16.

Shozawa, A.; Suto, C. Hemocytes of *Pomacea canaliculata*: I. Reversible Aggregation Induced by Ca^{2+}. *Dev. Comp. Immunol.* **1990**, *14*, 175–184.

Sigel, A.; Sigel, H.; Sigel, R. K. O. Eds. *Metal Ions in Lifes Sciences, Vol. 5, Metallothioneins and Related Chelators*; RSC Publishing, Cambridge, 2009.

Silva-Castiglioni, D.; Oliveira, G. T.; Buckup, L. Metabolic Responses of *Parastacus defossus* and *Parastacus brasiliensis* (Crustacea, Decapoda, Parastacidae) to Hypoxia. *Comp. Biochem. Physiol. A* **2010**, *156*, 436–444.

Sinclair, B. J.; Ferguson, L. V.; Salehipour-Shirazi, G.; MacMillan, H. A. Cross-tolerance and Cross-talk in the Cold: Relating Low Temperatures to Desiccation and Immune Stress in Insects. *Integr. Comp. Biol.* **2013**, *53*, 545–556.

Singh, N. P.; McCoy, M. T.; Tice, R. R.; Schneider, E. L. A Simple Technique for Quantitation of Low Levels of DNA Damage in Individual Cells. *Exp. Cell Res.* **1988**, *175*, 184–191.

Siwela, A. H.; Nyathi, C. B.; Naik, Y. S. A Comparison of Metal Levels and Antioxidant Enzymes in Freshwater Snails, *Lymnaea natalensis*, Exposed to Sediment and Water Collected from Wright Dam and Lower Mguza Dam, Bulawayo, Zimbabwe. *Ecotoxicol. Environ. Saf.* **2010**, *73*, 1728–1732.

Skála, V.; Černíková, A.; Jindrová, Z.; Kašný, M.; Vostrý, M.; Walker, A.J.; Horák, P. Influence of *Trichobilharzia regenti* (Digenea: Schistosomatidae) on the Defence Activity of *Radix lagotis* (Lymnaeidae) Haemocytes. *PLoS ONE* **2014**, *9*, e111696.

Sminia, T. Structure and Function of Blood and Connective Tissue Cells of the Fresh Water Pulmonate *Lymnaea stagnalis* Studied by Electron Microscopy and Enzyme Histochemistry. *Z. Zellforsch.* **1972**, *130*, 497–526.

Sminia, T. Haematopoiesis in the Freshwater Snail *Lymnaea stagnalis* Studied by Electron Microscopy and Autoradiography. *Cell Tissue Res.* **1974**, *150*, 443–454.

Sminia, T. Gastropods. In *Invertebrate Blood Cells Volume 1*; Ratcliffe, N. A., Rowley, A. F., Eds.; Academic Press: London, 1981; pp 191–232.

Sminia, T.; Barendsen, L. A Comparative Morphological and Enzyme Histochemical Study on Blood Cells of the Freshwater Snails *Lymnaea stagnalis, Biomphalaria glabrata,* and *Bulinus truncatus. J. Morphol.* **1980**, *165*, 31–39.

Sminia, T.; van der Knaap, W. P. W.; Kroese, F. G. M. Fixed Phagocytes in the Freshwater Snail *Lymnaea stagnalis. Cell Tissue Res.* **1979**, *196*, 545–548.

Sminia, T.; Van der Knaap, W.; Van Asselt, L. Blood Cell Types and Blood Cell Formation in Gastropod Molluscs. *Dev. Comp. Immunol.* **1983**, *7*, 665–668.

Smith, B. S. Sexuality of the American Mud Snail *Nassarius obsoletus* (Say). *Proc. Malac. Soc. Lond.* **1971**, *39*, 377–378.

Snyder, M. J. Cytochrome P450 Enzymes in Aquatic Invertebrates: Recent Advances and Future Directions. *Aquat. Toxicol.* **2000**, *48*, 529–547.

Sokolova, I. M. Energy-limited Tolerance to Stress as a Conceptual Framework to Integrate the Effects of Multiple Stressors. *Integr. Comp. Biol.* **2013**, *53*, 597–608.

Sokolova, I. M.; Lannig, G. Interactive Effects of Metal Pollution and Temperature on Metabolism in Aquatic Ectotherms: Implications of Global Climate Change. *Clim. Res.* **2008**, *37*, 181–201.

Sokolova, I. M.; Frederich, M.; Bagwe, R.; Lannig, G.; Sukhotin, A. A. Energy Homeostasis as an Integrative Tool for Assessing Limits of Environmental Stress Tolerance in Aquatic Invertebrates. *Mar. Environ. Res.* **2012**, *79*, 1–15.

Song, X.; Zhang, H.; Wang, L.; Zhao, J.; Mu, C.; Song, L.; Qiu, L.; Liu, X. A Galectin with Quadruple-domain from Bay Scallop *Argopecten irradians* is Involved in Innate Immune Response. *Dev. Comp. Immunol.* **2011**, *35*, 592–602.

Sousa, W. P.; Gleason, M. P. Does Parasitic Infection Compromise Host Survival Under Extreme Environmental Conditions? The Case for *Cerithidea californica* (Gastropoda: Prosobranchia). *Oecologia* **1989**, *80*, 456–464.

Souza, S. D.; Andrade, Z. A. On the Origin of the *Biomphalaria glabrata* Hemocytes. *Mem. Inst. Oswaldo Cruz* **2006**, *101*, 213–218.

Stang-Voss, C. Zur Ultrastruktur der Blutzellen wirbelloser Tiere. III. Über die Haemocyten der Schnecke *Lymnea stagnalis* L. (Pulmonata). *Z. Zellforsch.* **1970**, *107*, 141–156.

Stange, D.; Sieratowicz, A.; Oehlmann, J. Imposex Development in *Nucella lapillus*— Evidence for the Involvement of Retinoid X Receptor and Androgen Signalling Pathways In Vivo. *Aquat. Toxicol.* **2012**, *106*, 20–24.

Stapley, J.; Reger, J.; Feulner, P. G. D.; Smadja, C.; Galindo, J.; Ekblom, R.; Bennison, C.; Ball, A. D.; Beckerman, A. P.; Slate, J. Adaptation Genomics: The Next Generation. TREE **2010**, *25*, 705–712.

Stefano, G. B.; Cadet, P.; Zhu, W.; Rialas, C. M.; Mantione, K.; Benz, D.; Fuentes, R.; Casares, F.; Fricchione, G. L.; Fulop, Z.; Slingsby, B. The Blueprint for Stress Can Be Found in Invertebrates. *Neuroendocrinol. Lett.* **2002**, *23*, 85–93.

Sterling, P. Principles of Allostasis: Optimal Design, Predictive Regulation, Pathophysiology and Rational Therapeutics. In *Allostasis, Homeostasis, and the Costs of Adaptation*; Schulkin, J., Ed. MIT Press: Cambidge, MA, 2003; pp 1–24.

Sterling, P. Allostasis: A Model of Predictive Regulation. *Physiol. Behav.* **2012**, *106*, 5–15.

Sterling, P.; Eyer, J. Allostasis: A New Paradigm to Explain Arousal Pathology. In *Handbook of Life Stress, Cognition and Health*; Fisher, S.; Reason, J., Eds. John Wiley: Chichester, 1988; pp 629–649.

Sternberg, R. M.; Gooding, M. P.; Hotchkiss, A. K.; LeBlanc, G. A. Environmental Endocrine Control of Reproductive Maturation in Gastropods: Implications for the Mechanism of Tributyltin-Induced Imposex in Prosobranchs. *Ecotoxicology* **2010**, *19*, 4–23.

Stickle, W. B.; Rice, S. D.; Moles, A. Bioenergetics and Survival of the Marine Snail *Thais lima* During Long-term Oil Exposure. *Mar. Biol.* **1984**, *80*, 281–289.

Storey, K. B.; Storey, J. M. Metabolic Rate Depression in Animals: Transcriptional and Translational Controls. *Biol. Rev.* **2004**, *79*, 207–233.

Strathmann R. R.; Strathmann, M. F. Oxygen Supply and Limits on Aggregation of Embryos. *J. Mar. Biol. Assoc. U.K.* **1995**, *75*, 413–428.

Sun, J.; Zhang, H.; Wang, H.; Heras, H.; Dreon, M. S.; Ituarte, S.; Ravasi, T.; Qian, P.-Y.; Qiu, J.-W. First Proteome of the Egg Perivitelline Fluid of a Freshwater Gastropod with Aerial Oviposition. *J. Proteome Res.* **2012**, *11*, 4240–4248.

Tallmark, B.; Norrgren, G. The Influence of Parasitic Trematodes on the Ecology of *Nassarius reticulatus* (L) in Gullmar Fjord (Sweden). *Zoon* **1976**, *4*, 149–154.

Teyke, T. Nitric Oxide, But Not Serotonin, is Involved in Acquisition of Food-attraction Conditioning in the Snail *Helix pomatia*. *Neurosci. Lett.* **1996**, *206*, 29–32.

Tielens, A. G.; Horemans, A. M.; Dunnewijk, R.; van der Meer, P.; van den Bergh, S. G. The Facultative Anaerobic Energy Metabolism of *Schistosoma mansoni* Sporocysts. *Mol. Biochem. Parasitol.* **1992**, *56*, 49–57.

Todgham, A. E.; Stilman, J. H. Physiological Responses to Shifts in Multiple Environmental Stressors: Relevance in a Changing World. *Integr. Comp. Biol.* **2013**, *53*, 539–544.

Travers, M. A.; da Silva, P. M.; Le Goïc, N.; Marie, D.; Donval, A.; Huchette, S.; Koken, M.; Paillard, C. Morphologic, Cytometric and Functional Characterisation of Abalone (*Haliotis tuberculata*) Haemocytes. *Fish Shellfish Immunol.* **2008a**, *24*, 400–411.

Travers, M. A.; Le Goic, N.; Huchette, S.; Koken, M.; Paillard, C. Summer Immune Depression Associated with Increased Susceptibility of the European Abalone, *Haliotis tuberculata* to *Vibrio harveyi* Infection. *Fish Shellfish Immunol.* **2008b**, *25*, 800–808.

Travers, M. A.; Le Bouffant, R.; Friedman, C. S.; Buzin, F.; Cougard, B.; Huchette, S.; Koken, M.; Paillard, C. Pathogenic *Vibrio harveyi*, In Contrast to Non-pathogenic Strains, Intervenes with the p38 MAPK Pathway to Avoid an Abalone Haemocyte Immune Response. *J. Cell. Biochem.* **2009**, *106*, 152–160.

Tunholi-Alves V. M.; Tunholi, V. M.; Castro N. R.; D'Oliveira Sant'ana, L.; Santos-Amaral L.; Martins de Oliveira, A. P.; Garcia, J.; Carvalho Thiengo, S.; Pinheiro, J.; Maldonado, A., Jr. Activation of Anaerobic Metabolism in *Biomphalaria glabrata* (Mollusca: Gastropoda) Experimentally Infected by *Angiostrongylus cantonensis* (Nematoda, Metastrongylidae) by High-performance Liquid Chromatography. *Parasitol. Intern.* **2014**, *63*, 64–68.

Uhlenbruck, G.; Steinhausen, G.; Cheesman, D. F. An Incomplete anti-B Agglutinin in the Eggs of the Prosobranch Snail *Pila ovata*. *Experientia* **1973**, *29*, 1139–1140.

Valavanidis, A.; Vlahogianni, T.; Dassenakis, M.; Scoullos, M. Molecular Biomarkers of Oxidative Stress in Aquatic Organisms in Relation to Toxic Environmental Pollutants. *Ecotoxicol. Environ. Saf.* **2006**, *64*, 178–189.

van Dam, H.; Wilhelm, D.; Herr, I.; Steffen, A.; Herrlich, P.; Angel, P. ATF-2 is Preferentially Activated by Stress-activated Protein Kinases to Mediate c-Jun Induction in Response to Genotoxic Agents. *EMBO J.* **1995**, *14*, 1798–1811.

van der Knaap, W. P. W.; Adema, C. M.; Sminia, T. Invertebrate Blood Cells: Morphological and Functional Aspects of the Haemocytes in the Pond Snail *Lymnaea stagnalis*. *Comp. Haematol. Int.* **1993**, *3*, 20–26.

Van Straalen, N. M. Ecotoxicology Becomes Stress Ecology. *Environ. Sci. Technol.* **2003**, *37*, 324A–330A.

Vasseur, P.; Cossu-Leguille, C. Biomarkers and Community Indices as Complementary Tools for Environmental Safety. *Environ. Int.* **2003**, *28*, 711–717.

Veenstra, J. A. Neurohormones and Neuropeptides Encoded by the Genome of *Lottia gigantea*, with Reference to Other Mollusks and Insects. *Gen. Comp. Endocrinol.* **2010**, *167*, 86–103.

Vergani, L. Metallothioneins in Aquatic Organisms: Fish, Crustaceans, Molluscs, and Echinoderms. In *Metal Ions in Lifes Sciences, Vol. 5, Metallothioneins and Related Chelators*; Sigel, A., Sigel H., Sigel R. K. O., Eds.; RSC Publishing: Cambridge, **2009**; pp 199–238.

Vergote, D.; Bouchut, A.; Sautiere, P. E.; Roger, E.; Galinier, R.; Rognon, A.; Coustau, C.; Salzet, M.; Mitta, G. Characterisation of Proteins Differentially Present in the Plasma of *Biomphalaria glabrata* Susceptible or Resistant to *Echinostoma caproni*. *Int. J. Parasitol.* **2005**, *35*, 215–224.

Vernberg, W. B.; Vernberg, F. J. Influence of Parasitism on Thermal Resistance of the Mud-flat Snail, *Nassarius obsoletus* Say. *Exp. Parasitol.* **1963**, *14*, 330–332.

Vorontsova, Y. L.; Slepneva, I. A.; Yurlova, N. I.; Glupov, V. V. Do Snails *Lymnaea stagnalis* have Phenoloxidase Activity in Hemolymph? *ISJ* **2015**, *12*, 5–12.

Waite, J. H. The Phylogeny and Chemical Diversity of Quinoned-tanned Glues and Varnishes. *Comp. Biochem. Physiol. B: Biochem. Mol. Biol.* **1990**, *156*, 491–496.

Wang, L.; Wang, L.; Huang, M.; Zhang, H.; Song, L. The Immune Role of C-type Lectins in Molluscs. *ISJ* **2011**, *8*, 241–246.

Wang, L.; Yue, F.; Song, X.; Song, L. Maternal Immune Transfer in Mollusc. *Dev. Comp. Immunol.* **2015**, *48*, 354–359.

Watson, S. N.; Wright, N.; Hermann, P. M.; Wildering, W. C. Phospholipase A2: The Key to Reversing Long-term Memory Impairment in a Gastropod Model of Aging. *Neurobiol. Aging* **2013**, *34*, 610–620.

Welker, A. F.; Moreira, D. C.; Campos, É. G.; Hermes-Lima, M. Role of Redox Metabolism for Adaptation of Aquatic Animals to Drastic Changes in Oxygen Availability. *Comp. Biochem. Physiol. A Mol. Integr. Physiol.* **2013**, *165*, 384–404.

Wendt, D. E. Effect of Larval Swimming Duration on Success of Metamorphosis and Size of the Ancestrular Lophophore in *Bugula neritina* (Bryozoa). *Biol. Bull.* **1996**, *191*, 224–233.

Whalen, K. E.; Sotka, E. E.; Goldstone, J. V.; Hahn, M. E. The Role of Multixenobiotic Transporters in Predatory Marine Molluscs as Counter-defense Mechanisms Against Dietary Allelochemicals. *Comp. Biochem. Physiol. C Toxicol. Pharmacol.* **2010**, *152*, 288–300.

Whalen, K. E.; Morin, D.; Lin, C. Y.; Tjeerdema, R. S.; Goldstone, J. V.; Hahn, M. E. Proteomic Identification, cDNA Cloning and Enzymatic Activity of Glutathione *S*-transferases from the Generalist Marine Gastropod, *Cyphoma gibbosum*. *Arch. Biochem. Biophys.* **2008**, *478*, 7–17.

Whitehead, A. Interactions Between Oil-spill Pollutants and Natural Stressors can Compound Ecotoxicological Effects. *Integr. Comp. Biol.* **2013**, *53*, 635–647.

Whitehead, A.; Crawford, D. L. Variation Within and Among Species in Gene Expression: Raw Material for Evolution. *Mol. Ecol.* **2006,** *15,* 1197–1211.

Widdows, J.; Donkin, P.; Salkeld, P. N.; Evans, S. V. Measurement of Scope for Growth and Tissue Hydrocarbon Concentrations of Mussels (*Mytilus edulis*) at Sites in the Vicinity of the Sullom Voe Oil Terminal: A Case Study. In *Fate and Effects of Oil in Marine Ecosystems*; Van den Brink, W. J., Kuiper, J., Eds.; Martinus Niijhof: Dordrecht, 1987; pp 269–277.

Widdows, J.; Donkin, P.; Brinsley, M. D.; Evans, S. V.; Salkeld, P. N.; Franklin, A.; Law, R. J.; Waldock, M. J. Scope for Growth and Contaminant Levels in North Sea Mussels, *Mytilus edulis. Mar. Ecol. Prog. Ser.* **1995,** *127,* 131–148.

Widdows, J.; Donkin, P.; Staff, F. J.; Matthiessen, P.; Law, R. J.; Allen, Y. T.; Thain, J. E.; Allchin, C. R.; Jones, B. R. Measurement of Stress Effects (Scope for Growth) and Contaminant Levels in Mussels (*Mytilus edulis*) Collected from the Irish Sea. *Mar. Environ. Res.* **2002,** *53,* 327–356.

Wieser, W. Metabolic End Products in Three Species of Marine Gastropods. *J. Mar. Biol. Assoc. UK* **1980,** *60,* 175–180.

Wilbrink, M.; Groot, E. J.; Jansen, R.; De Vries, Y.; Vermeulen, N. P. E. Occurrence of a Cytochrome P-450 Containing Mixed-function Oxidase System in the Pond Snail *Lymnaea stagnalis. Xenobiotica* **1991a,** *21,* 223–233.

Wilbrink, M.; Vand de Merbel, N. C.; Vermeulen, N. P. E. Glutathione-*S*-Transferase Activity in the Digestive Gland of the Pond Snail *Lymnaea stagnalis. Comp. Biochem. Physiol. C: Toxicol. Pharmacol.* **1991b,** *99,* 185–189.

Williams, E. A.; Degnan, S. M. Carry-over Effect of Larval Settlement Cue on Postlarva Gene Expression in the Marine Gastropod *Haliotis asinina. Mol. Ecol.* **2009,** *18,* 4434–4449.

Williams, G. A.; De Pirro, M.; Cartwright, S.; Khangura, K.; Ng, W. C.; Leung, P. T. Y.; Morritt, D. Come Rain or Shine: The Combined Effects of Physical Stresses on Physiological and Protein-level Responses of an Intertidal Limpet in the Monsoonal Tropics. *Funct. Ecol.* **2011,** *25,* 101–110.

Winberg, G. G. Rate of Metabolism and Food Requirements of Fishes. *Transl. Ser. Fish. Res. Board Can.* **1960,** *194,* 1–202.

Wingfield, J. C. Modulation of the Adrenocortical Response to Stress in Birds. In *Perspectives in Comparative Endocrinology*; Davey, K. G.; Peter, R. E.; Tobe, S. S., Eds.; National Research Council Canada: Ottawa, 1994; pp 520–528.

Wingfield, J. C.; Breuner, C.; Jacobs, J. Corticosterone and Behavioural Responses to Unpredictable Events. In *Perspectives in Avian Endocrinology*; Harvey, S.; Etches, R. J., Eds.; Journal of Endocrinology Ltd.: Bristol, 1997; pp 267–278.

Winterbourn, C. C. Reconciling the Chemistry and Biology of Reactive Oxygen Species. *Nat. Chem. Biol.* **2008,** *4,* 278–286.

Wo, K. T.; Lam, P. K. S.; Wu, R. S. S. A Comparison of Growth Biomarkers for Assessing Sublethal Effects of Cadmium on a Marine Gastropod, *Nassarius festivus. Mar. Pollut. Bull.* **1999,** *39,* 165–173.

Wojtaszek, J.; Poloczek-Adamowicz, A.; Adamowicz, A.; Fuks, U.; Dzugaj, A. Cytomorphometry and Seromucoid Concentration in the Hemolymph of Selected Snail Species. *Zool. Pol.* **1998,** *43,* 87–101.

Wright, B.; Lacchini, A. H.; Davies, A. J.; Walker, A. J. Regulation of Nitric Oxide Production in Snail (*Lymnaea stagnalis*) Defence Cells: A Role for PKC and ERK Signalling Pathways. *Biol. Cell* **2006,** *98,* 265–278.

Wu, X. J.; Sabat, G.; Brown, J. F.; Zhang, M.; Taft, A.; Peterson, N.; Harms, A.; Yoshino, T. P. Proteomic Analysis of *Schistosoma mansoni* Proteins Released During In Vitro Miracidium-to-sporocyst Transformation. *Mol. Biochem. Parasitol.* **2009,** *164,* 32–44.

Xu, C.; Li, C. Y.; Kong, A. N. Induction of Phase I, II and III Drug Metabolism/Transport by Xenobiotics. *Arch. Pharm. Res.* **2005,** *28,* 249–268.

Xu, Q.; Guo, L.; Xie, J.; Zhao, C. Relationship Between Quality of Pearl Cultured in the Triangle Mussel *Hyriopsis cumingii* of Different Ages and Its Immune Mechanism. *Aquaculture* **2011,** *315,* 196–200.

Yamaguchi, K.; Furuta, E.; Nakamura, H. Chronic Skin Allograft Rejection in Terrestrial Slugs. *Zool. Sci.* **1999,** *16,* 485–495.

Yang, J.; Wang, L.; Zhang, H.; Qiu, L.; Wang, H.; Song, L. C-type Lectin in *Chlamys farreri* (CfLec-1) Mediating Immune Recognition and Opsonization. *PLoS ONE* **2011,** *6,* e17089.

Yawetz, A.; Manelis, R.; Fishelson, L. The Effects of Aroclor 1254 and Petrochemical Pollutants on Cytochrome P450 from Digestive Gland Microsomes of Four Species of Mediterranean Molluscs. *Comp. Biochem. Physiol. C: Toxicol. Pharmacol.* **1992,** *103C,* 607–614.

Yonow, N.; Renwrantz, L. Studies on the Haemocytes of *Acteon tornatilis* (L.) (Opistobranchia: Acteonidae). *J. Moll. Stud.* **1986,** *52,* 150–155.

Yoshino, T. P.; Vasta, G. R. Parasite—Invertebrate Host Immune Interactions. *Adv. Comp. Environ. Physiol.* **2006,** *24,* 125–167.

Yoshino, T. P.; Boyle, J. P.; Humphries, J. E. Receptor–Ligand Interactions and Cellular Signalling at the Host–Parasite Interface. *Parasitology* **2001,** *123 Suppl.,* S143–S157.

Yoshino, T. P.; Dinguirard, N.; Kunert, J.; Hokke, C. H. Molecular and Functional Characterization of a Tandem-repeat Galectin from the Freshwater Snail *Biomphalaria glabrata,* Intermediate Host of the Human Blood Fluke *Schistosoma mansoni. Gene* **2008,** *411,* 46–58.

Yousif, F.; Blahser, S.; Lammler, G. The Cellular Responses in *Marisa cornuarietis* Experimentally Infected with *Angiostrongylus cantonensis. Parasitol. Res.* **1980,** *62,* 179–190.

Yue, F.; Zhou, Z.; Wang, L.; Ma, Z.; Wang, J.; Wang, M.; Zhang, H.; Song, L. Maternal Transfer of Immunity in Scallop *Chlamys farreri* and Its Transgenerational Immune Protection to Offspring Against Bacterial Challenge. *Dev. Comp. Immunol.* **2013,** *41,* 569–577.

Zahoor, Z.; Davies, A. J.; Kirk, R. S.; Rollinson, D.; Walker, A. J. Nitric Oxide Production by *Biomphalaria glabrata* Haemocytes: Effects of *Schistosoma mansoni* ESPs and Regulation Through the Extracellular Signal-regulated Kinase Pathway. *Parasite Vectors* **2009,** *2,* 18.

Zanette, J.; Goldstone, J. V.; Bainy, A. C.; Stegeman, J. J. Identification of CYP Genes in *Mytilus* (mussel) and *Crassostrea* (oyster) Species: First Approach to the Full Complement of Cytochrome P450 Genes in Bivalves. *Mar. Environ. Res.* **2010,** *69 Suppl.,* S1–S3.

Zarubin, T.; Han, J. Activation and Signaling of the p38 MAP Kinase Pathway. *Cell Res.* **2005,** *15,* 11–18.

Zelck, U. E.; Janje, B.; Schneider, O. Superoxide Dismutase Expression and H_2O_2 Production by Hemocytes of the Trematode Intermediate Host *Lymnaea stagnalis* (Gastropoda). *Dev. Comp. Immunol.* **2005,** *29,* 305–314.

Zelck, U. E.; Gege, B. E.; Schmid, S. Specific Inhibitors of Mitogen-activated Protein Kinase and PI3-K Pathways Impair Immune Responses by Hemocytes of Trematode Intermediate Host Snails. *Dev. Comp. Immunol.* **2007,** *31,* 321–331.

Zhang, H. M.; Zhunge, H. X.; Wang, Y. F.; Gong, W.; Lu, X. B.; Huang, L. H. Studies on Haemocytes of *Oncomelania hupensis. Chin. J. Parasitol. Parasit. Dis.* **2007a,** *25,* 114–115.

Zhang, S. H.; Zhu, D. D.; Chang, M. X.; Zhao, Q. P.; Jiao, R.; Huang, B.; Fu, J. P.; Liu, Z. X.; Nie, P. Three Goose-type Lysozymes in the Gastropod *Oncomelania hupensis*: cDNA Sequences and Lytic Activity of Recombinant Proteins. *Dev. Comp. Immunol.* **2012**, *36*, 241–246.

Zhang, S. M.; Zeng, Y.; Loker, E. S. Characterization of Immune Genes from the Schistosome Host Snail *Biomphalaria glabrata* that Encode Peptidoglycan Recognition Proteins and Gram-negative Bacteria Binding Protein. *Immunogenetics* **2007b**, *59*, 883–898.

Zhou, J.; Cai, Z. H.; Zhu, X. S.; Li, L.; Gao, Y. F. Innate Immune Parameters and Haemolymph Protein Expression Profile to Evaluate the Immunotoxicity of Tributyltin on Abalone (*Haliotis diversicolor supertexta*). *Dev. Comp. Immunol.* **2010**, *34*, 1059–1067.

INDEX

T - #0195 - 160425 - C424 - 229/152/19 - PB - 9781774635261 - Gloss Lamination